# EVOLUTION OF SIZE EFFECTS IN CHEMICAL DYNAMICS
## Part 1

ADVANCES IN CHEMICAL PHYSICS

VOLUME LXX

# EVOLUTION OF SIZE EFFECTS IN CHEMICAL DYNAMICS
# Part 1

ADVANCES IN CHEMICAL PHYSICS
VOLUME LXX

*Edited by*

**I. PRIGOGINE**

University of Brussels
Brussels, Belgium
and
University of Texas
Austin, Texas

*and*

**STUART A. RICE**

Department of Chemistry
and
The James Franck Institute
The University of Chicago
Chicago, Illinois

AN INTERSCIENCE® PUBLICATION
**JOHN WILEY & SONS**
NEW YORK • CHICHESTER • BRISBANE • TORONTO • SINGAPORE

An Interscience® Publication



**Library of Congress Cataloging Number: 58-9935**

ISBN 0-471-62784-4

Printed in the United States of America

10 9 8 7 6 5 4 3 2 1

# CONTRIBUTORS TO VOLUME LXX
## Part 1

R. Stephen Berry, The James Franck Institute and the Department of Chemistry, The University of Chicago, Chicago, Illinois

Paul Brumer, Chemical Physics Theory Group, Department of Chemistry, University of Toronto, Toronto, Ontario, Canada

Peter M. Felker,* Arthur Amos Noyes Laboratory of Chemical Physics, California Institute of Technology, Pasadena, California

R. B. Gerber, Department of Physical Chemistry and The Fritz Haber Research Center for Molecular Dynamics, The Hebrew University of Jerusalem, Jerusalem, Israel

Joshua Jortner, School of Chemistry, Tel Aviv University, Ramat Aviv 69978, Israel

Jan Kommandeur, Laboratory for Physical Chemistry, The University of Grönigen, Nijenborgh, Gröningen, The Netherlands

Jeffery L. Krause, The James Franck Institute and the Department of Chemistry, The University of Chicago, Chicago, Illinois

R. D. Levine, The Fritz Haber Research Center for Molecular Dynamics, The Hebrew University, Jerusalem, Israel

R. A. Marcus, Arthur Amos Noyes Laboratory of Chemical Physics, California Institute of Technology, Pasadena, California

Shaul Mukamel, Department of Chemistry, The University of Rochester, Rochester, New York

Mark A. Ratner, Department of Chemistry, Northwestern University, Evanston, Illinois

Stuart A. Rice, The James Franck Institute and the Department of Chemistry, The University of Chicago, Chicago, Illinois

Moshe Shapiro, Department of Chemical Physics, Weizmann Institute of Science, Rehovot, Israel

* *Present Affiliation:* Department of Chemistry, University of California, Los Angeles, Los Angeles, California

DAVID J. TANNOR,* The James Franck Institute and the Department of Chemistry, The University of Chicago, Chicago, Illinois

DAVID M. WARDLAW, Department of Chemistry, Queen's University, Kingston, Ontario, Canada

AHMED H. ZEWAIL, Arthur Amos Noyes Laboratory of Chemical Physics, California Institute of Technology, Pasadena, California

* *Present Affiliation:* Department of Chemistry, Illinois Institute of Technology, Chicago, Illinois

# INTRODUCTION

Few of us can any longer keep up with the flood of scientific literature, even in specialized subfields. Any attempt to do more and be broadly educated with respect to a large domain of science has the appearance of tilting at windmills. Yet the synthesis of ideas drawn from different subjects into new, powerful, general concepts is as valuable as ever, and the desire to remain educated persists in all scientists. This series, *Advances in Chemical Physics*, is devoted to helping the reader obtain general information about a wide variety of topics in chemical physics, which field we interpret very broadly. Our intent is to have experts present comprehensive analyses of subjects of interest and to encourage the expression of individual points of view. We hope that this approach to the presentation of an overview of a subject will both stimulate new research and serve as a personalized learning text for beginners in a field.

ILYA PRIGOGINE
STUART A. RICE

# CONTENTS

# EVOLUTION OF SIZE EFFECTS IN CHEMICAL DYNAMICS Part 1

ADVANCES IN CHEMICAL PHYSICS

VOLUME LXX

# LEVEL STRUCTURE AND DYNAMICS FROM DIATOMICS TO CLUSTERS*

JOSHUA JORTNER

*School of Chemistry, Tel-Aviv University, Ramat Aviv 69978, Israel*

R. D. LEVINE

*The Fritz Haber Research Center for Molecular Dynamics, The Hebrew University, Jerusalem 91904, Israel*

STUART A. RICE

*Department of Chemistry and The James Franck Institute, The University of Chicago, Chicago, Illinois 60637, USA*

## CONTENTS



## I. INTRODUCTION

In our 1981 review,[1] we suggested that analyses using a microscopic description of dynamics would increasingly influence our interpretation of the properties of large systems; that the merging of the microscopic and macroscopic methods of analysis would lead to an understanding of the gradual transition from the behavior characteristic of the intramolecular dynamics of

* Work supported by the U.S.–Israel Binational Science Foundation, The U.S. Army European Research Office, The U.S. Air Force Office of Scientific Research (AFOSR), and the U.S. National Science Foundation.

an isolated small molecule to that characteristic of the statistical dynamics of a condensed phase. In particular, processes on surfaces[2] and interfaces and in molecular clusters were expected to provide the bridge whereby one could examine electronic structure, vibrational and electronic excitations, and intramolecular dynamics in large but finite systems[3] as a function of the size of the system.

This volume provides an account of some of the results of studies that illustrate the expectations previously expressed. Indeed, our understanding of the transition from small- to large-system dynamics has been enriched to such an extent that the basic results are already to be found in primers of chemical reactivity.[4] In the same period we have also witnessed a qualitative improvement in our knowledge of the dynamics of small molecules. As is always the case, much of the progress that has been made is derived from new experimental techniques used in an imaginative fashion. The most important new experimental techniques use various combinations of lasers and supersonic beams to explore chemically interesting questions. For smaller systems, our understanding of chemical reactivity and, in particular, our ability to control and probe collision processes, has been advanced significantly. We consider first the impact of new spectroscopic techniques[5–8] on our knowledge of intramolecular dynamics, including the new theoretical techniques that are being developed to examine the large amounts of data produced by high-resolution spectroscopy at chemically significant levels of excitation. We then examine the level structure and properties of clusters. The dynamics of nuclear motion, the nature of electronic excitations, and the electronic–vibrational relaxation phenomena in nonmetallic clusters, provide a wealth of novel information[3] that helps bridge the different interpretations of the dynamics of isolated molecules and condensed phases.

## II. HIGH-RESOLUTION SPECTROSCOPY AND INTRAMOLECULAR DYNAMICS

Direct high-overtone excitation[6,7,9–15] of a molecule reaches states with local-mode character that carry significant transition strength from the ground state. Much of the extant theoretical effort has been directed at computing the corresponding level structure,[16–23] the nature of the optically prepared state,[10,24,25] and the ensuing intramolecular dynamics.[26–35] Brute force quantum-mechanical computations of the spectrum run into the difficulty that the size of the Hamiltonian matrix required to achieve convergence for these high-lying states taxes the abilities of present day computers. Semiclassical methods of analysis clearly have an advantage in this case. It remains to be seen, however, how well the procedure of adiabatic switching[35] can "zero in" on the eigenstate of interest for Hamiltonians of realistic complexity. Many of the theoretical studies have been carried out on model systems for which key terms such as Coriolis coupling are neglected, even though said coupling has been clearly implicated[5,36–39] in intramolecular energy redistribution.[40,41] This is also the case for the algebraic approach,[22,42–44] which even in zeroth order treats the oscillations as anharmonic and accounts for Fermi[43] and for Darling–Dennison[42] (or, equivalently,[45,46] for local-mode) coupling between the anharmonic oscillators. However, the algebraic method has not yet been generalized to include vibration–rotational coupling, nor has it been applied to realistic spectra.

Spectroscopic studies at even higher levels of excitation on the ground-state potential energy surface is possible by an indirect route. In stimulated emission pumping[5] (SEP) an initial state is laser pumped to a specific, single, rotation–vibration level of an electronically excited state. A second laser stimulates the emission from the electronically excited state down to an excited level of the ground electronic state. The SEP spectrum[5,47,48] of $C_2H_2$, for example, is quite different from that obtained using direct overtone excitation.[6,11] The low-resolution ($\sim$0.3 cm$^{-1}$) SEP spectrum at about 26,500 cm$^{-1}$ shows a series of features with a density of about 0.6 per cm$^{-1}$. Higher-resolution spectroscopy reveals that each feature is a clump of lines with a density of about 8 per cm$^{-1}$ (Fig. 1). The usual spectroscopic methods of assignment of lines are unlikely to be useful in practice (and possibly even in principle).[49,50] Statistical tests based on near-neighbor spacing distributions[5] can be applied[5,51] but are complicated by level congestion. A new approach[52] is to examine the "survival probability" $P(t)$, that is,[16,53] the square modulus of the Fourier transform of the spectrum $I(\omega)$,

$$P(t) = \left| \int I(\omega) \exp(-i\omega t)\, dt \right|^2 \tag{1}$$

For a "stick spectrum" of $\mu$ lines where $y_n$ is the intensity of the $n$th transition at the frequency $\omega_n$, $I(\omega) = \sum_{n=1}^{N} y_n \delta(\omega - \omega_n)$, there are two contributions to $P(t)$:

$$P(t) = \sum_n y_n^2 + \sum_{n \neq m} \sum_m y_n y_m^* \exp[i(\omega_n - \omega_m)t] \tag{2}$$

Upon taking an ensemble average, the first term, which contributes only at the origin, yields $N^2 \langle y^2 \rangle$, where $\langle y^2 \rangle$ is the averaged intensity. The average value of the second term depends in an essential way on the correlation between positions of the different lines. On the assumptions that (1) the distributions of intensity and of positions are independent and (2) that the probability to find another line at a distance $\Delta\omega$ from the first line depends

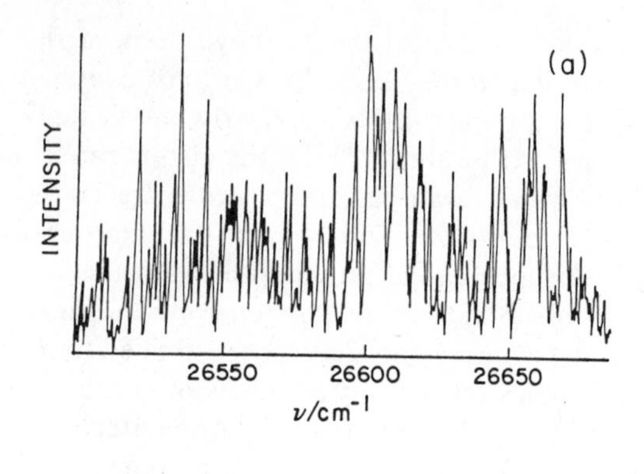


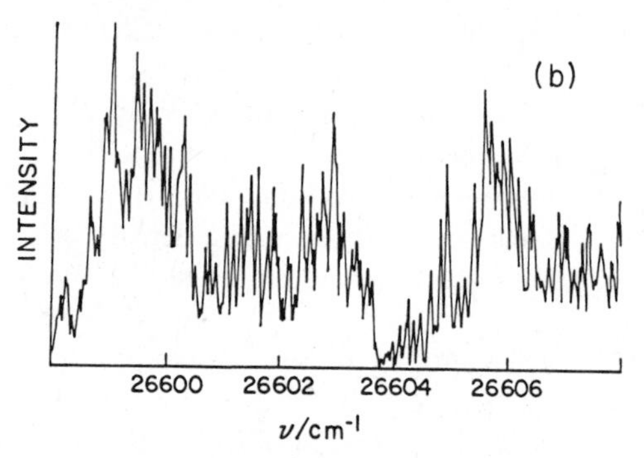


**Figure 1.** SEP spectrum of $C_2H_2$ at about 26,500 $cm^{-1}$ of vibrational excitation in the ground electronic state. [Adapted from J. P. Pique, Y. Chen, R. W. Field and J. L. Kinsey, *Phys. Rev. Lett.* **58**, 475 (1987).] (*a*) Low-resolution ($\sim 0.3$ $cm^{-1}$) spectrum; (*b*) high-resolution ($\sim 0.05$ $cm^{-1}$) spectrum. Each feature in (*a*) is resolved into a series of lines. The overall "envelope" variation in intensity in (*b*) is the "clump" structure seen in (*a*).

only on the frequency difference $\Delta\omega$ (and not on the absolute positions), the second term of Eq. (2) yields $N\langle y^2\rangle[1 - Gb_2(t)]$, where $b_2(t)$ is the Fourier transform of the cluster function[54] $Y_2(\omega)$.* From the computation of $P(t)$ one can therefore extract the value of $G \equiv \langle y\rangle^2/\langle y^2\rangle$ (provided the number, $N$, of lines has been counted) and the $t$ dependence of $b_2(t)$. We expect $b_2(t)$ to decline from unity with a width, $t_c$, which is inverse to the frequency range of level correlations. For an anharmonic but "regular" spectrum, the level positions are uncorrelated. In the limit when the corresponding classical motion is chaotic, the correlation range is the mean spacing, so that $t_c = \rho/c$,

* The cluster function is defined[54] such that for an infinite spectrum with unit mean spacing $\frac{1}{2}(1 - Y_2(\omega_n, \omega_m))\, d\omega_n\, d\omega_m$ is the probability of finding two transitions in the intervals $\omega_n$, $\omega_n + d\omega_n$ and $\omega_m$, $\omega_m + d\omega_m$.

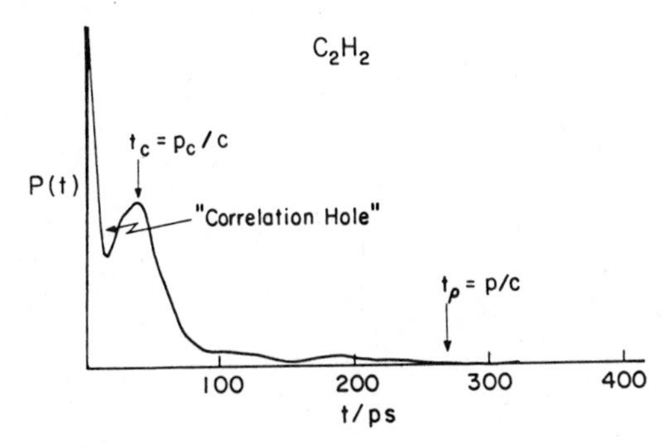


**Figure 2.** The survival probability, $P(t)$, computed as an average over several clumps. (Same source as Fig. 1.) The overall decline of $P(t)$ with $t$ is due to the frequency dependence of the envelope of the clump structure.

where $\rho$ is the density of states. In general, $t_c = \rho_c/c$, where $\rho_c$ is the density of correlated states. Figure 2 shows the results for $P(t)$ for the high-resolution spectrum of $C_2H_2$. The overall decline of $P(t)$ with time is due to the envelope shape of the clump of states. The correlation hole following the fast decay [the first term in (2)] is evident but is not very deep, since $G < 1$, $t_c \simeq 45$ ps, and $\rho_c \simeq \rho/6$. At a somewhat higher energy even more lines are strongly correlated.

In Fig. 2 we have considered the properties of the distribution of line positions irrespective of their intensities. Should we not also examine the distribution of intensity, irrespective of position? This can be done by dividing the ordinate in Fig. 1 into bins and counting the number of transitions whose intensity falls within any given bin. The results[55] are shown in Fig. 3 together with a fit to the functional form[56]

$$P(y) = y^{(\nu/2)-1} \exp\left(\frac{-\nu y}{2\langle y\rangle}\right) \Big/ \left(\frac{2\langle y\rangle}{\nu}\right)^{\nu/2} \Gamma\left(\frac{\nu}{2}\right) \tag{3}$$

for which $G = \langle y\rangle^2/\langle y^2\rangle = \nu/(\nu + 2)$. The value of $\nu$ is sensitive to the precise choice of the base line of zero intensity. However, sensibly independent of this choice, $\nu \geq 3$ and so $G \geq 0.6$. For a quantal spectrum where the corresponding classical motion is chaotic, one expects[56,57] $\nu = 1$ or $G = \frac{1}{3}$. This is not the case for $C_2H_2$ at 26,500 $cm^{-1}$, as is also evident from the analysis of Fig. 2. This is, however, the case for the spectrum[12] of $NO_2$ at 17,000 $cm^{-1}$, as shown in Fig. 4.

Another route for accessing the ground potential energy surface at energies at or even above dissociation[58,59] is via a nonradiative transition from an excited electronic state. One would then expect fluctuations in the lifetimes of the predissociating states much as the transition strengths fluctuate for the

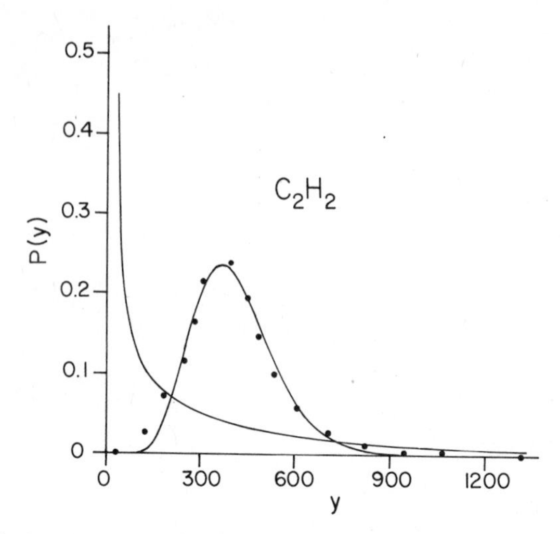


**Figure 3.** Statistical analysis of the intensity distribution of the high-resolution spectrum of $C_2H_2$ at about 26,500 $cm^{-1}$, including nearly 4000 lines. (Adapted from Ref. 55.) The solid line is the maximum entropy distribution (cf. Ref. 56) given by Eq. (3) with $\nu = 3.2$.

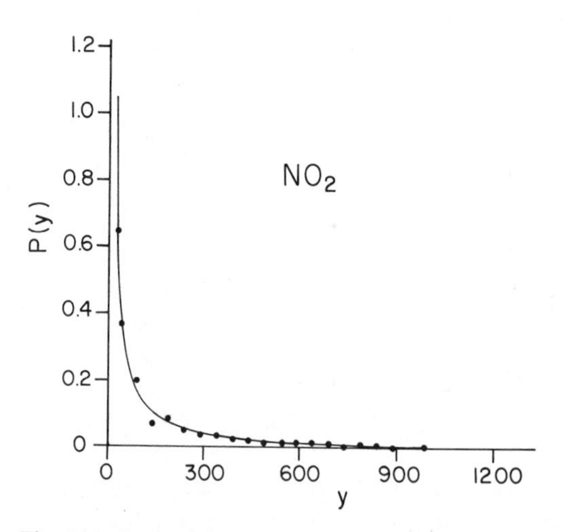


**Figure 4.** Same as Fig. 3 but for $NO_2$ (data of Ref. 12). Here the fit by Eq. (3) shows that no constraints are needed ($\nu \simeq 1$) so that the fluctuations are as random as possible. (Adapted from Ref. 55.)

bound-state spectra. For $D_2CO$, an order of magnitude variation in rates is observed[58] for states within a 0.2 $cm^{-1}$ energy range at about 28,000 $cm^{-1}$. For $SiH_2$ fluorescence lifetimes of individual ro-vibronic states at 17,000 $cm^{-1}$ vary[59] by almost three orders of magnitude and can be well fitted[60] by a distribution of the type (3) with $n = 1$. An essentially statistical intensity distribution is also found[60] for benzene in the "channel three" region.[61]

## III. LASER STUDIES OF CHEMICAL REACTIVITY

That lasers have played a key role as promoters and as probes of chemical reactions is well known and extensively documented.[1,4,7,62–72] In many of these applications the laser is employed as an intense, nearly monochromatic, light source whose characteristics ensure species selectivity, a well-characterized spectroscopy, and adequate intensity for multiphoton processes. Some possible applications, notably "laser-assisted collisions"[73,74] and transition-state spectroscopy,[75,76] are yet in their infancy, but the extant studies already suggest considerable promise for influencing and probing chemical reactions.

In recent experiments additional properties of lasers have been used in an essential way and these have opened up new avenues for exploring and understanding chemical reactivity. The first of these properties is the seemingly obvious spatial collimation characteristic of laser light. Doppler shift measurements[77–80] have now successfully been used to deconvolute spectral line shapes and also to measure the speed distribution of reaction products along the light beam direction (Fig. 5). Another property of laser radiation that is beginning to contribute to our understanding of steric aspects of chemical

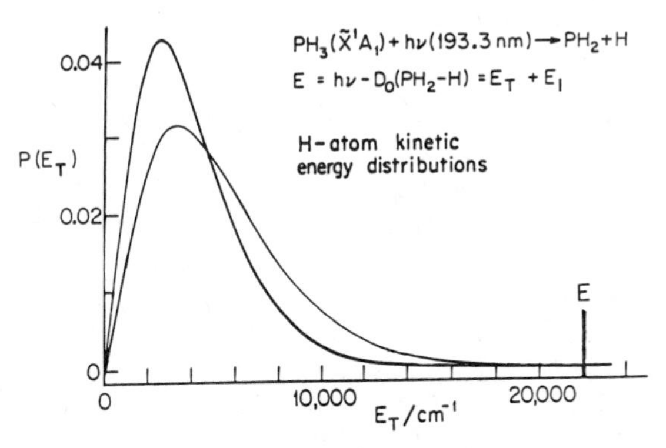


**Figure 5.** H atom kinetic energy distribution from the photolysis of $PH_3$ at 193.3 nm as deduced from Doppler profiles (using Lyman-$\alpha$ radiation). The two curves bound the range of acceptable fit to the measured profiles. [Adapted from Z. Xu, B. Koplitz, S. Buelow, D. Baugh, and C. Wittig, *Chem. Phys. Lett.* **127**, 534 (1986).]

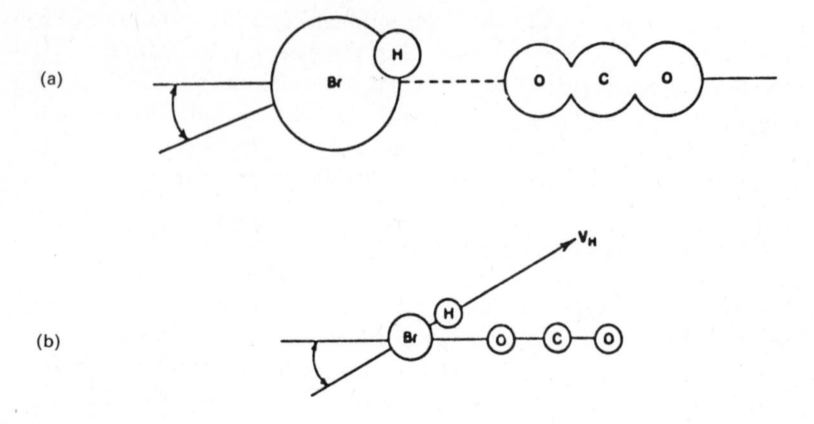


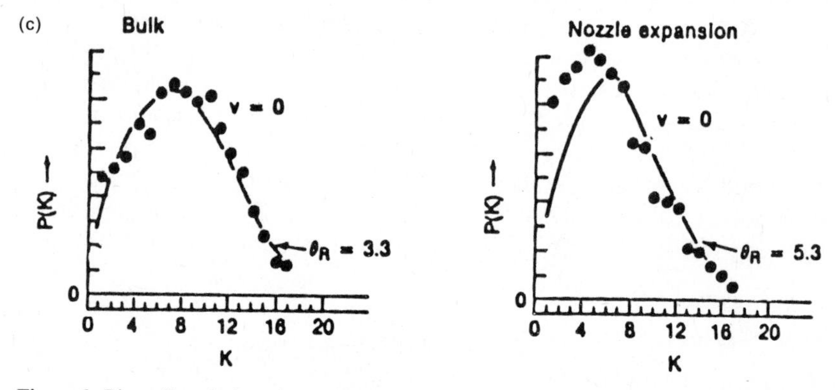


**Figure 6.** Photodissociation of the $CO_2HBr$ van der Waals molecule. (*a*) The proposed structure of the adduct. At the minimum all atoms are collinear but the restoring force for the bending (as shown) is small. (*b*) At the instant of dissociation. The H atom is directed into a narrow cone about the OCO axis. (*c*) The measured OH ($v = 0$) rotational distribution under bulk conditions and from the photolysis of the adduct. The solid curve is computed assuming a linear rotational surprisal and the $\theta_R$ values shown. The OH distribution from the dissociation of the adduct is rotationally cold. [Adapted from S. Buelow, M. Noble, H. Radhakrishnan, H. Reisler, C. Wittig and G. Hancock, *J. Phys. Chem.* **90**, 1015 (1986).]

reactions is polarization.[81] The polarization of a light beam can be used to prepare aligned reagents and also to probe the alignment of the products.[82–88] Stereoselectivity can also be studied by photodissociation of a van der Waals adduct prepared in a supersonic beam.[89,90] Figure 6 shows the proposed structure of $CO_2HBr$ with the hydrogen atom adjacent to the oxygen. Upon photolysis of HBr, the velocity vector of the H atom is confined to a cone about the OCO axis. The OH vibrational distribution from the reaction

$$\mathrm{OCO + H \rightarrow CO + OH}$$

is different (i.e., colder) than that observed when the velocity vector of the hot H atom is randomly oriented (Fig. 6). In the past, lasers have been used to select the initial states of the reagents and to probe the states of the products formed, which are the "scalar" characteristics of the reaction dynamics. With the newer experiments, vector characteristics are being selected and/or probed.

Another aspect of chemical reactivity that is receiving considerable attention is the reaction of electronically excited species[91–97] Questions of orbital stereoselectivity[92,96,97] are likely to keep both experimentalists and theorists busy for some time to come.

Laser pulses of short duration have been used to prepare, in bulk samples, a large initial concentration of reactive atoms or radicals.[98–100] More recently, by using a second probe laser pulse, delayed with respect to the initiation pulse, it is found possible to monitor reaction products formed after a single collision of the reactive species. The recent progress[101–103] in experimental studies of the $H + H_2$ exchange reaction is due to this technique. It is finally possible (Fig. 7) to report agreement between experiments and for this fundamental chemical reaction.

Unimolecular dissociation can also be studied in this fashion,[104] Fig. 8. The first pulse provides the energy and the second probes the dissociation products in real time. To achieve the time resolution sufficient to monitor the unimolecular dissociation, both pulses need to be ultrashort. The absolute unimolecular rate constant into any given internal state of the product can be so measured. Relative rates of reaction from an initial state into a set of final states have been available for some time since they are provided by the measurement of the products' state distribution. With the new techniques, the magnitude of the rate can be measured. Such results for neutrals (and also measurements of lifetimes of metastable multiphoton produced ions[105,106]) will test in detail our theoretical understanding of unimolecular reactions. The fundamental concept that a bottleneck separates the energy rich molecule from the products needs to be reexamined.[107,108] Nor is it going to be enough to stand, *a la* RRKM, at the bottleneck and count phase points as they cross.

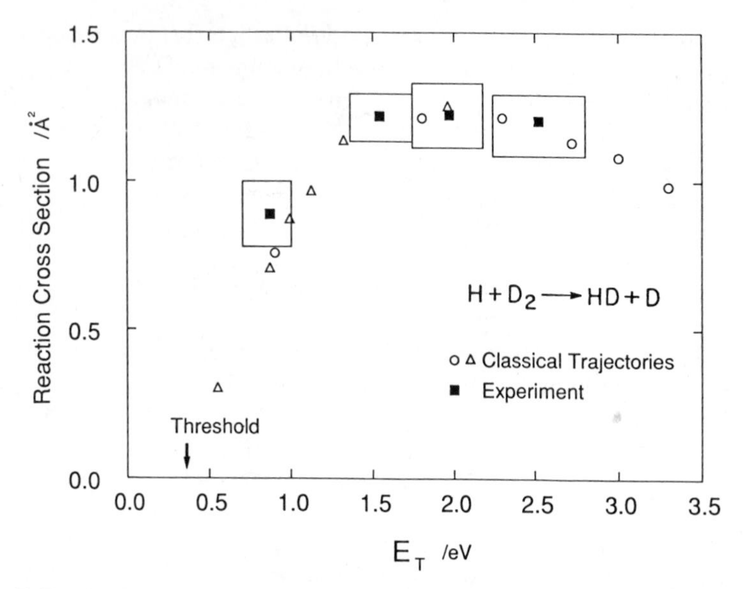


**Figure 7.** Two laser measurements of the translational energy dependence of the reaction cross section for $H + D_2 \rightarrow HD + D$. Rectangles: experimental results [K. Tsukiyama, B. Katz, and R. Bersohn, *J. Chem. Phys.* **84**, 1934 (1986) and private communication]. Points: Classical trajectory computations on the best available *ab initio* potential energy surface [by N. C. Blais and D. G. Truhlar (triangles) and by I. Schechter (circles)].

Rather, it will be necessary to include the exit valley interactions, so as to account for the partitioning into specific final states.

Theorists have also considered the yet to be tapped capabilities of lasers and have recently centered attention on coherence effects. Consider the probability to observe a set of final states $\rho_f$ following optical excitation of an initial state $|i\rangle$ by a weak field. This can be written as

$$P_{fi} = \lim_{T \to \infty} \frac{1}{T} \int_0^T \mathrm{Tr}[\rho_f \rho_i(t)]\, dt \tag{4}$$

where

$$\rho_i(t) = U(t)\mu|i\rangle\langle i|\mu^\dagger U^\dagger(t) \tag{5}$$

with $U(t)$ the evolution operator of the unperturbed molecule and $\mu$ is the dipole operator. Introducing a set of energy eigenstates for the molecule,

$$U(t)|n\rangle = \exp(-iE_n t/\hbar|n\rangle \tag{6}$$

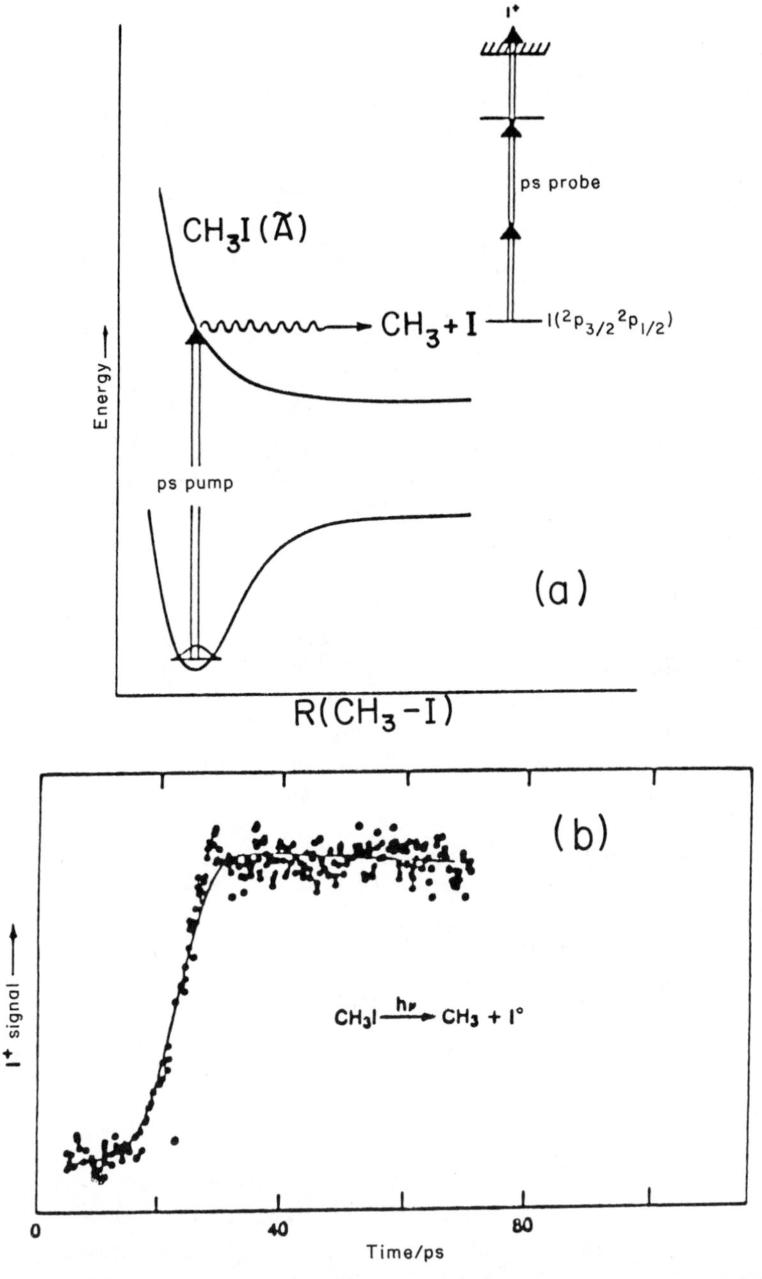


**Figure 8.** A picosecond laser experiment to determine the photodissociation lifetime of $CH_3I^*$. [Adapted from J. L. Knee, L. R. Khundar, and A. H. Zewail, *J. Chem. Phys.* **83**, 1996 (1985).] (*a*) An outline of the experiment: a picosecond laser pulse pumps $CH_3I$ to the repulsive $\tilde{A}$ state. The I (or I*) atoms are detected as ions using a picosecond laser pulse for resonance-enhanced multiphoton ionization. (*b*) The $I^+$ ion signal. Deconvolution yields a lifetime of $\leq 0.5$ ps.

one can write (4) as

$$P_{fi} = \lim_{T\to\infty} \frac{1}{T} \int_0^T \sum_n \sum_m \langle n|\rho_f|m\rangle \exp[i(E_n - E_m)t/\hbar] \langle m|\rho_i|n\rangle \tag{7}$$

and, after integration (which cancels all off-diagonal terms in the $T \to \infty$ limit), we obtain[109]

$$P_{fi} = \sum_n \langle n|\rho_f|n\rangle \langle n|\rho_i|n\rangle \tag{8}$$

where $\rho_i = \rho_i(0)$. The result (8) is purely diagonal and carries on phase information. To achieve photoselectivity, that is, to be able to influence the magnitude of $P_{fi}$, it is necessary to remove one or more of the restrictions imposed in the derivation of (8). The conditions used in the derivation of (8) are: (1) The external field is weak, so that the excitation can be described by first-order perturbation theory. This need not be the case.[28] (2) The molecule evolves unperturbed following excitation. This can be modified by using a coherent sequence of pulses.[109–111] For example,[110] suppose the first ultrashort pulse promotes the molecule to an excited electronic state surface; so far, as in (4). However, rather than being allowed to dissociate from the upper state, a second laser pulse brings the molecule back down to the ground potential energy surface. By controlling the ultrashort delay between the two pulses, one determines the extent of propagation on the upper surface. Different propagation times can (Fig. 9) lead to a selective dissociation pathway on the lower potential energy surface.

Another theoretical possibility[112] is to use a coherent superposition of initial states. Very schematically, one would write

$$|i\rangle = C_1|1\rangle + C_2|2\rangle \tag{9}$$

so that four terms contribute to $\langle n|\rho_i|n\rangle$:

$$\begin{aligned}\langle n|\rho_i|n\rangle = |\langle n|\mu|1\rangle|^2 C_1^2 + |\langle n|\mu|2\rangle|^2 C_2^2 + \langle n|\mu|1\rangle\langle 2|\mu|n\rangle C_1 C_2^* \\ + \langle n|\mu|2\rangle\langle 1|\mu|n\rangle C_1^* C_2.\end{aligned} \tag{10}$$

By controlling the phase relations in the initial state (i.e., the phase difference between $C_1$ and $C_2$) and the relative intensity of the two lasers used to photolyze the eigenstates $|1\rangle$ and $|2\rangle$, one can influence the branching factors into different product states (Fig. 10).

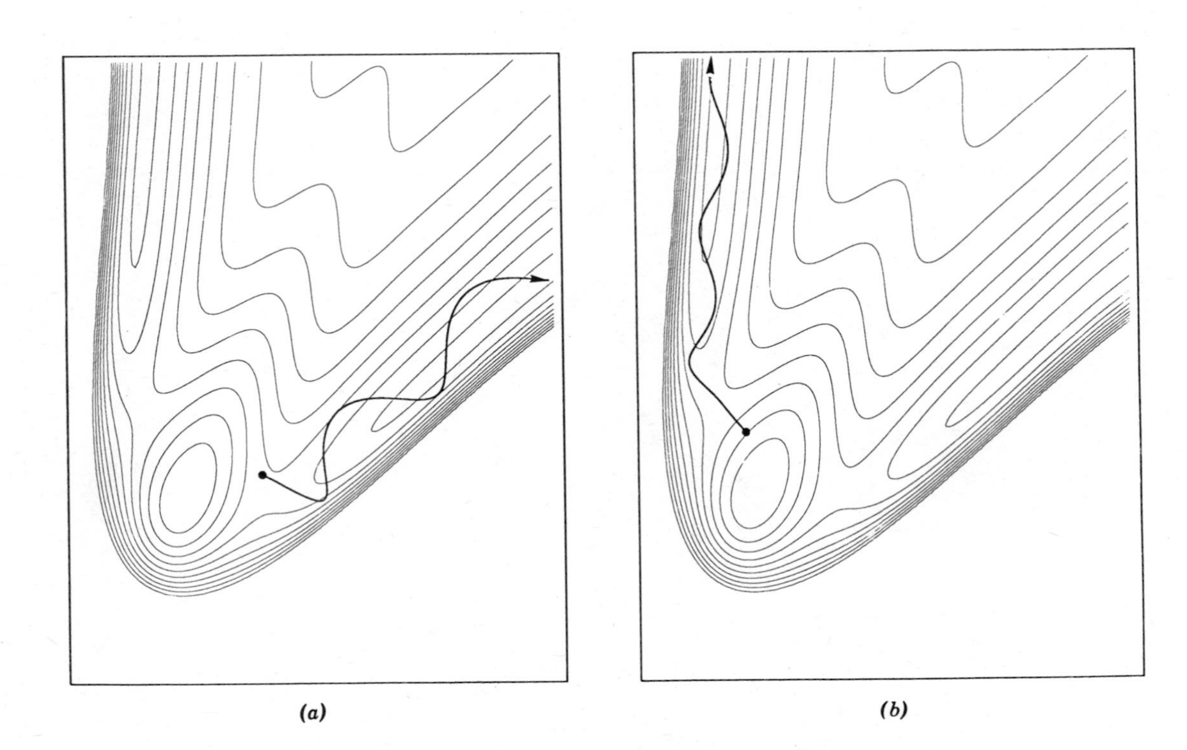


**Figure 9.** Selectivity of photodissociation controlled using a coherent pulse sequence. [Adapted from D. J. Tannor, R. Kosloff, and S. A. Rice, *J. Chem. Phys.* **85**, 5805 (1986).] The system is pumped at $t = 0$ to an excited electronic state surface on which it propagates for a time $T$ and is then brought down to the ground surface. Two classical trajectories are shown for (*a*) $T = 600$ au, (*b*) $T = 2100$ au. The computations are for vertical up and down transitions. Parts (*c*) and (*d*) show the corresponding quantum-mechanical results. The different plots are for (*c*): 0, 800, 1000 au propagation on the ground-state surface and (*d*): 0, 1000, 1200 au propagation on the ground-state surface.

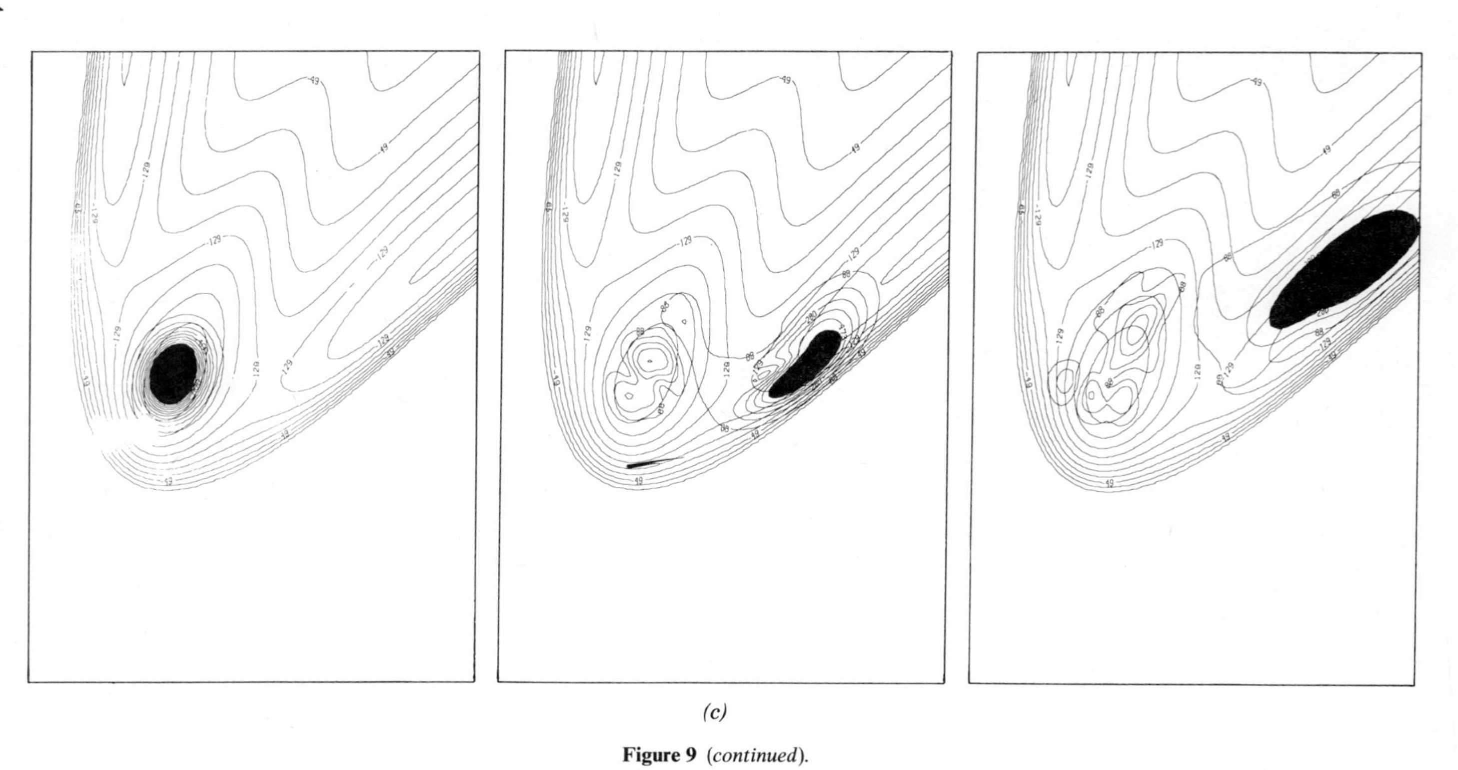

*(c)*

**Figure 9** *(continued)*.

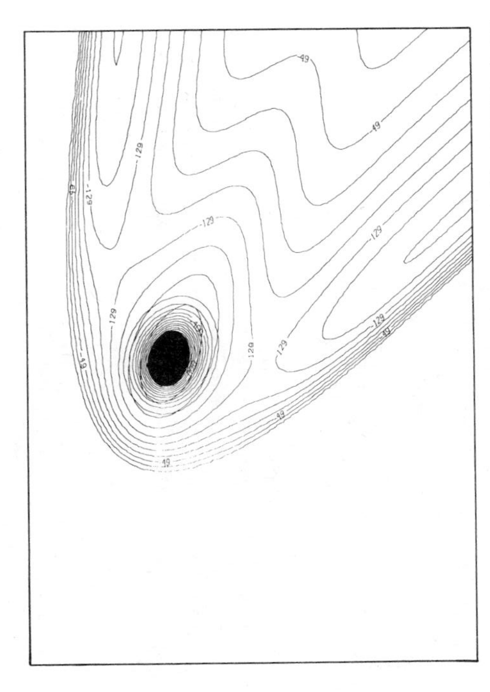

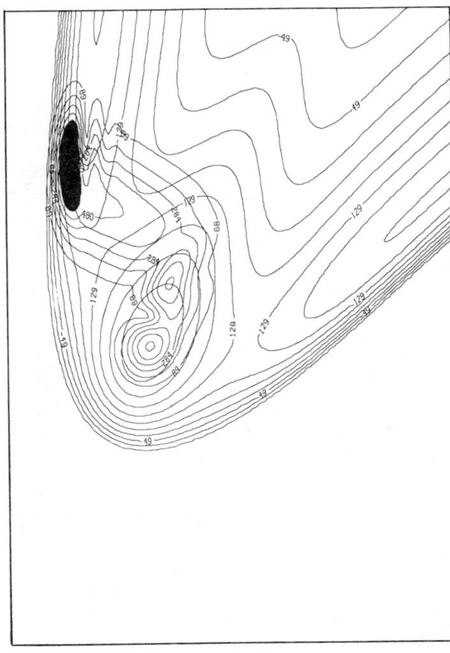

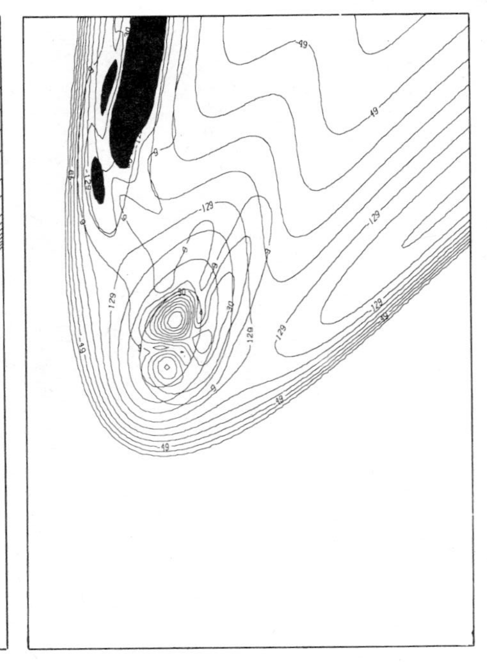

*(d)*

**Figure 9** *(continued)*.

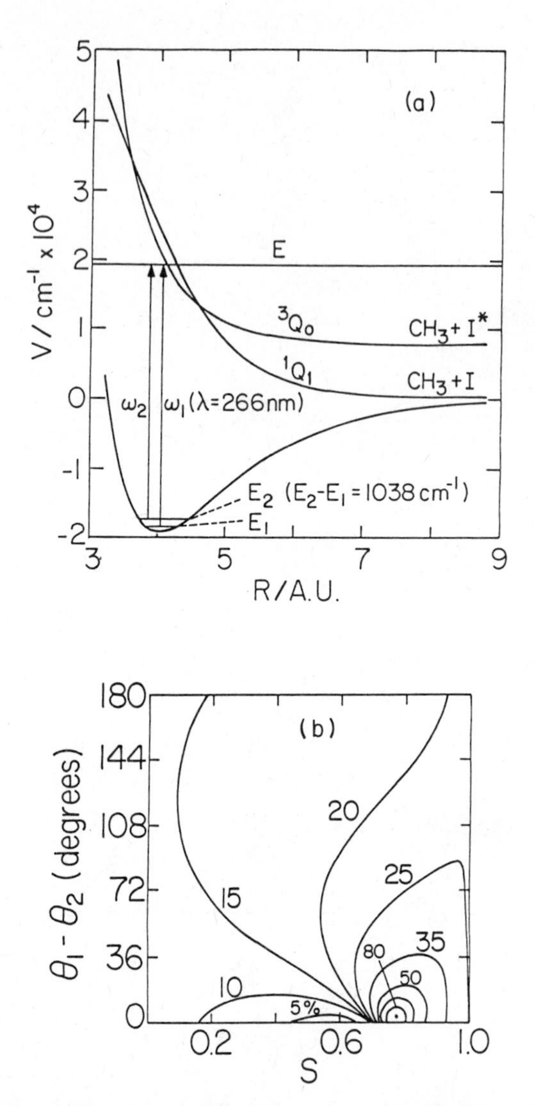


**Figure 10.** Photodissociation of an initial coherent linear superposition of states in $CH_3I$. [Adapted from P. Brumer and M. Shapiro, *Chem. Phys. Lett.* **123**, (1986).] (*a*) A cut of the potential energy surface along the dissociation coordinate. The figure shows the two bound states [cf. (9)] that are photodissociated (to the same total energy). (*b*) Contour plot of the yield of ground state I atoms as a function of the relative intensity (*S*) of the two photodissociating lasers and of the phase difference.

## IV. LEVEL STRUCTURE AND DYNAMICS OF CLUSTERS

A realization of the gradual transition from the intramolecular dynamics characteristic of finite molecular systems to that characteristic of infinite systems, that is, condensed phases, emerges from the study of the properties of clusters.[113–116] As has been repeatedly emphasized, the level structure of

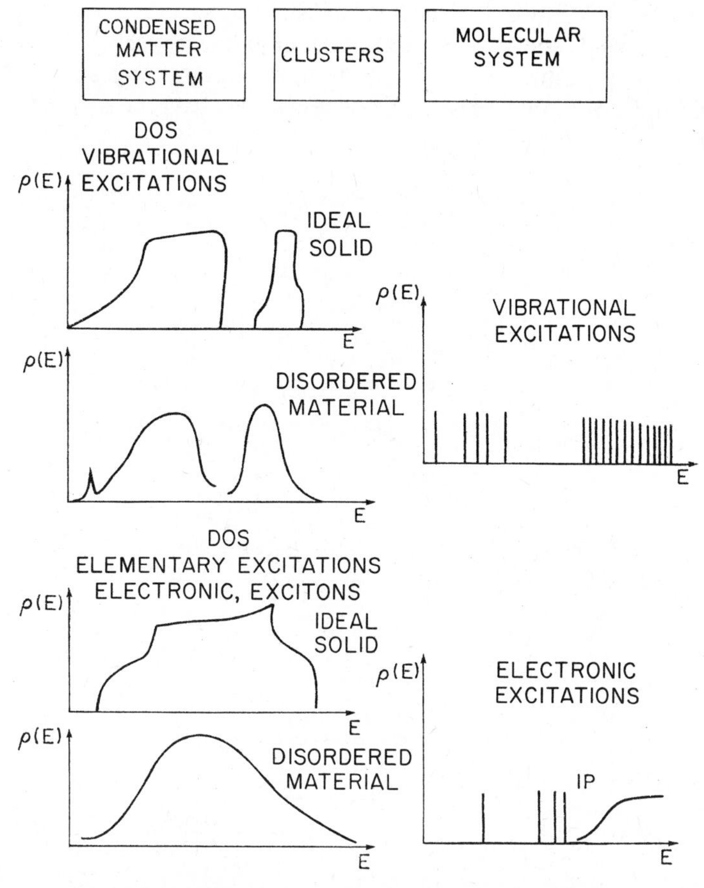


**Figure 11.** Density of states (DOS) of vibrational and electronic excitation in condensed matter systems and in molecular systems. The DOS for electronic and vibrational excitation in condensed matter systems is *continuous*, and is characterized by van Hove topological singularities for ordered structures and by exponential Mott tails for disordered materials. The DOS for electronic excitation of molecular systems is *discrete* below the first ionization potential. In molecular systems the vibrational DOS is discrete, while in large molecules a quasicontinuum of vibrational states exists at high energies.

elementary excitations, for example, phonons and excitons, in clusters cannot be described using only the traditional concepts of molecular physics and/or condensed matter physics (Fig. 11). Rather, the geometric structure, level structure, and internal dynamics of clusters are intermediate between those of isolated molecules and condensed matter. In 1977, when cluster physics and chemistry were in their embryonic stages, the study of clean and isolated clusters was viewed as the "theoretician's dream."[117] This dream has become a reality! The remarkable progress in the technology of supersonic jets and cluster beams[118,119] provides isolated clusters amenable to experimental interrogation, while the combination of beam and laser techniques permits probing of the dynamics of nulcear motion, the nature of electronic states, electronic and vibrational relaxation, as well as chemical processes in clusters.

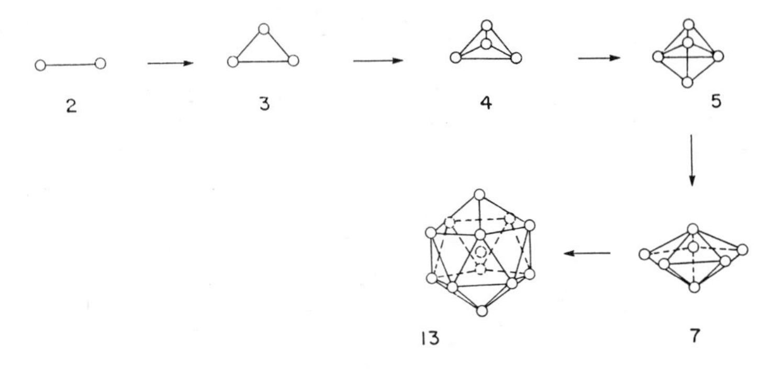


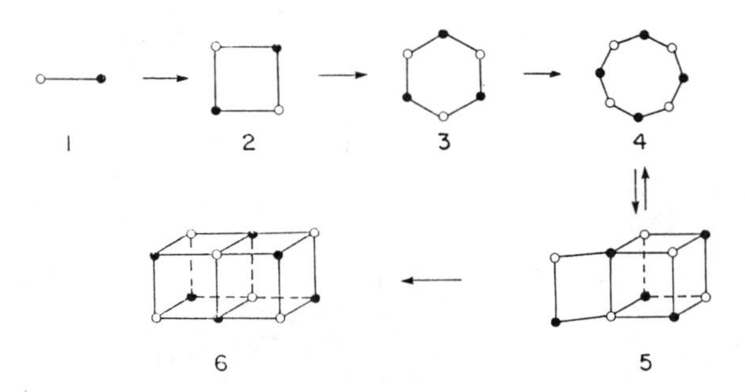


**Figure 12.** Structures of small rare-gas and ionic clusters (Refs. 125–129).

There are some compelling reasons for the study of isolated clusters. First, they provide a vehicle for the exploration of a wealth of energetic, dynamic, and chemical phenomena in large finite systems whose size can be continuously varied, thus providing the means for the exploration of the evolution of size effects in chemical dynamics. Second, these systems allow the study of microscopic aspects of macroscopic phenomena, such as nucleation, adsorption, desorption, and catalysis. Third, a variety of astrophysical applications rest on cluster physics and chemistry, including the formation mechanism and properties of cosmic dust.[121] Furthermore, clusters in the interstellar medium may be responsible for some fascinating phenomena in astrophysical spectroscopy;[122,123] for example, the undefined ultraviolet and visible absorption bands in the range 443–800 nm[123] may result from the existence of inorganic clusters, while graphite clusters may give rise to large polycyclic aromatic hydrocarbons which presumably are responsible for some infrared interstellar emission bands in the range 3.3–11 $\mu$m.[122]

Small clusters ($n = 2$–$100$) may be van der Waals molecules that are held together by "weak interactions" in rare gas, molecular, and hydrogen-bonded systems, as well as ionic and covalent structures where the chemical bonding corresponds to "strong interactions." The structures of small clusters at low temperature (Fig. 12) are different from the corresponding solid-state structures.[124–127] The cluster structure is well characterized at 0 K; at nonzero temperature the vibrational motion gives rise to the coexistence of several dominant structures (Fig. 13), which undergo isomerization.[128,129] Such isomerization within a small cluster can be thought of as the precursor of melting[120] in large clusters. All of the above pertain to chemically pure clusters. The properties of heteroclusters are also of considerable interest, as a guest molecule of different species may act as a microscopic probe for the nuclear motion and the electronic excitations of the cluster. In this context, van der Waals molecules composed of a single large aromatic molecule bound to a rare-gas atom[130–153] (Fig. 14), whose excited-state energetics (Fig. 15) have been thoroughly explored provide basic information on the microscopic characteristics of solvent perturbations. Furthermore, they provide insight into the energetics and dynamics of vibrational motion of rare gases adsorbed on microsurfaces of graphite, providing a bridge between cluster and surface phenomena.[154]

Large clusters ($n > 10^2$) exhibit a gradual evolution of the corresponding solid-state structure at low temperature. Three unique characteristics of large clusters, which differ from those of the bulk material, should be noted. First, a large surface/volume ratio can enhance the importance of microscopic catalysis[155,156] on clusters. Second, quantum size effects,[157] which show up via the dependence of some specific properties, for example, electric and magnetic susceptibilities, on the size and shape of the cluster, are of consider-

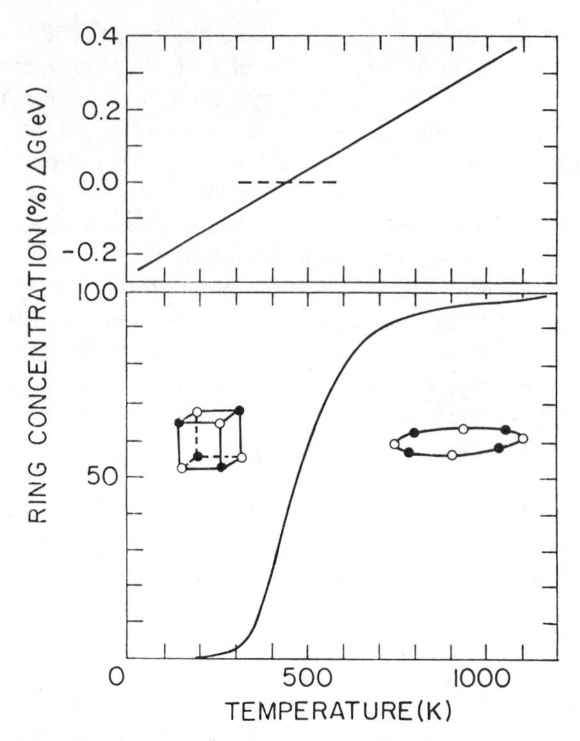


**Figure 13.** Free energy difference between the cubic and the ring-shaped cluster of $(NaCl)_4$ (top), and the temperature dependence of the relative concentration of the ring-shaped structure (bottom). Data from Refs. 128 and 129.

able interest for the elucidation of the properties of metallis clusters[157,158] and of excess electron states in nonmetallic clusters.[159] Third, large clusters exhibit a dependence of some intensive thermodynamic properties on the cluster size.[124,160,161]

Quantum size effects in large metallic clusters are related to the central, and as yet unresolved, issue of the nature of the insulator–metal transition in finite systems.[162] In this context, the discreteness of the electronic level structure of a cluster is crucial, with the level spacing between the electronic states being[157] $\delta \simeq 4E_F/3N$, where $E_F$ is the Fermi energy and $N$ is the number of filled states. In order to attain a level spacing of 1 K, the particle should have a radius of 70 Å. The theory of the electronic properties of metallic clusters rests on the statistical character of the energy-level distribution, with the spacing between the energy levels in the vicinity of the Fermi energy being taken as a random variable.[157,158] The statistical nature of the energy spectrum presumably originates from the microscopic variations in the roughness

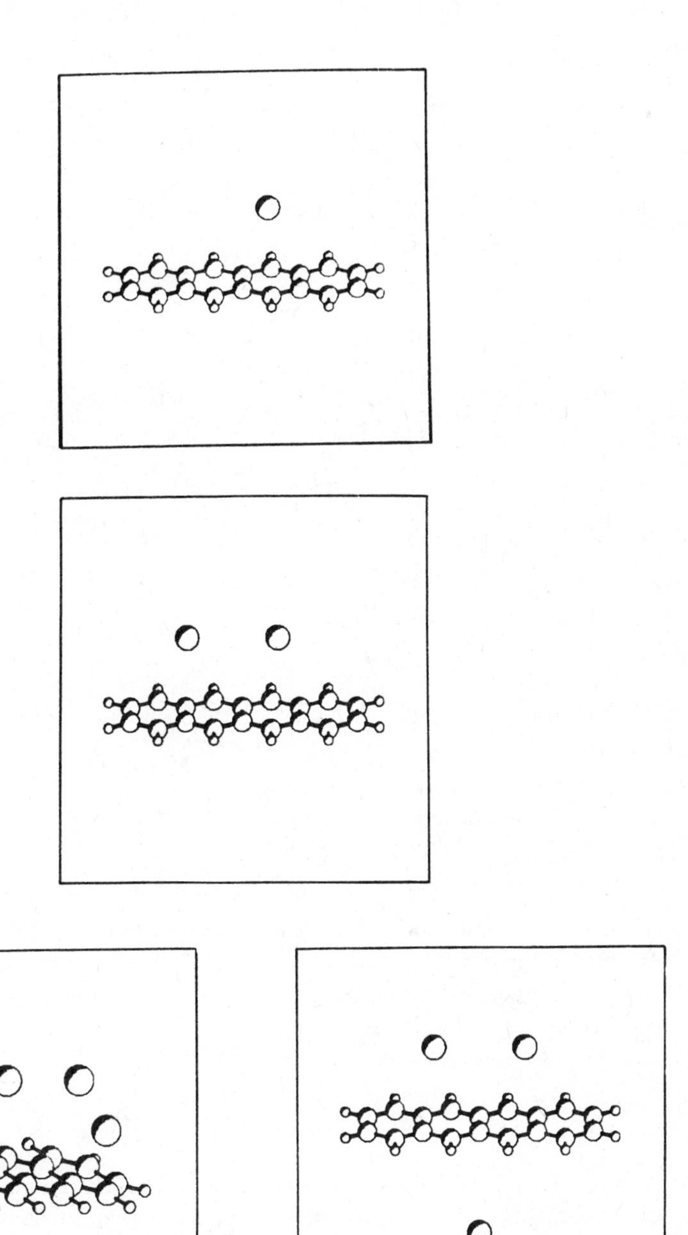

**Figure 14.** Structures of tetracene: $Ar_n$ ($n = 1, 2, 3$) complexes (Ref. 136).

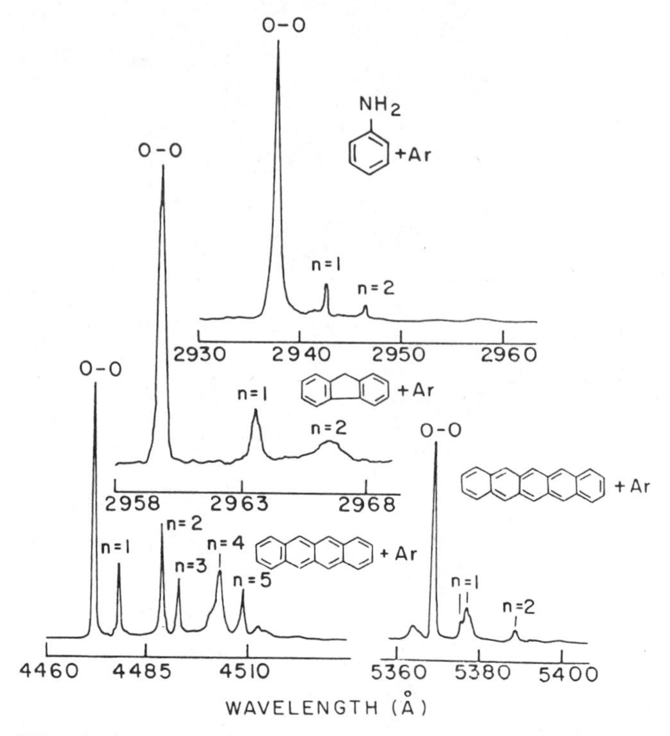


**Figure 15.** LIF excitation spectra of $S_0 \rightarrow S_1$ electronic origin (0–0) of several bare molecules M, and $M \cdot Ar_n$ complexes in supersonic expansions of Ar (Refs. 131, 132, 136). The coordination numbers, $n$, of the $M \cdot Ar_n$ complexes are marked.

of the surface, which gives rise to a random potential.[157,158] The spectrum of an ensemble of particles of the same size, characterized by a mean level spacing $\delta$ in the vicinity of $E_F$, can be described by the transformation properties of the Hamiltonian, being specified by one of the following pair distribution functions for adjacent level spacing[158]:

$$P_2^a(\Delta) = \Omega_a \Delta^{-1}(\Delta/\delta)^a \exp[-B_a(\Delta/\delta)^2] \tag{11}$$

with $a = 0$, 1, 2, and 4 indicating Poisson, orthogonal, unitary, and sympathetic distributions, while $\Omega_a$ and $B_a$ are numerical constants.[158] The most popular distribution, corresponding to the orthogonal, Wigner, type,[164] seems to apply in the limit of weak magnetic fields and small spin–orbit coupling.[158] The electronic spectrum of a metal cluster has much in common with that of highly excited vibrational states of a large polyatomic molecule, as was alluded to in Section II. In both cases a statistical description of the

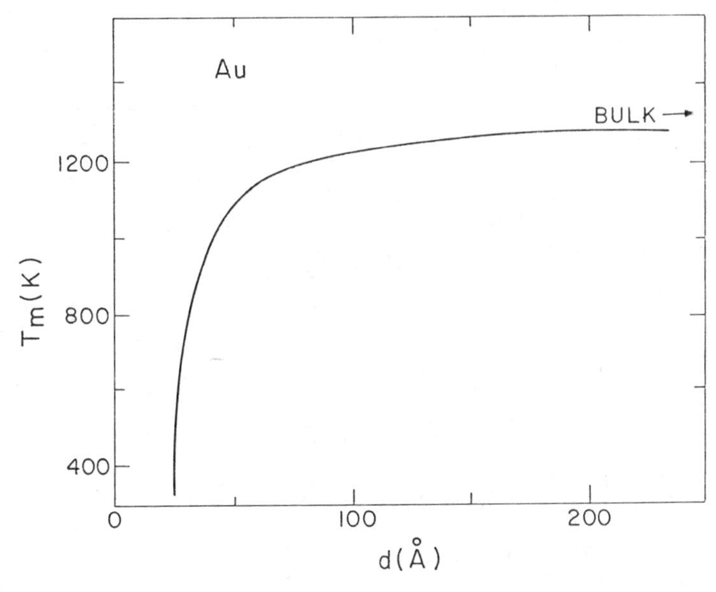


**Figure 16.** Size dependence of the melting temperature of gold clusters supported on a substrate (Ref. 167).

level structure is required. It should be noted, however, that for vibrational excitations of a molecule the statistical nature of the spectrum originates from instrinsic "random" intramolecular (anharmonic) interactions, as in the case of "random" nuclear interactions.[163] On the other hand, the level structure of a metallic particle can be traced to boundary perturbations and may not be representable by a random-matrix description.[163,165] Mapping techniques,[165] that rest on the equivalence of the eigenvalue problems for a simple operator (e.g., the Hamiltonian) in a complicated domain and a complicated operator in a simple domain may be useful in this context; their extension to the description of the time evolution of a complicated system will be of considerable interest both for molecular and cluster dynamics.

Disorder can be incorporated into clusters by either introducing local defects, for example, vacancies or dislocations, or by considering the effects of thermal excitation of the nuclear vibrations. Consider the melting of small clusters.[166–175] It was predicted in 1909[166] that the melting temperature of a droplet decreases with decreasing size. This prediction has been borne out experimentally[167] for metal clusters (Fig. 16). The applicability of simple thermodynamic arguments[161] is questionable, as phase transitions in a finite system are expected to be smeared out by fluctuations. An alternative analysis describes the melting of clusters in terms of Lindeman's hypothesis,[168,169]

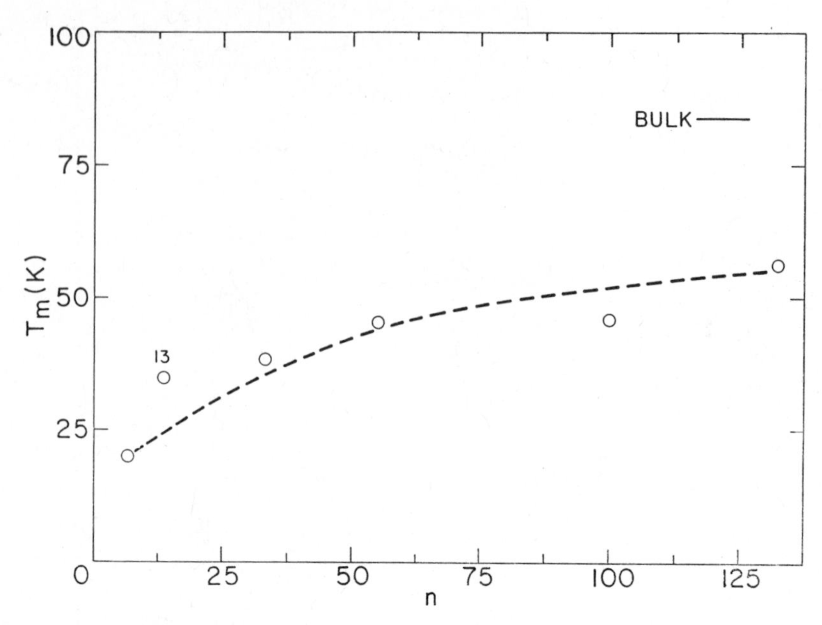


**Figure 17.** Melting temperatures, $T_m$, of $Ar_n$ ($n = 7$–$133$) microclusters and small clusters from molecular dynamics computer simulations. $T_m$ is obtained from the break in the caloric curve (Refs. 171–174).

which rests on the notion of the onset of large amplitude nuclear motion. Yet another is the dynamic description of melting.[170,174,175] Small van der Waals clusters at finite temperature have been shown, by molecular dynamics computer simulations,[171–175] to undergo a "transition" from a rigid low-temperature form to a nonrigid high-temperature form.[170] An attempt has been made[170] to associate these two forms of a finite cluster with two phases of an infinite sample of bulk material on the basis of several of the physical properties of the system, for example, a break in the caloric equation of state (Fig. 17), the specific heat, the mean coordination number, the radial distribution function, the bond length fluctuation, the power spectrum, the density of vibrational states, and the diffusion coefficient. All these attributes, with the exception of the last three, pertain both to melting and to cluster isomerization. Indeed, the melting of small clusters may be synonymous with fast isomerization between several distinct nuclear configurations. Molecular dynamics simulations of alkali-halide clusters[175] have elucidated some of the size dependence of the kinetics and dynamics of these transformations. For small clusters, distinct diffusionless isomerization is exhibited. Intermediate size clusters reveal hierarchal kinetics, with isomerization preceding the onset of the diffusion, while a genuine "sharp" melting and solid-liquid coexistence

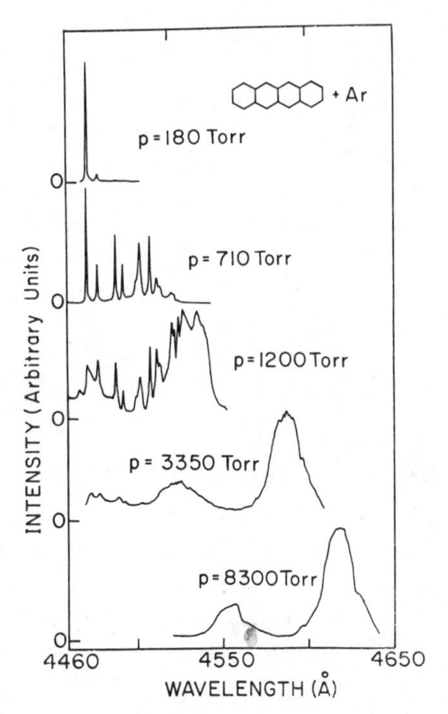


**Figure 18.** LIF excitation spectra of tetracene seeded in a supersonic expansion of Ar, exhibiting the transition from the isolated molecule (at $p = 180$ torr) to the molecule solvated in an Ar cluster ($p = 8300$ torr). Data from Ref. 153.

prevails for relatively large clusters. Such size dependences of the sequences of physical processes in clusters signal the coalescence of time scales for intrawell and interwell dynamics. The observed gradual transition from isomerization to melting in clusters establishes a connection between thermodynamic and kinetic size effects; its further exploration should help us understand the dynamics of phase transitions in finite systems.

Experimental exploration of the initial and final states of cluster isomerization can rely on structural determinations.[127] What can be learned from the interrogation of vibrational[176] or electronic excitations[120,153] of a probe molecule in a cluster? Inhomogeneous line broadening, due to the appearance of isomers or coexisting phases, within the same cluster, provides a spectroscopic signal for isomerization or melting. We note that anomalous line broadening in the infrared spectrum of $SF_6Ar_n$[176] has been attributed, on the basis of Monte Carlo simulations,[177] to isomerization. Intracluster inhomogeneous broadening is exhibited in the $S_0 \rightarrow S_1$ electronic spectrum of the tetracene molecule in an Ar cluster[120,153] (Figs. 18 and 19). The red shift of

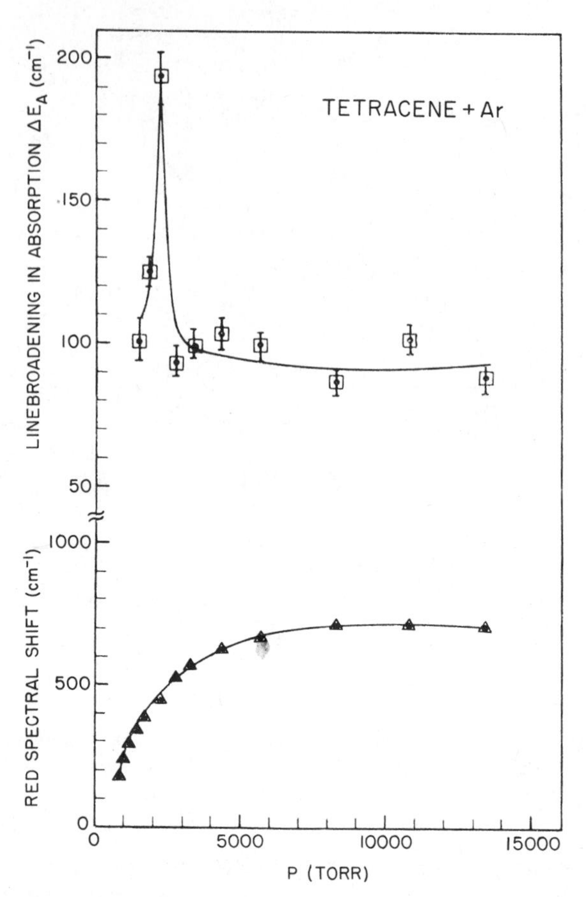


**Figure 19.** Spectral shifts relative to the 0–0 transition of the bare molecule, and the widths (FWHM), of the 0–0 transition of tetracene in Ar clusters (Ref. 153).

the spectrum of the large molecule (Fig. 18), which is dominated by dispersive interactions, illustrates the continuous modification of its optical properties upon solvation. The corresponding linewidth (Fig. 19), which exhibits a marked anomaly, may be influenced by cluster melting.

The processes of energy acquisition, storage and disposal in clusters are of considerable interest in their own right and also for the interpretation of similar processes in finite systems. Consider vibrational energy excitation of an intramolecular vibration of a molecule in a cluster, or of a cluster intermolecular mode(s), which can be accomplished by collisional excitation, photoselective vibrational excitation, electronic excitation followed by intramolecular radiationless transitions or exciton trapping.[178] In charged clusters vibrational excitation can originate from ionization followed by hole trapping (rare gases)[179–182] and from electron attachment (alkali halides).[159,183] Vibrational energy relaxation falls into two categories: (1) Reactive vibrational predissociation,[138,184,185] which results in V-T, V-R, and V-V exchange; (2) nonreactive vibrational energy redistribution, resulting in energy exchange between an intramolecular vibration and low-frequency intermolecular modes.[131] The existing extensive experimental and theoretical studies[138,184–193] of vibrational predissociation of small van der Waals molecules, for example, $HeI_2$, $(Cl_2)_2$, and $(N_2O)_2$, have focused attention on the reactive channel. A central conclusion emerging from these theoretical studies,[184] which is supported by experiment,[138,185] is that statistical theories of the rate of a unimolecular reaction fail when applied to vibrational predissociation of small van der Waals molecules. It is, of course, implied, that there is some mode selectivity in the reaction. Information regarding vibrational energy degradation in larger clusters comes from molecular dynamics computer simulations of the dissociation of ground-state $Ar_n$ ($n = 4$–$6$), which can be accounted for in terms of the statistical theory of unimolecular reactions, and which imply rapid vibrational energy randomization. Sequential vibrational energy redistribution, followed by dissociation, also seems to be ubiquitous in larger van der Waals molecules, for example, $(C_2H_4)_2$, $(NH_3)_2$, $(C_6H_6)_2$, and $(HF)_n$ ($n = 2$–$6$).[187–193] The difference between the dephasing lifetimes of some van der Waals molecules ($\approx 10^{-12}$ $s^{-1}$), obtained from line broadening, and their vibrational predissociation lifetimes ($> 10^{-9}$ s),[193] comes from the contribution of the nonreactive vibrational redistribution to the dephasing.[194] This situation is analogous to relaxation associated with channel 3 of benzene,[61] where intrastate vibrational energy redistribution in the $S_1$ manifold precedes $S_1 \rightarrow S_0$ internal conversion.

Information on size effects in intramolecular dynamics of clusters is emerging. The available data are concerned with the relationship between excited-state energies and dynamics of some large clusters (Fig. 19), and show the cluster size dependence of vibrational relaxation of an excitation of a large organic molecule in a rare-gas cluster.[120,153] Another interesting phenomenon involves the "transition" from reactive to nonreactive relaxation with increasing the cluster size. For a given excitation, reactive vibrational predissociation is characteristic to vibrational energy flow in small clusters, while in condensed phases the typical behavior is nonreactive vibrational energy redistribution. Recent molecular dynamics calculations of the dynamics of highly excited $Xe_2^*$ excimers in $Xe_2Ar_{11}$ and $Xe_2^*Ar_{53}$ clusters demonstrate the "transition" from reactive molecular-type behavior in small clusters to solid-state nonreactive behavior in a large cluster.[178]

Recently, clusters have been used as reagents for chemical reactions.[155,156,195–204] Of considerable interest are those chemical reactions

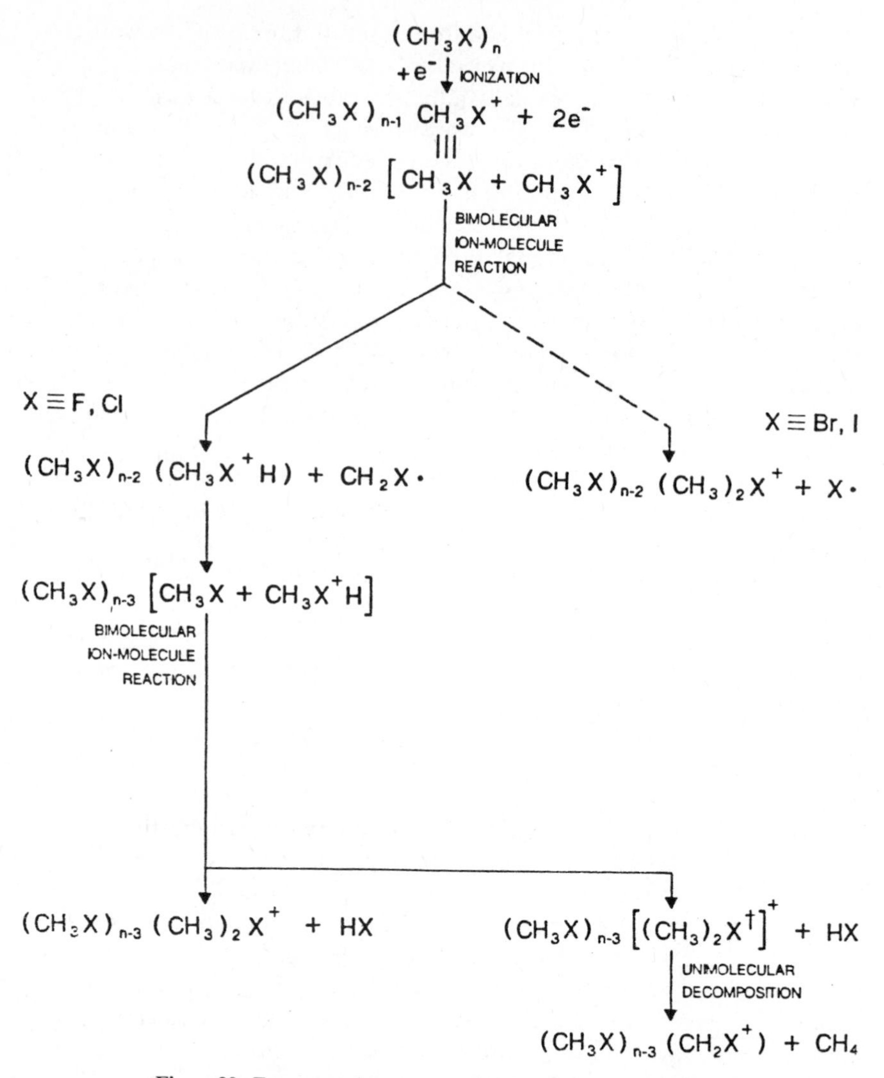


**Figure 20.** Fragmentation pattern of $(CH_3F)_n$ clusters (Ref. 205).

of the cluster which differ from molecular processes.[205] The fragmentation patterns of $(CH_3F)_n$ clusters ionized by an electron beam[205] (Fig. 20) provide novel information regarding microscopic solvent effects on covalent bond reformation and the extent of energy transfer from a reactive intermediate to the microscopic solvent environment of the cluster. There are, moreover, other exciting directions for further research. In particular, the exploration of chemical size effects in clusters is of considerable interest. In this context, one should enquire: What is the minimum size of a cluster which is required to induce a specific chemical process? The influence of chemical size effects is documented in the following systems:

1. The "critical" size of a polar cluster required to attach an electron in a localized stable state,[206,207]

$$(H_2O)_n + e \rightarrow (H_2O)_n^-, \qquad n \geq 11$$

$$(NH_2)_n + e \rightarrow (NH_3)_n^-, \qquad n \geq 35$$

2. The minimum size of a polar cluster which induces excited-state proton transfer from an aromatic naphthol[208]

$$A - OH^*(NH_3)_n \rightarrow A - O^- NH_4^+(NH_3)_{n-1}, \qquad n \geq 3$$

3. Microscopic catalysis on metal clusters,[155,156,209–211]

$$M_nD_2 \rightarrow M_n(D + D)$$

being a typical chemisorption process where M is a transition metal, exhibits a dramatic dependence on the cluster size, with the detailed reactivity pattern very different for each metal.[210,211] For Co Nb clusters the onset of reactivity is exhibited at $n = 3$, $Co_6 - Co_9$ and $Nb_8 - Nb_9$ clusters are comparatively inert, while a second onset of reactivity occurs for $Co_{10}$ and $Nb_{10}$. For $Fe_n$ clusters[209,211] two onsets of reactivity have been observed, at $n = 6$ and $n = 17$, and the reactivity is correlated with the ionization potentials of the clusters for high $n$.[209]

## References


1. J. Jortner and R. D. Levine, in *Photoselective Chemistry* (J. Jortner, R. D. Levine, and S. A. Rice, eds.), Wiley, New York, 1981, p. 1.
2. B. Pullman, J. Jortner, A. Nitzan, and R. B. Gerber (eds.), *Dynamics on Surfaces*, Reidel, Dordrecht, 1984.
3. J. Jortner and B. Pullman, *Large Finite Systems*, D. Reidel Publishing Company, Dordrecht, 1987.
4. R. D. Levine and R. B. Bernstein, *Molecular Reaction Dynamics and Chemical Reactivity*, Clarendon Press, Oxford, 1987.
5. C. E. Hamilton, J. L. Kinsey, and R. W. Field, *Ann. Rev. Phys. Chem.* **37**, 493 (1986).
6. M. J. Berry, *R. A. Welch Foundation Conferences on Chemical Research* **XXVII**, 133 (1984).
7. F. F. Crim, *Ann. Rev. Phys. Chem.* **35**, 657 (1984).
8. R. E. Smalley, *Ann. Rev. Phys. Chem.* **34**, 129 (1983).
9. K. V. Reddy, D. F. Heller, and M. J. Berry, *J. Chem. Phys.* **76**, 2814 (1982).
10. K. K. Lehmann, G. Scherer, and W. Klemperer, *J. Chem. Phys.* **77**, 2853 (1982).
11. K. K. Lehmann, G. Scherer, and W. Klemperer, *J. Chem. Phys.* **79**, 1369 (1983).

12. K. K. Lehmann and S. L. Coy, *J. Chem. Phys.* **83**, 3290 (1985).
13. D. W. Chandler, W. E. Farneth, and R. N. Zare, *J. Chem. Phys.* **77**, 4447 (1982).
14. J. M. Jasinski, J. K. Frisoli, and C. B. Moore, *J. Chem. Phys.* **79**, 1312 (1983).
15. J. W. Perry and A. H. Zewail, *J. Chem. Phys.* **80**, 5333 (1984).
16. E. J. Heller, *Acc. Chem. Res.* **14**, 368 (1981); N. De Leon and E. J. Heller, *J. Chem. Phys.* **78**, 4005 (1983).
17. D. W. Noid, M. L. Koszykowski, and R. A. Marcus, *Ann. Rev. Phys. Chem.* **32**, 267 (1981).
18. G. Hose and H. S. Taylor, *Chem. Phys.* **84**, 375 (1984).
19. C. Jaffe and W. P. Reinhardt, *J. Chem. Phys.* **77**, 5191 (1982).
20. J. G. Leopold and D. Richards, *Rev. Mod. Phys.* (to be published).
21. K. M. Christoffel and J. M. Bowman, *Chem. Phys. Lett.* **85**, 220 (1982); R. M. Roth, R. B. Gerber, and M. A. Ratner, *J. Phys. Chem.* **87**, 2376 (1983).
22. O. S. van Roosmalen, F. Iachello, R. D. Levine, and A. E. L. Dieperink, *J. Chem. Phys.* **79**, 2515 (1983).
23. B. Maessen and M. Wolfsberg, *J. Chem. Phys.* **80**, 4651 (1984); H. Romanowski, J. M. Bowman, and L. B. Harding, *J. Chem. Phys.* **82**, 4155 (1985).
24. K. M. Christoffel and J. M. Bowman, *J. Phys. Chem.* **85**, 2159 (1981); S. K. Gray, *Chem. Phys.* **75**, 67 (1983).
25. T. A. Holme and J. S. Hutchinson, *J. Chem. Phys.* **84**, 5455 (1986).
26. J. S. Hutchinson, J. T. Hynes, and W. P. Reinhardt, *J. Phys. Chem.* **90**, 3528 (1986); E. L. Sibert, J. T. Hynes, and W. P. Reinhardt, *J. Chem. Phys.* **81**, 1135 (1984).
27. D. J. Tannor, M. Blanco, and E. J. Heller, *J. Phys. Chem.* **88**, 6240 (1984).
28. S. Mukamel, *J. Phys. Chem.* **88**, 832 (1984); S. Mukamel and K. Shan, *Chem. Phys. Lett.* **117**, 489 (1985).
29. J. S. Hutchinson, *J. Chem. Phys.* **82**, 22 (1985).
30. B. A. Waite, S. K. Gray, and W. H. Miller, *J. Chem. Phys.* **78**, 259 (1983).
31. G. A. Voth, and R. A. Marcus, *J. Chem. Phys.* **84**, 2254 (1986).
32. K. N. Swamy, W. L. Hase, B. C. Garret, C. W. McCurdy, and J. F. McNutt, *J. Phys. Chem.* **90**, 3517 (1986).
33. M. S. Child, *Accts. Chem. Res.* **45**, 451 (1985).
34. A. A. Stuchebrukhov, M. V. Kuzmin, V. N. Bagratashvili, and V. S. Letokhov, *Chem. Phys.* **107**, 429 (1986).
35. R. T. Skodje, F. Borondo, and W. P. Reinhardt, *J. Chem. Phys.* **82**, 4611 (1985); B. R. Johnson, *J. Chem. Phys.* **83**, 1204 (1985); T. P. Grozdanov, S. Saini, and H. S. Taylor, *J. Chem. Phys.* **84**, 3246 (1986).
36. E. Riedle, H. J. Neusser, and E. W. Schlag, *J. Phys. Chem.* **86**, 4847 (1982).
37. N. L. Garland, E. C. Apel, and E. K. C. Lee, *Chem. Phys. Lett.* **B95**, 209 (1983).
38. H. L. Dai, C. L. Korpa, J. L. Kinsey, and R. W. Field, *J. Chem. Phys.* **82**, 1688 (1985).
39. C. S. Parmenter, *Faraday Discuss. Chem. Soc.* **75**, 7 (1983).
40. T. Uzer, *Chem. Phys. Lett.* **110**, 356 (1984); M. L. Sage, *J. Chem. Phys.* **84**, 1565 (1986).
41. R. B. Shirts, *J. Chem. Phys.* **85**, 4949 (1986).
42. I. Benjamin, R. D. Levine, and J. L. Kinsey, *J. Phys. Chem.* **87**, 727 (1983).
43. O. S. van Roosmalen, I. Benjamin, and R. D. Levine, *J. Chem. Phys.* **81**, 5986 (1984).
44. I. Benjamin, R. H. Bisseling, R. Kosloff, R. D. Levine, J. Manz, and H. R. H. Schor, *Chem. Phys. Lett.* **116**, 255 (1985).
45. K. K. Lehmann, *J. Chem. Phys.* **79**, 1098 (1983).
46. R. D. Levine and J. L. Kinsey, *J. Phys. Chem.* **90**, 3653 (1986).
47. E. Abramson, R. W. Field, D. Imre, and J. L. Kinsey, *J. Chem. Phys.* **83**, 453 (1985).
48. J. P. Pique, Y. Chen. R. W. Field and J. L. Kinsey, *Phys. Rev. Lett.* **58**, 475 (1987).
49. I. C. Percival, *Adv. Chem. Phys.* **36**, 1 (1977).

50. E. B. Stechel and E. J. Heller, *Ann. Rev. Phys. Chem.* **35**, 563 (1984).
51. R. L. Sundberg, E. Abramson, J. L. Kinsey, and R. W. Field, *J. Chem. Phys.* **83**, 466 (1985).
52. L. Leviandier, M. Lombardi, R. Jost, and J. P. Pique, *Phys. Rev. Lett.* **56**, 2449 (1986).
53. P. Pechukas, *J. Phys. Chem.* **88**, 4823 (1984).
54. F. J. Dyson, *J. Math. Phys.* **3**, 405 (1962); M. L. Metha, *Random Matrices*, Academic Press, New York, 1967.
55. J. P. Pique, R. D. Levine, R. W. Field, and J. L. Kinsey (unpublished).
56. Y. Alhassid and R. D. Levine, *Phys. Rev. Lett.* **57**, 2879 (1986); J. Brickmann, Y. M. Engel, and R. D. Levine, *Chem. Phys. Lett.* (to be published).
57. E. J. Heller and R. L. Sundberg, in *Chaotic Behavior in Quantum Systems* (G. Casati, ed.), Plenum, New York, 1985.
58. D. R. Guyer, W. F. Polik, and C. B. Moore, *J. Chem. Phys.* **84**, 6519 (1986).
59. J. W. Thoman, J. I. Steinfield, R. I. McKay, and A. E. W. Knight, *J. Chem. Phys.* (to be published).
60. R. D. Levine, in this volume.
61. U. U. Schubert, E. Riedle, H. J. Neusser, and E. W. Schlag, *J. Chem. Phys.* **84**, 6182 (1986).
62. A. Ben-Shaul, Y. Haas, K. L. Kompa, and R. D. Levine, *Lasers and Chemical Change*, Springer Verlag, Berlin, 1981.
63. R. B. Bernstein, *Chemical Dynamics via Molecular Beam and Laser Techniques*, Clarendon Press, Oxford 1982.
64. A. H. Zewail (ed.), *Photochemistry and Photobiology*, Harwood, Schur, 1983.
65. R. Vetter and J. Vigue (eds.), *Recent Advances in Molecular Reaction Dynamics*, Edition CNRS, Paris, 1986.
66. S. R. Leone, *Science* **227**, 889 (1985).
67. W. D. Lawrence, C. B. Moore, and H. Petek, *Science* **227**, 895 (1985).
68. W. M. Jackson and A. B. Harvey (eds.), *Lasers as Reactants and Probes in Chemistry*, Howard University Press, Washington, 1985.
69. V. S. Letokhov, *Nonlinear Laser Chemistry*, Springer Verlag, Berlin, 1983.
70. H. Reisler and C. Wittig, *Ann. Rev. Phys. Chem.* **37**, 307 (1986).
71. S. H. Lin (ed.), *Advances in Multiphoton Processes* (to be published).
72. S. H. Lin, Y. Fujimura, J. J. Neusser, and E. W. Schlag, *Multiphoton Spectroscopy of Molecules*, Academic Press, New York, 1984.
73. J. K. Ku, D. W. Setser, and D. Oba, *Chem. Phys. Lett.* **109**, 429 (1984).
74. T. C. Maguire, P. R. Brooks, and R. F. Curl, Jr., *Phys. Rev. Lett.* **50**, 1918 (1983).
75. H. J. Foth, J. C. Polanyi, and H. H. Telle, *J. Phys. Chem.* **86**, 5027 (1982).
76. D. Imre, J. L. Kinsey, A. Sinha, and J. Krenos, *J. Phys. Chem.* **88**, 3956 (1984).
77. W. P. Moskowitz, B. Steward, R. M. Bilotta, J. L. Kinsey, and D. E. Pritchard, *J. Chem. Phys.* **80**, 5496 (1984).
78. R. Schmiedl, H. Dugan, W. Meier, and K. H. Welge, *Z. Phys. A* **403**, 137 (1982).
79. G. E. Hall, N. Sivakumar, P. L. Houston, and I. Burak, *Phys. Rev. Lett.* **56**, 1671 (1986).
80. Z. Yu, B. Koplitz, S. Buelow, D. Baugh, and C. Wittig *Chem. Phys. Lett.* **127**, 534 (1986).
81. R. N. Zare, *Ber. Bunsenges Phys. Chem.* **86**, 422 (1982).
82. R. Vasudev, R. N. Zare, and R. N. Dixon, *J. Chem. Phys.* **80**, 4863 (1984).
83. I. Nadler, D. Mahgerefteh, H. Reisler, and C. Wittig, *J. Chem. Phys.* **82**, 3885 (1985).
84. P. Andresen, G. S. Ondrey, B. Titze, and E. Rothe, *J. Chem. Phys.* **80**, 2548 (1984).
85. M. Dubs, U. Brühlmann, and J. R. Huber, *J. Chem. Phys.* **84**, 3106 (1986).
86. G. E. Hall, N. Sivakumar, and P. L. Houston, *J. Chem. Phys.* **84**, 2120 (1986); P. L. Houston, *J. Phys. Chem.* (to be published).
87. S. Klee, K. H. Gericke, and F. J. Comes, *J. Chem. Phys.* **85**, 40 (1986).
88. M. P. Docker, A. Hodgson and J. P. Simons, *Discuss. Faraday Soc.* **82**, (1986).

89. W. H. Breckenridge, C. Jouvet, and B. Soep, *J. Chem. Phys.* **84**, 1443 (1986).
90. S. Buelow, M. Noble, G. Radharkrishnan, H. Reisler, C. Wittig, and G. Hanock, *J. Phys. Chem.* **90**, 1015 (1986).
91. W. H. Breckenridge and H. Umemoto, *J. Chem. Phys.* **80**, 4168 (1984).
92. C. T. Rettner and R. N. Zare, *J. Chem. Phys.* **77**, 2416 (1982).
93. H. J. Yuh and P. J. Dagdigian, *J. Chem. Phys.* **79**, 2086 (1983).
94. R. Vetter, *Laser Chem.* **6**, 149 (1986).
95. H. Schmidt, P. S. Weiss, J. M. Mestdagh, M. H. Covinsky, and Y. T. Lee, *Chem. Phys. Lett.* **118**, 539 (1985).
96. M. O. Hale, I. V. Hertel, and S. R. Leone, *Phys. Rev. Lett.* **53**, 2296 (1984); I. V. Hertel, H. Schmidt, A. Bahring, and E. Meyer, *Rep. Prog. Phys.* **48**, 375 (1985).
97. M. H. Alexander and P. J. Dagdigian, *J. Chem. Phys.* **83**, 2191 (1985).
98. P. L. Houston, *Adv. Chem. Phys.* **47**, 625 (1981).
99. S. R. Leone, *Ann. Rev. Phys. Chem.* **35**, 109 (1984).
100. J. Wolfrum, *Laser Chem.* **6**, 125 (1986).
101. E. E. Marinero, C. T. Rettner, and R. N. Zare, *J. Chem. Phys.* **80**, 4142 (1984).
102. D. P. Gerrity and J. J. Valentini, *J. Chem. Phys.* **81**, 1298 (1984); **82**, 1328 (1985).
103. K. Tsukiyama, B. Katz, and R. Bersohn, *J. Chem. Phys.* **84**, 1934 (1986).
104. J. L. Knee, L. R. Khundar, and A. H. Zewail, *J. Chem. Phys.* **83**, 1966 (1985).
105. H. J. Neusser, H. Kühlewind, U. Boesl, and E. S. Schlag, *Ber. Bunsenges Phys. Chem.* **89**, 276 (1985).
106. J. L. Durant, D. M. Rider, S. L. Anderson, F. D. Proch, and R. N. Zare, *J. Chem. Phys.* **80**, 1817 (1984).
107. M. J. Davis and S. K. Gray, *J. Chem. Phys.* **84**, 5389 (1986); S. K. Gray, S. A. Rice, and M. J. Davis, *J. Phys. Chem.* **90**, 3476 (1986).
108. C. Wittig, I. Nadler, H. Reisler, M. Noble, J. Catanzarite, and G. Radhakrishnan, *J. Chem. Phys.* **83**, 5581 (1985); J. Troe, *J. Chem. Phys.* **85**, 1708 (1986).
109. K. S. J. Nordholm and S. A. Rice, *J. Chem. Phys.* **61**, 213 (1974); **61**, 768 (1974).
110. D. J. Tannor and S. A. Rice, *J. Chem. Phys.* **83**, 5013 (1985); D. J. Tannor, R. Kosloff, and S. A. Rice, *J. Chem. Phys.* **85**, 5805 (1986).
111. E. T. Sleva, M. Glasbeek, and A. H. Zewail, *J. Phys. Chem.* **90**, 1232 (1986).
112. P. Brumer and M. Shapiro, *Discuss. Faraday Soc.* **82** (1986).
113. Proceedings of International Meeting on Small Particles and Inorganic Clusters, *J. Phys. (Paris) Colloque* **C-2**, 38 (1977).
114. Proceedings of the Second International Meeting on Small Particles and Inorganic Clusters, Lausanne, 1980, *Surface Sci.* **106** 1–608 (1981).
115. Proceedings of Bunsengesellschaft Discussion Meeting on Experiments on Clusters, Koningstein, 1983, *Ber. Bunsenges, Phys. Chem.* **88** (1984).
116. Proceedings of the Third International Meeting on Small Particles and Inorganic Clusters, Berlin 1985, *Surface Sci.* **156**, I-1072 (1985).
117. J. Friedel, *J. Phys.* (*Paris*) **38**, C2, 1 (1977).
118. E. W. Becker, K. Bier, and W. Henkens, *Z. Physik.* **146**, 33 (1956).
119. O. F. Hagena and W. Obert, *J. Chem. Phys.* **56**, 1793 (1972).
120. J. Jortner, *Ber. Bunsenges. Phys. Chem.* **88**, 188 (1984).
121. B. Donn, J. Hecht, R. Khanna, J. Nuth, D. Stranz, and A. B. Anderson, *Surface Sci.* **106**, 576 (1981).
122. Leger, L. d'Hendecourt, and N. Boccara (eds.), *Polycyclic Aromatic Hydrocarbons and Astrophysics* NATO ASI Series Vol. 191, D. Reidel Company, Dordrecht, 1987.
123. G. H. Herbig, *Astrophys. J.* **196**, 129 (1975).
124. H. P. Blates, *J. Phys.* (*Paris*) **38**, C2, 153 (1977).

125. J. Barker, *J. Phys.* (*Paris*) **38**, C2, 37 (1977).
126. M. R. Hoare, *Adv. Chem. Phys.* **40**, 49 (1979).
127. J. Farges, M. F. de Feraudy, B. Raoult, and G. Torchet, *J. Chem. Phys.* **78**, 5067 (1983).
128. T. P. Martin, *Phys. Rep.* **95**, 169 (1983).
129. U. Landman, D. Scharf, and J. Jortner, *Isomerization and Melting in Alkali Halide Clusters* (unpublished).
130. A. Amirav, U. Even, and J. Jortner, *Chem. Phys. Lett.* **67**, 9 (1979).
131. A. Amirav, U. Even, and J. Jortner, *J. Chem. Phys.* **75**, 2489 (1981).
132. A. Amirav, U. Even, and J. Jortner, *J. Phys. Chem.* **85**, 309 (1981).
133. A. Amirav, U. Even, and J. Jortner, *J. Chem. Phys.* **67**, 1 (1982).
134. S. Leutwyler, U. Even, and J. Jorner, *Chem. Phys. Lett.* **86**, 439 (1982).
135. S. Leutwyler, U. Even, and J. Jortner, *J. Chem. Phys.* **79**, 5769 (1983).
136. M. J. Ondrechen, Z. Berkovitch-Yellin, and J. Jortner, *J. Am. Chem. Soc.* **103**, 6586 (1981).
137. U. Even, A. Amirav, S. Leutwyler, M. J. Ondrechen. Z. Berkovitch-Yellin, and J. Jortner, *J. Farad. Discuss. Chem. Soc.* **73**, 153 (1982).
138. D. H. Levy, *Advances in Chemical Physics*, Wiley Interscience, New York, 1981, Vol. 47, p. 323.
139. T. R. Heyes, W. Wenke, H. L. Selzle, and E. W. Schlag, *Chem. Phys. Lett.* **77**, 19 (1980).
140. W. E. Henke, W. Yu, H. L. Selzle, E. W. Schlag, D. Witz, and S. H. Lin, *Chem. Phys.* **92**, 187 (1985).
141. I. Raitt, A. M. Griffiths, and P. A. Freedman, *Chem. Phys. Lett.* **80**, 225 (1981).
142. A. M. Griffiths, and P. A. Freedman, *Chem. Phys.* **63**, 469 (1981).
143. A. Amirav and J. Jortner, *Chem. Phys.* **85**, 19 (1984).
144. A. Amirav, M. Sonnenschein, and J. Jortner, *J. Phys. Chem.* **88**, 199 (1984).
145. S. Leutwyler, A. Schmelzer, and R. J. Meyer, *Chem. Phys.* **79**, 4385 (1983).
146. S. Leutwler, *J. Chem. Phys.* **81**, 5480 (1984).
147. S. Leutwyler, *Chem. Phys. Lett.* **107**, 284 (1984).
148. J. C. Kettley, T. F. Palmer, and J. P. Simmons, *Chem. Phys. Lett.* **115**, 40 (1985).
149. M. M. Doxtadler, P. M. Gulis, S. A. Schwarz, and M. R. Topp, *Chem. Phys. Lett.* **112**, 483 (1984).
150. J. Jortner, U. Even, S. Leutwyler, and Z. Berkovitch-Yellin, *J. Chem. Phys.* **78**, 309 (1981).
151. U. Even, J. Jortner, and Z. Berkovitch-Yellin, *Canad. J. Chem.* **63**, 2073 (1985).
152. A. Amirav, U. Even, and J. Jortner, *Chem. Phys. Lett.* **72**, 16 (1980).
153. A. Amirav, U. Even, and J. Jortner, *J. Phys. Chem.* **86**, 3345 (1982).
154. S. Leutwyler and J. Jortner (unpublished).
155. M. D. Morse and R. E. Smalley, *Ber. Bunsenges. Phys. Chem.* **88**, 208 (1984).
156. R. L. Whetten, D. M. Cox, D. J. Tevor, and A. Kaldor. *Surface Sci.* **156**, 8 (1985).
157. R. Kubo, *J. Phys.* (*Paris*) **38**, C2, 270 (1977).
158. W. P. Halperin, *Rev. Mod. Phys.* **58**, 553 (1986).
159. U. Landman, D. Scharf, and J. Jortner, *Phys. Rev. Lett.* **54**, 1860 (1985).
160. J. P. Borel, *Surface Sci.* **106**, 1 (1981).
161. W. M. Hasegawa, M. Watabe, and K. Hoshino, *Surface Sci.* **106**, 11 (1981).
162. R. Kubo, *J. Phys. Soc. Japan* **17**, 975 (1962).
163. T. A. Brody, J. Flores, J. B. French, P. A. Mello, A. Pandy, and S. S. M. Wong, *Rev. Mod. Phys.* **53**, 385 (1981).
164. E. P. Winger, *Ann. Math.* **53**, 36 (1951); **62**, 548 (1955).
165. J. Barojas, E. Cota, E. Blaister-Barojas, J. Flores, and P. A. Mello, *Ann. Phys.* (*N.Y.*) **107**, 95 (1977).
166. P. Pawlow, *Z. Phys. Chem.* **66**, 545 (1909).
167. D. A. Buffat and J. P. Borel, *Phys. Rev. A* **13**, 2289 (1976).

168. P. R. Couchaman and C. L. Ryan, *Phil. Mag. A* **37**, 369 (1978).
169. K. Hoshino and S. Shimamura, *Phil. Mag. A* **40**, 137 (1979).
170. G. Natanson, F. Amar, and R. S. Berry, *J. Chem. Phys.* **78**, 399 (1983).
171. C. L. Briant and J. J. Burton, *J. Chem. Phys.* **63**, 2045 (1975).
172. J. B. Kaelberer and R. D. Etters, *J. Chem. Phys.* **66**, 3233 (1977).
173. R. D. Etters and J. B. Kaelberer, *J. Chem. Phys.* **66**, 5112 (1977).
174. J. Jellinek, T. L. Beck, and R. S. Berry, *J. Chem. Phys.* **84**, 2783 (1986).
175. J. Luo, U. Landman, and J. Jortner, in *Physics and Chemistry of Small Clusters*, (P. Jena, ed.) Plenum Press, New York, 1986.
176. T. E. Gough, D. G. Knight, and G. Scoles, *Chem. Phys. Lett.* **97**, 155 (1983).
177. D. Eichenauer and R. J. Le Roy, *Phys. Rev. Lett.* (to be published).
178. D. Scharf, J. Jortner, and U. Landman, *Chem. Phys. Lett.* **126**, 495 (1986).
179. H. Haberland, *Surface Sci.* **156**, 305 (1985).
180. J. J. Saenz, J. M. Soler, and N. Garcia, *Surface Sci.* **156**, 121 (1985).
181. J. J. Saenz, J. M. Soler, N. Garcia, and O. Echt, Chem. *Phys. Lett.* **109**, 71 (1984).
182. E. E. Polymeropoulos and J. Brickmann, *Surface Sci.* **156**, 563 (1985).
183. U. Landman, R. N. Barnett, C. L. Cleveland, D. Scharf, and J. Jortner, *J. Phys. Chem.* (to be published).
184. J. A. Beswick and J. Jortner, *Adv. Chem. Phys.* **47**, 363 (1981).
185. K. C. Janda, *Adv. Chem. Phys.* **60**, 201 (1985).
186. J. W. Brody and J. D. Doll, J. Chem. Phys. **73**, 2767 (1980).
187. M. F. Vernon, J. M. Lisy, A. S. Kwok, D. J. Krajnovich, A. Tramer, Y. R. Shen, and Y. T. Lee, *J. Phys. Chem.* **85**, 3327 (1981).
188. T. E. Gough, R. E. Miller, and G. Scoles, *J. Chem. Phys.* **69**, 1588 (1978).
189. M. P. Casassa, D. S. Bomse, J. L. Beauchamp, and K. C. Janda, *J. Chem. Phys.* **72**, 6805 (1980).
190. M. A. Hoffbauer, K. Liu, C. F. Giese, and W. R. Gentry, *J. Chem. Phys.* **78**, 5567 (1983).
191. M. J. Howard, S. Burdenski, C. G. Fiese, and W. R. Gentry, *J. Chem. Phys.* **80**, 4137 (1984).
192. A. Mitchell, M. J. McAuliffe, C. F. Giese, and W. R. Gentry, *J. Chem. Phys.* **83**, 4271 (1985).
193. R. D. Johnson, S. Burdenski, M. A. Hoffbauer, C. F. Giese, and W. R. Gentry, *J. Chem. Phys.* **84**, 2624 (1986).
194. W. R. Gentry (private communication).
195. J. C. Whitehead and R. Grice, *Discussions Faraday Soc.* **55**, 320 (1973).
196. D. L. King, D. A. Dixon, and D. R. Herschbach, *J. Am. Chem. Soc.* **96**, 3328 (1974).
197. R. B. Behrens Jr., A. Freedman, R. R. Herm, and T. P. Parr, *J. Chem. Phys.* **63**, 4622 (1975).
198. A. Gonzalez Urena, R. B. Bernstein, and G. R. Phillips, *J. Chem. Phys.* **62**, 1818 (1975).
199. D. J. Wren and M. Menzinger, *Chem. Phys.* **66**, 85 (1982).
200. J. Nieman and R. Naaman, *Chem. Phys.* **90**, 407 (1984).
201. A. J. Stace and A. K. Shukla, *J. Phys. Chem.* **86**, 865 (1982).
202. N. Nishi, K. Tamamoto, H. Shinohara, U. Nagashima, and T. Okyama, *Chem. Phys. Lett.* **122**, 599 (1985).
203. Y. Oono and C. Y. Ng, *J. Am. Chem. Soc.* **104**, 4752 (1982).
204. A. J. Stace, *J. Am. Chem. Soc.* **107**, 755 (1985).
205. J. F. Garvey and R. B. Bernstein, *Chem. Phys. Lett.* **126**, 394 (1986).
206. H. Haberland, H. G. Schindler, and D. R. Worsnop, *J. Phys. Chem.* **88**, 3903 (1984).
207. J. V. Coe, D. R. Worsnop and K. H. Bowen, *J. Chem. Phys.* (to be published).
208. O. Cheshnovsky and S. Leutwyler, *Chem. Phys. Lett.* **121**, 1 (1985).
209. R. L. Whetten, D. M. Cox, D. J. Trevor, and A. Kaldor, *Phys. Rev. Lett.* **54**, 1494 (1985).
210. M. E. Geusic, M. D. Morse, and R. E. Smalley, *J. Chem. Phys.* **82**, 590 (1985).
211. M. D. Morse, M. E. Geusic, J. R. Heath, and R. E. Smalley, *J. Chem. Phys.* **83**, 2293 (1985).

# INDEPENDENT AND COLLECTIVE BEHAVIOR WITHIN ATOMS AND MOLECULES

R. STEPHEN BERRY AND JEFFREY L. KRAUSE

*Department of Chemistry and the James Franck Institute, The University of Chicago, Chicago, Illinois 60637*

## CONTENTS



## I. INTRODUCTION

The traditional model of atomic structure has been the quantum-mechanical analogue of a solar system, with electrons perturbing each other only a little from their individual states of definite energy and angular momentum. On the other hand, the traditional model for molecular structure has been the quantum-mechanical counterpart of balls held together by springs in a fairly rigid, well-defined structure. These two pictures are so different and lend themselves to such different computational methods that they have remained as separated, almost unrelated fields within chemical physics.

Beginning less than 10 years ago, the independent-particle model for atoms was challenged, first for a specific set of rather exotic states of helium and, more recently, for the ground and ordinary excited states of the alkaline earth atoms Be, Mg, Ca, Sr, and Ba. Evidence has been building that the quantization in these two-electron and quasi-two-electron atoms corresponds to collective, moleculelike behavior, rather than to independent-particle-like behavior.

The emerging picture is one in which the quantum-mechanical equivalents of the constants of motion for the two valence electrons in these atoms are like those associated with the near-rigid rotations, bending vibrations, and stretching vibrations we normally associate with linear triatomic molecules. These new results bring into question the range of validity of the nearly-independent-particle model, the quantum-mechanical counterpart of Bohr's planetary model, for atoms with more than one valence electron.

In this chapter we review the recent history of and evidence for collective, moleculelike behavior of valence electrons in atoms and indicate some of the questions that will have to be explored in order to resolve the question of how well the electrons in atoms are described by independent-particle or collective models. We then turn the question around and ask whether atoms in a molecule could, under suitable circumstances, display independent-particle behavior, with their own one-particle angular momenta behaving like near-constants of the motion. The larger question that emerges is then one of whether few-body systems—the valence electrons of an atom, the atoms that constitute a small polyatomic molecule, and perhaps others such as the nucleons in a nucleus, all of which have heretofore seemed nearly unrelated—share characteristics to the extent that we can devise a unifying picture of the dynamics of few-body systems that will expose their commonalities as well as their obvious differences.

## II. ORIGINS OF THE MOLECULAR PICTURE OF ATOMS

The collective, moleculelike model for atoms grew from the interpretation of states of doubly excited helium, He**. The absorption spectrum of helium gas at energies of about 57–64 eV, well above the first ionization threshold, exhibits a series of striking, rather sharp features[1] that converge to the second ionization threshold, corresponding to $He^+$ (2*s* or 2*p*) + *e*. These features were attributed to autoionizing states—transient states or resonances with lifetimes long enough to justify assignments with quantum numbers like those of bound states—which, in this case, correspond to one electron bound to an *excited* $He^+$ ion. Thus, in these states of He**, the total energy is enough to form $He^+$(1*s*) + *e*, but, in the state produced by the particular excitation process, that total energy is shared by the two electrons, with one in the second shell (principal quantum number $n = 2$) and the other in that shell or some higher one.

The interpretation of these autoionizing excited states even from the outset recognized the inadequacy of the independent-particle picture.[2] To account for the number of observed series and their intensities, it was necessary to invoke strong mixing of configurations, each with its own individual-particle quantum numbers. To obtain even a minimally satisfactory description of the observed spectra, one had to use a superposition of at least two configurations (antisymmetrized products of one-electron orbital functions) such as this combination for the lowest-energy doubly excited state of He (designated in the independent-particle picture as "$2s^2$"):

$$\psi(\text{“}2s^{2}\text{”}\,^{1}S^{e}) \sim \psi_{2s}(1)\psi_{2s}(2) + x\psi_{2p}(1)\psi_{2p}(2), \tag{1}$$

where $\psi$ is the wavefunction for the two-electron system and $\psi_{nl}(j)$ indicates the one-electron orbital of electron $j$ with principal quantum number $n$ and orbital quantum number $l$. The numerical parameter $x$ lies between $\pm 1$ and indicates by its deviation from zero how badly the independent-particle model is spoiled.

Several approaches were pursued in the process of finding an interpretation more physically intuitive than the somewhat hollow, *ex post facto* interpretation of Eq. (1), that is, that the independent-particle picture represented by a single configuration is "spoiled" by electron–electron correlation, particularly by angular correlation, because the second configuration leaves the radial distribution relatively unaffected but changes the angular distribution.

Accuracy of calculation has not been a problem for some time. For example, elaborate calculations based on many variational terms in a Hylleraas representation have been carried out by Bhatia and co-workers,[3] and highly accurate complex-scaling calculations have been performed by Ho.[4] A representation based on collective "hyperspherical" coordinates

$$R \equiv \sqrt{r_1^2 + r_2^2},\ \alpha \equiv \arctan(r_1/r_2) \text{ and } \theta_{12},$$

where $\theta_{12}$ is the angle between the one-electron position vectors $\mathbf{r}_1$ and $\mathbf{r}_2$, was developed rigorously[5–7] and continues to be used[8] to help complete the interpretation.* Recently, Frey and Howard have analyzed the effects of

* Incautiously, in our view, because of the lack of justification of an "adiabatic" approximation frequently made in the separation of variables in this approach. The so-called "adiabatic approximation" is suggested by its mathematical convenience, but this use of the hyperspherical representation contradicts both physical intuition and assumptions regarding the treatment of "fast" and "slow" variables made in the Born–Oppenheimer approximation, which is supposed to be the model for this separation of variables. In the "adiabatic" hyperspherical approach, the coordinate $R$ is held constant and solutions are found corresponding to eigenvalues and eigenfunctions for the other variables $\alpha$ and $\theta_{12}$. These eigenvalues, depending parametrically on $R$, provide an "effective potential" for the Schrödinger equation for behavior with respect to $R$. However, the motion with respect to $R$ is the *fastest*, rather than the slowest among all the degrees of freedom; that is, the level spacings associated with excitations in the $R$ variable are the largest spacings in the system's spectrum. In short, the hyperspherical method badly needs an analysis specifying necessary and sufficient conditions for the validity of its "adiabatic" approximation, in which $R$ is treated as if it were the "slowest" variable rather than the "fastest." Reference 9 seems to be the only source to address this point.

nonadiabatic couplings in such calculations for the ground states of He and $H^-$ and found them to be quite important.[9] A group-theoretical approach was found[10-12] whose starting point approximates helium with two independent hydrogenlike electrons and whose next step is the introduction of electron–electron repulsion as a strong perturbation. The wavefunctions based on that group-theoretical work were analyzed to elucidate how the electrons are spatially correlated.[13] This chapter focuses on the inferences derived from the group-theoretical approach (but not on the group-theoretical analysis itself); all but the most recent work on the hyperspherical approach has been reviewed recently.[7] One connection between the hyperspherical and group-theoretical methods was made by Lin and Macek[14] between one of the earliest, functionally limited group-theoretical approaches,[11] a hydrogenic-based configuration interaction expansion, and functions derived by the adiabatic hyperspherical method. The inferences made in this work may be quite dependent on the very limited basis set of Herrick and Sinanoğlu, and are not necessarily characteristic of the collective quantization inherent in the group-theoretical approach. A more general connection has been given recently by Watanabe and Lin,[15] which may develop enough to expose fully the relationships between the two representations for two-electron systems, even for two electrons in an effective core potential, as is often used to describe alkaline earth atoms. A somewhat different molecular model for the doubly excited states of He has been proposed recently by Feagin and Briggs.[16] They describe these states in terms of the molecular of adiabatic potential curves scaled from those of the $H_2^+$ molecular ion. Finally, by beginning with the classical equations of motion, Klar has recently derived the equilibrium configurations for two-electron atoms and shown that semiclassical quantization of the rotation yields two series of rotational levels converging to the double-ionization threshold.[17]

## III. THE COLLECTIVE MOLECULELIKE PICTURE OF He**

The group-theoretical work based on the electron–electron interaction breaking the especially high symmetry of one electron attached to a point nucleus led to the emergence of the molecular picture. This work was inspired by the discovery by Wulfman[10] and Herrick and Sinanoğlu[11] that the mixing coefficients induced by Coulomb repulsion in the doubly excited states of He could be predicted by group-theoretical techniques. Kellman and Herrick[18] then recognized that the designations and energy level spacings of some of the states of He** seem to define terminating rotor series analogous to those long known for nuclei.[19] Next, a supermultiplet pattern[20] provided a phenomenological organization of all the levels designated by equal values of the principal quantum numbers $n_1$ and $n_2$. This work culminated with the identification of

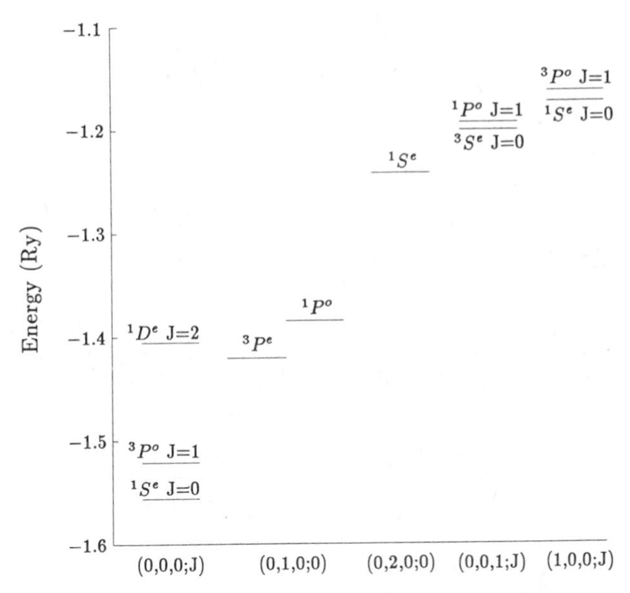


**Figure 1.** Energy levels of the doubly excited helium atom with principal quantum numbers $n_1 = n_2 = 2$, and the four lowest levels with $n_1 = 2$, $n_2 = 3$. At left, all the levels are shown on a common scale; across the diagram, the same levels are shown with quantum numbers below indicating the kind of excitation as $(v_1, v_2, v_3; J)$, with $v_1$ being the number of quanta in the symmetric stretching mode, $v_2$ being the number of quanta in the bending mode, $v_3$ being the number of quanta in the antisymmetric stretching mode, and $J$ being the number of quanta of (near) rigid rotation.

the supermultiplet structure equivalent to that of a linear rotor–vibrator composed of two electrons on opposite sides of the heavy nucleus, an $e^-$–$He^{2+}$–$e^-$ structure analogous to a linear H–C–H molecule.[21] All the states of He** with $n_1 = n_2$ can be described in terms of excitations of a doubly degenerate bending vibration associated with the $e^-$–$He^{2+}$–$e^-$ angle and the rotations of this three-body system. The energy levels of He** for $n_1 = n_2 = 2$ and some for $n_1 = 2, n_2 = 3$ are shown in Fig. 1, on a single scale at the left, and arrayed in the rotor–vibrator pattern in the main figure.

If the rotor–vibrator model of He** is correct, we reasoned, then the spatial distributions of electron probability density *in the internal* $e^-$–$He^{2+}$–$e^-$ *coordinate system* should exhibit specific relationships implied by the model. A series of these distributions is shown in Fig. 2–4. The figures are generated from well-converged variational wavefunctions $\psi(\mathbf{r}_1, \mathbf{r}_2)$ as follows.[22,23] From $\psi(\mathbf{r}_1, \mathbf{r}_2)$ for a state of interest, one constructs $|\psi(\mathbf{r}_1, \mathbf{r}_2)|^2$, a density in terms of six independent variables; then one integrates over three of the six variables, specifically the three Euler angles specifying the orientation of the three-body system in space. The remaining reduced density is a function of only three variables; several reasonable choices of these variables offer themselves, such as hyperspherical coordinates $R, \alpha, \theta_{12}$; the normal mode coordinates, $r_1 + r_2$, $r_1 - r_2$, $\theta_{12}$; or the "local mode" coordinates that we choose, namely, $r_1$, $r_2$, and $\theta_{12}$. The three-variable reduced density $\rho(r_1, r_2, \theta_{12})$ is readily converted to the *conditional* probability density $\rho(r_2, \theta_{12}|r_1 = \xi)$, the probability distribution for the distance $r_2$ of electron 2 from the nucleus and the angle $\theta_{12}$, provided electron 1 is at the distance $\xi$ from the nucleus. It is important to realize that such distributions are simply convenient forms—reduced densities—to which the full six-variable densities can be reduced for purposes of interpretation and visualization. The conditional probability distributions shown here are all based on well-converged variational wavefunctions. Well-converged functions must be used because, as is well known, incompletely converged variational wavefunctions based on energy minimization may yield very good expectation values of energy, yet represent other properties, such as spatial distributions of electron density, inaccurately enough to be misleading.

As a first example the states of any pure rotor series should all have maximum probability for $r_1 = r_2$ and $\theta_{12} = \pi$, and the probability densities for members of a given rotor series should have very similar spatial distributions in their internal coordinate system (called the "intrinsic coordinate system" in the context of nuclear physics). The total wavefunctions of different rotor states in any series should differ primarily only in the parts that describe the rotation of the figure axis in space; these parts do not affect the distributions in their internal coordinate systems. At a higher level of approximation, the distributions for the states of a given series may be expected to differ a little because of centrifugal distortion; such differences, of course, are apparent in the internal coordinate system.

The left-hand column of Fig. 2 shows just such spatial distributions of probability density for the rotor series of He** with $n_1 = n_2 = 2$. In the traditional independent-particle representation, these would be designated $2s^2$ $^1S^e$, $2s2p$ $^3P^o$, and $2p^2$ $^1D^e$; for convenience, we shall sometimes refer to them in this way, but with quotation marks. The three distributions in the column, all drawn for the most probable value of $r_1$, are similar but differ in the increasing angular spread, as the angular momentum increases.

A second implication of the collective, moleculelike model concerns the states with excitation in the bending mode. The bending mode of the linear rotor–vibrator is of course doubly degenerate. Hence, there are two independent states with one quantum of bending, which should be very similar to one another and should differ markedly from the state with no excitation: as with all first excited vibrational states, their probability distributions should be zero at the potential minimum where the density of the unexcited state is a maximum. In this case, the implication is that the two first excited bending states

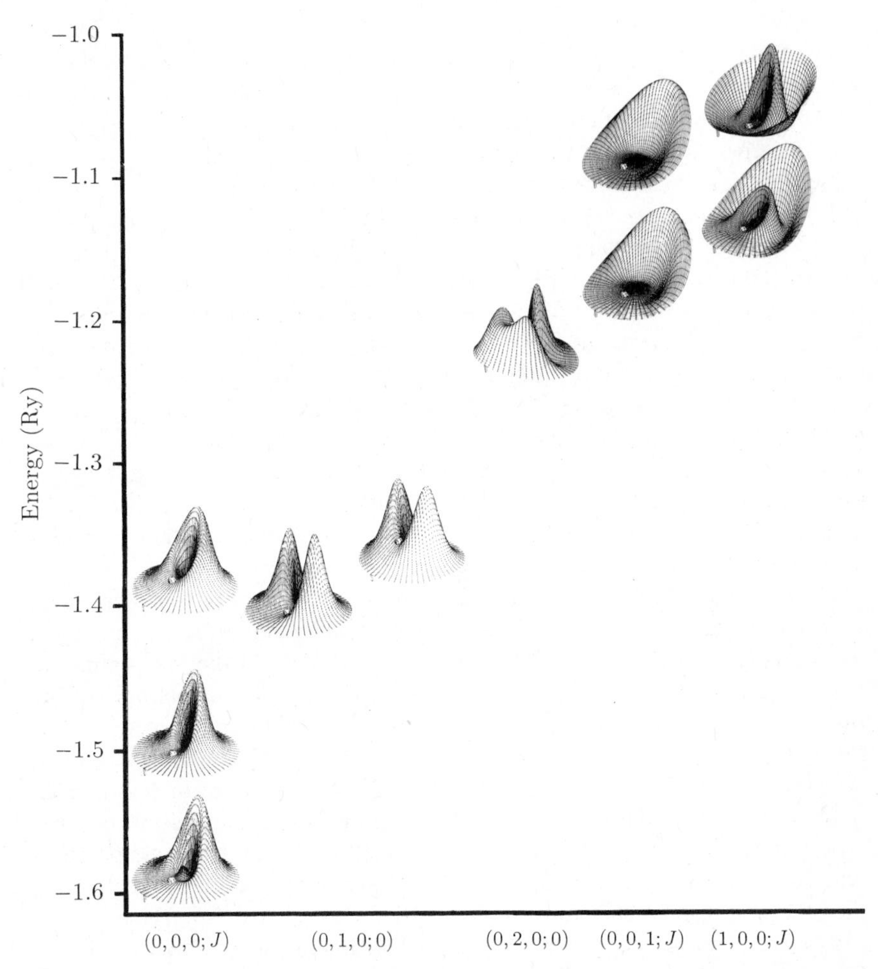


**Figure 2.** Conditional probability distributions for the states of doubly excited helium with $n_1 = n_2 = 2$, and the four lowest levels with $n_1 = 2, n_2 = 3$, all with one electron at approximately its most probable distance from the nucleus. The distributions are drawn in a cylindrical polar representation. The nucleus is located at the center of each distribution, and the position of the fixed electron is denoted by a small sphere. The location of $\theta_{12} = 0$ is denoted by a small vertical mark. These distributions are the probability densities in the spatial coordinates $r$ and $\theta$ of finding one electron, if the other electron is located at the position of the sphere. The maximum value of $r$ in each figure is 8 bohr. The quantum numbers are given below, as in Fig. 1, and the pattern in the same as in that figure. Where the states have nearly the same energy, such as the stretching states, the vertical positions of the distributions in this figure with respect to energy are only approximate. The distributions are taken from the calculations of Ref. 23. Note the strong similarities of the pair $(0, 0, 1; 0)(0, 0, 1; 1)$ and the pair $(1, 0, 0; 0)(1, 0, 0; 1)$ corresponding to antisymmetric and symmetric stretching mode states with $J = 0$ and 1. The similarities are strongly persuasive of the validity of the collective molecular picture.

should have similar energies (not identical because of the differences in Coriolis interactions) and zeros where $r_1 = r_2$ and $\theta_{12} = \pi$. The Herrick–Kellman supermultiplet picture assigns the "$2s2p$" $^1P^o$ and "$2p^2$" $^3P^e$ states as this pair. In the independent-particle picture, or in any other picture yet devised except the collective molecular, there is no particular reason for a relation between these states; in the collective molecular picture, these two states differ only in the way they are affected by Coriolis interaction, which couples bending with the antisymmetric stretching mode. The conditional probability distributions for these states are shown in Fig. 2 as the states with $v_2$, the number of bending quanta, equal to 1 and no rigid-rotor quanta ($J = 0$). The two distributions have the expected zeros at $\theta_{12} = \pi$ and maxima $\theta_{12} \simeq 0.55\pi$. Moreover, the two distributions are quite similar, differing primarily because the $^1P^o$ state does not exhibit the node at $\theta_{12} = 0$, which is mandatory for the $^3P^e$ state.

The one other state in Fig. 2 for which $n_1 = n_2 = 2$, the "$2p^2$" $^1S^e$, corresponds to the state with two quanta of bending excitation whose angular momenta exactly cancel so the state has zero net angular momentum, a maximum in its probability density at $\theta_{12} = \pi$, and a zero and then a second maximum at smaller values of $\theta_{12}$.

Kellman and Herrick had conjectured that He** might have states corresponding to excitation of collective, normal stretching modes. Such states have been identified[23]: the "$2s3s$" $^3S^e$ state is the first state of the antisymmetric stretching mode (and can be thought of as a "giant dipole resonance" of the two-electron system), and the corresponding "$2s3s$" $^1S^e$ state corresponds to the first excited state of the symmetric stretching mode. These are also shown for He** in Fig. 2. The triplet is constrained to look more or less like an antisymmetric stretching mode even in the pure $2s3s$ configuration because such a state must have a zero at $r_1 = r_2$ and the exclusion principle prevents one electron from being in the region of small $r$, that is, the $2s$ region, when the other electron is there, and similarly when one electron is in the region of large $r$, the $3s$ region, the other electron must be in the region of small $r$. Hence, the antisymmetric character of the vibration is already induced by the exclusion principle in a single-configuration representation. This state is an example of a state for which two quite different models, the independent particle and the collective molecular, generate roughly the same description for what seem to be quite different reasons. Such a case is of little use for choosing between models, but is of course a reassurance of the possible validity of either.

## IV. THE PHYSICAL BASIS OF COLLECTIVE, MOLECULELIKE BEHAVIOR

In the ground state and singly excited states of helium, the electrons have a high probability of penetrating into the region near the nucleus where their kinetic energies are large and, more specifically, large relative to the electron–electron repulsion even when the electrons are quite close together. Hence, the electrons can pass one another without undergoing very strong scattering, and the angular momenta of the individual electrons are moderately well preserved. This is apparent in how the conditional probability distributions for such independent-particle states come to take the form of (squares of) only slightly perturbed spherical harmonics.[22]

If both electrons are promoted out of the 1*s* shell, the exclusion principle (or, more strictly, the condition of orthogonality of wavefunctions) prevents either from having a high probability of being in the region near the nucleus and hence from having a high kinetic energy. The consequence is that electron–electron repulsion is much more important for He** than for the ground state and the singly excited states of helium. In particular, the probability of finding the electrons close together is vastly lower for He** than for the states of lower energy, angular correlations are extremely important for He**, and the behavior of the states reflects these correlations by exhibiting collective rather than independent-particle characteristics, in the main.

If the nuclear charge of He** could be increased, the binding energy and kinetic energy of the electrons would increase correspondingly, and one would expect the behavior of the electrons to become more independent particlelike. However, as the nuclear charge $Z$ increases and the electrons become more hydrogenic, the energy separations of one-electron states with the same $n$ but different $l$ decrease, so that they can intermix more readily. In reality, one can increase $Z$ by comparing He** with $(Li^+)^{**}$ and $(Be^{2+})^{**}$ and other two-electron ions of higher charge. Some states of $(Li^+)^{**}$ and $(Be^{2+})^{**}$ are slightly more independent particlelike than the corresponding states of He**, but as $Z$ is made still larger, for example, in $(Ne^{8+})^{**}$, some states, such as the "$2p^2$" $^1D^e$ become very clearly independent particlelike, while the "$2s^2$" $^1S^e$ and "$2s2p$" $^3P^o$ remain correlated,[23] as Wulfman and co-workers, based on an analysis of group-theoretical constants of the motion, showed they should.[24] In other words, some but not all states of the two-electron systems remain strongly collective for all values of $Z$.

## V. THE ALKALINE EARTH ATOMS AND ALKALI NEGATIVE IONS

The doubly excited states of helium are amenable to theoretical studies, and some of those states can be studied spectroscopically. Nevertheless, by any reasonable criterion, these transient states are rather exotic. They illustrate a phenomenon of atomic structure that was not heretofore expected, and, in so doing, open our minds to thinking of atomic structure in more general terms. They force us to ask whether other atoms, expecially atoms with two valence electrons, in mundane, accessible states, exhibit collective, moleculelike behavior to an extent large enough to influence their observable properties significantly.

The alkaline earth atoms—Be, Mg, Ca, Sr, and Ba—are the natural species to consider first. The important question here is whether the states of the two *valence* electrons are better described by collective, rotation–vibration quantization or by independent-particle quantization. Difficulties with the latter have been discussed.[6,8b,25] The new issue is whether moleculelike quantization is much more nearly free of its own problems.

To probe the collective or independent-particle character of the valence electrons in these atoms, it suffices to use well-converged variational wavefunctions for just the valence electrons in the effective field of the closed-shell core electrons. (The robustness of any conclusions can be checked by comparing results based on different choices of core potentials and different sets of basis functions for the expansions.) Conditional probability distributions for the *ground* and *bound excited states* of the alkaline earth atoms show that these species are much more like linear triatomic molecules than like quantum analogues of solar systems.[26,27] Although the zero-point amplitudes of their "vibrations" are larger than for He**, the alkaline earth atoms exhibit the angular correlations characteristic of moleculelike collective behavior even in their ground states, as shown in Fig. 3.

Figure 3 shows conditional probability distributions for the ground states of the alkaline earth atoms with the fixed electron at several different positions. This figure illustrates the point that for these states, angular correlation dominates the distributions until one electron gets very far (a very improbable distance) from the nucleus. These states are also, as the figure shows, correlated radially as well as angularly. That is, the distributions tend to be peaked about the most probable distance from the nucleus. Since the distributions do not change radically in the radial direction as one electron moves further from the nucleus, the radial effects are not dramatic. One effect of radial correlation in the rotor states is the centrifugal distortion, which mixes stretching and rotational motions together to expand[27] or contract[23] the distributions with increasing rotational excitation.

Radial correlation is more important in the bending states. This is best understood in terms of Coriolis coupling, which mixes bending and rotational motions. Evidence for this coupling is found, for example, in the energy splitting of the states with one quantum of bending vibration. Were there no Coriolis coupling, these states would be degenerate. The stretching vibrations are inherently radial effects. This can be seen, too, in how quickly the stretching states make the transition to independent-particlelike behavior. With angular correlation comparatively unimportant, as one electron moves near the nucleus, the other electron behaves as if it were in the field of a nucleus with charge $Z - 1$.

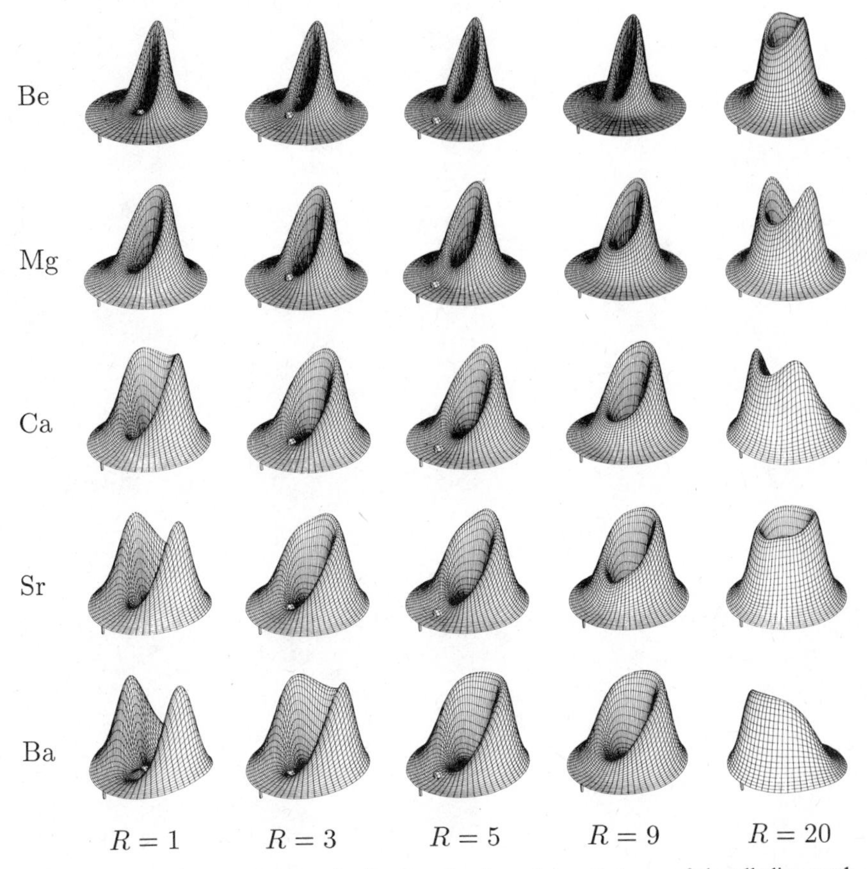


**Figure 3.** Conditional probability distributions for the valence electrons of the alkaline earth atoms in their ground states, for various values of the distance between the nucleus and one valence electron. The results are taken from calculations of Ref. 27.

We see, then, that although the original Herrick–Kellman model for doubly excited states was concerned primarily with angular correlation, the molecular model inherently includes radial as well as angular correlations. Similarly, while earlier work in our group considered the particles-on-a-sphere problem,[28] which includes purely angular effects, and then particles-on-concentric-spheres,[29] which reflects consequences of radial motion on the angular correlations, the complete six-dimensional calculations used to present the conditional probability distributions in this work make no assumptions regarding the relative importance of angular versus radial correlations.

The array of the states of the alkaline earth atoms analogous to those of Fig. 2 is shown in Fig. 4 (the $^1S^e$ state with two bending quanta is omitted). The position of one electron is set at its most probable value for this figure.

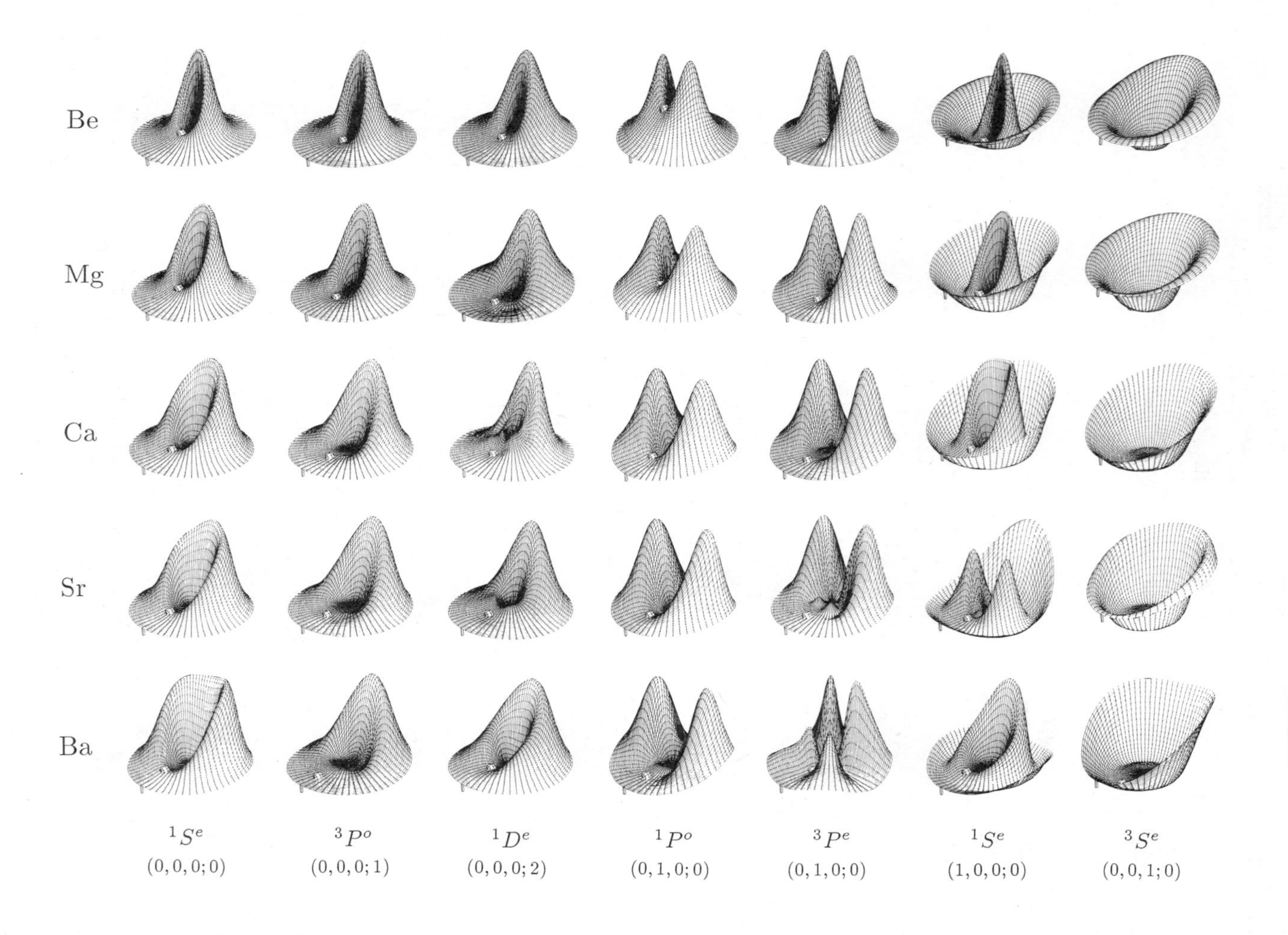
Be
Mg
Ca
Sr
Ba
$^1S^e$
(0, 0, 0; 0)
$^3P^o$
(0, 0, 0; 1)
$^1D^e$
(0, 0, 0; 2)
$^1P^o$
(0, 1, 0; 0)
$^3P^e$
(0, 1, 0; 0)
$^1S^e$
(1, 0, 0; 0)
$^3S^e$
(0, 0, 1; 0)

The forms of the distributions of almost all the states in Fig. 4 correspond closely to those in Fig. 2. The overall pattern is one of rotor states, states of vibrational excitation, and combinations of rotational and vibrational excitations. One of the most striking features of this set of distributions, which is not apparent from the figure itself, is that in several instances quite different atomic orbital configurations—that is, independent-particle representations—give rise to nearly the same spatial distributions. For example, the $^1D^e$ second excited rotor state of Be is dominated by the $2p^2$ configuration and of Mg, by the $3p^2$ configuration, but the corresponding $^1D^e$ states of Ca, Sr, and Ba are dominated, respectively, by the $4s3d$, $5s4d$, and $6s5d$ configurations. Nevertheless, these states have, as the figures show, virtually the same spatial distributions of probability density.

The one notable exception is the "$5d^2$" $^3P^e$ state of Ba that would be one of the partner states of the first excited bending vibration in this atom. This state has neither the single-lobed distribution expected for a state with one quantum of bending vibration nor the symmetric two-lobed distribution of a pure $d^2$ independent-particle state; it is just between these two extremes. This is a sort of antithesis of the first excited states of the antisymmetric stretching mode, the $ns(n+1)s$ $^3S^e$ states, which are well described by both the collective and independent-particle approaches; in this $^3P^e$ state, neither picture is adequate.

The negative ions of the alkali atoms have stable ground states and all but $Li^-$ are presumed to have bound "$np^2$" $^3P^e$ states.[30] The ground states of these ions have conditional probability distributions very much like those of the alkaline earths, and their first excited $^3P^e$ states are much like the corresponding states of the alkaline earths, except that they extend to large values of $r$.[31] This is true even for the $^3P^e$ state of $Li^-$, which lies above the energy of Li in its ground state, plus a free electron with no kinetic energy.

The limits of collective quantization in atoms are as yet quite unknown. In light atoms such as He** or Be, the shell structure corresponding to principal quantum numbers is clearly marked even though the quantization corresponds to collective rather than independent-particle behavior. In heavy atoms such as Sr and Ba, the shell structure of the valence electrons seems to be blurred because excited states associated with one set of principal quantum numbers

◀ **Figure 4.** Conditional probability distributions for the valence electrons in the ground states and other low-lying states of the alkaline earth atoms, with one electron at approximately its most probable distance from the nucleus. The states shown are those corresponding to the states of He** shown in Fig. 2, that is, those corresponding to the moleculelike quantum numbers $(v_1, v_2, v_3; J)$ shown there are at the bottom of this figure. The dominant configurations are not necessarily the same for all alkaline earths in a given column; for example, the $^1D^e$ rotor states of Be and Mg are predominantly $2p^2$ and $3p^2$, respectively, while those of Sr and Ba are predominantly $5s4d$ and $6s5d$, respectively. This figure is adapted from Ref. 27.

overlap and interweave with states having different apparent principal quantum numbers. In the heavier atoms, there are also states, such as the "$6s5d$" $^3D^e$ of Ba, which do not seem to fit naturally into the collective molecular picture. The complications involved in interpreting the Ba spectrum, in fact, make it difficult to determine the assignment of states in the molecular model. The states we chose, based on the examination of conditional probability distributions, have, for example, the unfortunate aspect that the state assigned as having two quanta of rotational excitation lies lower in energy than the state with one quantum of excitation. On the other hand, Kellman assigns a few of the states of Sr and Ba differently than we have; his assignments of the states are based on their positions in the energy-level diagram.[32] Where differences occur, as with the first $^1D^e$ rotor states, the criteria of position on the energy scale and of spatial distributions lead to conflicting assignments. Clearly, work of a more quantitative nature is necessary to sort these questions out. Recently, steps in this direction have been taken by calculating expectation values of $\mathbf{p}_1 \cdot \mathbf{p}_2$ as a measure of the form and degree of correlation in these atoms,[33] and by calculating suitable overlaps (see below).

These problems aside, the present picture is thus one in which many states of atoms, both common and exotic, must henceforth be thought of as collective; some states such as the ground and singly excited states of He, clearly fit the traditional independent-particle model; and most of the elements of the Periodic Table must now be studied carefully in many states to see how they should be interpreted to fit into a coherent scheme of excitations. Then it may become possible to infer, before doing elaborate calculations, whether any particular atomic state is represented better by an independent-particle model or a collective model, and to test the efficiency of collective basis sets for accurate computations.

## VI. INDEPENDENT-PARTICLE BEHAVIOR IN MOLECULES?

One can turn the question around in this way: If the electrons in atoms can sometimes exhibit moleculelike, collective quantization instead of independent-particle behavior, could the atoms in a molecule, such as the hydrogens in $H_2O$, $NH_3$, or $CH_4$, exhibit independent-particle behavior instead of their well-known collective rotor–vibrator excitations? One kind of independent-particle behavior is now well established for light atoms in molecules, namely, local stretching-mode behavior.[34] When as many as four or five quanta of excitation are put into the stretching motions of hydrogen in $H_2O$, the most likely state to be excited is one in which only one O–H bond carries the energy of excitation.[35] Another more familiar kind of localization of energy is the excitation of a small chromophore in a much larger molecule. However, neither of these is quite the kind of independent-particle behavior of electrons in the ground state of helium. If there are vibration–rotation states for which the full analogy holds, they must be characterized by angular momentum quantum numbers for the individual hydrogen atoms. The transition for collective behavior to independent-particle behavior in the water molecule would be characterized by a transition from the collective rotor–vibrator quantization to independent-particle quantization. The former is expressible in terms of the rotor quantum numbers (very approximately just $J$ and $K$ of the symmetric rotor), the symmetric stretching quantum number $v_1$, the bending quantum number $v_2$, and the antisymmetric stretching quantum number $v_3$. Independent-particle quantization is characterized by a stretching quantum number $v_j$ for each atom $j$, an angular momentum quantum number $l_j$ for each atom, and a total angular momentum quantum number $J$, due to the coupling of the $l_j$'s. The transition between these extremes would be seen as a change in the pattern of the level spacings. The most dramatic aspect of this change would probably be the transformation of the bending motion into a rotational motion.

What kind of sequence of states might exhibit this transition? The most obvious would be states highly excited in a *local* stretching mode, in order that one hydrogen can get out of the way of the other, and excited enough in the bending mode to allow one hydrogen to pass the other, through $\theta_{12} = 0$. Roughly, for $H_2O$, this would require about seven or eight quanta of a local stretching mode and about five or six quanta of bending.

Can such states exist in the spectrum of bound states of the water molecule? The answer is still ambiguous, for two reasons. The first is that methods are *almost* but not quite developed to the point of yielding spectroscopically useful predictions (or retrodictions) of rotation–vibration levels high enough to test a model or check an observed spectrum of rotation–vibration levels—provided an adequately accurate potential surface is available. The second reason is that there so far appears to be no effective potential surface that has the required accuracy over a range of configuration space large enough to encompass the regions critical to this collective-to-independent transition.

Incidentally, even if no independent-particle bound states exist for $H_2O$ or other small molecules, there could be transient resonance states above the dissociation limit with independent-particle character; they may just be more difficult to diagnose and observe than stable bound states.

## VII. CONCLUDING REMARKS

The existence of collective, moleculelike states of two-electron and quasi-two-electron atoms now seems well established. The extent to which atoms with three or more valence electrons show similar behavior is not yet known; that problem is now an obvious and pressing one. The two-electron systems have been amenable to a graphical analysis that yields clear results relatively easily. To explore three-electron atoms and beyond will require new tools of analysis.[25b,25c] Perhaps the most obvious is projection of accurate wavefunctions, which can now be constructed (except in the vicinity of *very* short electron–electron distances), onto simple rotor–vibrator functions, in which the parameters of the rotor–vibrator functions such as moments of inertia and force constants are treated as variational parameters to maximize the overlap.[36] This approach will even be useful for the two-electron systems, as a way to quantify the degree of collective character of a state, just as the projection of an accurate function into its dominating independent-particle configuration is a quantification of the degree of validity of the independent-particle model.

The existence of collective, moleculelike states in atoms suggests the tantalizing possibilities of a more unified picture of few-body systems at the atomic level. In such a picture, the degree of independent-particle or collective character would be determinable from the force laws of the component particles and the energy and other quantum numbers of the state in question, without requiring either elaborate computations and after-the-fact analyses like those that brought us to this point or the assumption of specific behavior and a corresponding model at the outset. Such a picture could put few-body systems of many kinds into a common, unifying context, and allow us to see generalizations that we have until now overlooked.

## Acknowledgments

The research on which our own group's contribution to this subject are based has been supported by the National Science Foundation. We wish to acknowledge the contributions of the students and associates who have collaborated on these problems: Paul Rehmus, Michael Kellman, Huoy-Jen Yuh, Gregory Ezra, and Grigory Natanson. One of us (RSB) also wishes to acknowledge the hospitality of the Aspen Center for Physics where this paper was first drafted.

## References

1. R. P. Madden and K. Codling, *Phys. Rev. Lett.* **10**, 516 (1963); R. P. Madden and K. Codling. *Astrophys. J.* **141**, 364 (1965).
2. J. W. Cooper, U. Fano, and F. Prats, *Phys. Rev. Lett.* **10**, 518 (1963).
3. A. K. Bhatia and A. Temkin, *Revs. Mod. Phys.* **36**, 1050 (1964); A. K. Bhatia, A. Temkin, and J. F. Pekeris, *Phys. Rev.* **153**, 177 (1967); A. K. Bhatia, P. G. Burke, and A. Temkin, *Phys. Rev. A* **8**, 21 (1973); **10**, 459(E) (1974); A. K. Bhatia, *Phys. Rev. A* **6**, 120 (1972); A. K. Bhatia and A. Temkin, *Phys. Rev. A* **11**, 2018 (1975); A. K. Bhatia, *Phys. Rev. A* **15**, 1315 (1977).
4. Y. K. Ho, *Phys. Rev. A* **23**, 2137 (1981); *Phys. Lett.* **79A**, 44 (1980); *J. Phys. B* **12**, 387 (1979).
5. J. Macek, *J. Phys. B* **1**, 831 (1968).
6. C. H. Greene, *Phys. Rev. A* **23**, 661 (1981).
7. U. Fano, *Rep. Prog. Phys.* **46**, 97 (1983).
8. (*a*) C. D. Lin, *Phys. Rev. A* **25**, 76 (1982); (*b*) *J. Phys. B* **16**, 723 (1983); (*c*) *Phys. Rev. A* **29**, 1019 (1984).
9. J. G. Frey and B. J. Howard *Chem. Phys.* **111**, 33 (1987).

10. C. Wulfman, *Chem. Phys. Lett.* **23**, 370 (1973).
11. D. R. Herrick and O. Sinanoğlu, *Phys. Rev. A* **11**, 97 (1975); O. Sinanoğlu and D. R. Herrick, *J. Chem. Phys.* **62**, 886 (1975); *J. Chem. Phys.* **65**, 850 (1976); D. R. Herrick, *Phys. Rev. A* **12**, 413 (1975).
12. C. Wulfman and S. Kumei, *Chem. Phys. Lett.* **23**, 367 (1973); O. Navaro and A. Freyere, *Mol. Phys.* **20**, 861 (1971); S. I. Nikitin and V. N. Ostrovsky, *J. Phys. B* **11**, 1681 (1978); S. I. Nikitin and V. I. Ostrovsky, *J. Phys. B* **9**, 3141 (1976).
13. P. Rehmus, M. E. Kellman, and R. S. Berry, *Chem. Phys.* **31**, 239 (1978).
14. C. D. Lin and J. H. Macek, *Phys. Rev. A* **29**, 2317 (1984).
15. S. Watanabe and C. D. Lin, *Phys. Rev. A* **34**, 823 (1986).
16. J. M. Feagin and J. S. Briggs, *Phys. Rev. Lett.* **57**, 984 (1986).
17. H. Klar, *Phys. Rev. Lett.* **57**, 66 (1986).
18. M. E. Kellman and D. R. Herrick, *J. Phys. B* **11**, L755 (1978).
19. C. F. A. de Shalit and H. Feshbach, *Theoretical Nuclear Physics, Vol. 1: Nuclear Structure.* Wiley, New York, 1974, Chap. VI, pp. 377ff.
20. D. R. Herrick and M. E. Kellman, *Phys. Rev. A* **21**, 418 (1980); D. R. Herrick, M. E. Kellman, and R. D. Poliak, *Phys. Rev. A* **22**, 1517 (1980).
21. M. E. Kellman and D. R. Herrick, *Phys. Rev. A* **22**, 1536 (1980).
22. P. Rehmus, C. C. J. Roothaan, and R. S. Berry, *Chem. Phys. Lett.* **58**, 321 (1978); P. Rehmus and R. S. Berry, *Chem. Phys.* **38**, 257 (1979); H.-J. Yuh, G. S. Ezra, P. Rehmus, and R. S. Berry, *Phys. Rev. Lett.* **47**, 497 (1981).
23. G. S. Ezra and R. S. Berry, *Phys. Rev. A* **28**, 1974 (1983).
24. C. Wulfman, *Phys. Rev. Lett.* **51**, 1159 (1983); Y. Ho and C. Wulfman, *Chem. Phys. Lett.* **103**, 35 (1983); C. F. Wulfman and R. D. Levine, *Phys. Rev. Lett.* **53**, 238 (1984).
25. (*a*) P. F. O'Mahony and C. H. Greene, *Phys. Rev. A* **31**, 250 (1985); (*b*) P. F. O'Mahony, *Phys. Rev. A* **32**, 908 (1985); (*c*) P. F. O'Mahony and S. Watanabe, *J. Phys. B* **18**, L239 (1985).
26. J. L. Krause and R. S. Berry, *Phys. Rev. A* **31**, 3502 (1985).
27. J. L. Krause and R. S. Berry, *J. Chem. Phys.* **83**, 5153 (1985).
28. G. S. Ezra and R. S. Berry, *Phys. Rev. A* **25**, 1513 (1982).
29. G. S. Ezra and R. S. Berry, *Phys. Rev. A* **28**, 1989 (1983).
30. D. W. Norcross, *Phys. Rev. Lett.* **32**, 192 (1974).
31. J. L. Krause and R. S. Berry, *Comments At. Mol. Phys.* **18**, 91 (1986).
32. M. E. Kellman, *Phys. Rev. Lett.* **57**, 1738 (1985).
33. J. L. Krause, J. D. Morgan III, and R. S. Berry, *Phys. Rev. A* **35**, 3189 (1987).
34. B. R. Henry, *Vib. Spectra Structure* **10**, 269 (1981).
35. R. T. Lawton and M. S. Child, *Mol. Phys.* **37**, 1799 (1979); M. S. Child and R. T. Lawton, *Chem. Phys. Lett.* **87**, 217 (1982).
36. J. E. Hunter III and R. S. Berry, *Phys. Rev. A* (in press, 1987).

# FLUCTUATIONS IN SPECTRAL INTENSITIES AND TRANSITION RATES

R. D. LEVINE

*The Fritz Haber Research Center for Molecular Dynamics, The Hebrew University, Jerusalem 91904, Israel*

## CONTENTS



## I. INTRODUCTION

Spectral intensities, lifetimes of excited and/or predissociating states, and other transition rates can all be computed as (squares) of off-diagonal matrix elements. Our purpose here is different than the usual approach. We seek to characterize not individual rates but the overall distribution of rates and the moments thereof. Clearly such an approach is only of interest when, for example, an optical probe can access many final states, that is, at higher levels of excitation. It is also relevant to larger systems where individual transitions are not of interest and only lower-resolution measurements need be interpreted.

As a specific problem consider an energy-rich polyatomic molecule that can dissociate. Transition-state theory[1] (in the RRKM framework) can be used to compute the rate constant as a function of excess energy, $k(E)$. The result is a smooth, monotonically increasing function of $E$. Say, now, the rates are measured for individual quantum states. Recent experimental results[2–5] show that such rates can vary widely over a narrow energy range. Such variations of the rates are also found in semiclassical and quantal model computations.[6,7] In theoretical studies, the specificity can be related to the detailed nature of the intramolecular dynamics. The observed experimental variations are also being interpreted in terms of special coupling effects such as Coriolis,[2] singlet–triplet mixing,[4] or limited coupling of slow vibrations to the reaction coordinate.[3,5,7] Does it therefore follow that in the limit of stronger coupling the fluctuations in the rates will disappear and the monotonic dependence on the total energy, as given by transition-state theory, will be the rule? Our point of view is that quantal fluctuations will remain even in the limit of fully chaotic classical dynamics. To be sure, the extent of fluctuation will diminish, but it will settle down to a finite, universal, limit. Moreover, even before that limit will be reached, there are general trends in these fluctuations that are worth exploring.

Fluctuations in spectral intensities[8–14] provide an even more detailed probe. Transition rates examine only the variation with the initial state (all final states being usually summed over). A spectrum can be studied for the distribution of intensities over all final states and, moreover, the initial state can be varied. The measurement of spectra at the required resolution for the excitation range of interest is not easy, but improved and novel techniques[15,16] are currently providing a wealth of data.

The study of fluctuations has its origins in nuclear physics.[17,18] There is a highly developed and well-reviewed approach based on the notion of an ensemble of Hamiltonians.[18–21] We shall not follow this approach for two reasons. The first is that our system of interest is a definite molecule whose Hamiltonian may not be precisely known to us but is, in principle, known and is known also in practice in computational studies. Thus what we seek to characterize is the distribution of intensities over the different final states for a given initial state in a given molecule. In the ensemble approach one computes the distribution of intensities for a particular transition (i.e., for given initial and final states) for different Hamiltonians in the admissible set. The second reason is a more practical one. Our intended applications are *not* only to such systems that are as statistical as the symmetry restrictions will allow. It is necessary to admit the possibility that the excited molecule will fail to uniformly sample the available phase space.* It is our intention to allow for the possibility of such deviations and indeed to use them as a diagnostic tool.

What we are examining is the signature, in the discrete quantum spectra of the onset of irregularity and eventually of chaos, in the corresponding classical system. There is a very extensive literature dealing with such questions.[9,22–29] We shall, however, approach it from the point of view of characterizing the distribution rather than that of trying to compute the spectra, semiclassically[23,30,31] or otherwise. We shall point out, however, the connection with dynamics and, in particular, with the semiclassical spectrum. This connection will serve to show that the smooth envelope of the spectrum against which we measure the fluctuation is generated by "short-time" dynamics. For transition rates, we shall find that the "zero-time" approximation[32–34] (i.e., classical transition-state theory[35,36]) provides the required smooth variation.

The essential statistical approach is introduced in Section II. To simplify the derivation, we consider the distribution of fluctuations against a flat background. With the hindsight of Section III, this will turn out to mean that we examine a portion of the spectrum (or a range of initial states, when considering rates) over a narrow enough energy interval such that the density of states is hardly varying. In other words, the results of Section II are strictly valid only for local fluctuations. The punchline, provided in Section III, is that the functional form for fluctuations is universal and as derived in Section II provided we examine the fluctuations with respect to the local, smooth, envelope. Section IV is a summary of the practical issues that arise in examining (experimental or computational) data and forges a link with the thermodynamic theory of fluctuations.

## II. THE DISTRIBUTION OF FLUCTUATIONS

### A. Strength Function and Sum Rules

A fully resolved discrete spectrum from the initial state $i$ is given by the strength function $S(E')$

$$S(E') = \sum_f |\langle f | T | i \rangle|^2 \delta(E' - E_f). \tag{2.1}$$

* One could conceivably model this in the ensemble approach by restricting the range of Hamiltonians in the ensemble.

Here $T$ is the transition operator, which is the dipole for ordinary optical transitions. The intensities of the individual transitions satisfy the sum rule

$$\sum_f |\langle f|T|i\rangle|^2 = \langle i|T^\dagger T|i\rangle. \tag{2.2}$$

The "closure" property, which is used to derive (2.2), can also be used to express the higher moments of the strength function in closed form, for example,

$$\int dE' E'^n S(E') \equiv \sum_f E_f^n |\langle f|T|i\rangle|^2 = \langle i|T^\dagger H^n T|i\rangle, \tag{2.3}$$

where $H|f\rangle = E_f|f\rangle$. Indeed, the strength function itself can be written as an expectation value over the initial state

$$S(E') = \langle i|T^\dagger \delta(E' - H)T|i\rangle. \tag{2.4}$$

In Section III we shall discuss the corresponding sum rules for a spectra at finite resolution.

For the purpose of taking the classical limit we consider a more symmetric strength function $Y(E, E')$, where the initial states at energy $E$ and the final states at the energy $E'$ are treated on equal footing:

$$Y(E, E') = \sum_i \sum_f |\langle f|T|i\rangle|^2 \delta(E - E_i)\delta(E' - E_f). \tag{2.5}$$

$Y(E, E')$ can be regarded as the strength function $S(E')$ "averaged" over all initial states at the energy $E$. In fact, (2.5) is just a sum over initial states and the mathematical average is $Y(E, E')/\rho(E)$ where $\rho(E)$,

$$\rho(E) \equiv \mathrm{Tr}\{\delta(E - H)\} = \sum_i \delta(E - E_i), \tag{2.6}$$

is the density of states. As in applications of information theory to collision dynamics,[37] one is interested in symmetric quantities, which are either $Y(E, E')$ or

$$\omega(E, E') = Y(E, E')/\rho(E)\rho(E'). \tag{2.7}$$

In Section II.H and III we shall consider not only spectral intensities but also rates of unimolecular processes. If the operator $T$ is interpreted as the transition operator, then $S(E')$ in (2.4) is (proportional to) the rate of transitions out of the particular state $i$ to all final states of an energy $E'$. In current terminology,[4] $S(E')$ is the "state-specific" rate. The rate averaged over all initial states of the same energy $E$ is

$$\langle S(E')\rangle = \sum_i \delta(E - E_i)S(E') \Big/ \sum_i \delta(E - E_i) = Y(E, E')/\rho(E). \qquad (2.8)$$

Summing over all final states, we can write, in the language of unimolecular reactions, $\int \langle S(E')\rangle \, dE' \equiv k(E)$, where $k(E)$ is the reaction rate constant for a microcanonical ensemble of reactants.[35,36] When transition-state theory is valid, $k(E) = N_{\ddagger}(E - E_0)/h\rho(E)$, where $N_{\ddagger}(E - E_0)$ is the number of states of the transitions state (i.e., excluding the reaction coordinate) at an energy up to $E - E_0$. As usual, $E_0$ is the minimal energy that needs to be present along the reaction coordinate. As is well known, $k(E)$ is a smoothly varying function of the energy in the limit where the density of states is high, so that $N_{\ddagger}(E - E_0)$ is practically a continuous function of $E$. This limiting behavior of the averaged strength function $\langle S(E')\rangle$ will be of particular importance in Section III.

### B. The Distribution of Intensities

Our purpose in this section is to derive the distribution of intensities irrespective of the energy of the final state. Operationally, this is carried out as follows. Consider a plot of $S(E')$ versus $E'$ (i.e., intensity versus frequency such as shown, for example, in Fig. 1 of the introductory chapter by Jortner, Levine, and Rice). Divide the intensity axis in bins. (As an aid to visualization consider lines parallel to the abscissa whose intersection with the ordinate marks the boundaries of the bins.) Now count how many transitions have intensity in any given bin. This is the experimental histogram giving the distribution of intensities. Each transition is counted not in terms of its energy but in terms of its intensity. The corresponding theoretical construct follows a similar argument. Rather than performing the summation over final states in the sum rules (2.2)–(2.4) in terms of their energy, sum over the values of the intensities. We shall regard (for convenience) the intensity $y$ as a continuous variable with a probability (density) function $P(y)$ such that the fraction of transitions in the interval $y$ to $y + \delta y$ is $P(y)\delta y$.

In terms of the distribution of intensities the sum rule (2.2) can be written as

$$\int dy\, y P(y) = \langle i|\, T^{\dagger} T \,|i\rangle, \qquad (2.9)$$

where $y$ is the intensity whose distribution is taken to be independent of the energy of the final state. The only other generally available input on $P(y)$ is its normalization

$$\int dyP(y) = 1. \tag{2.10}$$

### C. The Distribution of Maximal Entropy

We now seek that distribution $P(y)$ that is characteristic for a discrete spectrum where the corresponding classical motion is chaotic. Such a distribution must still satisfy the conditions (2.9) and (2.10), but these do not suffice to determine a unique $P(y)$. In the regular regime, where selection rules operate, a few transitions are expected to be strong and many others will be very weak. As the corresponding classical motion becomes increasingly chaotic, there is less variation among adjacent eigenstates. The strength distribution will then become more uniform. In the limit we seek that distribution where the total strength [i.e., area equals $S(E)$ which equals $\langle i|T^\dagger T|i\rangle$] is most uniformly distributed[12]. The required distribution is thus one of maximal entropy, subject to the constraints (2.9) (i.e., a given total strength) and (2.10).

To implement the procedure of maximal entropy, we need to write down the entropy. Here care need be exercised since the usual expressions for the entropy are only valid for that distribution which, in the absence of all but the normalization constraint, is uniform. We suggest that this "natural" distribution is not necessarily $P(y)$. Consider first the case where all bound states and the transition operator are real. Then $y = x^2$ where $x$ is real. In the absence of any constraints we require that the "natural" distribution remain unchanged under an orthogonal change of the basis set. The distribution that is unchanged by a rotation of $\mathbf{x}$ is the distribution uniform in $x$. The alternative, a distribution uniform in $y$, will not stay uniform after a rotation of the basis.

It follows that for *real* transition amplitudes we wish to maximize

$$S = -\int dxP(x)\ln P(x) \tag{2.11}$$

subject to

$$\int dxx^2P(x) \equiv \langle x^2\rangle (= \langle i|T^\dagger T|i\rangle) \tag{2.9$'$}$$

and to

$$\int dxP(x) = 1. \tag{2.10$'$}$$

Standard Lagrange multipliers procedure leads to the result[12]

$$P(x) = (2\pi\langle x^2\rangle)^{-1/2}\exp(-x^2/2\langle x^2\rangle). \tag{2.12}$$

For a given total strength $\langle x^2\rangle$, the distribution (2.12) is the most entropic (and hence the most uniform one, cf. Section C.3 of Ref. 37). The direct proof of this claim is to consider some alternative distribution $Q(x)$ that is normalized and has the same integrated strength

$$\int dx x^2 Q(x) = \int dx x^2 P(x) = \langle x^2\rangle. \tag{2.13}$$

Now, using Jensen's inequality,

$$0 \leq \int dx Q(x) \ln[Q(x)/P(x)], \tag{2.14}$$

and (2.13) which implies on using (2.12) that

$$\int dx Q(x) \ln P(x) = \int dx P(x) \ln P(x).$$

$$0 \leq \int dx Q(x) \ln[Q(x)/P(x)], \tag{2.15}$$

$$= -\int dx P(x) \ln P(x) - \left(-\int dx Q(x) \ln Q(x)\right). \quad \text{Q.E.D.}$$

One can show that the result, Eq. (2.15), that (2.12) is the distribution of maximal entropy implies also that $P(x)$ is "more uniform" than $Q(x)$.

In general, if $y = g(x)$ and if $\{x_i\}$ is the set of real roots of $y = g(x_i)$, then the probability density of $y$ is

$$P(y) = \sum_i P(x_i)/|dg(x_i)/dx|. \tag{2.16}$$

For the special case $y = x^2$ and $P(x)$ given by (2.12),

$$P(y) = (2\pi\langle y\rangle)^{-1/2} y^{-1/2} \exp(-y/2\langle y\rangle). \tag{2.17}$$

The intensity distribution given by (2.17) has been derived[12] for real transition amplitudes. It is referred to as the Porter–Thomas distribution,[17] and has been extensively studied in nuclear physics. It is the distribution of transition strengths given by the ensemble method. The results of Section III will reaffirm (2.17) but with one important qualification. The variable $y/\langle y\rangle$ in (2.17) need be understood as the ratio of the actual intensity of the transition to the value of the smooth spectral envelope at the energy of the transition. If the envelope is flat, we recover (2.17) as is.

It is the proof that (2.17) is the correct result but in a local sense that is one of the central results of this chapter.

Figure 1 shows a computed[13] spectrum* and the resulting intensity distribution for an initial state at an energy where large-scale classical chaos prevails. Clearly, (2.17) accounts well for the results. A fit of (2.17) to measured[16] intensity fluctuations in $NO_2$ is shown in Fig. 4 of the first chapter, by Jortner, Levine, and Rice.

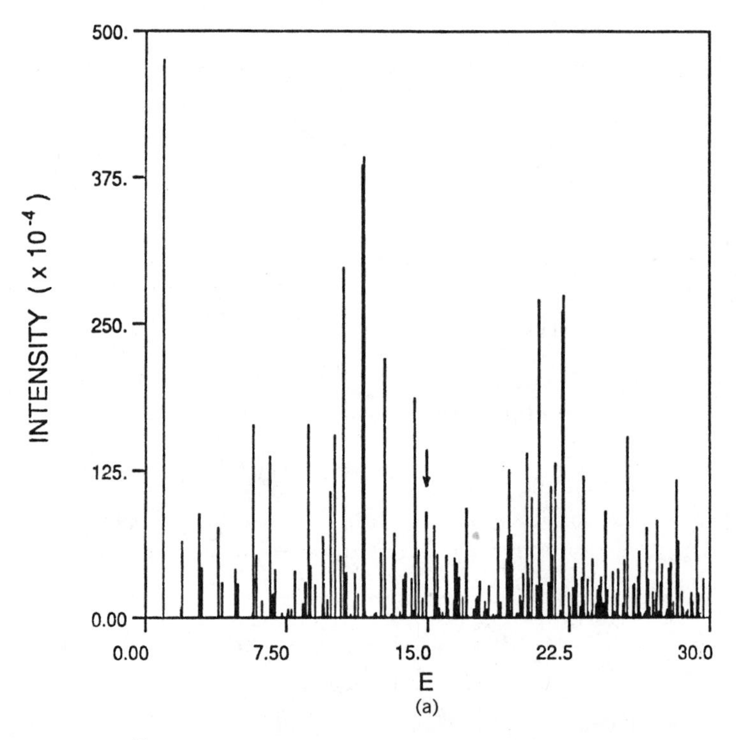


**Figure 1.** Computed[13] discrete stick spectrum (*a*) and the fit by (2.17) (solid line) to the distribution of fluctuations in that spectrum, (*b*). The initial state is at an energy (indicated by an arrow in (*a*)) above the onset of large-scale chaos in the classical dynamics of the Hamiltonian. The accessed final states span a wide range of energies about *E*, and there are no very strong or many very weak transitions.

* The computation is for a model system with two vibrational coordinates. The spectrum is between two different electronic states.

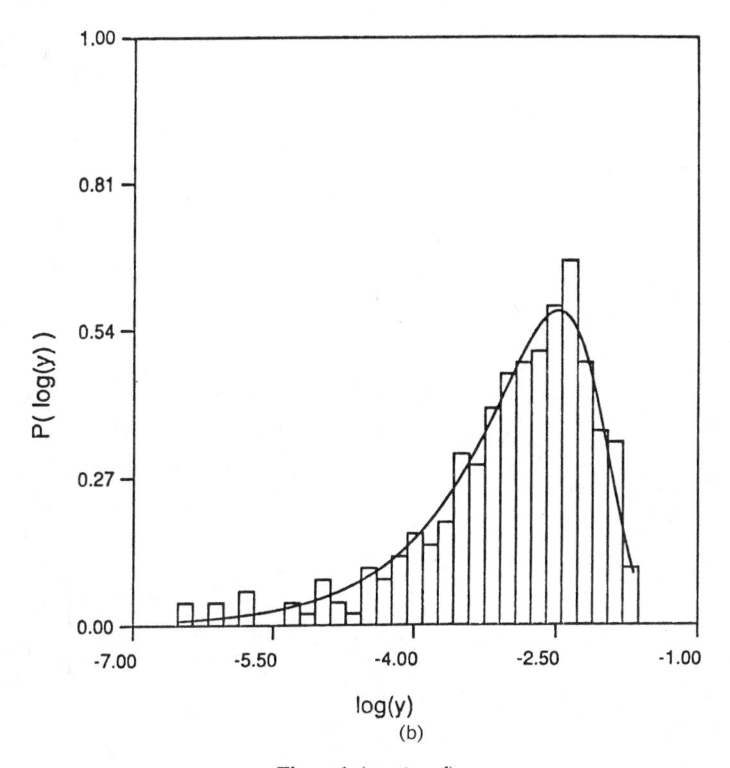


**Figure 1** *(continued)*.

A close fit to (2.17) is often not the case as can be seen in Fig. 2 [where (2.17) is the dashed line]. In particular, in the transition from regular to chaotic regimes, the corresponding discrete quantum spectra show more extensive fluctuations. The example of Fig. 2 is the first of several which will emphasize that, unfortunately, it is in the region of low intensities where the deviations from (2.17) are most noted. Therefore, we must turn to exploring the possible origins of such deviations, beginning with the more obvious possibilities.

In Section III we introduce the maximum entropy formalism.[37] Results such as (2.17) and others below can then be derived within a unified framework. The price of a more general approach is that it is more cumbersome and hence it is delayed to the end.

### D. Complex Transition Amplitude

In the more general case, the transition amplitude may be complex, $y = x_1^2 + x_2^2$, where $x_1$ and $x_2$ are the real and imaginary parts of the amplitude. In this case, in the absence of any constraints the distribution of $y$ is uniform. To prove this put $x_1 = r\cos\theta$, $x_2 = r\sin\theta$ so that $y = r^2$. Now, since the

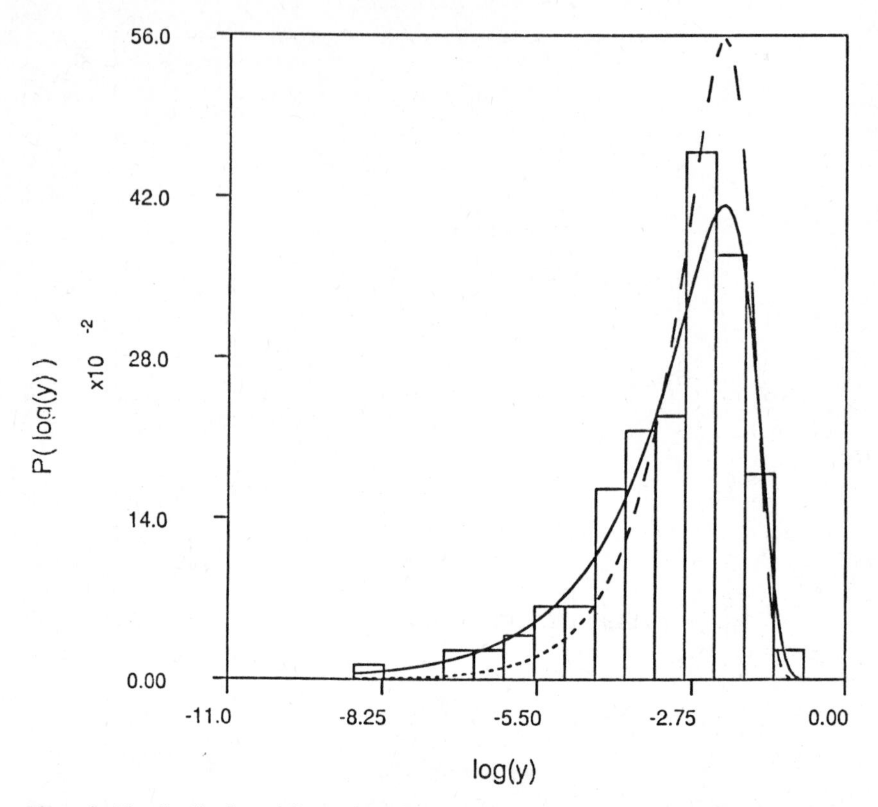


**Figure 2.** The distribution of fluctuations for the same initial state shown in Fig. 1 but to an excited electronic state that is not much displaced with respect to the ground state. (Adapted from Ref. 13.) Owing to the weaker coupling, the fit by (2.17), shown as a dashed line, is no longer very accurate for the weaker transitions. The solid line is (2.25) with $v = 0.63$.

distribution in either $x$ is uniform, $P(y)\delta y$ is proportional to the area of a ring bounded by two circles of radii $r$ and $r + \delta r$, respectively,

$$P(y)\delta y \propto 2\pi r \delta r = \pi \delta r^2 = \pi \delta y. \tag{2.18}$$

Hence the natural distribution in $y$ is uniform, and the entropy of $P(y)$ is $-\int dy P(y) \ln P(y)$. Imposing the sum rule

$$\langle y \rangle = \int dy y P(y) \tag{2.9'}$$

as a constraint, leads to a distribution of maximum entropy

$$P(y) = (\langle y \rangle)^{-1} \exp(-y/\langle y \rangle). \tag{2.19}$$

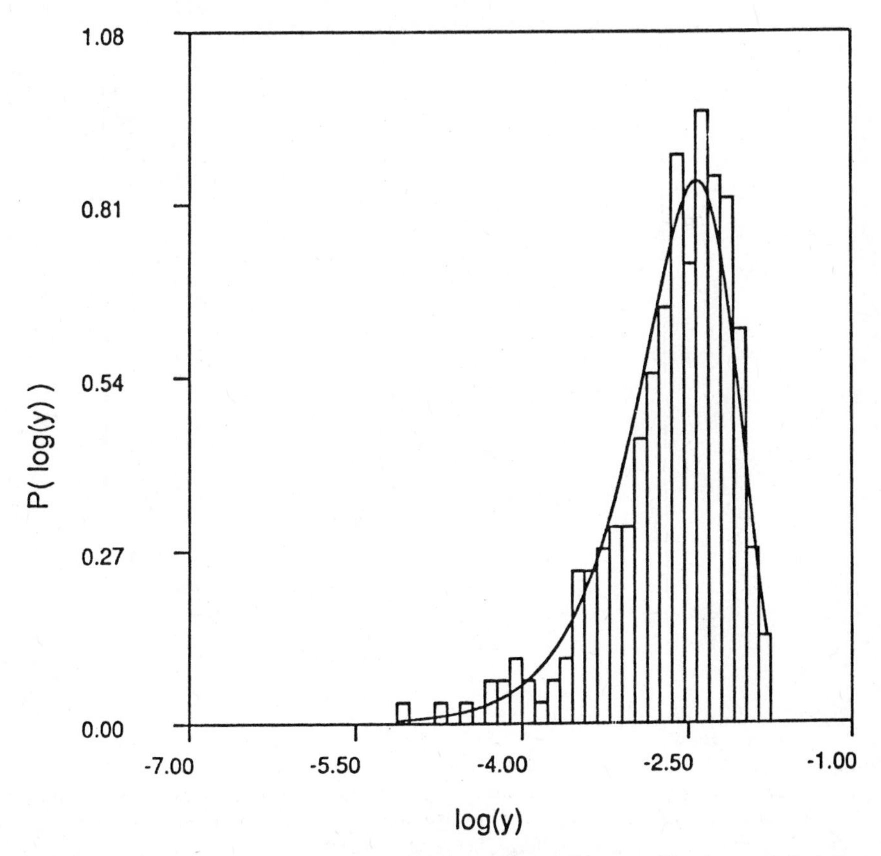


**Figure 3.** Distribution for fluctuations under the same conditions as Fig. 1 except that the transition amplitude is complex. The solid line is the fit by (2.19).

Figure 3 shows a fit of (2.19) to computational results generated using a non-Hermitian $T$ operator,* such that its Hermitian and skew-Hermitian parts have the same variance.

The difference between (2.17) and (2.19) is most notable for $y$ values

* It is important to emphasize that in deriving (2.19) the only constraint that was imposed was a given total transition strength, $\langle y \rangle$, as implied (2.9). The values of $\langle x_1^2 \rangle$ and of $\langle x_2^2 \rangle$ are not individually given by (2.2). Only their sum is specified. If we do insist that $\langle x_1^2 \rangle$ and $\langle x_2^2 \rangle$ are imposed as distinct constraints, then $P(x_1)$ and $P(x_2)$ are given by (2.12). $P(y)$ has then to be determined as a convolution. Prof C. Schlier evaluated the convolution integral analytically. When $\langle x_1^2 \rangle = \langle x_2^2 \rangle$ the result for $P(y)$ is Eq. (2.19). A special case where it would make physical sense to impose $\langle x_1^2 \rangle$ and $\langle x_2^2 \rangle$ as separate constraints is when the transition operator $T$ is not Hermitian but is normal $TT^\dagger = T^\dagger T$. Then $T$ has a unique decomposition into a Hermitian ($T_1$) and skew-Hermitian ($T_2$) parts $T = T_1 + iT_2$ where $T_1$ and $T_2$ commute. Then $\langle x_1^2 \rangle = \langle i | T_1^\dagger T_1 | i \rangle$ and $\langle x_2^2 \rangle = \langle i | T_2^\dagger T_2 | i \rangle$. See also the Appendix.

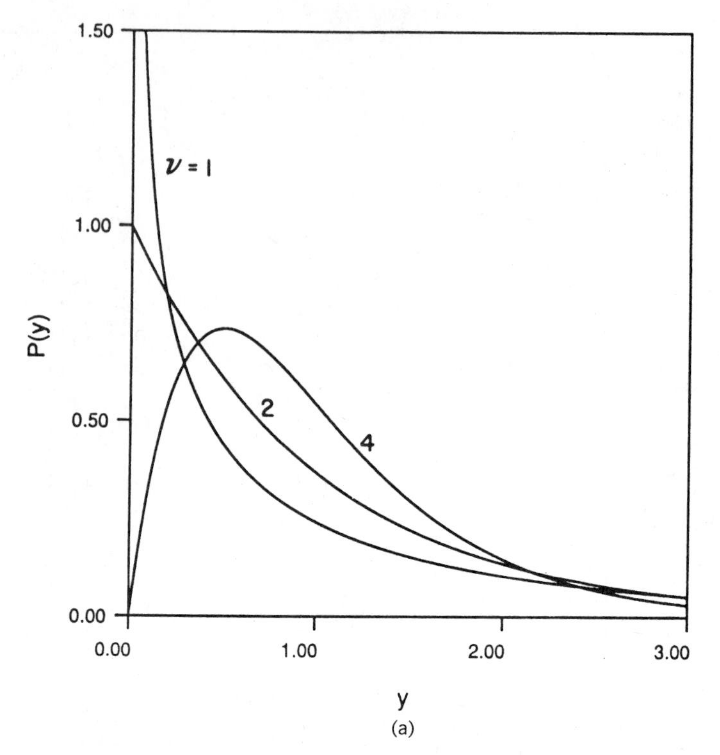


**Figure 4.** (*a*) Plot of $P(y)$, Eq. (2.25), versus $y$ for $v = 1, 2, 4$. Note that $P(y)$ is peaked for finite $y$ only for $v > 2$ and that the width diminishes (i.e., fluctuations are reduced) as $v$ increases. Such a plot underemphasizes the range $y < \langle y \rangle$. To spread out this important region it proves convenient to plot (as in Fig. 1–3) $P(\log y)$ versus $\log y$ where

$$P(\log y) = P(y)/(d \log y/dy)$$
$$= y^{v/2} \exp(-vy/2\langle y \rangle) \Big/ \left(\frac{2\langle y \rangle}{v}\right)^{v/2} \Gamma\left(\frac{v}{2}\right).$$

Such a plot is shown in (*b*) for $v = 1, 2, 4$. Note that it is always peaked at $y = \langle y \rangle$. In nuclear physics, the plots shown are often $P(y)$ versus $\log y$. This is not a consistent representation but does have the advantage that the fall-off as $y \to 0$ is more moderate, as can be seen in (*c*).

significantly below the average intensity $\langle y \rangle$, Fig. 4. This is the very region most susceptible to experimental noise. The drawing of the results versus a logarithmic scale in $y$, Fig. 4, offers some magnification of this critical region. Even high quality data suffer from this problem of the need for highest accuracy for the lowest of intensities, cf. Fig. 1.2, and 3 of the Jortner, Levine, and Rice chapter (which shows a fit to the measured[15] intensity fluctuation in $C_2H_2$).

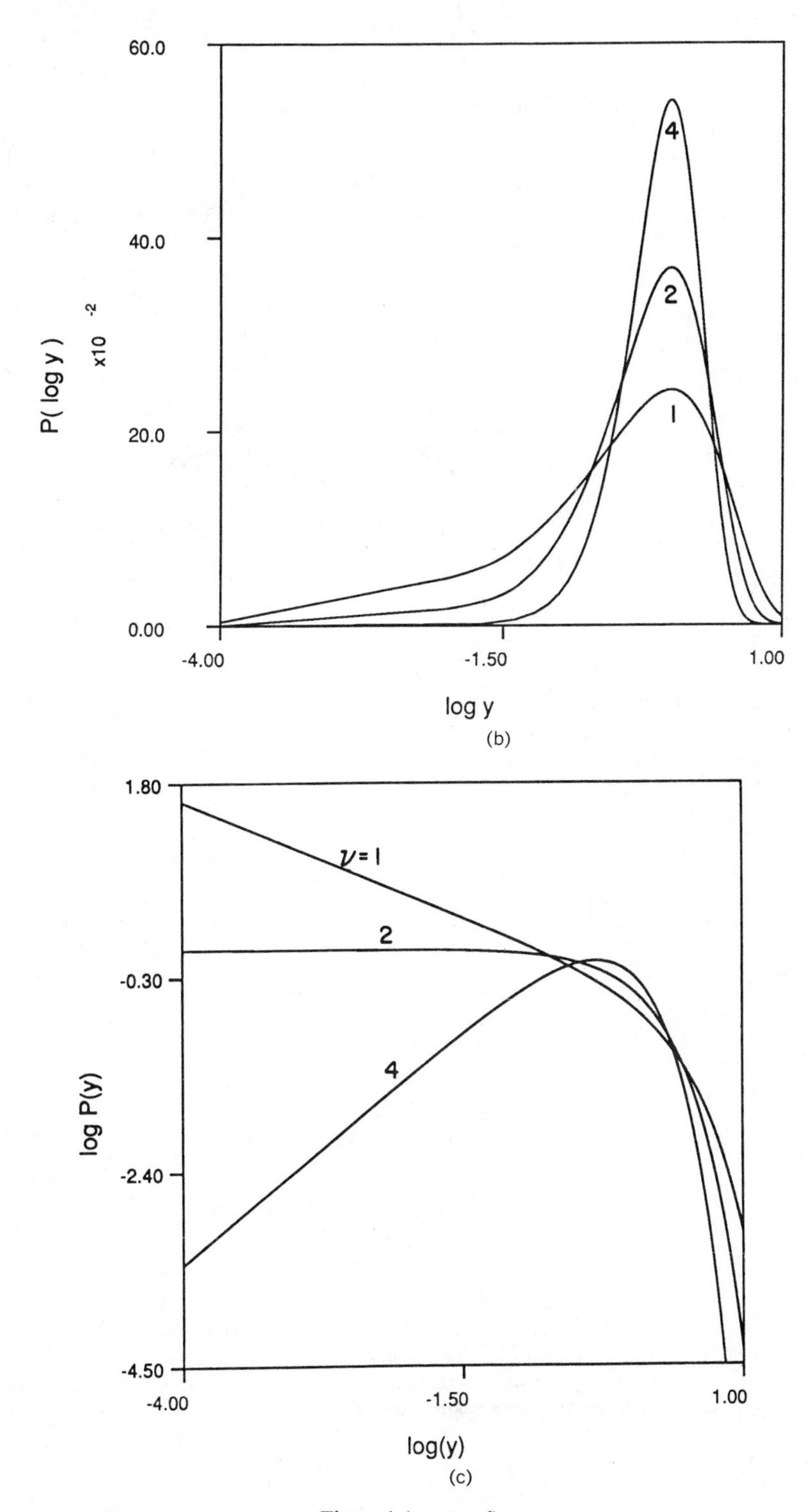


**Figure 4** (*continued*).

### E. The Cumulative Distribution

In our own work we found the sensitivity to small intensities to be a real problem.* One possible solution is to consider not $P(y)$ but the cumulative distribution function that is the fraction of all transitions with intensity up to $y$. In actual applications we note that for either experimental or computational data, the distribution of intensities is generated as a histogram. Say the intensity axis is divided into 1 bins of equal size and $P_n$ is the fraction of transitions with intensities in the $n$th bin. The histogram is now rearranged

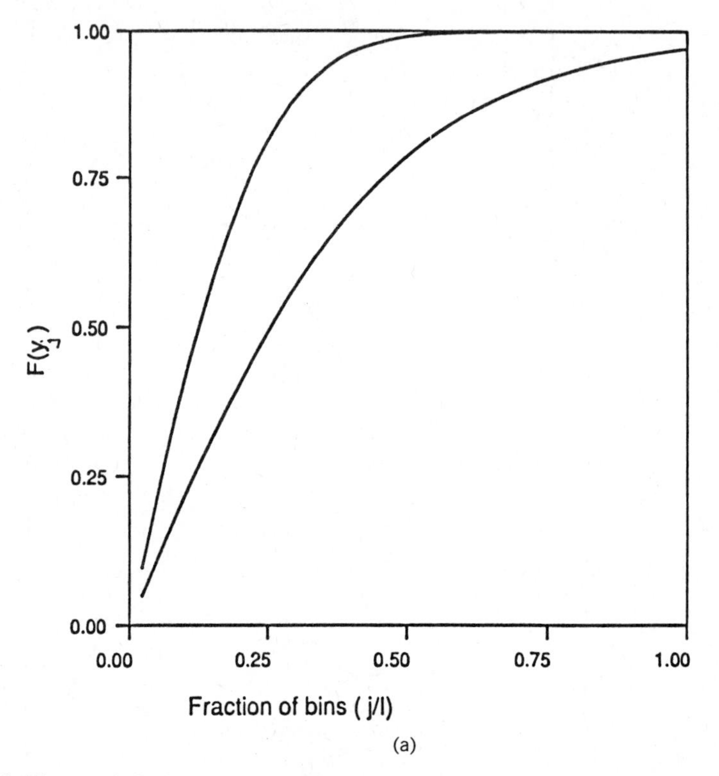


(a)

**Figure 5.** The cumulative intensity distribution determined from the observations[2,38] and computed for (2.17) at the same value of $\langle y \rangle$ as the observed spectrum of $C_6D_6$. (*a*) For the $14_0^1$ band; (*b*) for the $14_0^1 1_0^1$ band that cannot be assigned.

* It affects also computational data since, apart from very special models, the eigenstates need be determined by diagonalizing the Hamiltonian in a basis set that is exceedingly large. As the basis size is increased, the very weakest transitions are (as is only to be expected on the basis of Murphy's law and other considerations) the slowest to converge.

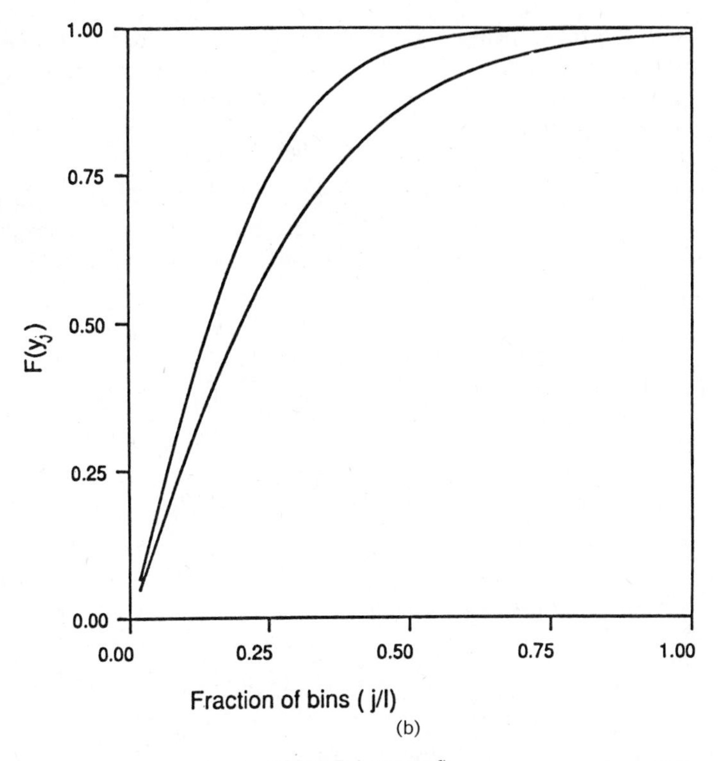


**Figure 5** *(continued)*.

so that each bin has a $P_n$ equal or less than that of the preceding bin. The new first bin has the highest fraction of transitions. The cumulative probability $F_j$,

$$F_j = \sum_{n=1}^{j} P_n, \tag{2.20}$$

is computed for the rearranged histogram so that $F_j$ increases as rapidly as possible towards its limit value $F_I = 1$.

The lower is the entropy of the intensity distribution, the more rapid is the increase of the cumulative probability.[37] Figure 5 compares the results for two bands in[2,38] $C_6D_6$ with the results for the distribution (2.17) at the same value of the total strength $\langle y \rangle$. For the band where quantum numbers can be assigned, each transition can be identified and no lines are missing, the cumulative distribution of the experimental results increases more rapidly than that for the most entropic intensity distribution. Not so for the band that is in the "channel three" region[2] where intramolecular energy transfer is quite dominant. A similar behavior is found also for computational results.[13]

As a practical point, it is necessary to mention that for the purpose of this comparison, the theoretical distribution [e.g., (2.17)] is also first converted into a histogram with the same bin boundaries as the experimental (or computational) data.

### F. The $\chi$-Square Distribution

Experimental and computational results often do deviate from the distributions of maximal entropy subject only to a given total strength. Consider the following simple modification. The spectrum can be regarded as the set of expectation values of the state $\rho_i$

$$\rho_i = T|i\rangle\langle i|T^\dagger/\mathrm{Tr}(T|i\rangle\langle i|T^\dagger). \tag{2.21}$$

Say now there is a good discrete symmetry $G$

$$[\rho_i, G] = 0, \tag{2.22}$$

then $\rho_i$ can be resolved into its components according to the representations of $G$. Any given transition probability $\langle f|\rho_i|f\rangle$ is then the sum of independent terms, or, in the simplest case

$$y = \sum_{j=1}^{\nu} x_j^2. \tag{2.23}$$

In (2.23) $\nu$ is the number* of independent components of $\rho_i$. When the distribution of the independent amplitudes is uniform, $P(y)\delta y$ is proportional to the volume bounded between two $\nu$-dimensional spheres of radii $r$ and $r + \delta r$, respectively, $y = r^2$. Hence (2.18) is replaced by

$$P(y)\delta y \propto r^{\nu-1}\delta r \propto r^{\nu-2}\delta y \tag{2.24}$$

or $P(y) \propto y^{\nu/2-1}$. Our two previous results correspond to $\nu = 1$ and 2, respectively. Note, however, that, in general, (2.23) represents a conjecture. We suspect there may be some symmetry that serves to partition the phase space into disjoint regions but we usually do not know for sure that such is the case.

The entropy of $P(y)$ is now given in terms of the deviance of $P(y)$ from $y^{\nu/2-1}$. The result of seeking the distribution of maximum entropy subject to a given total strength, $\langle y\rangle$, cf. (2.15), leads to

$$P(y) = y^{\nu/2-1}\exp(-\nu y/2\langle y\rangle)\Bigg/\left(\frac{2\langle y\rangle}{\nu}\right)^{\nu/2}\Gamma\left(\frac{\nu}{2}\right). \tag{2.25}$$

* Note that the general case is $y = \sum_j \rho_j x_j^2$. This is further discussed in the Appendix.

The distribution (2.25) is known in statistics[39] as a "$\chi$-square" distribution with $\nu$ degrees of freedom. It is usually derived as the distribution of $y$ as given by (2.23) when each one of the $\nu$ $x_j$'s is independent with a gaussian distribution of mean zero and unit variance. To bring (2.25) to this standard form note that the mean of $y$ is $\nu\langle x^2\rangle$ assuming all the $x$'s to have the same variance. Hence the variable $t = y\nu/\langle y\rangle$ will bring (2.25) to the standard form in statistics

$$P(t) = t^{\nu/2-1}\exp(-t/2)\Big/2^{\nu/2}\Gamma\left(\frac{\nu}{2}\right). \tag{2.26}$$

Sometimes $t$ is denoted by $\chi^2$.

The distribution (2.25) [of which (2.19) and (2.17) are special cases corresponding to $\nu = 1$ and 2, respectively] becomes narrower with increasing $\nu$, Fig. 4. This can also be seen analytically,

$$\langle (y/\langle y\rangle)^r\rangle = \left(\frac{2}{\nu}\right)^r\frac{\Gamma(r+\nu/2)}{\Gamma(\nu/2)}, \tag{2.27}$$

and, in particular,

$$\langle (y - \langle y\rangle)^2\rangle = 2\langle y\rangle^2/\nu. \tag{2.28}$$

For a given overall strength (i.e., given $\langle y\rangle$), the variance decreases with increasing $\nu$ as $2/\nu$.

The decline in the variance implies that the fluctuations are reduced. The intensities tend to be more tightly clustered about their mean value $\langle y\rangle$ when $\nu$ is higher. Unfortunately, inherent experimental averaging (i.e., a finite resolution) will have the very same effect. This is intuitively obvious and the mathematical reasoning concurs: Say the fully resolved spectrum is indeed most entropic and of the form (2.17), that is, with $\nu = 1$. Let the finite-resolution observed spectrum be such that each recorded transition is a sum of $\nu$ unresolved lines. Then (2.25) expresses the observed intensity and the observed distribution will be $\chi$-square with $\nu$ degrees of freedom. The intensity distribution for the SEP spectrum[15] of $C_2H_2$ and the fit to (2.25) is shown in Fig. 3 of the Jortner, Levine, and Rice chapter.

The derivation of (2.25) assumed that [cf. (2.23)] the $\nu$ components of $y$ are independent. If we follow the nuclear physics practice[21,40] of defining an effective $\nu$ by (2.29)

$$\frac{2}{\nu_{\text{eff}}} = \frac{\langle (y - \langle y\rangle)^2\rangle}{\langle y\rangle^2}, \tag{2.29}$$

then one can show that

$$1 \leq \nu_{\text{eff}} \leq \nu, \tag{2.30}$$

where the minimal value results when the $\nu$ components are completely correlated (see the Appendix).

Support for the functional form (2.25) is also provided by computational studies. The value of $\nu$ that is so obtained is not necessarily an integer [cf. (2.29); see also Section II G). Nonintegral values of $\nu$ can also be due to the effects of finite samples and to the inherent uncertainty in the value of $\nu$ [cf. Eq. (2.40)]. There is however another approach where the parameter $\nu$ can vary continuously, as follows: The most general form for the entropy of $P(y)$, as given by the grouping property,[37,41] is

$$S = -\int P(y)[\ln P(y) - S(y)]. \tag{2.31}$$

Here $S(y)$ is the entropy at a given value of $y$, associated, say, with the distribution among the $\nu$ $x_j$'s for a specified $\sum x_j^2$. If there are no additional constraints, then we obtain, by maximizing (2.31), the "natural" or prior distribution $P^0(y) \propto \exp[S(y)]$. In general

$$P(y) \propto \exp[S(y) + \text{additional constraints}]. \tag{2.32}$$

Hence $S(y)$ is as much a constraint on $P(y)$ as, say, the given total strength $y$. One sometimes writes loosely $S(y) = \ln V$, where $V$ is the available volume in phase space. If there are no other constraints, this volume is uniformly sampled.

For a dimension-bearing variable, one generally expects $\delta S \propto \delta y/y$ or $S(y) \propto \ln y$. Hence we can impose $\langle \ln y \rangle$ as a constraint on the maximization of the entropy. This will lead to a $\chi$-square distribution where now $(\nu - 1)/2$ is the Lagrange multiplier for this additional constraint. The value of $\nu$ is now to be determined as usual, by equating the value of $\langle \ln y \rangle$ as determined from (2.25)

$$\langle (\ln(y/\langle y \rangle)) \rangle = \psi\left(\frac{\nu}{2}\right) - \ln\left(\frac{\nu}{2}\right) \to -\frac{1}{\nu} - \frac{1}{3\nu^2} + O(\nu^{-4}) \tag{2.33}$$

to the average value of $\ln y$ as computed for the data. In (2.33), $\psi(x)$ is the $\psi$ function[42] $\psi(x) = d \ln \Gamma(x)/dx$, and the asymptotic expansion is quite accurate. A graph of the right-hand side of (2.33) versus $\nu$ is shown in Fig. 6.

The reasoning behind the choice of $\nu$ as a variable suggests that $\nu \geq 1$. In

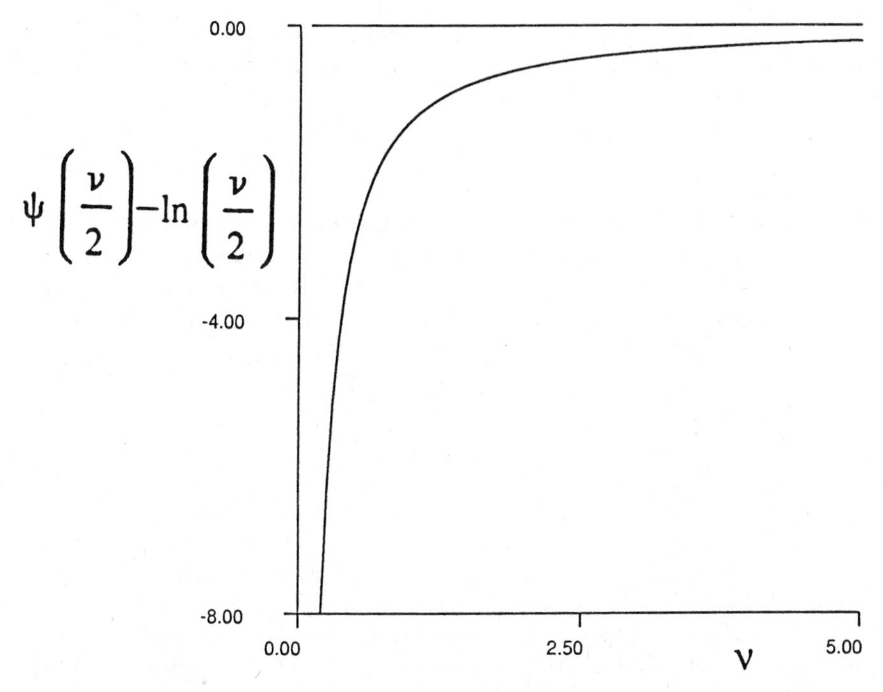


**Figure 6.** A plot of $\psi(\nu/2) - \ln(\nu/2)$ versus $\nu$ for use with Eq. (2.33).

Section III we shall refine the procedure so that the secular variation of the intensity with energy can be incorporated. One can then also analyze data when the classical motion is more regular. For computational studies one often finds that that $\nu \leq 1$ in that regime, Fig. 2 (and Fig. 8). That is physically reasonable in that the extent of the fluctuation diminishes as $\nu$ increases [cf. (2.28)] and one expects more fluctuations in the more regular regime.

### G. Determining the Number of Degrees of Freedom

Given an experimental (or computed) distribution of intensities $P(y)$, we determine the values of $\langle y \rangle$ and $\nu$ by minimizing $H$,

$$H = \int dy P(y) \ln[P(y)/P^T(y)], \tag{2.34}$$

where $P^T(y)$ is a $\chi$-square distribution with trial values of $\langle y \rangle$ and $\nu$ that are varied so as to minimize $H$. This is not quite the same as a least square fit, but will produce the theoretically correct distribution, namely, the one of maximal entropy with the same values of $\langle y \rangle$ and $\langle \ln y \rangle$ as the experimental distribution.[41]

Alternatively, $\langle y \rangle$ can be computed directly from the data. The situation is more delicate with respect to $\langle \ln y \rangle$. If we compute it from the raw data,

$$\langle \ln y \rangle = \sum_f \ln y_f \Big/ \sum_f 1, \tag{2.35}$$

(where summation is over all transitions) or if we draw a histogram of the data first and compute $\langle \ln y \rangle$ from the histogram, we find that the use of (2.33) can lead to different values of $v$. The reason is, of course, that $\ln(y/\langle y \rangle)$ emphasizes the very weak transitions. This is a real problem, which we have already alluded to. The nuclear physicists[21] often determine $v$ from the variance of $y$, cf. (2.29),

$$v = 2\langle y \rangle^2 / \langle (y - \langle y \rangle)^2 \rangle. \tag{2.36}$$

This centers attention on the width of the distribution rather than on its behavior for small $y$. As a pragmatic measure, the use of (2.36 ) is fine. Unfortunately, in principle, it is $\langle \ln y \rangle$ that determines $v$, and unless the sample is large and relatively noise free, (2.33) and (2.36) will not yield the same value for $v$ even when (2.25) is the true distribution.* We thus *emphasize* that the proper procedure is either to minimize $H$ or to use (2.35) to compute $\langle \ln y \rangle$ and then use (2.33) to determine $v$. We have verified that the two routes give practically identical results. The use of (2.33) is the recommended procedure for those not interested in the minimal value† of $H$.

Since $(v - 1)/2$ can be regarded as a Lagrange multiplier, its error limits can be estimated.[43–45] There are three origins for the uncertainty in the value of $v$. The first is the effect of finite sample size.[46] This turns out to be often smaller than the other two which are due to noise in the data and to the inherent uncertainty in Lagrange multipliers.

The average squared fractional error in the data, $s^2$, is defined by[43]

$$s^2 \equiv \sum_f y_f (\delta y_f / y_f)^2. \tag{2.37}$$

The variance in $\ln(y/\langle y \rangle)$ can be computed from (2.25),

$$\Delta^2(\ln(y/\langle y \rangle)) \equiv \int dy [\ln(y/\langle y \rangle) - \langle \ln(y/\langle y \rangle) \rangle]^2 P(y) = \psi'\left(\frac{v}{2}\right). \tag{2.38}$$

* We have verified that this is the case by numerical experiments; (2.36) is not an accurate way for estimating $v$.

† Min $H$ is a measure of the quality of the fit. It should be below $s^2$ for a fit to within the noise limit in the data.

Here $\psi'(x) = d\psi(x)/dx$ is the derivative of the $\psi$ function, which has the expansion[42]

$$\psi'\left(\frac{\nu}{2}\right) = \frac{2}{\nu - 1}\left(1 - \frac{1}{3(\nu - 1)^2}\right) + O[(\nu - 1)^{-5}]. \tag{2.39}$$

Assuming that $\langle y \rangle$ is accurately known, the uncertainty in $\nu$ is[44]

$$\delta\left(\frac{\nu}{2}\right) \leq \frac{s}{\Delta[\ln(y/\langle y \rangle)]}. \tag{2.40}$$

If the data are grouped into $N$ bins, then $s^2 \simeq 1/N$. Typically, the experimental noise makes $s^2$ much larger. Given $s$ and $\nu$ one computes from (2.40) the possible range of values $\pm\delta\nu$ of $\nu$ about its mean.

### H. Lifetimes and Transition Rates

The intensities in an optical spectra can be regarded as transition rates. Hence, the distributions under consideration apply equally well, for example, to the distribution of predissociation rates. Indeed, it is in this context that such fluctuations were first considered in nulear physics.

Figure 7 compares the observed[4] distribution of lifetimes of ro-vibronic states of $SiH_2$ with excess energy of about 17,000 $cm^{-1}$ with the distribution $P(\tau) = P(y)/(d\tau/dy)$ with $\tau = 1/y$ and $\langle \tau \rangle = 1/\langle y \rangle$,

$$P(\tau) = \tau^{-(\nu/2+1)} \exp\left(-\frac{\nu\langle \tau \rangle}{2\tau}\right)\langle \tau \rangle^{\nu/2} \Big/ \left(\frac{2}{\nu}\right)^{\nu/2} \Gamma\left(\frac{\nu}{2}\right). \tag{2.41}$$

The fit with $\nu = 1$ is quite accurate for the data shown in Fig. 7. There is thus no evidence for deviation from the most entropic behavior in this system.[47]

In the classically chaotic limit, rate constants for specific quantum states are therefore expected to extensively fluctuate about their monotonic average $k(E)$ (cf. also Section III E)

$$\langle (k - k(E))^2 \rangle = 2k^2(E)/\nu. \tag{2.42}$$

For $\nu = 1$, the case shown in Fig. 7, the standard deviation of the fluctuations, is $(2)^{1/2}$ in units of the mean!

Off-diagonal matrix elements, in general, are also expected to exhibit similar fluctuations, and numerical studies[48] verify this expectation.

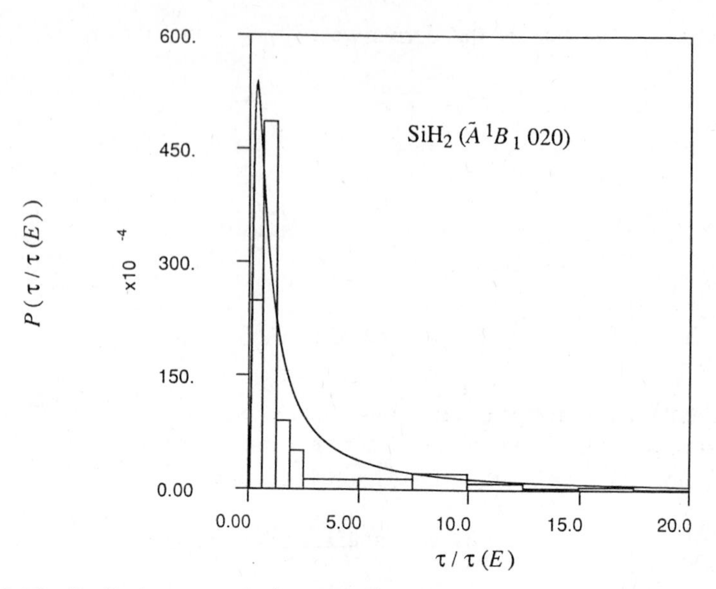


**Figure 7.** The distribution of lifetimes[3] in $SiH_2(\tilde{A}^1B_1\,020)$ versus $\tau/\tau(E)$. The solid line is the fit by (2.41) with $v = 1$. The histogrammic representation of the data uses nonuniform bins so as to emphasize the distribution at low $\tau$'s. [The most probable $\tau$ for $v = 1$ is at $\tau(E)/3$.] All 222 measured levels were included in the fit but 14 levels have $\tau > 20\tau(E)$ and hence are not included in the drawing.

### I. Discussion

The distribution we have considered is that of the intensities or other transition rates, irrespective of their energy. As long as we are examining only a narrow energy interval, this is quite reasonable. Otherwise, however, there is a secular variation reflecting essentially the energy dependence of the "short-time dynamics" approximation. What we really expect is to have fluctuations about the local average. In other words, $\langle y \rangle$ would vary as we scan a wide energy interval. It turns out that the secular variation in $\langle y \rangle$ requires only a technical change. The argument is (2.17) or (2.19) or (2.25) is not $y/\langle y \rangle$ but $y$ scaled by the local average of $y$. Proving this result is the subject of Section III.

## III. THE DISTRIBUTION OF SPECTRA

### A. The Envelope

A discrete stick spectrum, normalized to a total strength of unity, is given by

$$S(E) = \sum_f p_f \delta(E_f - E), \tag{3.1}$$

where*

$$p_f = y_f \Big/ \sum_f y_f = y_f / N\langle y \rangle, \tag{3.2}$$

with $N$ as the number of lines in the spectrum $N = \sum_f 1$. At finite resolution such a spectrum will be smooth and can be written as the envelope

$$S^0(E) = \sum_f p_f A(E_f - E). \tag{3.3}$$

Here $A(E_f - E)$ is the line shape function that is of unit area and reduces to a $\delta$ function as its width shrinks. Thus

$$\int dES(E) = \int dES^0(E) = 1. \tag{3.4}$$

We shall forego for the moment the question of how to specify the line shape function.

As in Section II A, also at finite resolution, one can write down sum rules, for example,

$$S^0(E) = \langle i | T^\dagger A(E - H) T | i \rangle / \langle i | T^\dagger T | i \rangle, \tag{3.5}$$

or

$$\int dEE^n S^0(E) = \langle i | T^\dagger \overline{H^n} T | i \rangle / \langle i | T^\dagger T | i \rangle, \tag{3.6}$$

where

$$\overline{H^n} = \int dEH^n A(E - H). \tag{3.7}$$

The denominators in (3.5) or (3.6) are due to the normalization $\int dES^0(E) = 1$ and equal the total strength.

At finite resolution, transitions at energies away from $E_f$ will contribute to the (normalized) strength $p_f^0$ at $E_f$,

* We ask the reader's permission to use the very same $\langle z \rangle$ symbol for the average of $z$ in two, quite distinct meanings. One, reserved for use in $\langle y \rangle$ only is an average over all transitions. The other is an average over fluctuations in the strength of a given transition.

$$p_f^0 = \sum_j p_j A(E_j - E_f). \tag{3.8}$$

A consistency condition is that the finite resolution spectrum of the finite resolution strengths be the initial finite resolution envelope

$$S^0(E) = \sum_f p_f^0 A(E_f - E). \tag{3.9}$$

It follows from (3.3) taken at $E_j$ that

$$\sum_j p_j A(E_j - E_f) = \sum_j p_j^0 A(E_j - E_f). \tag{3.10}$$

We note in passing that the self-consistency condition (3.10), implies on using (3.8) and the symmetry of the line shape $A(E_j - E_f) = A(E_f - E_j)$ that $\sum_f p_f p_f^0 = \sum_f p_f^{0^2}$, Hence

$$0 \leq \sum_f (p_f - p_f^0)^2 = \sum_f p_f^2 - \sum_f p_f^{0^2}. \tag{3.11}$$

The averaged spectrum $p_f^0$ is thus more uniform than the actual spectrum, as is to be expected. The proper fraction

$$F = \sum_f p_f^{0^2} \Big/ \sum_f p_f^2 \tag{3.12}$$

is the $F$ parameter of Heller,[8,9] if we identify $p_f^0$ with his "stochastic limit."

In Section III E we shall explore the dynamical significance of the envelope function and identify it as a "short-time dynamics" approximation. As a practical issue we shall also explore classical mechanics approximations for $S^0(E)$.

### B. Fluctuations

The essential point is that the envelope does not determine a unique, high-resolution, spectrum. Seemingly, that is wrong. We have after all the set of equations

$$p_f^0 = \sum_j p_j A(E_j - E_f) \tag{3.8}$$

one for each transition. Can we not invert them so as to determine the $p_j$'s? The answer is no. Indeed, our very consistency condition (3.10) implies that there is more than one stick spectrum that gives rise to the same envelope.* There is thus a whole class of spectra that gives rise to the same envelope.

We cannot infer a unique spectrum from a given envelope. We thus do the next best thing—infer the distribution of $p_j$'s that is consistent with the given envelope.

Let $P(p_j)$ be the distribution of the intensity of the $j$th line taken by considering all the spectra that have the same envelope. Since each individual spectrum satisfies (3.8), so does their average

$$p_f^0 = \sum_j A(E_j - E_f) \int dp_j p_j P(p_j). \tag{3.13}$$

Equation (3.13) will be imposed as a set of constraints (one for each transition) in determining $P(p_j)$ by the procedure of maximum entropy.

### C. The Probability of a Spectrum

The probability of an entire spectrum will be determined by the procedure of maximum entropy.[49] The purpose is to show that under the constraints considered so far, including the various sum rules, the distribution of intensities of different lines are independent of one another. When considered as a function of the ratio of the actual intensity to its mean, local value, all transitions have the same distribution, which is $\chi$-square in the general case. Hence, one can group intensities together into bins, and consider the distribution of intensity irrespective of line position *provided* one takes proper account of the variation of the envelope with energy. The technical backup for these statements and the specific conditions under which they will fail are the subjects of this subsection.

The set of line strengths will be denoted $\mathbf{p}$ with $P(\mathbf{p})$ being the distribution of intensities for an entire spectrum. The constraints are: (1) normalization

$$\int d\mathbf{p} P(\mathbf{p}) = 1; \tag{3.14}$$

(2) a given overall strength that is equivalent, cf. (3.2), to normalization of the

* An analogous situation may be more familiar. Let $p_j$ be a nonthermal distribution of the internal states of, say, diatomic molecules diluted in an excess of monoatomic buffer gas. Owing to collisions, the distribution $p_j$ will relax to a final, unique equilibrium distribution. Let $p_j^0$ be that equilibrium distribution $A_{fj}$ [$\equiv A(E_f - E_j)$] is the evolution matrix. It converts any initial distribution to a unique final distribution *and* it leaves the equilibrium distribution invariant $p_f^0 = \sum_j A_{jf} p_j^0$. Given the unique final equilibrium distribution $p_f^0$, we have no way whatever of knowing which was the initial nonequilibrium distribution which relaxed to it.

intensities

$$\sum_j \int d\mathbf{p} p_j P(\mathbf{p}) = 1; \tag{3.15}$$

and (3) a given envelope, (3.13),

$$p_f^0 = \sum_j A(E_j - E_f) \int d\mathbf{p} p_j P(\mathbf{p}), \qquad f = 1, 2, \ldots, N. \tag{3.13}$$

The entropy of the distribution is

$$S = -\int d\mathbf{p} P(\mathbf{p}) [\ln P(\mathbf{p}) - S(\mathbf{p})], \tag{3.16}$$

where $S(\mathbf{p})$ is the entropy of a particular spectrum. If only normalization is imposed as a constraint (with the Lagrange multiplier $\lambda_0$), then the maximum of $S$ obtains for

$$P^0(\mathbf{p}) = \exp[S(\mathbf{p}) - \lambda_0] \tag{3.17}$$

with the normalization condition

$$\exp(N\lambda_0) = \int d\mathbf{p} \exp[S(\mathbf{p})]. \tag{3.18}$$

In the general case we expect (cf. Section II F) $S(\mathbf{p}) \propto \ln \mathbf{p}$ so that $P^0(\mathbf{p})$ factorizes into a product

$$P^0(\mathbf{p}) = \prod_j P^0(p_j). \tag{3.19}$$

Next we impose two constraints, normalization of $P(\mathbf{p})$ and a given total strength, (3.15). The latter is a single constraint on the entire spectrum, and we denote its Lagrange multiplier by $\lambda$. The maximum of $S$ is found for

$$P(\mathbf{p}) = \exp\left(S(\mathbf{p}) - \lambda_0 - \lambda \sum_j p_j\right) = \prod_j P(p_j), \tag{3.20}$$

where

$$P(p_j) = P^0(p_j) \exp(-\mu_0 - \lambda p_j). \tag{3.21}$$

Here $\mu_0$ has been defined by

$$\exp(\mu_0) = \int dp_j P^0(p_j) \exp(-\lambda p_j) \tag{3.22}$$

and the value of $\lambda$ is to be determined from the constraint equation (3.15). Take the most general case $P^0(p_j) \propto p_j^{\nu/2-1}$. Then, using (3.21) and (3.22)

$$\langle p_j \rangle \equiv \int dp_j p_j P(p_j) = \nu/2\lambda \tag{3.23}$$

so that the constraint condition (3.15) gives $\lambda = \nu N/2$. Using the definition (3.2) of $p_j$ in (3.21) leads to the final answer

$$P(y_j) = \left(\frac{\nu y_j}{\langle y \rangle}\right)^{\nu/2-1} \exp(-\nu y_j/2\langle y \rangle) \Big/ 2^{\nu/2}\Gamma\left(\frac{\nu}{2}\right). \tag{3.24}$$

That is precisely our previous result (2.25). Considering an entire spectrum has, so far, failed to establish any correlation between the distributions of different lines nor to provide any new results.

We now impose not only a given overall strength but also a given envelope. The average *local* intensity is thus also specified. This requires the introduction of the set of $N$ constraints (3.13), each of which is assigned the Lagrange multiplier $\gamma_f$. The distribution of maximal entropy subject to the three constraints (1)–(3) is

$$P(\mathbf{p}) = P^0(\mathbf{p}) \exp\left(-\lambda_0 - \lambda \sum_j p_j - \sum_f \gamma_f \sum_j A(E_f - E_j) p_j\right). \tag{3.25}$$

It is convenient to introduce $\mu_j$ by

$$\mu_j \equiv \lambda + \sum_f \lambda_f A(E_f - E_j). \tag{3.26}$$

Then

$$P(\mathbf{p}) = \prod_j P(p_j) \tag{3.27}$$

with

$$P(p_j) = P^0(p_j) \exp(-\mu_{0j} - \mu_j p_j). \tag{3.28}$$

Even with a specified envelope, the distribution of intensities of different lines factorizes as a product of independent distributions. Now, however, each line has its own Lagrange multiplier $\mu_j$, whereas when only the total strength was specified, the distribution of intensities of different lines [cf. (3.21)] had a common multiplier, $\lambda$, for all the $p_j$'s.

The value of the $N$ $\mu_j$'s is to be determined from the $N$ constraints (3.13). $\mu_{0j}$ is a function of $\mu_j$ as determined by the normalization condition. From (3.28) and the normalization of $P(p_j)$

$$\exp(\mu_{0j}) = 1/\mu_j, \qquad \mu_j = \nu/2\langle p_j\rangle. \tag{3.29}$$

Hence our final result

$$P(p_j) = \left(\frac{\nu p_j}{\langle p_j\rangle}\right)^{\nu/2-1} \exp(-\nu p_j/2\langle p_j\rangle)\Big/ 2^{\nu/2}\Gamma\left(\frac{\nu}{2}\right) \tag{3.30}$$

normalized such that

$$\int_0^\infty P(p_j) d(\nu p_j/\langle p_j\rangle) = 1 \tag{3.31}$$

and

$$\langle (p_j - \langle p_j\rangle)^2\rangle = 2\langle p_j\rangle^2/\nu. \tag{3.32}$$

All transitions fluctuate with a common distribution *provided* that the variable is not the intensity itself but the intensity divided by its local average, $p_j/\langle p_j\rangle$. It is essential however to note that this presupposes that the number $\nu$ of degrees of freedom is the same along the entire spectrum. This is certainly true when the underlying classical motion is chaotic over the same range in energy (when $\nu = 1$ or 2). However, it need not be true over a wide energy range where the character of the classical motion can change.

Computational studies,[13] Fig. 8, show that for the same system but at very different energy ranges the value of $\nu$ will differ. Specifically, it will be higher when the classical behavior is more chaotic.

What about our sum rules, should they not be imposed as constraints? The answer is that the envelope must also satisfy the sum rules, and since the distribution is chosen among the set that reproduces the envelope, the sum rules primarily govern the specification of the latter.

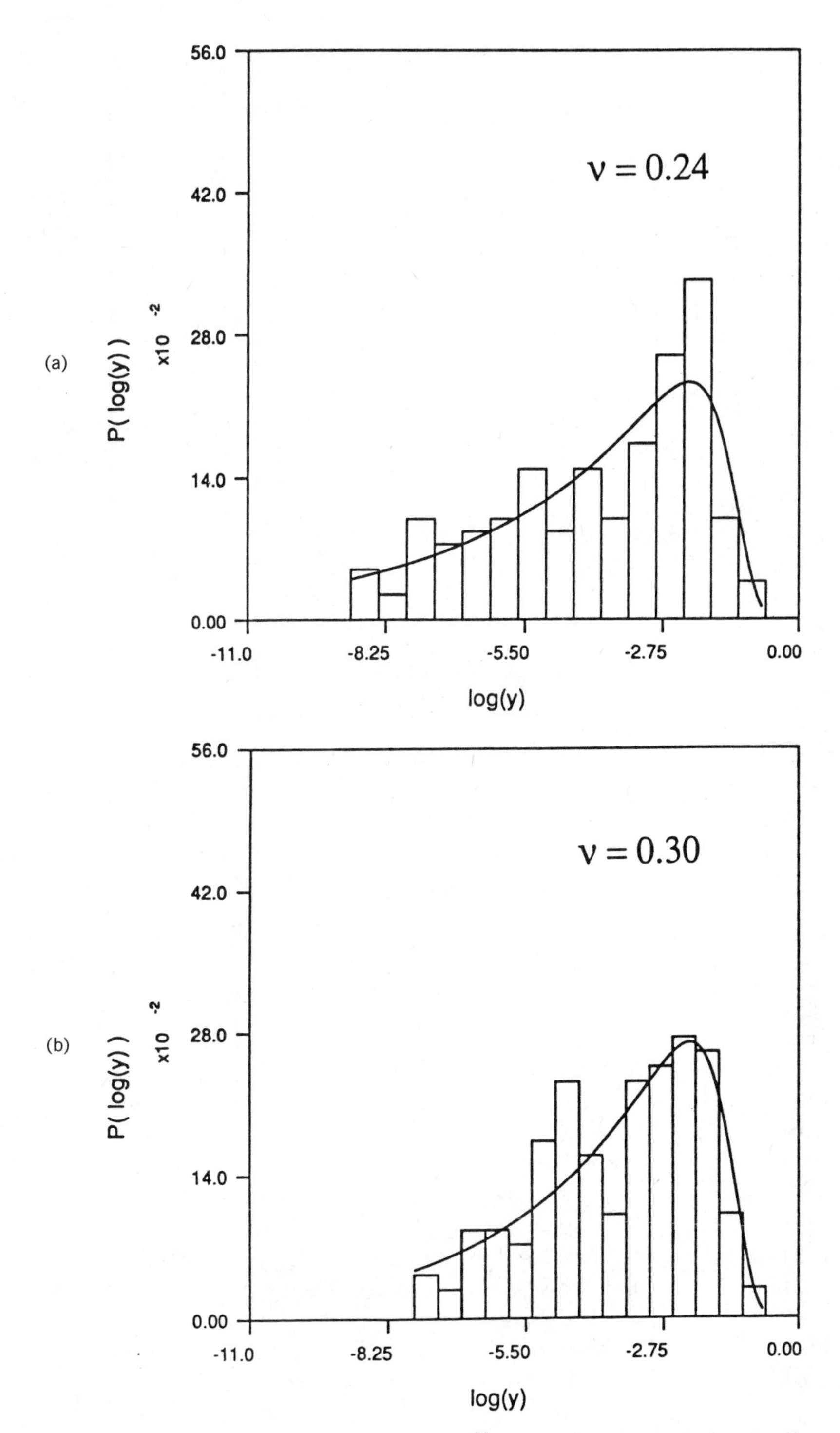


**Figure 8.** Distribution of intensities for computed[13] spectra when the classical motion does not indicate large-scale chaos. The fit shown is to (3.30) with the values of $\nu$ as indicated. The very low $\nu$'s correspond to initial states in the regular regime and to weak coupling.

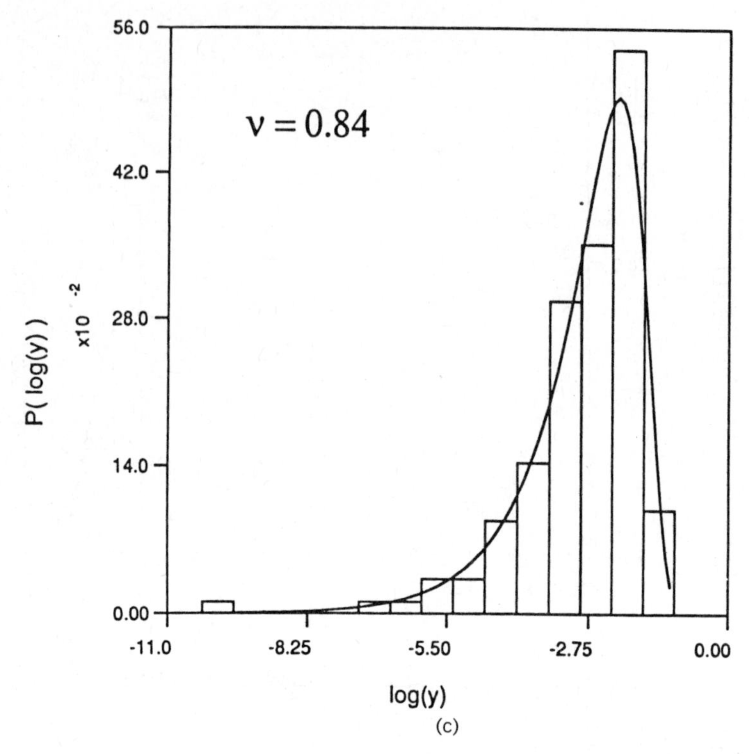

(c)

**Figure 8** (*continued*).

### D. The Effective Number of States in the Spectrum

We consider the number of final states that are effectively coupled by the transition operator to the initial state. The expectation is that when the number is far smaller than the number of available states, then selection rules must be operative, adjacent states will be quite different, and fluctuations will be large. Such behavior is interpreted as the underlying classical dynamics being regular so that the system, being confined to an invariant torus, fails to sample the available phase space. If, however, many final states can be accessed (and, as we shall argue, there are no very large differences in intensity), then fluctuations will be small. This will be taken as the signature of irregular and ultimately chaotic classical dynamics.

There is a natural definition of the effective number $N_e$ of final states when the spectrum is uniform. Then $p_f = 1/N_e$ for all the lines that carry any strength. When the intensities of different lines do differ, we follow Shannon[50]

$$\log N_e = -\sum_{f=1}^{N} p_f \log p_f, \tag{3.33}$$

where summation is over all the possible $N$ final states. One readily verifies that $N_e \leq N$, with equality if and only if all possible states are accessed with equal probability. Any deviation from this overall uniformity (including $p_f = 1/N_e$, $N_e < N$) leads to a lower value for $N_e$.

In general, the envelope is not uniform either. Let there be $N_0$ available final states, determined by considering all the spectra that are consistent with the given envelope. Then

$$-\log(N_e/N_0) = \sum_f p_f \log(p_f/p_f^0) \equiv DS. \tag{3.34}$$

The right-hand side of (3.34) is, by definition, the entropy deficiency, $DS$. It is nonnegative and vanishes if and only if $p_f = p_f^0$. The fraction of the available final states that are actually accessed is

$$N_e/N_0 = \exp(-DS), \tag{3.35}$$

a result familiar from other applications.[51]

The inequality[52]

$$\sum_f a_f \log x_f \leq \log\left(\sum_f a_f x_f\right) \tag{3.36}$$

can be used to offer an approximate version of (3.34) when $\sum_f p_f^2$ is not very much below unity;

$$N_e/N_0 \simeq \sum_f p_f^0 \Big/ \sum_f p_f^2 \equiv F. \tag{3.37}$$

The right-hand side of (3.37) is, by definition, the $F$ parameter of Heller. The information theoretic definition of $N_e/N_0$, as given by $\exp(-DS)$, is somewhat underestimated by $F$ as can be seen from Fig. 9 or, analytically, from (3.43).

Even in the chaotic limit the effective number of states $N_e$ will be below the available number $N_0$. The reason is the fluctuations of the intensities about their mean values. Fluctuations imply lower entropy or, equivalently, lower $N_e$. To estimate the importance of fluctuation in reducing the available volume in phase space, we use (3.12) with $\langle p_f \rangle = p_f^0$. Then, using (3.32),

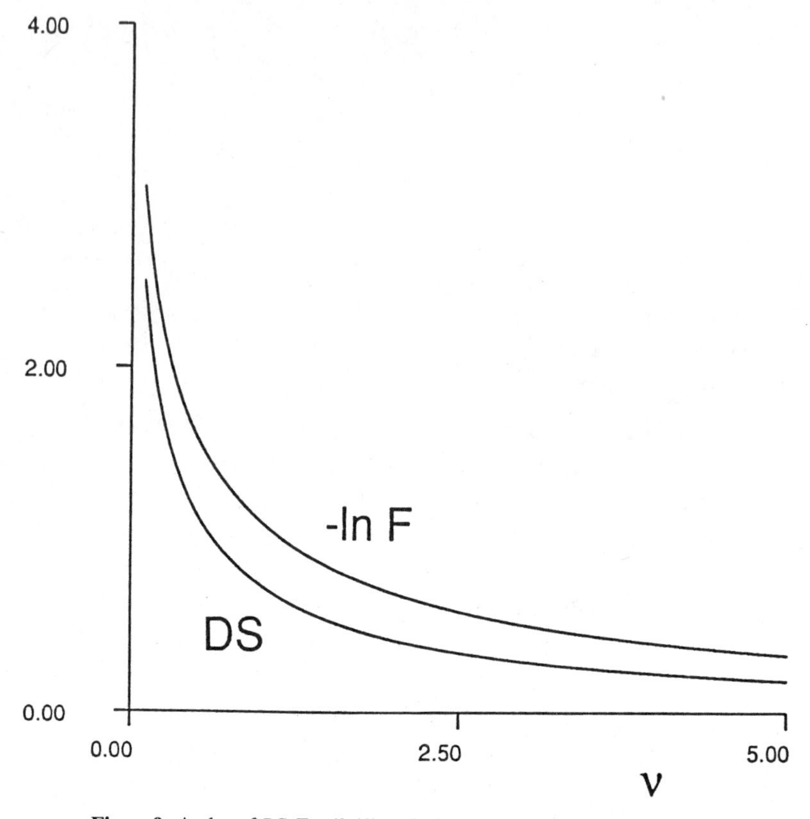


**Figure 9.** A plot of $DS$, Eq. (3.43) and of $-\ln F = \ln[(\nu + 2)/\nu]$ versus $\nu$.

$$F = \sum_f \langle p_f \rangle^2 \Big/ \sum_f \langle p_f^2 \rangle = \frac{\nu}{\nu + 2}. \tag{3.38}$$

Thus, for $\nu = 1$, fluctuations reduce $N_e$ to be about* one-third of $N_0$.

In obtaining the estimate (3.38) for $F$ we have replaced the average of $p_f$ over the set of spectra consistent with a given envelope with an energy average over a particular spectrum. This replacement is certainly useful for it offers a practical route to the evaluation of $\langle p_f \rangle$. In general, $F$ as defined by (3.12) is given as $F = G/G^0$, where

$$G = \left(\sum_f y_f\right)^2 \Big/ N \sum_f y_f^2 \quad (\simeq N_e/N), \tag{3.39}$$

* A more precise result based on computing $DS$ from $\nu = 1$ yields $N_e \simeq N_0/2$.

$$G^0 = \left(\sum_f y_f^0\right)^2 \Big/ N \sum_f y_f^{0^2} \quad (\simeq N_0/N), \tag{3.40}$$

and $p_f = y_f/\sum_f y_f$ and similarly for $p_f^0$ ($\sum y_f = \sum y_f^0$ by construction). In the approach of Section II where the envelope was effectively taken to be uniform, $G \equiv \langle y \rangle^2/\langle y \rangle^2 = v/(v+2)$.

The numerical value of $DS$ (like that of $F$) requires a knowledge of the spectrum. Here, too, we can use the result that $P(p_f/\langle p_f \rangle)$ is a universal function to replace the averaging of $\ln(p_f/\langle p_f \rangle)$ over all transitions of a particular spectrum by an average over the fluctuation

$$\begin{aligned} DS &= \sum_f p_f \ln(p_f/p_f^0) \\ &= \sum_f p_f^0(p_f/p_f^0)\ln(p_f/p_f^0) \\ &= \int_0^\infty dt(t/v)\ln(t/v)P(t), \end{aligned} \tag{3.41}$$

where $t = vp_f/\langle p_f \rangle$ and we have put $p_f^0 = \langle p_f \rangle$ as in (3.38). The integral can be readily evaluated using (3.30) and changing the number of degrees of freedom of $P(t)$ from $v$ to $v+2$. Use of the identity

$$\psi(z+1) = \psi(z) + \frac{1}{z} \tag{3.42}$$

leads to the compact result

$$\begin{aligned} DS &= \psi\left(\frac{v}{2}\right) - \ln\left(\frac{v}{2}\right) + \frac{2}{v} \\ &= \ln\left(\frac{1}{F}\right) + \psi\left(\frac{v}{2}+1\right) - \ln\left(\frac{v}{2}+1\right). \end{aligned} \tag{3.43}$$

$DS$ and $F$ are plotted versus $v$ in Fig. 9. In particular for $v = 1$ and 2, $DS \simeq 0.73$ [$\exp(-DS) \simeq 0.5$] and 0.42 [$\exp(-DS) = 0.66$], respectively. That $\exp(-DS) \geq F = v/(v+2)$, as can be seen from the graphs in Fig. 9, follows from (3.43).

### E. The Envelope Function and the Classical Limit

We have so far considered the smooth envelope as an average, (3.8), of the observed spectrum. The computation of the envelope from dynamical considerations is of interest not only for the purpose of analyzing data but also for tracing the dynamical origin of the fluctuations.

To proceed to the classical limit[53] it is convenient to start from the symmetric strength function $Y(E, E')$ introduced to (2.5), which can be written[54] as an expectation value over a microcanonical distribution of initial states at the energy $E$:

$$Y(E, E') \equiv \sum_i \sum_f |\langle f|T|i\rangle|^2 \delta(E - E_i)\delta(E' - E_f) \tag{3.44}$$
$$= \mathrm{Tr}\{\delta(E - H)T^\dagger \delta(E' - H)T\},$$

while [cf. (2.4)] a particular spectrum is given by

$$S(E') = \mathrm{Tr}\{|i\rangle\langle i|\, T^\dagger \delta(E' - H)T\}. \tag{3.45}$$

Now use the standard procedure[55] of replacing a delta function by a Fourier integral to get

$$Y(E, E') = (2\pi)^{-1} \int_{-\infty}^{\infty} dt \exp(i\omega t)\, \mathrm{Tr}\{\delta(E - H)T^\dagger(t)T(0)\} \tag{3.46}$$

where $T^\dagger(t)$ is the Heisenberg picture operator ($\hbar = 1$)

$$T^\dagger(t) = \exp(iHt)T^\dagger \exp(-iHt), \tag{3.47}$$

and $\omega = E' - E$. In (3.46) we can proceed to the classical limit[23,30] by defining the correlation function $\langle C(t)\rangle$

$$\langle C(t)\rangle \equiv \mathrm{Tr}\{\delta(E - H)T^\dagger(t)T(0)\}. \tag{3.48}$$

The average in (3.48) denotes a microcanonical expectation value.

The classical limit of (3.48) is a phase space integral*

$$\langle C(t)\rangle = \lim_{T\to\infty} \frac{1}{2T} \left\langle \int_{-T}^{T} T(t + \tau)T(\tau)\, d\tau \right\rangle \tag{3.49}$$

where $T(t)$ is now the classical (dipole or otherwise, as appropriate) function and averaging is over a classical microcanonical ensemble of initial conditions for the trajectories used to compute $T(t)$. The Fourier transform of (3.49) is

* This classical limit is different than that implied by Ref. 8 where a particular localized initial state, for which a classical limit can be taken is used.

the classical strength function

$$\langle C(\omega)\rangle = (2\pi)^{-1}\int_{-\infty}^{\infty} dt \exp(i\omega t)\langle C(t)\rangle, \tag{3.50}$$

and cf. (2.8),

$$\langle S(E')\rangle = \langle C(E' - E)\rangle/\rho(E). \tag{3.51}$$

In the terminology of random variables $\langle C(\omega)\rangle$ is the power spectrum of the variable $T(t)$. It is only under special circumstances,[56] however, that the power spectrum can be written as the average random power

$$\langle C(\omega)\rangle = \lim_{T\to\infty} (2T)^{-1}\left\langle\left|\int_{-T}^{T} T(t)\exp(i\omega t)\,dt\right|^2\right\rangle. \tag{3.52}$$

One can argue that equality in (3.52) will obtain in the ergodic limit,[56] but since our interest is often in systems not necessarily ergodic, (3.49) will be the procedure of choice.

The semiclassical evaluation of the averaged strength function can be discussed using periodic orbit theory[57] as applied to the similar (but simpler) problem of evaluating the average density of states.[53,58] One then finds

$$\langle C(t)\rangle \approx \sum_n C_n \exp[i\phi_n(t)] \tag{3.53}$$

where $\phi_n(t)$ is the action integral for a *closed* orbit (i.e., one that returns to its initial conditions at the time $t$). The primary variation of $\langle C(\omega)\rangle$ with the energy $E'$ of the final state is then due to the shortest closed paths. The smooth envelope of the spectrum is thus obtained by following the dynamics up to the first recurrence time. Longer times are required to build in the finer structure oscillations that reflect interferences between closed classical trajectories. The fluctuations in the spectrum can thus be viewed in the same way as other types of interference oscillations about the classical limit quantum mechanics.[59,60] The most familiar case is the interference pattern in the angular distribution for elastic scattering.*[61]

One expects that the "zero-time" limit will reproduce classical statistical mechanics.[53] In this limit, the distribution of intensities is uniform and varies

* Recall that even for that simple problem the amplitude of the interference oscillations does not vanish in the $h \to 0$ limit. It is only the frequency of oscillations that becomes more rapid in that limit.[61]

only with the energy $E$ of the initial state. When considering rates of unimolecular processes, this limit suffices, for it is the variation with $E$ that is of interest. Recent experiments[62,63] are however heralding the measurement of state-to-state unimolecular rates that would be amenable to the analysis of their fluctuation with both initial and final states.

### F. Lifetimes and Transition Rates

The fluctuations of decay rates of quantum states with variation in the initial level have already been considered in Section II H. Now we are in a better position to understand the derivation. Consider a group of initial levels in a narrow energy interval $\delta E$ about the energy $E$. Let $k_i$ be the decay rate of the $i$th level. Then

$$\sum_i{}' k_i \Big/ \sum_i{}' 1 \equiv \langle k \rangle = \langle k \rangle(E). \tag{3.54}$$

Summation in (3.54) is restricted to levels in the narrow energy interval and $\langle k \rangle(E)$, the analogue of the mean total strength function,

$$\langle k \rangle(E)/\rho(E) \equiv \int dE' Y(E, E')/\rho(E), \tag{3.55}$$

can be computed from the classical statistical approximation.

The procedure of Section III C will yield the analogue of (3.24)

$$P(k_j) = [\nu k_j/k(E)]^{\nu/2-1} \exp[-\nu k_j/2k(E)]/2^{\nu/2}\Gamma\left(\frac{\nu}{2}\right), \tag{3.56}$$

with a variance [in units of $\langle k \rangle^2(E)$, cf. (2.42)] of $2/\nu$. Such extensive fluctuations are indeed evident in recent measurements[2-5] and computations,[6,7,64,65] and an example is given in Fig. 7.

For systems with a continuous spectrum, the concept of "chaos" is not defined in classical mechanics,[66] since at least some trajectories will escape from the interaction region, as $t \to \infty$. One could introduce a definition based on the notion of "cantori," which act as barriers in phase space[67] and which can serve[68] to provide an analogue of a transition-state configuration for systems whose time evolution is constrained. Our own preferred interpretation is based on the discussion of Section III E. The fluctuations in rates are with respect to the "zero-time" limit of the dynamics. When $\nu = 1$, it is not necessary to propagate the system for a finite time in order to be able to predict the rates. The variation in rates are just the inherent fluctuations about the conventional, statistical $\langle k \rangle(E)$. When $\nu \neq 1$, there are specific features that do require dynamical input which can only be provided by integrating the equations of motion.

## IV. SUMMARY

Fluctuations in quantum transition amplitudes have been discussed as diagnostics of the dynamics. Despite the simplicity of the results and, in particular, the "universal" character of the distribution of fluctuations, much remains to be done. Many of the open problems have been recognized through the analysis of (experimental or computational) data. Hence we provide a summary (or "handbook") of the way we have processed the data in the hope that it can be improved upon and/or applied to additional cases by others.

### A. Handbook

Given a spectrum, the first step is the determination of the envelope. For computational data one can perform the required microcanonical average for the short-term closed trajectories, Section III E. Otherwise, or for experimental data, one needs to Fourier transform the spectrum so as to find the first recurrence time[8,69–70] $\tau_c$. This corresponds to the long-range variation with energy of $\langle S(E)\rangle$ and is expected[58] from periodic orbit theory to be of the order of the density of states $\rho(E)$. Truncating the Fourier transformed spectrum at $\tau_c$, one transforms back to get $S^0(E) \equiv \langle S(E)\rangle$. We have found that, in practice, the definition (3.3) with a line shape which is gaussian with a variance $\sigma^2$, $\sigma = \sigma_0/\rho(E)$, is good enough. The constant $\sigma_0$ is selected by ensuring that the self-consistency condition (3.10) is reasonably satisfied. Usually $\sigma_0$ between $\pi$ and $3\pi$ will do, and the results are not very sensitive to the precise choice*, see Fig. 10.

Next, histograms of the deviations of the intensities from the local value of the envelope [i.e., $S(E_f)/\langle S(E_f)\rangle$] are drawn. The theoretical distribution is quite peaked; hence, linear bins of equal length in intensity are not always reasonable. Bins of equal probability would be best.[71] Failing that, we bin so that within each bin the distribution is uniform.[72] The fit to the theoretical distribution is using the quality of fit function as discussed in Section II G. A shortcut is to use (2.36), but it is not recommended as it fails to properly weight the weak transitions. The recommended shortcut is to use (2.33) with $\langle \ln y\rangle$ computed using (2.35). The error limits on $v$ are placed using (2.40). Given $v$, the fraction of sampled phase space can be computed from (3.38) or (3.43).

* An unreasonably small value of $\sigma_0$ will make the envelope follow the actual spectrum too closely and hence yield an unreasonably high value for $v$ (i.e., fluctuations are suppressed). We find that past a certain reasonable threshold, and practically up to $\sigma_0 \to \infty$ (which corresponds to a uniform envelope), the value of $v$ does not vary much. Hence, the zeroth approximation of a uniform envelope is not unreasonable.

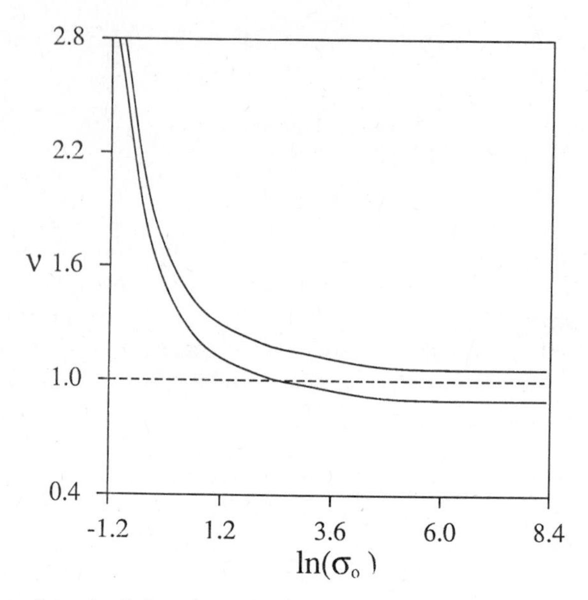


**Figure 10.** The values of $\nu \pm \delta\nu$ computed using (2.33) and (2.40) as a function of the width parameter $\sigma_0$ of the Gaussian lineshape function $A(E' - E)$ introduced in (3.3). The spectrum used as input has $\nu \simeq 1$.

## B. Fluctuations

The measured (or computed) spectra that are the subject of this chapter are reproducible (to within the stated error limits). Why then do we discuss "fluctuations." The simple answer is that the fluctuations are the variations in the intensities of different transitions of the same spectrum. They are the seemingly erratic changes of the intensity from one final state to another. The fluctuations are about the smooth envelope. An exact computation would reproduce the fluctuation. Only a statistical approximation will miss them and produce a smooth spectrum. The term "fluctuations" is thus used by analogy to refer to deviations from an equilibrium limit. Indeed that is exactly what we have done in Section III. The "probability of a spectrum" as discussed therein can be interpreted as the probability of a fluctuation. Indeed, the classical theory of fluctuations[73] can be derived along similar lines. Should we then not obtain a gaussian distribution in the intensities? Not quite. The gaussian (or Einstein) distribution is only valid for fluctuations near to the equilibrium limit.[73] In that limit $[(p - p^0)^2 < p^{0^2}]$ the $\chi$-square distribution, (2.25) or (3.30), does reduce to a gaussian of the same width [cf. (3.32)]:

$$P(p) \to (2\pi)^{-1/2} \exp\left(-\frac{\nu}{4}(p - p^0)/p^{0^2}\right). \tag{4.1}$$

This, however, is a poor approximation for the lower values of $\nu$, since the most probable value of $p$ is at $(\nu - 2)p^0/\nu$ (for $\nu \geq 2$) and equals zero otherwise. For $\nu \to \infty$ we therefore recover familiar results. But for $\nu$ of the order of unity, the region $(p - p^0)^2 < (p^0)^2$ is a typical (cf. Fig. 4a) so that for the more probable fluctuations the standard gaussian approximation is a poor one. The one essential difference with the usual statistical mechanics approach to fluctuations is that for the systems of interest here, the number of degrees of freedom is small.

**Acknowledgment**

I thank Y. M. Engel for assistance with the computations and drawings and Y. Alhassid and J. L. Kinsey for discussions. This work was supported by the U.S. Air Force Office of Scientific Research under Grant AFOSR 86-0011 and the Stiftung Volkswagenwerk. The Fritz Haber Research Center is supported by the Minerva Gesellschaft für die Forschung, mbH, Munich, BRD.

## APPENDIX: THE $\chi$-SQUARE DISTRIBUTION

It is important to spell out the limitations on the derivation of the distribution (2.25) of fluctuations. Consider the most general initial state, which is $\sum_{ij} |i\rangle \rho_{ij} \langle j|$. The initial state is pure if the rank of the matrix $\boldsymbol{\rho}$ is unity. Otherwise, it is a mixture. The transition intensity to the final state $f$ is $y = \sum_{ij} x_i^* \rho_{ij} x_j$, where $x_j = \langle j| T |f\rangle$. $y = \mathbf{x}^\dagger \boldsymbol{\rho} \mathbf{x}$ is then a quadratic form where the amplitudes $\mathbf{x}$ have a gaussian probability density

$$P(\mathbf{x}) \propto \exp(-\tfrac{1}{2}\mathbf{x}^\dagger \boldsymbol{\rho} \mathbf{x}). \tag{A.1}$$

One can then show that $y$ has a $\chi$-square distribution with $\nu$ degrees of freedom [i.e., (2.25)] where $\nu$ is the rank of the matrix $\boldsymbol{\rho}$.

A different result obtains if the amplitudes $\mathbf{x}$ are *independently* distributed:

$$P(\mathbf{x}) \propto \prod_{i=1}^{n} \exp(-x_i^2/2\langle x_i^2 \rangle). \tag{A.2}$$

That is, for each component there is a separate sum rule:

$$\langle x_i \rangle = 0, \quad \langle x_i^2 \rangle = \sum_f |\langle f| T |i\rangle|^2. \tag{A.3}$$

Such will be the case when the observed transition is an unresolved superposition of independent transitions. Putting $\tilde{\rho}_{ij} = \rho_{ij}(\langle x_i^2 \rangle \langle x_j^2 \rangle)^{1/2}$, we can write

$$y = \sum_{ij} \tilde{x}_i^* \tilde{\rho}_{ij} \tilde{x}_j, \tag{A.4}$$

where the $\tilde{x}_i$'s are independent gaussian variables of unit variance. The matrix $\rho_{ij}$ can be diagonalized so that

$$y = \sum_{i=1}^{v} \tilde{\rho}_i \tilde{z}_i^2, \tag{A.5}$$

where the $\tilde{z}_i$ are also gaussian variables of unit variance, the $\tilde{\rho}_i$'s are the nonzero eigenvalues of $\tilde{\boldsymbol{\rho}}$, and $v$ ($v \leq n$), is the rank of $\tilde{\boldsymbol{\rho}}$. The cumulants of the distribution of the variable $\alpha y$ (where $\alpha$ is a constant) are found to be[39]

$$\kappa_m = 2^{m-1}(m-1)! \sum_{i=1}^{v} (\alpha \tilde{\rho}_i)^m. \tag{A.6}$$

In particular,

$$\langle \alpha y \rangle = \kappa_1 = \alpha \sum_{i=1}^{v} \tilde{\rho}_i = \alpha \sum_{i=1}^{n} \rho_{ii} \langle x_i^2 \rangle \tag{A.7}$$

and

$$\langle \alpha^2 (y - \langle y \rangle)^2 \rangle = \kappa_2 = 2 \sum_{i=1}^{v} (\alpha \tilde{\rho}_i)^2. \tag{A.8}$$

The cumulants of (2.26), where the variable is $t = vy/\langle y \rangle$, are

$$\kappa_m = 2^{m-1}(m-1)!v. \tag{A.9}$$

Hence the variable $\alpha y$ has the $\chi$-square distribution (2.26) with $v$ degrees of freedom if and only if every $\tilde{\rho}_i$ in (A.5) is equal to $1/\alpha$. As a special case consider a spectrum that is experimentally not fully resolved, so that each observed intensity is a sum of $v$ transitions $y = \sum_i y_i$. This corresponds to (A.5) with $\tilde{\rho}_i = \langle z_i^2 \rangle$ with all $v\langle z_i^2 \rangle$'s being equal. It follows that the variable $y/\langle z^2 \rangle$ has the distribution (2.26).

Taking $\alpha = 1/\langle y \rangle$ in (A.8)

$$\langle (y - \langle y \rangle)^2 \rangle / \langle y \rangle^2 = 2 \sum_{i=1}^{v} p_i^2 \tag{A.10}$$

where $p_i = \tilde{\rho}_i / \sum_i \tilde{\rho}_i$, $\sum_i p_i = 1$. Hence, if we follow the nuclear physics practice[21,40] of defining an effective number $v_{\text{eff}}$ of degrees of freedom by equating $2/v_{\text{eff}}$ to the left-hand side of (A.10) [cf. (2.29)],

$$1 \leq \nu_{\text{eff}} \leq \nu, \tag{A.11}$$

where the minimal value is when all transitions are correlated for a pure initial state. If the observed intensity is a sum of $\nu$ transitions of the same spectrum, then we have the previously discussed case where $\nu_{\text{eff}} = \nu$.

When the intensity is a superposition of independent transitions, each with its own sum rule, the distribution is *not* $\chi$-square. The cumulants are however given by (A.6) and, in particular

$$\frac{1}{\nu_{\text{eff}}} = \sum_{i=1}^{\nu} p_i^2 \leq 1, \tag{A.12}$$

with $p_i = \langle z_i^2 \rangle / \sum_{i=1}^{\nu} \langle z_i^2 \rangle$. As expected, such a superposition will tend to reduce the fluctuations but the distributions will not be quite of the form (2.25).

## References


1. R. D. Levine and R. B. Bernstein, *Molecular Reaction Dynamics and Chemical Reactivity*. Oxford University Press, New York, 1987.
2. U. Schubert, E. Riedle, H. J. Neusser, and E. W. Schlag, *J. Chem. Phys.* **84**, 6519 (1986).
3. D. R. Guyer, W. F. Polik, and C. B. Moore, *J. Chem. Phys.* **84**, 6519 (1986).
4. J. W. Thoman, Jr., J. I. Steinfeld, R. McKay, and A. E. W. Knight, *J. Chem. Phys.* **86**, 5909 (1987).
5. H. L. Dai, R. W. Field, and J. L. Kinsey, *J. Chem. Phys.* **82**, 1606 (1985).
6. See, for example, K. N. Swamy, W. L. Hase, B. C. Garret, C. W. McCurdy, and J. F. McNutt, *J. Phys. Chem.* **90**, 3517 (1986).
7. B. A. Waite, S. K. Gray, and W. H. Miller, *J. Chem. Phys.* **78**, 259 (1983), and references therein.
8. E. J. Heller, *J. Chem. Phys.* **72**, 1337 (1980).
9. E. B. Stechel and E. J. Heller, *Ann. Rev. Phys. Chem.* **35**, 563 (1984).
10. V. Buch, M. A. Ratner, and R. B. Gerber, *Mol. Phys.* **46**, 1129 (1982).
11. P. Brumer and M. Shapiro, *Adv. Chem. Phys.* **60**, 371 (1985).
12. Y. Alhassid and R. D. Levine, *Phys. Rev. Lett.* **57**, 2879 (1986).
13. J. Brickmann, Y. M. Engel, and R. D. Levine, *Chem. Phys. Lett.* **137**, 441 (1987).
14. S. Mukamel, J. Sue, and A. Pandey, *Chem. Phys. Lett.* **105**, 134 (1984).
15. C. E. Hamilton, J. L. Kinsey, and R. W. Field, *Ann. Rev. Phys. Chem.* **37**, 493 (1986).
16. K. K. Lehmann and S. L. Coy, *J. Chem. Phys.* **83**, 3290 (1985).
17. C. E. Porter and R. G. Thomas, *Phys. Rev.* **104**, 483 (1956).
18. C. E. Porter, *Statistical Theory of Spectra: Fluctuations*. Academic Press, New York, 1965.
19. S. S. M. Wong, *Nuclear Statistical Spectroscopy*. Oxford University Press, New York, 1986.
20. M. Carmeli, *Statistical Theory and Random Matrices*. Deckker, New York, 1983.
21. T. A. Brody, J. Flores, J. B. French, P. A. Mello, A. Pandey, and S. S. M. Wong, *Rev. Mod. Phys.* **53**, 385 (1981).
22. E. J. Heller and R. L. Sundberg, in *Chaotic Behavior in Quantum Systems* (G. Casati ed.). Plenum Press, New York, 1985.
23. D. W. Noid, M. L. Koszykowski, and R. A. Marcus, *Ann. Rev. Phys. Chem.* **32**, 267 (1981).
24. S. A. Rice, *Adv. Chem. Phys.* **47**, 117 (1981).
25. P. Brumer, *Adv. Chem. Phys.* **47**, 201 (1981).

26. P. Pechukas, *J. Phys. Chem.* **88**, 4823 (1984).
27. R. Ramaswamy and R. A. Marcus, *J. Chem. Phys.* **74**, 1385 (1981).
28. A. Peres, *Phys. Rev. A* **30**, 504 (1984).
29. K. M. Christoffel and P. Brumer, *Phys. Rev. A* **33**, 1309 (1986).
30. D. M. Wardlaw, D. W. Noid, and R. A. Marcus, *J. Phys. Chem.* **88**, 536 (1984).
31. E. J. Heller, *Acc. Chem. Res.* **14**, 368 (1981).
32. W. H. Miller, S. D. Schwartz, and J. W. Tromp, *J. Chem. Phys.* **79**, 4889 (1983).
33. P. G. Wolynes, *Phys. Rev. Lett.* **47**, 968 (1981); E. C. Behrmann, G. A. Jongeward, and P. G. Wolyness, *J. Chem. Phys.* **83**, 668 (1985).
34. D. Thirumalai and B. J. Berne, *J. Chem. Phys.* **79**, 5029 (1983); J. D. Doll, *J. Chem. Phys.* **81**, 3536 (1984); R. E. Wyatt, *Chem. Phys. Lett.* **121**, 301 (1985).
35. W. L. Hase, *Acc. Chem. Res.* **16**, 258 (1983).
36. E. Pollak, in *Theory of Chemical Reaction Dynamics* (M. Baer, ed). CRC Press, Florida, 1985.
37. R. D. Levine, *Adv. Chem. Phys.* **47**, 239 (1981).
38. E. Riedle, H. J. Neusser, and E. W. Schlag, *J. Phys. Chem.* **86**, 4847 (1982).
39. See, for example, M. G. Kendall and A. Stuart, *The Advanced Theory of Statistics.* Hafner, New York, 1958.
40. L. Wilets, *Phys. Rev. Lett* **9**, 430 (1962).
41. R. D. Levine and J. L. Kinsey, in *Atom Molecule Collision Theory* (R. B. Bernstein, ed.). Plenum Press, New York, 1979.
42. See, for example, M. Abramowitz and I. A. Stegun, eds., *Handbook of Mathematical Functions.* Dover, New York, 1965.
43. Y. Alhassid and R. D. Levine, *Chem. Phys. Lett.* **73**, 16 (1980).
44. J. L. Kinsey and R. D. Levine, *Chem. Phys. Lett.* **65**, 413 (1979).
45. R. D. Levine, *The Theory and Practice of the Maximum Entropy Formalism in Maximum Entropy and Bayesian Methods in Applied Statistics* (J. H. Justice, ed.). Cambridge University Press, Cambridge, 1986.
46. Y. Tikochinsky and R. D. Levine, *J. Math. Phys.* **25**, 2160 (1984).
47. Y. M. Engel, R. D. Levine, J. W. Thoman, J. I. Steinfeld, and R. McKay, *J. Chem. Phys.* **86**, 6561 (1987).
48. M. Feingold and A. Peres, *Phys. Rev. A* **34**, 591 (1986); M. Feingold and Y. Alhassid (unpublished).
49. R. D. Levine, *J. Phys. A* **13**, 91 (1980).
50. C. E. Shannon, *Bell Syst. Tech. J.* **27**, 379 (1948).
51. R. D. Levine, "Statistical Dynamics," in *Theory of Reactive Collisions* (M. Baer, ed.). CRC Press, Florida, 1984.
52. R. Ash, *Information Theory.* Interscience, New York, 1961.
53. W. H. Miller, *Adv. Chem. Phys.* **25**, 69 (1974); *Science* **233**, 171 (1986).
54. Y. Alhassid and R. D. Levine (unpublished).
55. See, for example, R. D. Levine, *Quantum Mechanics of Molecular Rate Processes.* Clarendon Press, Oxford, 1969, Section 3.7.
56. See, for example, A. Papoulis, *Probability, Random Variables and Stochastic Processes.* McGraw Hill, New York, 1965.
57. M. Gutzwiller, *J. Math. Phys.* **12**, 343 (1971).
58. R. Balian and C. Bloch, *Ann. Phys.* **85**, 514 (1974).
59. W. H. Miller, *Acc. Chem. Res.* **4**, 161 (1971).
60. See also E. J. Heller, *J. Chem. Phys.* **68**, 2066 (1978); **68**, 3891 (1978).
61. R. B. Bernstein, *Adv. Chem. Phys.* **10**, 75 (1966).
62. N. F. Scherer, F. E. Doany, A. H. Zewail, and J. W. Perry, *J. Chem. Phys.* **84**, 1932 (1986).

63. M. P. Casassa, A. M. Woodward, J. C. Stephenson, and D. S. King, *J. Chem. Phys.* **85**, 6235 (1986).
64. Y. Y. Bai, G. Hose, C. W. McCurdy, and H. S. Taylor, *Chem. Phys. Lett.* **99**, 342 (1983); R. E. Wyatt, G. Hose, and H. S. Taylor, *Phys. Rev. A* **28**, 815 (1983).
65. K. C. Kulander, J. Manz, and H. H. R. Schor, *J. Chem. Phys.* **82**, 3088 (1985); R. H. Bisseling, R. Kosloff, J. Manz, F. Mrugala, J. Romelt, and G. Weichselbaumer, *J. Chem. Phys.* (to be published).
66. V. I. Arnold and A. Avez, *Ergodic Problems of Classical Mechanics*, Benjamin, New York, 1968.
67. R. S. Mackay, J. D. Meiss, and I. C. Percival, *Physica D* **13**, 55 (1984); D. Bensimon and L. P. Kadanoff, Physica D **13**, 82 (1984).
68. S. K. Gray, S. A. Rice, and M. J. Davis, *J. Phys. Chem.* **90**, 3470 (1986).
69. P. Pechukas, *Chem. Phys. Lett.* **86**, 553 (1982).
70. L. Leviander, M. Lombardi R. Jost, and J. P. Pique, *Phys. Rev. Lett.* **56**, 2449 (1986); J. P. Pique, Y. Chen, R. W. Field, and J. L. Kinsey, *Phys. Rev. Lett.* **58**, 475 (1987).
71. M. B. Faist, *J. Chem. Phys.* **65**, 5427 (1976).
72. E. A. Gislason and E. M. Goldfield, *J. Chem. Phys.* **80**, 701 (1984).
73. See, for example, L. D. Landau and E. M. Lifshitz, *Statistical Physics*. Pergamon Press, New York, 1960.

# SELF-CONSISTENT-FIELD METHODS FOR VIBRATIONAL EXCITATIONS IN POLYATOMIC SYSTEMS

R. B. GERBER

*Department of Physical Chemistry and The Fritz Haber Research Center for Molecular Dynamics, The Hebrew University of Jerusalem, Jerusalem 91904, Israel*

MARK A. RATNER

*Department of Chemistry, Northwestern University, Evanston, Illinois 60201*

## CONTENTS



## I. INTRODUCTION

This chapter is devoted to recent developments in a class of approximation methods that aim at describing the energy-level structure and dynamics of energy transfer in vibrationally highly excited polyatomic systems. The methods that are the subject of the present study are the self-consistent-field (SCF) approximations, in which framework each vibrational mode is described as moving in an effective field obtained by averaging the full potential function of the system over the motions of all the other modes. SCF methods provide an enormous simplification in treating the dynamics of polyatomic vibrations in the anharmonic, coupled mode regime, an advantage that grows rapidly as the number of degrees of freedom increases. This simplification is gained, however, at the expense of introducing approximations, which must be assessed and analyzed. The focus here will be on examining very recent improvements of the available SCF methods, on discussing the physical basis for the validity of the approximation, and on presenting some applications that illustrate the power of the approach and the possibilities it offers.

The motivation for the introduction of SCF methods, as well as of other approaches to the dynamics of several coupled vibrational modes, stems largely from progress in high-resolution spectroscopic studies of highly excited vibrations in polyatomic systems. This is particularly the case for the developments in SCF theory discussed here. For example, Field and co-workers,[1] Lehmann and co-workers,[2] and Bailly and co-workers[3] have shown that small polyatomics, such as $H_2CO$, $O_3$, HCN, $C_2H_2$, $CO_2$, exhibit regular, simple vibrational energy-level sequences well into the strongly anharmonic regime, at least for stretching mode excitations. The existence of such regular level sequences calls for the development of simple dynamical methods that can be used conveniently to assign the levels, interpret the modes and quantum numbers with which they are associated, and relate them as directly as possible to the coupled-mode potential energy surface of the system. In a sense, what seems desirable for dealing with the simple energy sequences of polyatomic vibrations is a scheme that extends and generalizes the notion of separability, and the normal-mode analysis, as familiar from the harmonic regime.[4] The improved SCF methods discussed in this article seem a promising tool for this purpose.

Quite a different experimental line of development that directly motivated much of the progress in SCF theory is the study of unimolecular dissociation dynamics of van der Waals clusters. Following the pioneering work by Levy and co-workers[5] on the vibrational predissociation of clusters such as $I_2He$, experimental work in this field grew very rapidly. The dissociation dynamics of many of these systems is qualitatively different from the statistical RRKM type, which overwhelmingly characterizes unimolecular decay of chemically bound polyatomic molecules. A dynamical approximation is needed to describe the dissociation of such systems, especially when the number of atoms excludes rigorous quantum-mechanical or classical calculations.

The first part of this chapter (Section II) deals with the SCF method for calculations of polyatomic vibrational energy levels. The aspect on which we focus is that the SCF approximation may strongly depend on the choice of the modes to which the SCF (generalized) separation is applied. Recent calculations have shown that a physically motivated choice of coordinates can yield much improved results. Probably the most useful development along this line is an approach that combines the SCF separation with a variational search (within a certain class of coordinates) for the modes in which the approximation is optimal. The previously mentioned methods, the physical arguments behind them, and examples which illustrate them are the core of Section 2.

The second part of the chapter (Section III) deals with the time-dependent self-consistent-field (TDSCF) method for studying intramolecular vibrational energy transfer in time. The focus is both on methodological aspects and on the application to models of van der Waals cluster systems, which exhibit non-RRKM type of behavior. Both Sections II and III review recent results. However, some of the examples and the theoretical aspects are presented here for the first time.

## II. SCF METHOD FOR VIBRATIONAL ENERGY LEVELS AND THE CHOICE OF COORDINATES

### A. Quantum-Mechanical and Semiclassical SCF

The quantum-mechanical SCF method for obtaining the vibrational energy levels is a direct adaptation of the Hartree approximation for electronic structrue calculations, which dates back to the early stages of quantum theory. The introduction of the method for vibrational modes is, however, rather recent and is due to Bowman and co-workers,[6,7] Carney et al.,[8] and Cohen et al.[9] The semiclassical version of the SCF, the SC-SCF method, proposed by Gerber and Ratner,[10] relies on the characteristically short de Broglie wavelengths typical of vibrational motions (as opposed to electronic ones) to gain some further simplification, but is otherwise based on the same physical considerations as the quantum-mechanical approximation. A brief review of the SCF and SC-SCF methods can be found in Ref. (11).

We describe first the quantum-mechanical SCF approximation. Consider a molecule of $N$ vibrational modes, whose Hamiltonian in a set of given Cartesian coordinates $q_1, \ldots, q_N$ is given by

$$H = -\sum_{i=1}^{N} \frac{\hbar^2}{2m_i} \frac{\partial^2}{\partial q_i^2} + V(q_1, \ldots, q_N), \tag{1}$$

where $m_i$ is the mass associated with the degree of freedom $q_i$, and $V$ is the potential energy function of the system. The SCF approximation makes the following *ansatz* for the eigenfunctions of the Hamiltonian (1):

$$\Psi(q_1, \ldots, q_N) = \prod_{i=1}^{N} \phi^{(i)}(q_i). \tag{2}$$

Applying the variational principle,

$$\delta\{\langle\Psi|H|\Psi\rangle/\langle\Psi|\Psi\rangle\} = 0 \tag{3}$$

to the ansatz (2), yields the single-mode SCF (or Hartree) equations:

$$h_i^{\text{SCF}}(q_i)\phi_n^{(i)}(q_i) = \varepsilon_n^{(i)}\phi_n^{(i)}(q_i), \tag{4}$$

where

$$h_i^{\text{SCF}}(q_i) = \frac{-\hbar^2}{2m_i}\frac{\partial^2}{\partial q_i^2} + U_i(q_i) \tag{5}$$

$$U_i(q_i) = \left\langle \prod_{j\neq i} \phi^{(j)}(q_j) \middle| V(q_1 \cdots q_n) \middle| \prod_{j\neq i} \phi^{(j)}(q_j) \right\rangle. \tag{6}$$

Substitution of (2) and (4)–(6) in the energy functional expression of Eq. (3) yields the following relation between the single-mode eigenvalues $\varepsilon_n^{(i)}$ and the total energy in the SCF approximation:

$$E_{n(1),\ldots,n(N)} = \sum_{i=1}^{N} \varepsilon_n^{(i)} + (1-N)\left\langle \prod_{j=1}^{N} \phi_{n(j)}^{(j)}(q_j) \middle| V \middle| \prod_{j=1}^{N} \phi_{n(j)}^{(j)}(q_j) \right\rangle, \tag{7}$$

where $n(i)$ is the quantum number in the $i$ mode.

In the SCF treatment, the modes are formally separable, each mode governed by a Hamiltonian $h_i^{\text{SCF}}(q_i)$ that depends explicitly on that mode only. But the effective potentials $U_i(q_i)$ in which each mode moves, the single-mode energies $\varepsilon_n^{(i)}$, and wavefunctions $\phi_n^{(i)}(q_i)$ are all self-consistently determined, hence some coupling between the modes is implicitly included. We refer to the latter as the "static" coupling between the modes, since each mode $i$ experiences an effective potential $U_i(q_i)$ generated by the averaged motions of all the other modes. This mean field does not, however, represent any effect of *energy transfer* between the modes. Obviously, the SCF approach described here is thus restricted to systems and states for which energy transfer between the modes can be neglected.

Semiclassical SCF makes the following additional simplifications.[10] The semiclassical Bohr–Sommerfeld quantization is applied to obtain the single-mode energies:

$$\int_a^b [2m_i(\varepsilon_n^{(i)} - U_i(q_i))]^{1/2} dq_i = (n + \tfrac{1}{2})\pi\hbar, \tag{8}$$

where $b$ and $a$ are the classical turning points associated with the integrand in (8). Also, the quantum-mechanical single-mode probability densities are replaced by their classical analogues:

$$|\phi^{(i)}(q_i)|^2 \approx \frac{1}{p_i(q_i)} C_i, \tag{9}$$

where $C_i$ is a normalization constant and $p_i$ is the classical momentum of mode $q_i$:

$$[C_i]^{-1} = \int_{a(i)}^{b(i)} \frac{dq_i}{p_i(q_i)}, \tag{10}$$

$$p_i(q_i) = [2m_i(\varepsilon_n^{(i)} - U_i(q_i))]^{1/2}. \tag{11}$$

The SCF potential is thus replaced in the SC-SCF by the classical average:

$$U_i(q_i) = \int_{a(1)}^{b(1)} \cdots \int_{a(N)}^{b(N)} V(q_i, \ldots, q_n) \prod_{j \neq i} \frac{dq_j}{p_j(q_j)} C_j. \tag{12}$$

Thus, the SC-SCF does not involve any calculation of single-mode wavefunctions. The single-mode energies are obtained by solution of the quantization condition (8), carried out simultaneously and self-consistently with the evaluation of the effective single-mode potentials of Eq. (12). The total energy in the SC-SCF approximation is

$$E_{n(1),\ldots,n(N)} = \sum \varepsilon_n^{(i)} - (N-1) \prod_{j=1}^{N} C_j \int_{a(1)}^{b(1)} \cdots \int_{a(N)}^{b(N)} V(q_1, \ldots, q_N) \frac{dq_j}{p_j(q_j)}. \tag{13}$$

The preceding expressions represent the simplest, primitive semiclassical limit of the SCF method. A more refined treatment may be required in some cases, for example, for systems with modes that have several disconnected classically allowed regions, with possible tunneling between the latter. Farrelly and Smith[12] gave an improved, uniform semiclassical SCF approximation for such systems.

On the whole, in most practical cases there is not a great difference between SCF and SC-SCF. Experience has shown that the two methods are nearly always comparable in accuracy since the errors due to the semiclassical approximation are small compared with the SCF errors. The SC-SCF is computationally more efficient, but in view of the effective algorithms now available for solving the single-mode Schrödinger equation, even the quantum SCF does not involve a major numerical effort. The SC-SCF seems to have an advantage in offering a simple, more transparent relation between the energy levels and the underlying potential energy surface. In fact, it was shown by Gerber et al.[13] that within the SC-SCF approximation measured vibrational energy levels of a polyatomic molecule can be directly inverted by a certain transform to yield the coupled-mode potential energy surface of the system (throughout the validity range of the method). This is an extension to polyatomic systems of the RKR inversion, which has long been known for diatomic molecules.[14] No such inversion is yet available for the quantum SCF.

Both the SCF and the SC-SCF have been applied to a substantial number of model systems and realistic molecular Hamiltonians[6–13,15–23] including cases such as $H_2O$ and its isotopic derivatives, $CO_2$, $O_3$, $H_2CO$, HCN, and so forth. Within their validity range, which will be discussed in the subsequent sections, the methods seem very useful, yielding typically an accuracy of better than 1% for individual excited energy levels. A major advantage of the SCF methods is that they are very useful for *assignment* of energy levels in the anharmonic regime: The fact that the SCF scheme provides a notion of (generalized) separability of modes and naturally interprets the quantum level structure associated with each degree of freedom (while allowing for static coupling between the modes) offers a simple quantitative way for analyzing the regular energy sequences recently observed in spectroscopic studies of molecules such as $O_3$, $H_2CO$, and HCN.[1–3] In addition to spectroscopic assignment, SCF, owing to its computational simplicity, provides an attractive tool for determining potential energy surfaces, including the anharmonic coupled-mode regime, by fitting spectroscopic energy-level data. This advantage is particularly strong in the case of SC-SCF, since, as was mentioned previously, within this approximation the potential energy surface can be determined from the full set of vibrational energy levels by direct inversion. A potential surface for the stretching modes of $CO_2$ was recently determined by this inversion,[19] and it compared very well with a potential energy surface determined by considerably more complicated trial and error fitting of the data.[3b] It seems desirable, however, to understand more profoundly the physical basis of the method and its regime of validity. This will be the topic of the forthcoming sections.

### B. SCF for Different Coordinate Systems

One of the main reasons for the good results obtained with the Hartree–Fock SCF method in electronic structure calculations for atoms and molecules is that the electrons keep away from each other due to the Pauli exclusion principle. This reduces the correlation between them, and provides a basis for the validity of the independent-particle model. The question arises as to the mechanisms that account for the validity of the SCF approximation in the vibrational case, which are obviously quite unrelated to the Pauli principle.

It turns out that the success of the vibrational SCF is closely related to and dependent on the coordinates one chooses. In fact, if one were to use the positions of the atomic nuclei themselves as coordinates in the SCF scheme, the approximation would fail completely since the hard repulsive forces between each pair of atoms give rise to strong correlations between the coordinates of these particles, clearly not included in the mean field treatment. Thus, the hard, repulsive part of the interatomic forces will break any SCF approximation based on separating mutually the motions of the different atoms. The successful SCF treatments of polyatomic vibrations apply the separation to *collective* coordinates of various types. The hard interatomic repulsions are then incorporated in the equations that describe the individual modes, rather than provide a correlation between different modes. Bond (or local) modes are already sufficiently "collective" in this respect, and interatomic repulsions are incorporated in the single-mode equations. The same holds for normal modes. The more a given choice of coordinates makes it possible to incorporate strong interaction terms between the atoms at the single-mode level, the better are these coordinates for the SCF approximation. It is intuitively attractive to use highly collective coordinates, that is, modes that involve simultaneous displacements of all or most atoms in the molecule. Such delocalized coordinates should provide better mutual screening, and correlations between them should average out to a small value.

In addition to the strength of the residual coupling between the modes, which is not included in the mean field potential, validity of the SCF approximation also depends on the frequency separation between the modes. This will be discussed further in Section II D, but the approximation obviously fails in cases of degeneracy, that is, resonance between modes. Buch[24] has shown that when the frequencies of the SCF modes are well separated and do not involve higher-order resonances, the correlations between the modes average out to a small residue over one or several single-mode oscillations. Thus, a good choice of coordinates for SCF is one that avoids correlation interactions between the modes and that does not involve near resonances between the SCF degrees of freedom.

It should be useful at this point to examine the importance of coordinate choice for SCF in the context of specific examples. Roth et al.[17] compared SCF results for local and for normal coordinates for nonbending models of water, and its isotopic variants, and of $CO_2$. The model Hamiltonian used for the stretching vibrations was of the following form in local coordinates:

$$\begin{aligned} H = {} & p_1^2/2\mu_1 + p_2^2/2\mu_2 + (\cos\phi)p_1p_2/\mu \\ & + D_1\{1 - \exp(-\beta_1 x_1)\}^2 + D_2\{1 - \exp(-\beta_2 x_2)\}^2, \end{aligned} \tag{14}$$

where $x_1, x_2$ are the distances between AB and BC, respectively, in the molecule

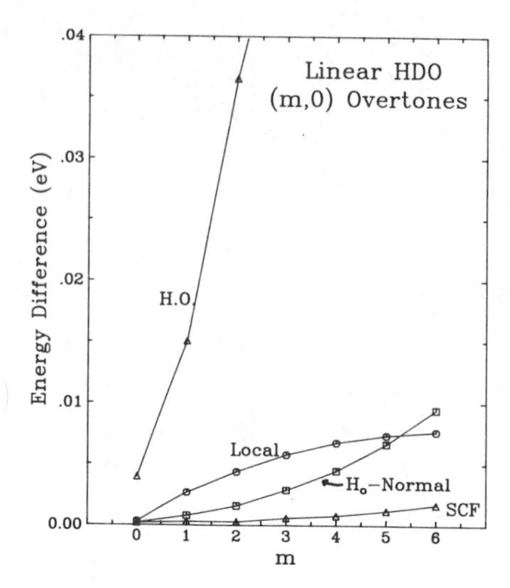


**Figure 1.** Difference in energy between approximate and exact results for the $(m, 0)$ overtones of linear HDO. In local modes, $m$ is the number of quanta in the O–D bond. The approximate results are uncoupled harmonic oscillators in normal modes (HO), uncoupled normal modes including all higher-order diagonal anharmonicities ($H_0$-normal), and SCF in normal modes. From Ref. 17.

A–B–C; $\phi$ is the bond angle; $\mu$ is the central atom mass; $\mu_1$ and $\mu_2$ are the reduced masses of AB and BC, respectively; $D_1$ and $\beta_1$ are the Morse parameters for the A ... B bond; $D_2$ and $\beta_2$ are the corresponding values for the B ... C bond; and $p_1$ and $p_2$ are the momenta conjugate to $x_1$, $x_2$. In local coordinates, the coupling between the modes is entirely in the mixed kinetic energy term. When the Hamiltonian is transformed to the normal coordinates representation, there are no mixed terms in the kinetic energy operator and the entire coupling between the modes is in the potential function. Figures 1 and 2 show the results obtained by Roth et al. for linear HDO. Figure 1 shows errors in the energy for $(m, 0)$ overtones of linear HDO given by several approximations: the harmonic oscillator (HO) treatment; uncoupled local modes; uncoupled normal modes (including anharmonicities to all order); and SCF in normal coordinates. For the Hamiltonian of Eq. (14), which involves only coupling of the type constant $p_1\, p_2$, SCF in local modes reduces to uncoupled local modes that are shown in the figure. Similar results are shown in Fig. 2 for the $(0, n)$ overtones of HDO. Not surprisingly, uncoupled local modes are found superior to uncoupled normal modes for the HO excitations even when the latter are used in a way that includes (uncoupled) anharmonicities. For the excitation of the heavier D atom, the uncoupled local modes and normal modes give virtually the same magnitude of error. However, both

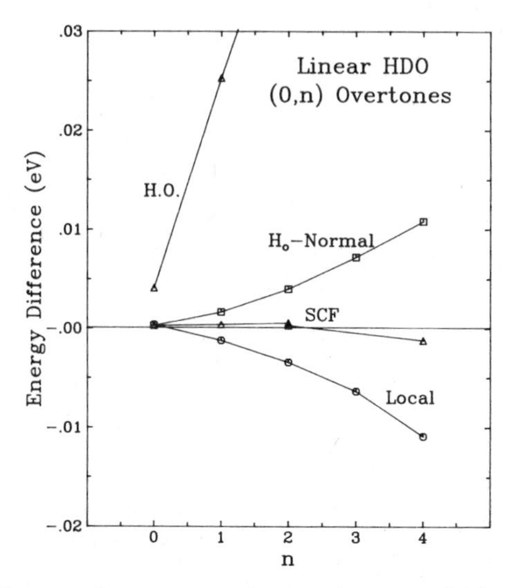


**Figure 2.** Difference in energy between approximate and exact results for the $(0, n)$ overtones of linear HDO. In local modes, $n$ is the number of quanta in the O–H bond. From Ref. 17.

for the $(m, 0)$ and the $(0, n)$ overtones, SCF in normal modes is seen to be much superior to SCF in local modes (which is the same as the uncoupled local modes approximation in this case), and, of course, also to the uncoupled normal modes model. Clearly, averaging over a relatively localized mode leaves a considerable residual correlation. In fact, in the preceding special case the mean field correction in local modes to the uncoupled part of the Hamiltonian vanishes. The mean field correction in normal modes evidently produces a large improvement over the uncoupled mode approximation, whatever the modes used for the latter.

Very recently, SC-SCF in hyperspherical coordinates was tested for the coupled stretching vibrations of $H_2O$ and $CO_2$.[25] The hyperspherical coordinates are highly collective and are thus expected to be very suitable for SCF. Also, for the states considered here the hyperspherical system provides a good frequency separation between the coordinates involved. For a linear, nonbending molecule A–B–C, the Hamiltonian of the stretching vibrations can be written as follows in hyperspherical coordinates (which are just plane polar coordinates in this case):

$$H = -\frac{\hbar^2}{2\mu_{\mathrm{BC}}}\left(\frac{1}{r}\frac{\partial}{\partial r}\left[r\frac{\partial}{\partial r}\right] + \frac{1}{r^2}\frac{\partial^2}{\partial \phi^2}\right) + V(r, \phi), \tag{15}$$

where $\mu_{\mathrm{BC}}$ is the reduced mass of B and C, and $r$ and $\phi$ are related to the normal stretching coordinates by

$$\phi = \tan^{-1}(y/x), \qquad r = x/\cos\phi, \tag{16}$$

and $x$ and $y$ are related to the atomic distance coordinates by

$$y = r_{BC}; \qquad x = \left(r_{AB} + \frac{m_B}{m_A + m_C} r_{BC}\right)\left(\frac{m_{A(B+C)}}{m_{BC}}\right)^{1/2}, \tag{17}$$

where $m_{A(B+C)}$ is the reduced mass of A with (BC). The SCF ansatz in polar (hyperspherical) coordinates is

$$\Psi(r, \phi) = \chi_m(r)\eta_n(\phi). \tag{18}$$

We refer to Ref. 25 for the detailed SCF and SC-SCF expressions. Table I, from Gibson et al.,[25] compares the results of hyperspherical SCF for various levels of $C^{18}O^{16}O$ with corresponding results for normal-mode SCF and with the exact values. Similar comparisons are shown in Table II for $H_2O$. The results show that, indeed, SCF separation in hyperspherical coordinates does

TABLE I
Computed Vibrational Eigenvalues (eV) for Two-Mode Model of $C^{18}O^{16}O$[a]

| State[b] | $E^{SCF}_{hyperspherical}$ [c] | $E_{exact}$ [d] | $E^{SCF}_{normal}$ [e] |
|---|---|---|---|
| 0, 0 | 0.2323 | | |
| | 0.2304 | 0.2309 | 0.2309 |
| | 0.2303 | | |
| 1, 0 | 0.3895 | | |
| | 0.3849 | 0.3849 | 0.3862 |
| 0, 1 | 0.5387 | | |
| | 0.5305 | 0.5306 | 0.5316 |
| 2, 0 | 0.5457 | | |
| | 0.5390 | 0.5377 | 0.5404 |
| 1, 1 | 0.6988 | | |
| | 0.6824 | 0.6808 | 0.6858 |
| 3, 0 | 0.7010 | | |
| | 0.6932 | 0.6894 | 0.6934 |
| 0, 2 | 0.8428 | | |
| | 0.8259 | 0.8248 | 0.8264 |

[a] Parameter values taken from Ref. 17.
[b] The quantum numbers refer, respectively, to the $r$ and $\phi$ functions.
[c] Reference 25.
[d] Reference 14.
[e] Reference 17.

TABLE II
Computed Vibrational Eigenvalues (eV) for Two-Mode Model of Linear $H_2O$[a]

| State[b] | $E^{SCF}_{hyperspherical}$[c] | $E_{exact}$[d] | $E^{SCF}_{normal}$[e] |
|---|---|---|---|
| 0, 0 | 0.4819 | | |
| | 0.4748 | 0.4745 | 0.4753 |
| 1, 0 | 0.9508 | | |
| | 0.9273 | 0.9200 | 0.9283 |
| 0, 1 | 0.9784 | | |
| | 0.9532 | 0.9467 | 0.9514 |
| 2, 0 | 1.4197 | | |
| | 1.3748 | 1.3536 | 1.3715 |
| 1, 1 | 1.4505 | | |
| | 1.3979 | 1.3719 | 1.3991 |
| 0, 2 | 1.4740 | | |
| | 1.4246 | 1.4093 | 1.4124 |

[a] Parameter values taken from Ref. 17.
[b] The quantum number refer, respectively, to the $r$ and $\phi$ functions.
[c] Reference 25.
[d] Reference 12.
[e] Reference 17.

yield better results than the corresponding approximation in normal coordinates. The improvement is much larger for $CO_2$ than for $H_2O$, since the low H/O mass ratio renders the $\phi$ motion effectively more localized than the corresponding behavior in $CO_2$. On the whole, the mutual screening of the hyperspherical modes is better than that of normal-mode coordinates, since the former are more delocalized. Hence, the hyperspherical coordinates are better for SCF purposes, at least for systems and states of the types considered here.

### C. Optimal-Coordinates SCF

The strong dependence of the SCF results on the coordinates that are being separated in this method suggests an important improvement in the SCF approach. One may seek to combine a search for coordinates that are best for the SCF, with the solution of the equations for the energies (and single-mode potentials). More specifically, a variational principle for the energy may be used to determine the best coordinates among a given total set of coordinate systems simultaneously with pursuing the solution for the single-mode energies.

Consider a one-parameter family of coordinate systems $q_1(\lambda), \ldots, q_n(\lambda)$, where $\lambda$ is a continuous parameter. In trying an ansatz of the SCF type, as in Eq. (2) for the wavefunction, one may also apply the variational principle of the energy functional to the coordinate parameter $\lambda$. We give here the results in the semiclassical limit, as presented in Eqs. (8)–(11), for the fixed coordinates case:

$$\int_a^b [2m_i(\varepsilon_n^{(i)}(\lambda) - U_i(q_i\{\lambda\})]^{1/2} dq_i = (n + \tfrac{1}{2})\pi\hbar, \tag{19}$$

$$U_i(q_i\{\lambda\}) = \underset{q_j, j \neq i}{\mathrm{Av}} [V(q_1, \ldots, q_n)] \quad \text{for } i = 1, \ldots, N, \tag{20}$$

$$\frac{\partial E_{n(1),\ldots,n(N)}}{\partial \lambda} = 0, \tag{21}$$

where the energy $E_{n(1),\ldots,n(N)}$ is related to the single-mode energies $\varepsilon_n^{(i)}(\lambda)$ as in Section 2.1; Av in Eq. (20) denotes the classical average over all the modes except $q_i$. The idea of improving SCF by optimizing coordinates was first proposed by Truhlar and co-workers,[15,20] by Lefebvre,[26] and by Moiseyev.[27] These studies were, however, confined to Cartesian coordinates, since these coordinates can conveniently be labeled by continuous transformation parameters. If $x$, $y$ form a Cartesian coordinate system, then any other Cartesian system can be obtained by the transformation

$$x' = x\cos\theta + y\sin\theta; \qquad y' = -x\sin\theta + y\cos\theta, \tag{22a}$$

which involves a single parameter $\theta$. In application to the symmetric stretch–bending interactions in $H_2O$, Thompson and Truhlar[20] have shown that the SCF in "optimal" Cartesian coordinates gives a mean error for the calculated energies that is some 40% lower than the corresponding error for the SCF in the normal modes. However, as we learned from the example showing the advantages of hyperspherical coordinates,[25] it is obvious that the set of Cartesian coordinate systems is too restrictive for actual applications.

Recently, Bacic et al.[28] have studied the bending–stretching energy levels of HCN by SCF approximation with modes optimized within the class of elliptical (spheriodal) coordinates. This example will be examined here in some detail since (1) the improvement obtained by coordinate optimization in this case is very large and (2) this case illustrated how intuitive considerations based on molecular geometry, the shape of the potential surface, and so forth, can be useful in choosing a suitable parametrized class of coordinates. Whenever the "trial set" of parametrized coordinates is physically well motivated, the OC-SCF procedure of Eqs. (19)–(21) can be expected to yield excellent results.

Bacic et al.[28] explored the excited bending levels of HCN, and the coupling of the bending mode to the C–H stretching vibration. The bending motion, in the highly excited levels of that mode, corresponds to the reaction coordinate for the HCN → HNC isomerization process. The study by Bacic et al. is the

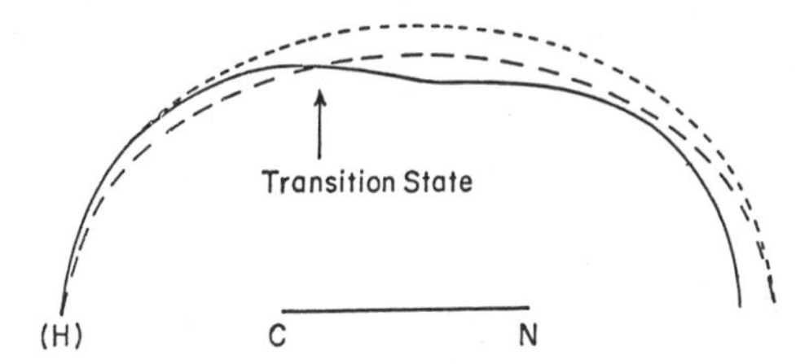


**Figure 3.** The minimum energy path for HCN: ——, exact; – – –, ellipse approximation with $a = 1.2237$ Å; ---, ellipse approximation with $a = 1.0631$ Å. $a$ is the half the interfocal distance. The 2 mep is for the potential surface of Murrell et al.[29]

first in which the SCF approximation was applied to highly excited bending states. Given the large mass separation between H and the C and N atoms, kinematic considerations suggest that the system should closely correspond to the two-center problem, and to the extent that the C and N centers may be treated as roughly similar, elliptical coordinates for the H motion seem a very natural choice. This intuitive argument is only very qualitative, of course, since the potential surface of the system does not necessarily correspond to an H atom interacting by central forces to the C and the N "centers." However, an inspection of the minimum energy path (mep) for the HCN $\rightleftarrows$ HNC process from the potential surface of Murrell et al.,[29] shown in Fig. 3, supports this intuitive suggestion. The form of the mep does correspond very roughly to an ellipse. One can obtain *quantitative* fitting of several parts of the mep by an ellipse, provided that different ellipses are used to fit different segments of the mep, as illustrated in Fig. 3. It will be seen that this directly bears on the fact that *optimal coordinates are state dependent, and may differ for states that probe different parts of the potential energy surface.* Several arguments suggest that "optimal coordinates" for excited bending states must be closely related to the mep: (1) The mep is obviously the preferred path for the HCN → HNC tunneling reaction. Coordinates that naturally describe motion along the mep should thus be advantageous for adequately describing tunneling. However, in bound systems tunneling rates and energy levels are related, and one expects that "good" coordinates for describing tunneling dynamics should also yield good results for the bending energy levels. (2) One expects as a general trend that the correlations between any two modes should increase as the total potential energy function becomes steeper. Inspection of the potential surface of Murrell et al.[29] suggests no reason for a different behavior in this case. Thus, motion in the vicinity of mep is expected to exhibit the lowest correlations.

These points all suggest that suitable modes for the system and states considered here should be the elliptical or prolate spheroidal coordinates, in which the H atom can be located in terms of its distances $r_1, r_2$ from two points on the CN axis. One works with the coordinates[30]

$$\xi = (r_1 + r_2)/2a, \quad 1 \le \xi < \infty, \tag{22b}$$
$$\eta = (r_1 - r_2)/2a, \quad -1 \le \eta \le 1,$$

where $2a$ is the distance between the foci.

Bacic et al.[28] *point out that the two reference points for the coordinate system should not necessarily be identified with the equilibrium positions of C and N, and in their treatment 2a is not the CN equilibrium distance, but a parameter to be determined for each state so as to optimize the SCF approximation.* This, indeed, takes advantage of the fact that the true mep is not an ellipse, and different elliptical fits (value of the parameter $a$) may be required to fit different parts of the mep. The Hamiltonian in the spheroidal coordinates is

$$H = \frac{\hbar^2}{2\mu_I a^2}\left\{\frac{1}{\xi^2 - \eta^2}\frac{\partial}{\partial_\xi}\left[(\xi^2 - 1)\frac{\partial}{\partial_\xi}\right]\right.$$
$$\left. + \frac{1}{\xi^2 - \eta^2}\frac{\partial}{\partial_\eta}\left[\left(1 - \eta^2\frac{\partial}{\partial_\eta}\right)\right]\right\} + V(\xi, \eta). \tag{23}$$

This Hamiltonian is an approximation to the true one, obtained by setting the CN moment of inertia to infinity (the more rigorous treatment is discussed in Ref. 28). In Eq. (23), $\mu_I$ denotes the reduced mass of H and CN, and $V$ is the potential surface. The CN vibration is ignored in the treatment, since its frequency is so high, and the full potential energy surface suggests that this mode is unlikely to play a significant role for the states considered here (and for the isomerization dynamics).

The OC-SCF equations in this case are

$$\int_{\xi_a}^{\xi_b} p_\xi(\xi)d\xi = (m + \tfrac{1}{2})\pi\hbar; \qquad \int_{\eta_b}^{1} p_\eta(\eta)d\eta = \tfrac{1}{2}(n + \tfrac{1}{2})\pi\hbar, \tag{24}$$

$$\frac{\partial}{\partial a}\varepsilon_m(n) = \frac{\partial}{\partial a}\varepsilon_n(m) = 0, \tag{25}$$

the symmetry of the integrand with respect to the $\eta$-domain is used in the second equation of (24). For the Hamiltonian (23) the total SCF energy equals the single-mode energies

$$E_{mn} = \varepsilon_m(n) = \varepsilon_n(m). \tag{26}$$

The single-mode momenta are given by

$$p_\xi(\xi) = [d_n(\xi)]^{-1/2}\left\{2\mu_I a^2\left[\varepsilon_m(n) - \left(\bar{V}_n(\xi) + \frac{\hbar^2}{2\mu_I a^2}\bar{t}_\eta(\xi)\right)\right]\right\}^{1/2}, \quad (27a)$$

$$p_\eta(\eta) = [d_m(\eta)]^{-1/2}\left\{2\mu_I a^2\left[\varepsilon_n(m) - \left(\bar{V}_m(\eta) + \frac{\hbar^2}{2\mu_I a^2}\bar{t}_\xi(\eta)\right)\right]\right\}^{1/2}. \quad (27b)$$

Here $\bar{V}_n(\xi)$, $\bar{V}_m(\eta)$ are the SCF potentials in the $\xi$ and $\eta$ modes, respectively, given by

$$\bar{V}_n(\xi) = \frac{2}{C_\eta}\int_{\eta_b}^{1}\frac{V(\xi,\eta)}{p_\eta(\eta)}d\eta; \quad \bar{V}_m(\eta) = \frac{1}{C_\xi}\int_{\xi_b}^{\xi_b}\frac{V(\xi,\eta)}{p_\xi(\eta)}d\xi; \quad (28)$$

and $\bar{t}_\eta(\xi)$, $\bar{t}_\xi(\eta)$, $d_m(\xi)$, $d_n(\eta)$, $C_\eta$, and $C_\xi$ are defined by

$$\bar{t}_\eta(\xi) = \frac{2}{\hbar^2 C_\eta}\int_{\eta_b}^{1} d\eta\frac{1-\eta^2}{\xi^2-\eta^2}p_\eta(\eta), \quad \bar{t}_\xi(\eta) = \frac{C_\xi^{-1}}{\hbar^2}\int_{\xi_a}^{\xi_b} d\xi\frac{\xi^2-1}{\xi^2-\eta^2}p_\xi(\xi), \quad (29)$$

$$d_m(\xi) = \int_{\eta_b}^{1}\frac{\xi^2-1\,d\eta}{(\xi^2-\eta^2)p_\eta(\eta)}, \quad d_m(\eta) = \int_{\xi_a}^{\xi_b}\frac{(1-\eta^2)\,d\xi}{(\xi^2-\eta^2)p_\xi(\xi)}, \quad (30)$$

$$C_\eta = 2\int_{\eta_b}^{1}\frac{d\eta}{p_\eta(\eta)}, \quad C_\xi = \int_{\xi_a}^{\xi_b}\frac{d\xi}{p_\xi(\xi)}. \quad (31)$$

Figure 4 shows the SCF results in spheroidal coordinates for the excited bending state (0, 4) as a function of the coordinate parameter $a$. For each of the states, the result is compared with the energy given by SCF in polar (hyperspherical) coordinates. Also shown are the results of a "bare-mode" approximation, a crude model which assumes for the $\xi$ mode a potential $V(\xi, \eta = 0)$, and similarly postulates a separate potential $V(\xi_{eq}, \eta)$ for the $\eta$ mode, without any self-consistency in the treatment of the two modes. It is evident from Fig. 4 that the physically motivated elliptical (spheroidal) SCF modes do better in this case than the hyperspherical coordinates. Also, the SCF correction gives an important improvement on the "bare-mode" results. Most important, coordinate *optimization*, that is, imposing condition, (25) yields a noticeably better result than the SCF energy in a spheroidal system that is not refined for the best $a$ value.

These findings are confirmed in an amplified way in the results for $(m, n) =$ (0, 16); (0, 18) shown in Fig. 5 from Bacic et al.[28] For these very high bending states (still, however, under the barrier for isomerization), the SCF in polar coordinates does so poorly compared with the spheroidal SCF that it is out of scale. Also the improvement gained by optimizing the coordinate parameter value $a$ is far greater for these highly excited levels than for the lower states

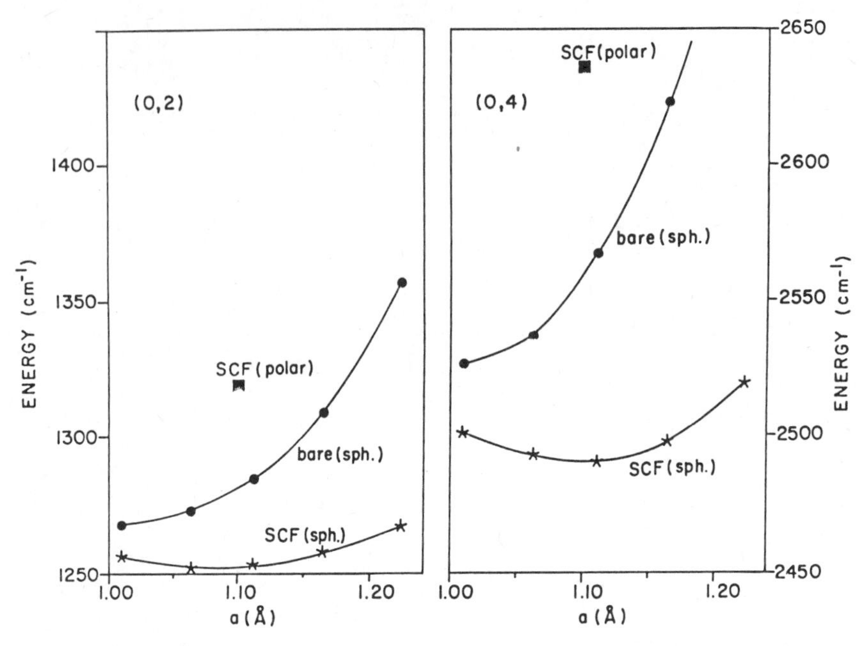


**Figure 4.** Optimal-coordinate behavior of the SCF energies of the (0, 2) and (0, 4) states of HCN. The spheroidal SCF results are plotted versus *a*, the spheroidal coordinates parameter (half the interfocal distance). Also shown are the bare-mode results versus the same parameter, and the polar coordinates SCF result.

of Fig. 4. *We conclude that good choice of coordinates for SCF is essentially always important for good results, and for the highly excited states especially it seems very important to use the optimal coordinates procedure.* That the optimal coordinates which give the best SCF depend on the state [there is a considerable difference between the *a* value which is optimal for (0, 2) and that which is best for (0, 18)] is very reasonable on physical grounds.

Another aspect of the optimal coordinates is shown in Fig. 6. This drawing gives the *locus* of the inner classical turning points $(\eta_t, \xi_t)$ for which the single-mode momenta (27) vanish, for all the bending states $(o, m)$ up to $m = 18$. It is evident that the set of turning points in the optimal coordinates follows very closely the minimum energy path. The same is not true for the turning points in polar SCF, which deviates considerably from the mep for the more excited levels, thus explaining why the results with these modes were relatively poor.

SCF is by no means the only framework in which "good coordinates" for coupled polyatomic vibrations can be defined, or pursued. Stefanski and Taylor[31] stressed the importance of coordinate system choice for simple inter-

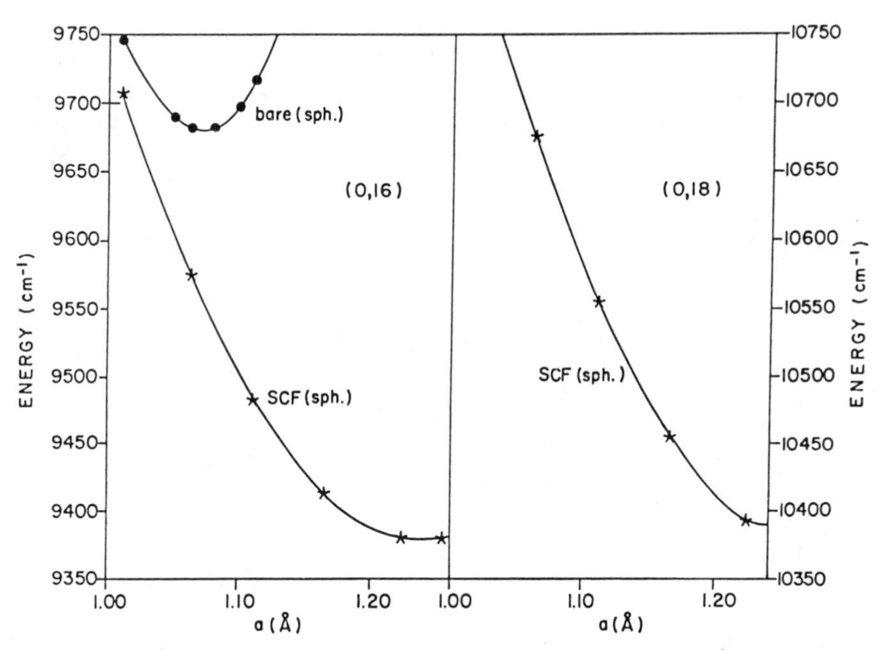


**Figure 5.** Optimal-coordinate behavior of the SCF energies of the (0, 16) and (0, 18) states of HCN. The spheroidal results are plotted versus $a$, the spheroidal coordinate parameter. In the case of (0, 16), the bare-mode results versus $a$ are also shown.

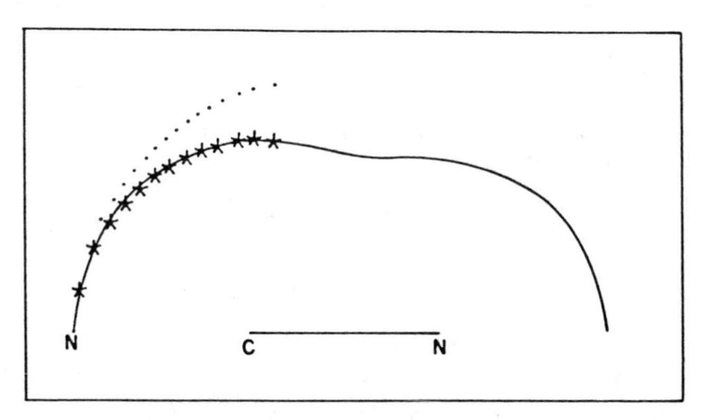


**Figure 6.** Location of SCF turning points with respect to the mep: ——, the mep for the potential surface of Murrell et al.[29]; ···, set of turning points at the barrier obtained from SCF in polar coordinates for bending states in the range $n = 0, \ldots, n = 20$; ∗∗∗, set of turning points at the barrier from SCF in optimal spheroidal coordinates for bending states in the range $n = 0, \ldots, n = 20$.

pretation of certain highly excited states of coupled oscillator systems, using the adiabatic approximation in their treatment. Sophisticated choices of coordinate systems to simplify the dynamics of wavepacket motions on various potential surfaces were proposed by Heller.[32] In OC-SCF the determination of good coordinates is an integral part of the method and has evidently been demonstrated as feasible in a realistic example. The present approach to improving SCF by optimizing the modes relies on physical intuition to find a parametrized family of coordinate systems, in the framework of which a variationally based procedure can then determine the best specific choice. Molecular geometry was the clue for the intuitive guess in the case of HCN. The question arises whether a more systematic method can be found to search for the good coordinates. This is obviously a very interesting, and formidable, topic for future research. In the meanwhile, on a pragmatic footing, we believe that there are excellent prospects of finding appropriate "trial classes" of coordinates by geometric and other considerations for some of the simple molecules that are actually at the focus of most experimental attention (e.g., $C_2H_2$, which is clearly related to HCN, $H_2CO$, $O_3$, etc.)

### D. Extensions for Nearly Degenerate States: CI-SCF

Degeneracies of the SCF states are an obvious cause for breakdown of the approximation in the form discussed in the previous sections. We discuss now an extension of the method that applies to such cases, that is, to resonances and near-resonances between SCF modes. Just as the vibrational SCF method is an adaptation of the Hartree approximation from electronic structure calculations, so is the generalization discussed here an application of the configuration interaction (CI) method, which uses for the wavefunctions a linear combination of the strongly interacting SCF states. Quantum CI for polyatomic vibrations was introduced by Bowman and co-workers,[7,21] the semiclassical version is due to Ratner et al.[33]

We now briefly describe the semiclassical configuration-interaction (SC-CI) approach, restricting attention for simplicity to a two-mode system:

$$H(q_1, q_2)\Psi_{m,n}(q_1, q_2) = E_{m,n}\Psi_{m,n}(q_1, q_2), \tag{32}$$

where the Hamiltonian $H$ is of the form of Eq. (1). The wavefunctions are expanded in a basis of single-mode SCF eigenfunctions:

$$\Psi_{m,n}(q_1, q_2) = \sum_{M,N} C_{MN}^{mn}\phi_M^{(1)}(q_1)\phi_N^{(2)}(q_2), \tag{33}$$

where the $\phi_M^{(i)}(q_i)$ are the eigenstates of $h_i^{\mathrm{SCF}}(q_i)$ as in Eq. (4). *Note that different SCF states are not orthogonal.* The eigenenergies $E_{m,n}$ are therefore solutions of the secular equation

$$\det[\mathbf{H} - E\mathbf{S}] = 0, \tag{34}$$

where

$$H_{MN,M'N'} = \langle \phi_M^{(1)} \phi_N^{(2)} | H | \phi_{M'}^{(1)} \phi_{N'}^{(2)} \rangle, \tag{35}$$

$$S_{MN,M'N'} = S_{M,M'}^{(1)} S_{N,N'}^{(2)}, \quad S_{M,M'}^{(1)} = \langle \phi_M^{(1)} | \phi_{M'}^{(1)} \rangle, \tag{36}$$

$S_{M,M'}^{(1)}$ is the overlap matrix between SCF states. A semiclassical, Fourier-component approximation can be given for such overlaps, the result being[33,34]

$$S_{M,M'}^{(1)} = \mathrm{Re}\left(\frac{1}{T}\int_0^T \exp\{i[\varepsilon_M^{(1)}(N) - \varepsilon_{M'}^{(1)}(N')]t/\hbar\}\gamma(t)\,dt\right), \tag{37}$$

where

$$\gamma(t) = \exp\left(-i\int_0^T \Delta V_{N,N'}(t')\,dt'/\hbar\right) \tag{38}$$

and

$$\Delta V_{N,N'}(t) = \bar{V}_N(q_1(t)) - \bar{V}_{N'}(q_1(t)). \tag{39}$$

The trajectory $q_1(t)$ is computed using the "average" Hamiltonian $h_{\mathrm{av}}^{\mathrm{SCF}}(1) = \frac{1}{2}[h_N^{\mathrm{SCF}}(1) + h_{N'}^{\mathrm{SCF}}(1)]$, and at the energy $\varepsilon$ for which the classical frequency $\omega(\varepsilon)$ equals the transition frequency $[\varepsilon_M^{(1)}(N) - \varepsilon_{M'}^{(1)}(N')]/\hbar$.

The (mean) vibrational period for the initial and final states involved is therefore $T = 2\pi/\omega(\varepsilon)$. Properties of such semiclassical matrix elements and their accuracy in different regimes is discussed by Buch et al.[34] The CI method is most efficient when the interaction potential $V(q_1, q_2)$ in the Hamiltonian can be conveniently expanded in the form

$$V(q_1, q_2) = \sum_{i,j} V_{ij} F_i(q_1) F_j(q_2). \tag{40}$$

In this case the matrix elements of $H$ can also be expressed in terms of single coordinate matrix elements of $F_j(q_i)$. The semiclassical expression for the matrix elements of $F_j(q_i)$ (between nonorthogonal SCF states) is[33]

$$\langle \phi_M^{(1)} | F_j | \phi_{M'}^{(1)} \rangle = \mathrm{Re}\left(\frac{1}{T}\int_0^T \exp\{i[\varepsilon_M^{(1)}(N) - \varepsilon_{M'}^{(1)}(N')]t/\hbar\} F_j(q_1(t))\gamma(t)\,dt\right), \tag{41}$$

where $q_1(t)$, $\gamma(t)$, and $T$ are defined as in Eq. (37). Equations (34)–(40) define the SC-CI scheme. An important question is whether one needs necessarily to use an SCF basis in the CI expansion Eq. (33). There is, in fact, evidence[33] that the SCF basis expansion converges much faster than an expansion using a "bare-mode" basis. In any case when a larger number of terms must be used in the expansion, and the secular equation[34] is of high dimensionality, the CI method becomes unattractive.

Thompson and Truhlar[35] have shown that Fermi resonances, which cannot be described by (single-term) SCF, can be very accurately treated by a small CI calculation involving only several states. Fermi resonances are thus an important class of cases where CI or SC-CI are effective. It will be useful in this respect to specialize the CI treatment to the (zero-order) twofold degeneracies in symmetric two-mode systems, such as the stretching vibrations of $H_2O$ in localized coordinates. Assuming that only the two strongly coupled SCF states contribute significantly to any eigenfunctions, the latter can be written as

$$\psi_{m,n}^{\pm} = \{\phi_m(q_1)\phi_n(q_2) \pm \phi_n(q_1)\phi_m(q_2)\}/\sqrt{2}. \tag{42}$$

Substituting this in the energy functional expression, and using the definition of the SCF single-mode Hamiltonian, one has

$$E_{m,n}^{\pm} = \varepsilon_m(n) + \varepsilon_n(m) - \langle \phi_m\phi_n | V | \phi_m\phi_n \rangle \pm \langle \phi_m\phi_n | V | \phi_n\phi_m \rangle, \tag{43}$$

where $\varepsilon_m(n)$ are the single-mode energies. The semiclassical approximation for the third term in Eq. (43) is given in Eq. (41). The "degeneracy splitting," as in the stretching vibration spectrum of $H_2O$, is given by

$$\Delta E_{m,n} = E_{m,n}^{+} - E_{m,n}^{-} = 2\langle \phi_m(q_1)\phi_n(q_2) | V(q_1, q_2) | \phi_n(q_1)\phi_m(q_2) \rangle. \tag{44}$$

A semiclassical estimate of the integral in (44) can be obtained from the stationary-phase approximation. Exchange degeneracy splitting is clearly a case where the CI approach results in a very simple, practically useful expression.

While an effort to carry out an SCF approximation in unsuitable coordinates may fail, and for such coordinates a large CI treatment may be required, the possibility remains that in good coordinates for the same problem SCF may yield adequate results without any CI correction. Even in dense energy manifolds unfavorable for SCF, coordinates may exist for which the correlations between modes are weak even on the scale of the small separations between the levels. SCF will then apply and CI will be unnecessary. Indeed, there is evidence that single SCF states can be found, virtually uncoupled, in a background of many other dense and strongly coupled states.[36] The question of good SCF states within an almost "statistical" background of other strongly coupled states is of much spectroscopic interest, since such states are expected to carry considerable oscillator strength and show characteristic selection rules for optical transitions. This topic merits considerable attention.

In conclusion, the advantage of the CI approach is that it can be made, in principle, numerically exact, and it offers a very systematic approach. The disadvantage of the scheme is that, as yet, no insight is available *a priori* as to how many CI terms are likely to be required in a calculation using a given system of coordinates. Also, the physical interpretation of a CI state, especially when it contains many terms, is much less transparent than that of an SCF state. Only in special cases, such as two-state Fermi resonances, can the CI approaches provide the simplicity and generality of SCF. Further research may point to some avenue for combining advantages of CI as a systematic algorithm with the physical insight and interpretation provided by the SCF.

## III. TDSCF AND INTRAMOLECULAR VIBRATIONAL ENERGY TRANSFER

### A. Quantum, Classical, and Semiclassical TDSCF

The SCF, or mean-field, approximation does not include the effect of energy transfer processes between the modes. The CI approach incorporates such effects in a time-independent framework, but as was noted in the previous section this method loses much of the simplicity and insight provided by the SCF model. The most natural extension of the SCF approximation that also describes energy transfer among the coupled modes in the system, and treats this effect by a mean-field approach, is the time-dependent self-consistent-field (TDSCF), or time-dependent mean-field, approximation.

Like the SCF method, the TDSCF has its origins in approaches to many-electron excitations[37–39] dating back to the early stages of quantum theory. In the last decade or so, this method has been used extensively in nuclear physics, especially in the context of heavy ion collisions.[40–44] The method proved very useful in describing a range of phenomena in many nuclear systems, including fission, fragmentation, compound nucleus formation, and collective excitation.[40] Applications of the method to intramolecular dynamics are rather recent. Heller[45] discussed the TDSCF approach [also referred to as TDH (time-dependent Hartree method)] in the context of time-dependent variational principles and argued that the scheme should be useful in molecular dynamics. Harris pointed out formal properties of the method in the context of vibrational spectroscopy.[46] Gerber et al. discussed the classical and semiclassical limits of the TDSCF approximation and their relation to the quantum version.[47] On that basis the authors proposed convenient adaptations of the method to intramolecular dynamics and discussed the physical properties and the limitations of TDSCF in the context of intramolecular dynamics.[47,48] Of particular interest for the present survey are the applications to energy transfer in non-RRKM van der Waals cluster system, for example, $I_2Ne$,[47] $I_2NeHe$,[49] $I_2[Ne]_N$, with $N$ up to 16,[50] and the evidence that TDSCF appears to work in both the strong- (RRKM) and the weak-coupling regimes of intramolecular dynamics.[51] In the last few years both the TDSCF approach and methods closely related to it have been extensively used in gas-phase collision problems and in molecule–surface scattering theory.[52–59].

To describe the TDSCF method here we shall consider only Hamiltonians with *pairwise* interactions between the modes, that is,

$$H = \sum_{i=1}^{N} h_i(q_1) + \sum_{i>j}^{N}\sum^{N} V_{ij}(q_i, q_j), \tag{45}$$

where $h_i(q_i)$ is some zero-order "bare" Hamiltonian for mode $i$. The restriction to pairwise interaction can easily be removed, and is kept here merely for technical convenience. To solve the time-dependent Schrödinger equation

$$i\hbar \frac{\partial \Psi(q_1, \ldots, q_N; t)}{\partial t} = H\Psi(q_1, \ldots, q_N; t). \tag{46}$$

The TDSCF approach uses the *ansatz*

$$\Psi(q_1, \ldots, q_N; t) = \prod_{i=1}^{N} \psi_i(q_i, t). \tag{47}$$

Substituting Eq. (47) into Eq. (46), multiplying both sides by $\prod_{i \neq j}^{N} \psi_i(q_i, t)$, and integrating over all $q_l$ with $l \neq j$, one finds after some algebra

$$i\hbar\dot{\psi}_j(q_j, t) + i\hbar\hat{\sigma}_j(t)\psi_j(q_j, t) = [\hat{h}_j^{\mathrm{SCF}}(q_j, t) + \hat{\varepsilon}_j(t)]\psi_j(q_j, t), \tag{48}$$

for $j = 1, \ldots, N$. In (48) the following definitions are used:

$$\hat{\sigma}_j(t) = \sum_{k \neq j} \sigma_{kk}(t), \quad \sigma_{kk} = \langle \psi_k | \dot{\psi}_k \rangle; \tag{49}$$

$$\hat{\varepsilon}_j(t) = \sum_{k \neq j} \varepsilon_k(t), \quad \varepsilon_k(t) = \langle \psi_k | H | \psi_k \rangle; \tag{50}$$

$$\hat{h}_j^{\mathrm{SCF}}(q_j, t) = h_j(q_j) + \sum_{p,l \neq j}\sum \bar{V}_{pl,pl}(t) + \sum_{l>j}\sum \bar{V}_{lj}(q_j, t) + \sum_{l>j}\sum \hat{U}_{jl}(q_j, t); \tag{51}$$

$$\hat{U}_{lj}(q_j, t) = \langle \psi_l | V_{lj}(q_l, q_j) | \psi_l \rangle; \tag{52}$$

$$\bar{V}_{pl,pl} = \langle \psi_p \psi_l | V_{pl} | \psi_p \psi_l \rangle. \tag{53}$$

Now introducing the phase-modified wavefunctions

$$\psi_k(q_k, t) = \phi_k(q_k, t) \exp\left[ -\left( \frac{i}{\hbar} \int_{t_0}^{t} \hat{\varepsilon}_k(t')\, dt' + \int_{t_0}^{t} \hat{\sigma}_k(t')\, dt' + (t - t_0) \sum_{p,l} \bar{V}_{pl,pl} \right) \right], \tag{54}$$

we obtain the TDSCF equations in the familar form

$$i\hbar \dot{\phi}_k(q_k, t) = h_k^{\text{SCF}}(q_k, t)\phi_k(q_k, t), \tag{55}$$

$$h_k^{\text{SCF}}(q_k, t) = h_k(q_k) + \bar{V}_k(q_k, t), \tag{56}$$

$$\bar{V}_k(q_k, t) = \sum_{l \neq k} \langle \phi_l | V_{lk}(q_l, q_k) | \phi_l \rangle. \tag{57}$$

The phase difference between the $\psi_k(q_k, t)$ of the ansatz and the $\phi_k(q_k, t)$ of the final TDSCF wavefunctions is physically unimportant, and will not affect the following development.

To consider the classical limit of the TDSCF equations, it is convenient to write the single-mode wavefunctions as

$$\psi_j(q_j, t) = A_j(q_j, t)\exp[iS(q_j, t)/\hbar], \tag{58}$$

where $A_j(q_j, t)$, $S_j(q_j, t)$ are real valued. As familiar from standard time-dependent quantum theory,[60] this leads to the following equations for the phase and amplitude functions:

$$\frac{\partial S(q_j, t)}{\partial t} + \frac{1}{2m_j}\left(\frac{\partial S_j}{\partial q_j}\right)^2 + V_j^{\text{SCF}}(q_j, t) = \frac{\hbar^2}{2m_j} \frac{\partial^2 A_j}{\partial q_j^2} \frac{1}{A_j(q_j, t)}, \tag{59}$$

$$m_j \frac{\partial A_j(x_j, t)}{\partial t} = -\left(\frac{\partial A_j}{\partial q_j}\right)\left(\frac{\partial S_j}{\partial q_j}\right) - \frac{1}{2} A_j(q_j, t) \frac{\partial^2 S_j}{\partial q_j^2}, \tag{60}$$

where $m_j$ is the mass associated with the mode $j$ in the "bare-mode" Hamiltonian $h_j(q_j)$ [see Eq. (45)]:

$$U_j^{\text{SCF}}(q_j, t) = V_j(q_j) + \bar{U}_j(q_j, t). \tag{61}$$

Here $V_j(q_j)$ is the bare single-mode potential in $h_j(q_j)$. The TDSCF "correction" to the single-mode potential thus becomes

$$\bar{U}_j(q_j, t) = \sum_{k \neq j} \int A_k^2(q_k, t) V_{jk}(q_j, q_k)\, dq_k, \tag{62}$$

where $V_{jk}$ is defined as in Eq. (45). Taking the classical limit $\hbar \to 0$ of Eq. (59), one has

$$\frac{\partial S_j^c}{\partial t} + \frac{1}{2m_j}\left(\frac{\partial S_j}{\partial q_j}\right)^2 + U_j^{\mathrm{SCF}}(q_j, t) = 0. \tag{63}$$

The equation[60] for the amplitude $A_j$ is formally $\hbar$-independent and thus does not change. The classical-limit amplitude function is obtained by solving Eq. (60), with $S_j^c$ substituted for $S(q_j, t)$. Equations (60)–(63) define the classical limit of the TDSCF, but rather than use trajectories, these equations lead directly to $A_k^2(q_k, t)$, the classical probability density, which is the analog of the quantum mechanical $|\phi_k(q_k, t)|^2$. For a stationary state in a time-independent potential $A_k^2(q_k, t) = A_k^2(q_k) = \text{const} \times [p_k(q_k)]^{-1}$, where $p_k(q_k)$ is the momentum of mode $k$ at the fixed energy of the system.

From a practical point of view the classical method using directly the probability density function is not convenient, and it is computationally preferable to use an approach that involves trajectory calculations. A derivation of such formulation can be made by starting from the quantum-mechanical TDSCF, and using semiclassical (gaussian) wavepackets. Here we merely quote the final result. In analogy to (62), the single-mode classical SCF potentials are given by

$$U_j^{\mathrm{SCF}}(q_j, t) = V_j(q_j) + \frac{1}{n} \sum_{k \neq j} \sum_{v=1}^{n} V_{jk}(q_j, q_k^{(v)}(t)). \tag{64}$$

Here the $q_k^{(v)}(t)$ are a set of single-mode trajectories for the $k$ mode. The index $v$ refers to the different initial conditions at some time $t = 0$ satisfied by each of these trajectories. The sampling over initial conditions in each mode (indexed by $k$) must be large enough to yield a converged average in Eq. (64). To obtain the trajectories, one solves the single-mode Hamilton's equations:

$$-\dot{p}_j = \frac{\partial h_j^{\mathrm{SCF}}(q_j, t)}{\partial q_j}, \tag{65a}$$

$$\dot{q}_j = p_j/m_j, \tag{65b}$$

for $j = 1, \ldots, N$, where $h_j^{\mathrm{SCF}}$ is the classical analogue of the self-consistent-field Hamiltonian of (56). The solution of Eq. (65) must be carried out simultaneously for all the modes, and consistently with the evaluation of the $U_j^{\mathrm{SCF}}(q_j, t)$, the single-mode potentials.

A classical TDSCF calculation thus proceeds as follows.[42] For each mode $j$, a set of initial values (consistent with the physical initial conditions) $q_j^{(1)}, \ldots, q_j^{(n)}$ is selected. One then solves simultaneously for all the set of single-mode trajectories $q_j^{(1)}(t), \ldots, q_j^{(n)}(t)$, pertaining to each of the modes $j = 1, \ldots, N$. We refer to the set of trajectories for each single mode evaluated by solving the classical TDSCF equations (64)–(65) as a "bundle of trajectories" for that mode. Classical TDSCF thus involves the self-consistent evaluation of the trajectory bundles for the various modes, the bundle for each mode being obtained from a potential that is an average over the bundles of the other modes. The complexity of a classical TDSCF calculation grows roughly linearly with $N$, the number of degrees of freedom, assuming the number of initial values per each mode is fixed. The complexity of full classical trajectory calculations grows exponentially with $N$ in the simpler algorithms, and in any case much faster than for TDSCF.[47]

The fact that TDSCF associates a separate Hamiltonian for each mode makes it straightforward to introduce a semiclassical TDSCF scheme in which some of the degrees of freedom are treated classically while the others are described quantum mechanically.[47,59] Other treatments which arbitrarily mix quantum-mechanical and classical degrees of freedom involve in general problems of consistency,[59] which are quite hard to resolve. The TDSCF offers a simple and very useful avenue for overcoming this, and many of the TDSCF calculations reported in the literature[47,55,56,59] are of this semiclassical type.

The foregoing discussion deals with TDSCF as introduced for fixed, preselected coordinates. Very recently, Kučar et al.[61] considered an optimal coordinates TDSCF approximation. For a two-mode model Hamiltonian expressed in terms of the Cartesian coordinates $(x, y)$, these authors used an ansatz of the type

$$\Psi(x, y, t) = R(t)\psi_1(x, t)\psi_2(y, t), \tag{66}$$

where $R$ is a rotation operator. This is tantamount to transforming the coordinates into: $x' = x\cos\theta + y\sin\theta$; $y' = -x\sin\theta + y\cos\theta$, where the transformation parameter $\theta$, determined variationally, is a function of time. The authors report excellent results with this coordinate-optimized TDSCF for several calculated variables as a function of time, including the expectation value of $p_x p_y$. (The SCF expectation value with the fixed coordinates $x$, $y$ for the product of momenta is of course zero.) It seems very reasonable to expect that coordinate-optimized TDSCF should indeed prove to be a much improved method over the fixed coordinate one, just as was found to be the case in the time-independent SCF. Insight needs to be developed as to the nature of suitable coordinate transformations to be optimized in this framework, and this should prove a useful avenue for future research.

### B. TDSCF for Weak and for Strong-Coupling (RRKM) Systems

The time-dependent potentials that act on each mode within the TDSCF approximation, allow for energy transfer out of and into any one of the modes. Although the TDSCF can therefore describe, in principle, intramolecular energy transfer the question must be answered whether, and for which type of systems, the amount and rate of energy transfer predicted by this method is correct. Considerable information has been gathered on this topic from numerical test calculations for model systems, for example, Refs.[47,49–51] One of the important conclusions that emerges, which is the topic of the present section, is that the TDSCF method seems to work well in the two experimentally established and physically very different regimes of intramolecular dynamics: *The approximation yields fairly good results for systems with relatively rapid intramolecular vibrational energy flow (i.e., RRKM type of dynamics), to which class most unimolecular processes belong.*[51] *At the same time TDSCF was found to work very well also for systems of relatively inefficient intramolecular transfer of the energy (weak-coupling dynamics)*, a behavior shown in vibrational predissociation of van der Waals clusters such as $I_2(v)Ne$.[5,47,49] We are unaware of any other dynamical approximation (as opposed, of course, to exact quantum or classical calculations) demonstrated to work in both opposing dynamical limits.

The first application of TDSCF to intramolecular dynamics was made to the vibrational predissociation process: $I_2(v)Ne \rightarrow I_2(v') + Ne$, which was studied by Gerber et al.[47] for the initial vibrational level $v = 25$ in the $^3\Pi$ electronic state of $I_2$. This and similar systems were previously studied by classical trajectories,[62] by quantum perturbation (distorted-wave) methods,[63] and by essentially exact coupled channel calculations.[64] This is therefore a suitable test case for the TDSCF, as far as dissociation dynamics of van der Waals cluster systems is concerned. The TDSCF calculations (and the reference calculations to which they were compared) were for a collinear model $I_2(v)Ne$. While the latter is almost definitely an inappropriate geometry for the real system, this should not affect the suitability of the model as a test case for assessing the validity of the TDSCF. The approximation was used in a version that mixes a quantum treatment for the $I_2$ vibration with a classical description of the I...Ne relative motion. The predissociation lifetime obtained from this semiclassical TDSCF (10 psec) was in good accord with results of classical trajectory calculations (12 psec) and with the quantum-mechanical calculations of Beswick and Jortner[63] (10 psec). The TDSCF calculations produced the propensity rule $v' = v - 1$ for the final state of $I_2$, also found in the quantum-mechanical results and a frequently observed empirical behavior in van der Waals predissociation systems. Finally, as the initial state was changed from $v = 25$ to $v = 28$, the lifetime dropped by about 30%, also in good agreement with the finding of Beswick and Jortner.[63,64] Evidently, the

TDSCF reproduces well at least some of the main observable properties for the system. More important still is the fact, which emerges from detailed analysis of the TDSCF solutions, that the approximation yields the correct physical mechanism for the dissociation process. As pointed out by Beswick and Jortner,[63,64] this is a highly non-RRKM system. Energy transfer between the I–I and the I...Ne mode occurs on the timescale of the dissociation process itself, and may thus be considered relatively efficient. Gerber et al.[47] found that in fact dissociation (and significant energy transfer) occur as a consequence of rather strong impulsive in-phase "collisions" between the Ne and the neighboring I atom. Until such a pure "collision" occurs, the effective coupling between the modes is rather weak. It is gratifying that the approximation is essentially able to reproduce the appropriate rate of these relatively rare in-phase "collisions."

The method was found equally successful for similar processes in more complicated van der Waals systems. Schatz et al.[49] studied vibrational predissociation processes of the type

$$XI_2(v)Y \rightarrow X + I_2(v')Y \rightarrow X + I_2(v'') + Y,$$

for $X$, $Y$ = He, Ne. The authors used collinear models for the complex, and compared "exact" classical dynamics with classical TDSCF. Good agreement was found between the exact classical and the TDSCF results for the dissociation lifetimes (e.g., in the case of $NeI_2Ne$, $\tau_{TDSCF} = 11$ psec, $\tau_{class} = 9$ psec). TDSCF also gives correctly the branching ratios, that is, the rate of $NeI_2He \rightarrow NeI_2 + He$ relative to $NeI_2He \rightarrow Ne + I_2He$. The dependence of the process on the rare gas is the same in TDSCF as in the classical results. Furthermore, both "exact" and mean-field dynamics predict a dynamical effect of the rare gases on each other (qualitatively shown by experimental results) although $X$ and $Y$ do not interact directly in the model used. The dissociation mechanism found in this case is qualitatively of the same type as in the triatomic complex $I_2X$ and is dominated by "hard" sudden collisions between the rare gas atom and an I neighbor. To illustrate the mechanism as reflected in the TDSCF trajectory-bundle approach, we consider Figs. 7 and 8 from Ref. 49, for the case of $HeI_2Ne$. Figure 7 shows the force acting on the He for two typical trajectories of the He...I mode, and for short times *before* significant coupling between the I–I and the He...I modes builds up (this is the so-called incubation period[49]). Only a single, relatively weak peak of force is seen for each trajectory, insufficient to effect dissociation. The displacement of the trajectories belonging to the bundle of the I–I mode are also shown in Fig. 7 for the early time interval of the process, and clearly the trajectories are out of phase, as in the initial state. A drastically different situation is revealed at a later time (Fig. 8): Now, as a consequence of phase-changing processes, the

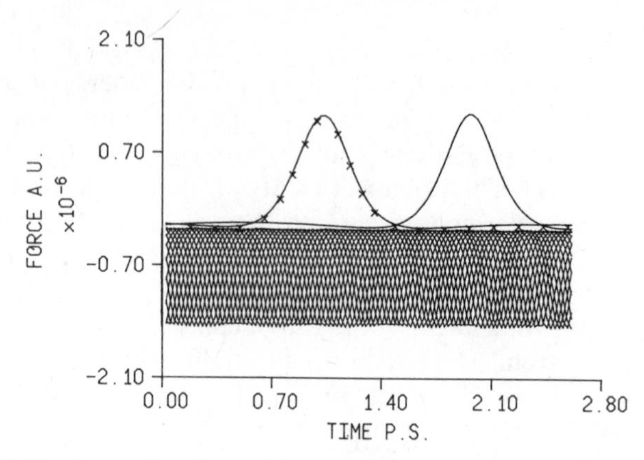


**Figure 7.** TDSCF trajectory bundle result for $HeI_2Ne$. The upper curves show the force on the He on two typical trajectories, while the lower curves show the 15 $I_2$ trajectory displacements. Short time behavior is shown, before the inphase "collision" occurs.

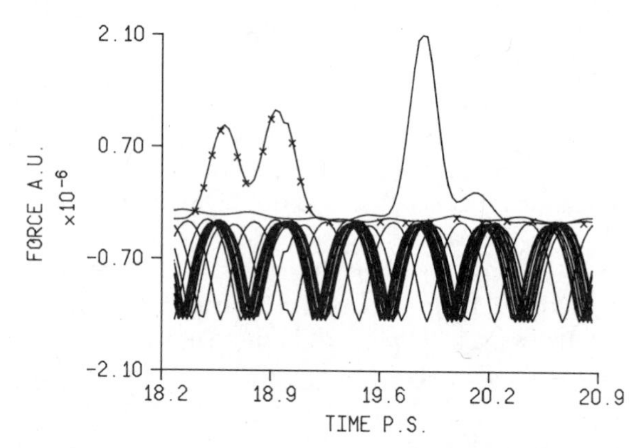


**Figure 8.** As in Fig. 7, but at longer times, after the induction period, when decay is occurring. Note that a dephasing-type process, from collisions with the gas atoms, has driven the $I_2$ trajectories largely into phase, resulting in larger force being exerted on the dissociating He.

trajectories of the I–I bundle are in phase, and this leads to a large force exerted on the I...He bond, causing dissociation. There is thus a period of weak coupling during which there is no significant energy flow, the He and I avoid a mutual hard collision. Then correlation builds up, the modes get in phase, and the intramolecular I...He encounter leading to dissociation takes place. It appears remarkable that the TDSCF is able to yield such an intricate mechanism.

Consider now the application of the TDSCF approximation to a process

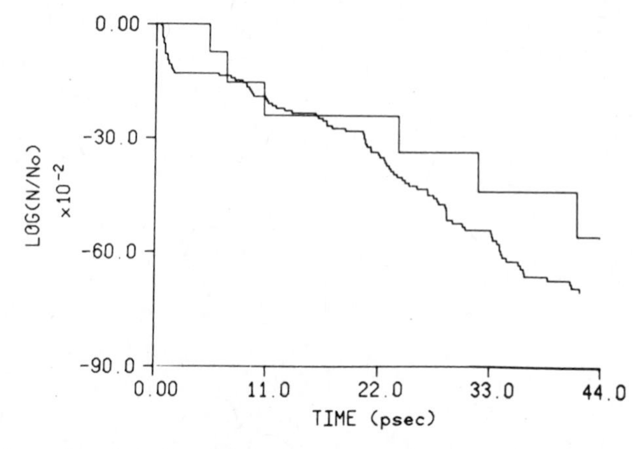


**Figure 9.** The number of nondissociated trajectories as a function of time. The coarse histogram is the TDSCF result. The finer histogram is for exact classical dynamics.

exhibiting a very different type of dynamics:

$$Ar_3 \to Ar_2 + Ar.$$

Buch et al.[51] carried out comparative classical trajectory and (trajectory-bundle) TDSCF calculations for initial states of $Ar_3$ with energies in each single mode below the dissociation threshold, but with a total energy sufficient for bond rupture. The calculations were carried out for a collinear model. As the classical trajectory as well as TDSCF results show, the dynamics of this system is essentially of the RRKM type, with a typical timescale for energy transfer between the two Ar...Ar modes that is fast compared with the dissociation lifetime. Analysis of the classical trajectories shows that resonance between the modes is the cause of the efficient transfer, but the amount of energy exchanged in a typical transfer event does not suffice for dissociation. Figure 9 shows plots of $\log[N(t)/N_0]$ versus time, as obtained from TDSCF and from exact classical dynamics; $N(t)$ is the number of nondissociated $Ar_3$ trajectories at time $t$ and $N_0$ is the number of trajectories used in the initial-state sampling. The results shown are for a case with initial energies $E_1 = E_2 = -0.15 \times 10^{-3}$ hartree in the two Ar...Ar modes. The lifetimes extracted from the mean slopes of the two curves are in very good agreement. $\tau_{\mathrm{TDSCF}} = 80.2$ psec, $\tau_{\mathrm{exact}} = 63.5$ psec. The TDSCF result deviates considerably from the exact classical one in the long-time regime, but the number of trajectories in this range is too small for meaningful statistics. In any case, the TDSCF approximation is found in this case to yield dissociation lifetimes to good accuracy for an RRKM type of system.

Eslava et al.[50] demonstrated in a recent study that the TDSCF approximation can predict the transition from weak- to strong-coupling dynamics as the initial excitation energy in increased in certain systems. The authors studied energy transfer from an initially vibrationally excited $I_2$ molecule embedded in a one-dimensional cluster of Ne atoms. Trajectory-bundle TDSCF results were reported for $I_2[Ne]_N$ with $N = 4$, 8, 16, and exact classical dynamics calculations were carried out in the $N = 4$ case. The process of $I_2$ vibrational relaxation in $I_2[Ne]_N$ was explored for timescales of up to $\sim 10^{-10}$ sec for initial vibrational states of $I_2$ in the range of $v = 25$–60. For $v \leq 35$, the short-time vibrational energy transfer from the I–I mode proceeds by the "impulsive" mechanism discussed earlier for small systems such as $I_2^{(v)}Ne$: The energy transfer depends on "hard" collisions between an I and a neighboring Ne atom. In this regime, resonances between the $I_2$ mode and cluster modes do not seem to occur, and therefore do not facilitate the energy transfer from the molecular vibration to the "solvent" modes. For initial state $v = 60$, the calculations of Eslava et al.[50] show a very different mechanism, with very rapid fluctuations of energy from and into the $I_2$ vibrational mode. The authors found evidence that energy transfer in this case involves resonances between the "internal" $I_2$ vibrations and some of the cluster modes. The change in mechanism of vibrational energy redistribution in $I_2[Ne]_N$ holds for all $N$ values studied ($N = 4, 8, 16$). For $N = 4$ the TDSCF result has been confirmed by an "exact" classical trajectory calculation. From these studies, in concert with the predissociation results just discussed, there is extensive evidence that the TDSCF approximation works well in the weak-coupling regime; some results show that it correctly describes the strong-coupling (RRKM) type of dynamics; and it seems at least reasonable to suggest that the method is valid also for the intermediate domain between the preceding two limits. Application of TDSCF to study intramolecular dynamics in the previously mentioned "intermediate regime," and for systems large enough to exclude exact classical calculations, appears to us as potentially very valuable, and it should be desirable to pursue it.

### C. Properties, Validity, and Limitations of TDSCF

It was previously argued that the TDSCF method can be valid in both the weak- and the strong-coupling regimes. This is of considerable importance for intramolecular dynamics where perturbative methods fail in the strong-coupling regime, and statistical approximations are not valid (at least as presently applied) in the weak-coupling cases. To this one should add several useful formal properties of the TDSCF method[40–43]: The total energy of the system is conserved in time within this approximation. Similarly, conservation of total angular momentum and time-reversal invariance are both obeyed. *In searching for the breakdown conditions of TDSCF* as applied to intramolecular dynamics, *the nature of the observables that one is attempting to calculate is of central importance.* Observable quantities associated with single-mode expectation values, for example the energy in a given mode as a function of time, are described well by TDSCF. Also, unimolecular lifetimes (which are physically related to the questions of when does a specific bond acquire a certain amount of energy or when does its length become larger than some critical value) are therefore well predicted by TDSCF. It is not surprising that the TDSCF, which assumes separability, should do best for single-mode expectation values and related observables: The difficulties are expected to appear in treating observables that involve correlations between modes. As previously mentioned, Kučar et al.[61] gave an example of a system with large fluctuations in time of the expectation values of $p_x p_y$ (the cross product of momentum), which TDSCF in fixed $(x, y)$ modes obviously cannot describe. This observable could, however, be well described by an "optimized coordinates" TDSCF, which uses the "best" linear combinations of $x$ and $y$, with time-dependent coefficients.[61] An important class of observables that inherently involves correlations between modes and cannot be described by TDSCF (at least is fixed coordinates) are detailed $S$-matrix elements, as pertinent to collisions or half-collisions[41–43]: Consider, for instance, an atom colliding with a target having two states $\chi_1(r_T)$, $\chi_2(r_T)$, where $r_T$ is the target coordinate. At asymptotic distances the atom can have kinetic energy $\varepsilon_{\text{kin}}^1 = E - \varepsilon_1^T$ or $\varepsilon_{\text{kin}}^2 = E - \varepsilon_2^T$ ($E$ is the total energy of the system, $\varepsilon_i^T$ is the target asymptotic energy), depending on the final state of the target. The wave-function of the system as $t \to \infty$ must thus contain the terms

$$\Psi(r_T, r_A, t \to \infty) = c_1 \chi_1(r_T)\eta_1(r_A) + c_2 \chi_2(r_T)\eta_2(r_A),$$

where $\eta_i$ is a wavefunction that describes the motion of the atom relative to the target, with asymptotic translational energy $\varepsilon_{\text{kin}}$, and the $c_i$ are constants. In TDSCF one has

$$\Psi^{\text{TDSCF}}(r_T, r_A, t \to \infty) = [d_1 \chi_1(r_T) + d_2 \chi_2(r_T)] \cdot [e_1 \eta_1(r_A) + e_2 \eta_2(r_A)],$$

where $d_i$, $e_i$ are constants. The "cross products" $\chi_1(r_T)\ \eta_2(r_A)$ are obviously unphysical and spurious[47] when occurring at asymptotic distances of the atom from the target. As a consequence of such spurious states, the TDSCF cannot produce a valid, detailed $S$ matrix in which the asymptotic variables corresponding to both collision partners (energies, etc.) are simultaneously specified.[41–43]

Several efforts were made by researchers in nuclear physics to formulate an improved mean-field theory, capable of yielding full state-resolved $S$-matrix elements in all variables, and free of the spurious asymptotic states and related difficulties.[41–43] Of these the approach by Griffin and co-workers[42] is probably the simplest although it is somewhat heuristic, while the methods of Levit[41] and Alhassid and Koonin[43] seem more rigorous yet harder to apply. None of these methods has yet proved to be the highly applicable tool that TDSCF is for nuclear processes, and to our knowledge they have not been applied at all to intramolecular dynamics. Thus, a practical extension of TDSCF, capable of producing consistently $S$-matrix elements, is certainly an open question, at least for molecular processes. On the other hand, such detailed $S$-matrix data are very rarely available for systems more complicated than a diatomic molecule colliding with an atom. More typically for the larger molecules, averaged or less detailed observables, such as lifetimes of dissociation or single-mode energy distribution, are the experimentally available quantities. The TDSCF is thus particularly geared for more averaged type of data, and for larger systems where detailed $S$-matrix information is unlikely to be relevant.

Another limitation of TDSCF of a very different type is related to the time domain. The difference between the exact Hamiltonian and the TDSCF one can be viewed as a perturbation that is neglected in the mean-field approximation, and the integrated effect of which is bound on the whole to grow in time. The fact that TDSCF can be in serious error when pursued for very long time scales has been noted for several cluster dissociation processes.[47,49,50] However, the long-tail time behavior was not of physical importance in the examples studied. Obviously this limitation of TDSCF is pertinent only to very slow processes (when in competition with much stronger channels).

Finally, a note is due regarding the initial states used in TDSCF. Obviously, (static) SCF eigenstates are stationary within the framework of the TDSCF, and will not show any nontrivial dynamical evolution in this approximation. (This is only provided we consider SCF and TDSCF for the same, fixed coordinates.) Under the exact Hamiltonian, static SCF states are, of course, not stationary and can evolve in a way that corresponds to a dynamical process such as energy transfer. The choice of the initial state in any theoretical calculation must reflect the preparation of that state in the corresponding experiment. Hence, any case where the initial state being prepared is believed to be extremely close to a (static) SCF eigenfunction cannot be pursued by TDSCF for the dynamics. Again, this does not seem a practically important limitation.

The preceding discussion of TDSCF and its limitations was confined to the mean-field approximation in fixed coordinates. Coordinate optimization should not only enhance the quantitative accuracy of the method, but could also remove some of the more qualitative difficulties. Unfortunately, the properties of coordinate-optimized TDSCF have not yet been discussed in the literature (see, however, Ref. 61), and this seems to us an essential route of development in the near future. In addition, extension of the simple TDSCF form of Eq. (47) either to contain several possible wavefunction products or to include a specific correlation factor should be investigated as a way to avoid the spurious states problem.

## IV. CONCLUDING REMARKS

In this chapter we analyzed several developments in SCF approaches to intramolecular dynamics, as pertinent both to the vibrational energy-level structure and to vibrational energy redistribution in polyatomic systems. The emphasis was on methodological progress, and on how the latter throws light on the physical basis for the validity of the method and on possible future applications. The main themes that emerged in the discussion can be summarized as follows.

First, the success of the SCF approximations depends greatly on the choice of modes that are being separated in the treatment. Physically well-motivated choices of the coordinates often lead to truly striking results. This has been amply demonstrated in SCF energy-level calculations, and there are early results along this line also for TDSCF descriptions of energy transfer dynamics. This shows that mutual screening of the modes that are being separated plays a central role in justifying the SCF approach to coupled vibrational degrees of freedom. The importance of choosing appropriate modes for mean-field approximation leads to optimal coordinates SCF (OC-SCF), in which the SCF equations are solved simultaneously and consistently with the search for the "best" coordinates for pursuing the approximation. The optimal coordinates so defined depend on the energy and the state, as is expected on physical grounds. The choice of improved coordinates, and in particular the variational optimization of the latter, appears a major step forward in both SCF and TDSCF theory.

Second, SCF methods have an important versatility in being applicable at the full quantum-mechanical, classical, or semiclassical levels. In the framework of TDSCF it is particularly useful that some degrees of freedom can be treated quantum mechanically while the other modes can be simultaneously and consistently described by classical dynamics. Such a mixed treatment is often very advantageous for systems with a considerable number of modes, since the main quantum effects can be described while still retaining most of the computational simplicity of the classical description.

A third important advantage of the SCF methods is that they provide a unified framework within which one can treat both energy-level calculations and dynamical processes. Having a unified language and closely related descriptions of spectroscopy and of energy transfer dynamics is attractive because both reflect on properties of the potential energy surface, and should best be used in conjunction to study the surface and its energetic and dynamical consequences.

Finally and perhaps most usefully, mean-field methods are computationally simple, and are also convenient for providing physical insight and interpretation. The advantages of simplicity make these methods promising tools for treating systems with a substantial number of degrees of freedom, and examples of such applications are already available.

However, the potential of SCF methods is by no means exhausted, nor are the approximations sufficiently understood. In particular, the question of how to approach the choice of "good" coordinate systems for the method is still largely open. It should be most useful to develop systematic guidelines regarding suitable coordinate families for different kinds of systems, molecular geometries, and so forth. The breakdown of the methods, especially in the case of TDSCF, must be better understood, and future work on this issue, both in formal terms and in establishing the physical considerations involved, seems crucial. Finally, one must look beyond SCF methods, and seek to develop schemes that have most of the simplicity of the mean-field approach, but which improve on the validity regime of the latter. Whatever is available at present along this line (e.g., the CI extension of the SCF treatment) does not seem satisfactory, and major progress is necessary on this topic.

We anticipate that SCF methods will grow in power and versatility, and will continue to prove extremely useful tools in the study of intramolecular dynamics.

## Acknowledgments

We are delighted to thank V. Buch, R. Roth, L. Gibson, Z. Bacic, L. Eslava, B. Barboy, and G. C. Schatz for close collaboration. The Fritz Haber Center at the Hebrew University is supported by the Minerva Gesellschaft für die Forschung, München, Federal Republic of Germany. We are grateful to the Chemistry Division of the NSF and to the Donors of the Petroleum Research Fund, administrated by the ACS, for partial support.

## References

1. D. E. Reisner, P. H. Vaccaro, C. Kittrell, R. W. Field, J. L. Kinsey, and H.-L. Dai, *J. Chem. Phys.* **77**, 575 (1982); D. E. Reisner, R. W. Field, J. L. Kinsey, and H.-L. Dai, *J. Chem. Phys.* **78**, 2817 (1983); E. Abramson, R. W. Field, D. Imre, K. K. Innes, and J. L. Kinsey, *J. Chem. Phys.* **83**, 453 (1985).
2. K. K. Lehmann, G. Scherer, and W. A. Klemperer, *J. Chem. Phys.* **77**, 2853 (1982); G. J. Scherer, K. K. Lehmann, and W. A. Klemperer, *J. Chem. Phys.* **78**, 2817 (1983).
3. (*a*) D. Bailly, R. Farrenq, G. Guelachivili, and C. Rossetti, *J. Mol. Spectros.* **90**, 74 (1981); (*b*) A. Chedin, J. Mol. Spectros. **76**, 430 (1979).
4. E. B. Wilson, Jr., J. C. Decius, and P. C. Cross, *Molecular Vibrations.* McGraw-Hill, New York, 1955.
5. K. E. Johnson, L. Wharton, and D. H. Levy, *J. Chem. Phys.* **69**, 2719 (1978); J. E. Kenny, K. E. Johnson, W. Sharfin, and D. H. Levy, *J. Chem. Phys.* **72**, 1109 (1982).
6. J. M. Bowman, *J. Chem. Phys.* **68**, 608 (1978).
7. J. M. Bowman, K. Christoffel, and F. Tobin, *J. Phys. Chem.* **83**, 905 (1979).
8. G. D. Carney, L. I. Sprandel, and C. W. Kern, *Adv. Chem. Phys.* **37**, 305 (1978).
9. M. Cohen, S. Greita, and R. P. McEachran, *Chem. Phys. Lett.* **60**, 445 (1979).

10. R. B. Gerber and M. A. Ratner, *Chem. Phys. Lett.* **68**, 195 (1979).
11. M. A. Ratner and R. B. Gerber, *J. Phys. Chem.* **90**, 20 (1986).
12. D. H. Farrelly and A. D. Smith, *J. Phys. Chem.* **90**, 1599 (1986).
13. R. B. Gerber, R. M. Roth, and M. A. Ratner, *Mol. Phys.* **44**, 1335 (1981).
14. C. J. H. Schutte, *Theory of Molecular Spectroscopy*. North Holland, Amsterdam, 1976, p. 280.
15. B. C. Garrett and D. G. Truhlar, *Chem. Phys. Lett.* **92**, 64 (1982).
16. D. Farrelly, R. M. Hedges, and W. P. Reinhardt, *Chem. Phys. Lett.* **96**, 599 (1983).
17. R. M. Roth, M. A. Ratner, and R. B. Gerber, *J. Phys. Chem.* **87**, 2376 (1983).
18. A. D. Smith, W.-K. Liu, and D. W. Noid, *Chem. Phys.* **89**, 345 (1984).
19. R. M. Roth, M. A. Ratner, and R. B. Gerber, *Phys. Rev. Lett.* **52**, 1288 (1984).
20. T. C. Thompson and D. G. Truhlar, *J. Chem. Phys.* **77**, 3031 (1982).
21. H. Romanowski, J. M. Bowman, and L. B. Harding, *J. Chem. Phys.* **82**, 4155 (1985).
22. P. Schatzberger, E. A. Halevi, and N. Moiseyev, *J. Phys. Chem.* **89**, 4691 (1985).
23. B. Barboy, G. C. Schatz, M. A. Ratner, and R. B. Gerber, *Mol. Phys.* **50**, 353 (1983).
24. V. Buch (unpublished).
25. L. L. Gibson, R. M. Roth, M. A. Ratner, and R. B. Gerber, *J. Chem. Phys.* **85**, 3425 (1986).
26. R. Lefebvre, *Int. J. Quant. Chem.* **23**, 543 (1983).
27. N. Moiseyev, *Chem. Phys. Lett.* **98**. 223 (1983).
28. Z. Bacic, R. B. Gerber, and M. A. Ratner, *J. Phys. Chem.* **90**, 3606 (1986).
29. J. N. Murrell, S. Carter, and L. O. Halonen, *J. Mol. Spectros.* **93**, 307 (1982).
30. H. Margenau and G. M. Murphy, *The Mathematics of Physics and Chemistry*. Van Nostrand, New York, 1956.
31. K. Stefanski and H. S. Taylor, *Phys. Rev. A* **31**, 2810 (1985).
32. E. J. Heller (private communication).
33. M. A. Ratner, V. Buch, and R. B. Gerber, *Chem. Phys.* **53**, 345 (1980).
34. V. Buch, M. A. Ratner, and R. B. Gerber, *Mol. Phys.* **42**, 197 (1981).
35. T. C. Thompson and D. G. Truhlar, *Chem. Phys. Lett.* **75**, 87 (1980).
36. M. A. Ratner and R. B. Gerber, in *Stochasticity and Intramolecular Energy Transfer*, Proceedings of the NATO Workshop, S. Mukamel and R. Lefebvre (eds.). Reidel, Dordrecht, Holland, 1987.
37. P. A. M. Dirac, *Proc. Camb. Phil. Soc.* **26**, 376 (1930).
38. A. D. Mclachlan and M. A. Ball, *Rev. Mol. Phys.* **36**, 844 (1964).
39. J. Linderberg and Y. Öhrn, *Propagators in Quantum Chemistry*. Academic Press, New York, 1972.
40. See, for instance, (*a*) P. Bonche, S. E. Koonin, and J. W. Negele, *Phys. Rev. C* **13**, 1226 (1976); (*b*) R. Y. Casson, R. K. Smith, and J. A. Marhun, *Phys. Rev. Lett.* **36**, 116 (1976); (*c*) J. W. Negele, S. E. Koonin N. Moller, J. R. Nix, and A. J. Sierle, *Phys. Rev. C* **17**, 1098 (1978).
41. S. Levit, *Phys. Rev. C* **21**, 1594 (1980).
42. J. J. Griffin, P. C. Lichtner, and M. Dworzecka, *Phys. Rev. C* **21**, 1351 (1980).
43. Y. Alhassid and S. E. Koonin, *Phys. Rev. C* **23**, 1590 (1981).
44. J. W. Negele, *Phys. Today* **38**, 24 (1985).
45. E. J. Heller, *J. Chem. Phys.* **64**, 63 (1976).
46. R. Harris, *J. Chem. Phys.* **72**, 1776 (1980).
47. R. B. Gerber, V. Buch, and M. A. Ratner, *J. Chem. Phys.* **77**, 3022 (1982).
48. R. B. Gerber, V. Buch, and M. A. Ratner, *Chem. Phys. Lett.* **91**, 173 (1982).
49. G. C. Schatz, V. Buch, M. A. Ratner, and R. B. Gerber, *J. Chem. Phys.* **79**, 1808 (1983).
50. L. A. Eslava, R. B. Gerber, and M. A. Ratner, *Mol. Phys.* **56**, 47 (1985).
51. V. Buch, R. B. Gerber, and M. A. Ratner, *Chem. Phys. Lett.* **101**, 44 (1983).
52. J. T. Muckerman, I. Rusinek, R. E. Roberts, and M. Alexander, *J. Chem. Phys.* **65**, 2416 (1976).

53. G. C. Schatz, *Chem. Phys.* **24**, 263 (1977).
54. B. Jackson and H. Metiu, *J. Chem. Phys.* **83**, 1952 (1985).
55. C.-Y. Lee, R. F. Grote, and A. E. DePristo, *Surface Sci.* **145**, 466 (1985).
56. A. M. Richard and A. E. DePristo, *Surface Sci.* **134**, 338 (1983).
57. C. Cerjan and R. Kosloff, *Phys. Rev. B.* **34**, 3832 (1986).
58. Z. Kirson, R. B. Gerber, A. Nitzan, and M. A. Ratner, *Surface Sci.* **137**, 527 (1984).
59. R. B. Gerber, R. Kosloff, and M. Berman, *Computer Phys. Rep.* (to be published).
60. A. Messiah, *Quantum Mechanics.* North-Holland, Amsterdam, 1962), Vol. 1, Chap. 6.
61. J. Kuĉar, H.-D. Meyer, and L.S. Cederbaum (unpublished).
62. S. B. Woodruff and D. L. Thompson, *J. Chem. Phys.* **71**, 376 (1979).
63. J. A. Beswick and J. Jortner, *J. Chem. Phys.* **68**, 2277 (1978); **71**, 376 (1979).
64. J. A. Beswick and J. Jortner, *J. Chem. Phys.* **69**, 512 (1978).

# SPECTROSCOPY AND PHOTODYNAMICS OF RELATIVELY LARGE MOLECULES

JAN KOMMANDEUR

*Laboratory for Physical Chemistry, The University of Groningen, Nijenborgh 16, 9747 AG Groningen, The Netherlands*

## CONTENTS



## I. INTRODUCTION

In a series on the evolution of size effects, it is inescapable that the term "size" be further investigated. This chapter reports on some spectroscopic and photodynamic properties of *relatively* large molecules. What do we mean by large, or even more explicitly, what do we mean by relatively large?

It will be shown that this terminology is critically dependent on the way one performs experiments. Properties that are taken as characteristic for large systems appear in photodynamics when one uses a broad laser as the excitation source, while properties usually taken as characteristic for small molecules are obtained when one uses a narrow laser. In this chapter we will define "relatively large" molecules as those molecules in which, with existing optical sources, both situations can be realized. We can then study the transition from "large" to "small" in one molecule just by changing the excitation source.

The experiments we will discuss have all been performed on isolated molecules in a supersonic nozzle and in a molecular beam. No extensive experimental details will be given, since the techniques are widely known.

Experience has shown that at present the "relatively large" situation is obtained for molecules consisting of some 10 atoms. In this chapter our experimental example will be pyrazine on which the vast majority of experiments have been carried out. It is a diazine (benzene in which two para-CH groups have been replaced by N); it is aromatic; and it was chosen for its size as well as for its absorption, which is accessible with today's lasers.

We are interested in the spectroscopy and the photodynamics of pyrazine, that is, in what happens after the molecule has been excited for a limited time by a (coherent) light source, as observed by either fluorescence or some other technique that probes the evolution in time of the originally excited state.

In Section II we will give a fairly simple but exact derivation of the equations that govern the behavior of the molecule in frequency and time space.

We will review in Section III the spectroscopy of the rotationless excited electronic state of pyrazine and in Section IV the time behavior of this state after various sorts of excitations. We will then be in a position to analyze the spectra in terms of conventional spectroscopy, and we can discuss the transition from small to large in one molecule.

In Section V we will discuss the effect of rotations on the time behavior and the quantum yield of pyrazine. In Section VI we will discuss some less conventional experiments, and finally, in Section VII we will draw some conclusions and make some attempts to generalize to other relatively large molecules.

## II. THEORY

The theory of the dynamics of molecules after optical excitation has been known for a very long time. In particular, we note the contributions of Bixon and Jortner,[1] Tramer and Voltz,[2] and Robinson and Langhoff.[3] The effects of the coherence widths of the light sources used has received less attention;[4] we therefore give a rather more detailed treatment here.

We are concerned with the dynamics of a molecule after optical excitation. If the state we excited is coupled only to the radiation field, we will only see the exponential decay to that field by spontaneous emission, and this is too well known to warrant our interest. It is therefore necessary that the state we excite is coupled to other states. We then arrive at the conventional model for "radiationless transitions": a "light" (optical transition probability carrying) state $|s\rangle$ coupled to a more or less dense manifold of background states $\{|k\rangle\}$. In pyrazine, $|s\rangle$ is the $^1B_{3u}$ state and $\{|k\rangle\}$ is the set of vibronic states arising from the $^3B_{3u}$ state. In addition, the state $|s\rangle$ may be coupled to another

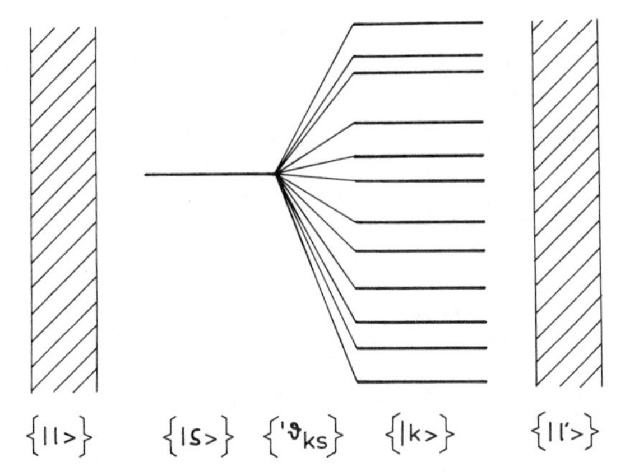


**Figure 1.** The model for radiationless decay from $|s\rangle$ to $\{|k\rangle\}$ with coupling constants $\{v_{ks}\rangle$, from $\{|k\rangle\}$ to $\{|l'\rangle\}$ and from $|s\rangle$ to $\{|l\rangle\}$.

manifold $\{|l\rangle\}$, which in pyrazine would be the set of vibronic states arising from the $^1A_g$ ground state ($|g\rangle$). Finally, the set $\{|k\rangle\}$ may also be coupled to (a different part of) $\{|l\rangle\}$. The various states and couplings are depicted in Fig. 1.

We start with the simplest case: $|s\rangle$ coupled to $\{|k\}$. Our Hamiltonian is

$$H = H_0 + \sum_k v_{sk} + \boldsymbol{\mu} \cdot \mathbf{E},$$

which has the following properties: $H_0$ generates the states $|g\rangle, |s\rangle, \{|k\rangle\}$ with orthonormal wavefunctions. The term $\sum_k v_{sk}$ connects only the states $|s\rangle$ and $\{|k\rangle\}$ with the coupling elements $\{v_{sk}\}$, while $\boldsymbol{\mu} \cdot \mathbf{E}$, the optical dipole transition, connects only the states $|g\rangle$ and $|s\rangle$.

We write the time-dependent wavefunction as a superposition of the states $|g\rangle, |s\rangle$, and $\{|k\rangle\}$:

$$\Psi(t) = \sum_n c_n(t)|n\rangle, \quad \text{where } n = |g\rangle, |s\rangle, \{|k\rangle\}. \tag{1}$$

We are interested in the part $c_s(t)$, which determines the behavior of the fluorescence in time or its Fourier transform $c_s(\omega)$, which will determine the absorption spectrum, since only the state $|s\rangle$ carries optical transition probability.

With the time-dependent Schrödinger equation

$$ih\frac{\partial\Psi(t)}{\partial(t)} = H\Psi(t),$$

we find with (1)

$$i h \sum_n |n\rangle \frac{\partial}{\partial t} c_n(t) = H \sum_n c_n(t)|n\rangle; \tag{2}$$

multiplying on the left with $|g\rangle$, $|s\rangle$, or $|k\rangle$, we obtain

$$i h \frac{\partial}{\partial t} c_g(t) = E_g c_g(t) + \mu E(t) c_s(t), \tag{3a}$$

$$i h \frac{\partial}{\partial t} c_s(t) = E_s c_s(t) + \mu E(t) c_g(t) + \sum_k v_{sk} c_k(t), \tag{3b}$$

$$i h \frac{\partial}{\partial t} c_k(t) = E_k c_k(t) + v_{sk} c_s(t), \quad \text{for all } k. \tag{3c}$$

We are interested in linear response, that is, all effects should be linear in $E(t)$, which means we should take the limit $E(t) \to 0$. Then we will have $c_g(t) \approx 1$ for all time and referring all frequencies to $w_g = 0$, we have $\partial c_g(t)/\partial t \approx 0$, which means we can neglect Eq. (3a).

To allow for spontaneous emission, we add an imaginary part to the energy of $|s\rangle$: $E_s \to E_s - i\gamma/2$; and for the electromagnetic field of the exciting (coherent) laser we write

$$\begin{aligned} E(t) &= E_0(t) \cos \omega_L t \\ &= \tfrac{1}{2} E_0(t)(e^{i\omega_L t} + e^{-i\omega_L t}). \end{aligned}$$

Substituting in (3b) we have

$$\frac{\partial}{\partial t} c_s(t) = \left(\frac{E_s}{ih} - \frac{\gamma}{2h}\right) c_s(t) + \frac{\mu E_0(t)}{ih} [\tfrac{1}{2}(e^{i\omega_L t} + e^{-i\omega_L t})] + \sum_k \frac{v_{sk}}{ih} c_k(t).$$

In the rotating frame we have

$$\begin{aligned} c_s(t) &= \hat{c}_s(t) e^{-i\omega_s t}, \quad \omega_s = E_s/h, \\ c_k(t) &= \hat{c}_k(t) e^{-i\omega_k t}, \quad \omega_k = E_k/h; \end{aligned}$$

and for simplification we take

$$\frac{\mu E_0(t)}{h} = \hat{\chi}(t), \quad \frac{v_{sk}}{h} \to v_{sk}, \quad \frac{\gamma}{h} \to \gamma.$$

Then, after some manipulation we have upon substitution

$$\frac{\partial}{\partial t}\hat{c}_s(t) = +\tfrac{1}{2}\gamma s\hat{c}_s(t) - i\hat{\chi}(t)\tfrac{1}{2}(e^{i(\omega_L+\omega_s)t} + e^{i(\omega_L-\omega_s)t}) - i\sum_k v_{sk}\hat{c}_k(t)e^{-i(\omega_k-\omega_s)t}$$

and similarly

$$\frac{\partial}{\partial t}\hat{c}_k(t) = iv_{sk}\hat{c}_s(t)e^{-i(\omega_s-\omega_k)t}.$$

The term $e^{i(\omega_L+\omega_s)t}$ oscillates very rapidly on any time scale we are interested in; therefore, we can replace it by its average, zero. Furthermore, we take

$$\omega_L - \omega_s = \omega_{Ls}$$

and

$$\omega_k - \omega_s = \omega_{ks}.$$

Then we have:

$$\frac{\partial}{\partial t}\hat{c}_s(t) = -\tfrac{1}{2}\gamma\hat{c}_s(t) - i[\tfrac{1}{2}\hat{\chi}(t)e^{-i\omega_{Ls}t}] - i\sum_k v_{sk}\hat{c}_k(t)e^{-i\omega_{ks}t} \tag{4a}$$

and

$$\frac{\partial}{\partial t}\hat{c}_k(t) = -iv_{sk}\hat{c}_s(t)e^{i\omega_{ks}t}. \tag{4b}$$

We now use the well-known Fourier transform property:

$$\int_{-\infty}^{+\infty} dt\, e^{i\omega t}\frac{\partial\hat{f}(t)}{\partial t} = i\omega f(\omega),$$

where $\hat{f}$ denotes the Fourier transform of $f$. Using this property and Fourier Transforming equations (4a) and (4b), we find

$$i\omega c_s(\omega) = -\tfrac{1}{2}\gamma c_s(\omega) - i\chi(\omega - \omega_{Ls}) - i\sum_k v_{sk}c_k(\omega - \omega_{ks}) \tag{5a}$$

and

$$-i\omega c_k(\omega) = -iv_{sk}c_k(\omega + \omega_{ks}).$$

Using the shift theorem $[\omega \rightarrow (\omega - \omega_{ks})]$ in the latter equation, we have

$$(\omega - \omega_{ks})c_k(\omega - \omega_{ks}) = v_{sk}c_s(\omega). \tag{5b}$$

Substituting (5b) into (5a) and solving for $c_s(\omega)$ yields

$$c_s(\omega) = \frac{\chi(\omega - \omega_{Ls})}{\omega - \sum_k \dfrac{v_{sk}^2}{\omega - \omega_{ks}} + \frac{1}{2}i\gamma}, \tag{6}$$

which is the central equation we need for our further considerations.

First note that apart from the requirement of linear response, Eq. (6) is the result of an exact derivation. As long as our system is represented by the appropriate part of the model of Fig. 1, all experiments in the limit of $E(t) \rightarrow 0$ should be represented by it.

Let us first study Eq. (6) in $\omega$ space. Clearly $c_s(\omega)$ is the amplitude of $|s\rangle$ as "prepared" by the laser $\chi(\omega - \omega_{Ls})$, which of course will have a certain coherence width around the frequency $\omega_{Ls}$. Further note that the denominator of Eq. (6) has poles at the frequencies $\{\omega_{ks}\}$, these poles being topped off by the radiative width $\gamma$ of $|s\rangle$. $c_s(\omega)$ has poles whenever $\omega = \sum_k v_{Ls}^2/(\omega - \omega_{ks})$, again topped off by the radiative width $\gamma$. At these poles $c_s(\omega)$ is maximal and we expect strong absorption; while at the poles of the denominator, $c_s(\omega)$ is minimal and we expect no, or at least very little, absorption.

If the width of $\chi(\omega - \omega_{Ls})$, that is, of the laser, is less than the spacing between the poles of $c_s(\omega)$, we can by scanning $\omega_{Ls}$, the central frequency of the laser, obtain a highly structured absorption spectrum. Apart from a very small correction due to $\gamma$, the maxima of this absorption spectrum correspond to the frequencies resulting from the diagonalization of the interaction matrix formed by $\omega_s$ and $\{\omega_k\}$ with the coupling elements $\{v_{sk}\}$. Apart from the interaction with the radiation field, these states (linear combinations of $|s\rangle$ and $\{|k\rangle\}$ have no further interaction. They are therefore called molecular eigenstates (MEs). Using a sufficiently narrow laser and eliminating all other sources of broadening (in particular the Doppler width), we can obtain an ME spectrum.

Obviously, using a broader laser, that is, with a width exceeding the separation of the poles of $c_s(\omega)$, we will upon scanning $\omega_{Ls}$ not obtain such a highly structured spectrum, the absorption rather presenting itself as a single broad line. It then looks as if there is just one single broad absorption.

Very roughly, one might think of the highly structured spectrum as a signature of a small molecule and of the broad spectrum as the signature of a large molecule. The latter may be further elucidated by considering Eq. (6) in the limit where the states $\{|k\rangle\}$ are so closely spaced that there is no thinkable laser that could probe between the poles of $c_s(\omega)$ or where other broadening mechanisms do not allow such probing. It is then convenient to consider $\{|k\rangle\}$ as a real continuum and replace the sum by an integral:

$$\sum_k \frac{v^2}{\omega - \omega_{ks}} \approx \langle v^2 \rangle \sum_k \frac{1}{\omega - \omega_{ks}} = \langle v^2 \rangle \int_{-\infty}^{+\infty} \frac{1}{\omega - \omega_k} \frac{dN}{d\omega_k} d\omega_k$$

$$= \langle v^2 \rangle \rho_k \int_{-\infty}^{+\infty} \frac{1}{\omega - \omega_k} d\omega_k,$$

which by integrating in the complex plane and using Cauchy's theorem yields

$$\sum_k \frac{v^2}{\omega - \omega_k} = -2\pi i \langle v^2 \rangle \rho_k,$$

where $\langle v^2 \rangle$ is the rms of the coupling and $\rho_k$ is the ($\omega$-independent) density of states $\{|k\rangle\}$. We can then write

$$c_s(\omega) = \frac{\chi(\omega)}{\omega + i(4\pi \langle v^2 \rangle \rho_k + \Gamma_r)/2},$$

which now yields a Lorentzian with width $\Gamma_{nr} + \Gamma_r$, where $\Gamma_{nr} = 4\pi \langle v^2 \rangle \rho$.

It would appear that this procedure can usually be carried out with confidence for a really dense set $\{|k\rangle\}$, which one would have in a large molecule. Therefore, a broad absorption might be taken as a signature of such a large system.

Let us now consider the behavior of $c_s$ in time space. $\hat{c}_s(t)$ can be obtained by Fourier transforming $c_s(\omega)$; $|\hat{c}_s(t)|^2$ is the magnitude of the amplitude of state $|s\rangle$. Since it is the state with the radiative property, $|\hat{c}_s(t)|^2$ determines the time behavior of the fluoresence.

Usually $c_s(\omega)$ is too complex to allow an analytical Fourier transformation, but some simple cases may be illustrative.

1. One state $|s\rangle$ not coupled to any others, except to the states of the radiation field. For the excitation we use a "white" laser, that is, $\chi(\omega - \omega_{Ls}) = C$, some constant, independent of $\omega$, denoting the strength of the interaction of molecule and light. Equation (6) becomes

$$c_s(\omega) = \frac{C}{\omega + i\gamma/2},$$

which is a Lorentzian. Its Fourier transform is an exponential:

$$\hat{c}_s(t) = ce^{-\frac{1}{2}\gamma t} \quad \text{and} \quad |\hat{c}_s(t)|^2 = C^2 e^{-\gamma t}.$$

The fluorescence decays with the radiative width, a result we would have expected.

2. Two equi-energetic states $|s\rangle$ and $|k\rangle$, coupled by $v_{sk}$. We now have for Eq. (6) (again using a "white" laser):

$$c_s(\omega) = \frac{C}{\omega - v^2/\omega + i\gamma/2}.$$

We take $\gamma \ll v$, and can then neglect the imaginary term:

$$c_s(\omega) \approx \frac{C}{\omega - v^2/\omega}$$

or

$$c_s(\omega) \approx \tfrac{1}{2}C\left(\frac{1}{\omega + v} + \frac{1}{\omega - v}\right),$$

$$\hat{c}_s(t) = FT(c_s(\omega)) = \tfrac{1}{2}C\{e^{-ivt} + e^{+ivt}\}$$

and

$$|\hat{c}_s(t)|^2 = \tfrac{1}{4}C^2(2 + e^{-i2vt} + e^{+i2vt}\},$$

$$|\hat{c}_s(t)|^2 = \tfrac{1}{2}C^2(1 + \cos 2vt).$$

The fluorescence oscillates in time with frequency $2v$. Of course, since we neglected the radiative width, it seems to oscillate forever. If we reintroduce this width, realizing that each molecular eigenstate in this symmetric case gets $\gamma/2$, we have

$$|\hat{c}_s(t)|^2 = \tfrac{1}{2}(1 + \cos 2vt)e^{-(1/2)\gamma t}.$$

The oscillation of the fluorescence is called a quantum beat. It should be noted that this quantum beat occurs in $|\hat{c}_s(t)|^2$, and therefore all other properties (as fluorescence-induced photoionization) that are determined by $|\hat{c}_s(t)|^2$ will also show this beat.

As is clear from the derivation we could also have determined the ME spectrum: in this simple case, two lines at a separation of $2v$ with a width $\gamma/2$ each, and Fourier transformed them directly. As long as the coherence width of the laser would have encompassed both states, we would have the quantum beat. On the other hand, if our laser was so narrow as to excite only one of the MEs, we would have simple exponential decay with twice the lifetime of the state $|s\rangle$.

In other words, using the broad laser we would find the quantum beat of frequency $2v$ in time, but not see it in frequency space. On the other hand, using the narrow laser we would not find the frequency $2v$ in time, but we would see it in frequency space as a separation between two absorptions.

What we have said for the laser also holds for the detection. If a monochromator would be so tuned as to allow only one component of the two MEs to be detected, we would not see a quantum beat, even though the excitation would be broad!

From our consideration of the broad band excitation of ME spectra and its subsequent observation of the difference frequency as a quantum beat, the generalization to a more complex ME spectrum is obvious:

$$\hat{c}_s(t) = \sum_i c_i e^{-i\omega_i t},$$

where $\omega_i$ is the frequency of the ME and $|c_i|^2$ is its absorption. $|\hat{c}_s(t)|^2$ will now contain all the frequency differences of the MEs excited and thus will show a complicated beat pattern. This pattern can be analyzed by Fourier transforming $|\hat{c}_s(t)|^2$ in its turn and the various frequency differences will display themselves on a $\omega$ scale. Clearly, if there are many of them, it is almost impossible to reconstruct the original ME spectrum; it becomes, at least, a complicated procedure. In this sense, directly measuring the absorption spectrum is a superior procedure.

When the $k$ manifold is very closely spaced, Eq. (6) can be transformed, as we have shown, into Eq. (7):

$$c_s(\omega) = \frac{\chi(\omega - \omega_{Ls})}{\omega + i(\Gamma_{nr} + \Gamma_r)/2}. \tag{7}$$

Now, in $\omega$ space $c_s(\omega)$ is a Lorentzian with a width of $(\Gamma_{nr} + \Gamma_r)$ and independent of the coherence width of the laser, as long as

$$\Delta\omega_c \leqq (\Gamma_{nr} + \Gamma_r).$$

Narrowing the laser will not yield any further information.

In time space $|\hat{c}_s(t)|^2$ is now a perfect exponential with decay rate $(\Gamma_{nr} + \Gamma_r)$, a clear signature of a large molecule with a dense $k$ manifold. Conversely, the observation of quantum beats is a signature of a small molecule. For a "relatively large" molecule we will have an intermediately dense manifold, which for a broad laser will show many beats. If the time detection of the system is not rapid enough to follow all these beats (the highest-frequency ones are of the order of the coherence width of the laser, which in time space may amount to a few picoseconds!), the fastest beats will be averaged out and, since they all start at $t = 0$, the result may look like an exponential. This would not be so had a narrower laser been used. We again find that a "relatively large" molecule may represent itself as a small or as a large molecule, depending on the type of experiment one performs.

It should be noted here that the use of Eq. (7) is hazardous. By assuming $\{|k\rangle\}$ to form a real continuum, one has implicitly assumed that the original photon energy put into the molecule is lost. This is clearly incorrect. Although the amplitude $c_s(\omega)$ has been so diluted over the dense background states that no fluorescence is observed, the photon energy is still in the molecule, be it in other degrees of freedom such as vibrations. There will be loss through infrared emission, but one would estimate that to occur on a much longer time scale. If, after excitation of a large isolated molecule showing a fluorescence decay in the nanosecond range, one would let it perform a chemical reaction some microseconds later, one would still expect the photon energy to manifest itself.

"Relatively large" molecules were treated by Lahmani et al.,[5] and since their theoretical results have been widely used in the literature, it is useful to summarize their work here.

They start from a number of assumptions:

1. The exciting source is "white," that is, a delta-pulse excitation in time is used. As we have stated, this amounts to replacing $\chi(\omega - \omega_{Ls})$ in Eq. (6) by a constant.
2. The coupling between $|s\rangle$ and $\{|l\rangle\}$ was assumed to be constant, or at least to vary little compared to its magnitude.
3. The density of $\{|k\rangle\}$ was assumed to be so high that a large number ($N \gtrsim 100$) of background levels was in interaction with the "light" state.
4. The quantum yield is 1, since no other irreversible decay than that to the radiation field is considered.

Under these conditions the singlet amplitude is distributed according to a Lorentzian distribution over the molecular eigenstates. Exciting with a broad (white) laser (or at least with a laser that completely spans the interaction width), one then sees in the fluorescence first the Fourier transform of the Lorentzian distribution, that is, an exponential decay. The density of $\{|k\rangle\}$ was, however, not taken to be so high as to dilute the singlet amplitude effectively to zero. It was taken to be "intermediate," which meant that each ME still had enough radiative probability so as to radiate independently, slower and therefore later. Exciting with a "white" laser therefore gives a fast and a slow decay:

$$I_f = A^+ \exp(-k^+ t) + A^- \exp(-k^- t).$$

Lahmani et al.[5] derived $A^+/A^- \approx N$, the number of levels participating in the singlet interaction, and in the simplest case $k^+ = 4\pi\langle v^2\rangle\rho_k$ and $k^- = \gamma_s^r/N$, where $\gamma_s^r$ is the radiative width of the singlet state.

Because of the intermediate nature of the density of $\{|k\rangle\}$, this theory has come to be known as intermediate level structure (ILS), and the fast decay observed is often called "dephasing." The theory appears to be completely correct, but conditions (1)–(4) are almost never completely fulfilled. The indiscriminate use of this theory for the interpretation of experimental results has therefore led to much confusion in the literature.

We need one more extension of Eq. (6). It may well be that both the states $|s\rangle$ and $\{|k\rangle\}$ are in their turn coupled to different states of a manifold $\{|l\rangle\}$ (see Fig. 1). For instance, if $|s\rangle$ is an excited singlet state $S_1$, $\{|k\rangle\}$ could denote the vibronic manifold of the electronic triplet state $T_1$ and $\{|l\rangle\}$ would then be the vibronic manifold of the ground state $S_0$.

The derivation of Eq. (6) can be straightforwardly extended to include these interactions, and as long as $|s\rangle$ and $\{|k\rangle\}$ do not decay to the same states of $\{|l\rangle\}$, the solution is simple:

$$c_s(\omega) = \frac{\chi(\omega - \omega_{Ls})}{\omega - \sum_k \dfrac{v_{ks}^2}{\omega - \omega_{ks} - \sum_l \dfrac{v_{kl}^2}{\omega - \omega_1}} - \sum_{l'} \dfrac{v_{kl'}^2}{\omega - \omega_{l'}} + \dfrac{i\gamma}{2}}, \tag{8}$$

and if we assume $\{|l\rangle\}$ to be a very dense manifold, we can replace the summations over $l$ and $l'$ by an imaginary term:

$$c_s(\omega) = \frac{\chi(\omega - \omega_{Ls})}{\omega - \sum_k \dfrac{v_{ks}^2}{\omega - \omega_{ks} + i\gamma_k/2} + i(\gamma_{nr}^s + \gamma_r^s)/2} \tag{9}$$

in the spirit of the derivation of Eq. (7), where $\gamma_k$ stands for the nonradiative decay of the states $\{|k\rangle\}$ to $\{|l\rangle\}$ and $\gamma_{nr}$ for that of $|s\rangle$ to $\{|l\rangle\}$. $\gamma_r$ is the radiative width of $|s\rangle$.

We will now review the experimental data for pyrazine, which illustrate many of the theoretical predictions made in this section.

## III. THE SPECTROSCOPY OF THE $P(1)$ ROTATIONAL MEMBER OF THE $^1B_{3u}(0–0)$ TRANSITION OF PYRAZINE

The vibronic spectrum of the first electronic transition ($^1A_g \rightarrow {}^1B_{3u}$) of pyrazine in a supersonic jet is given in Fig. 2. We focus our attention on the origin. Figure 3 shows the spectrum of the origin under much higher resolution, where the rotational lines are clearly resolved. We note here that for this parallel band only the $P(1)$ and the $R(0)$ transitions consist of single rotational lines. Because of the $\Delta J = \pm 1$ and $\Delta K = 0$ selection rules, all other $P(J)$ or $R(J)$ transitions consist of $(2J - 1)$, respectively, $(2J + 1)$ rotational lines, resulting from transitions between different sets of rotational states. Because of this "pile-up," we will in this section limit our discussion to the $P(1)$ member.

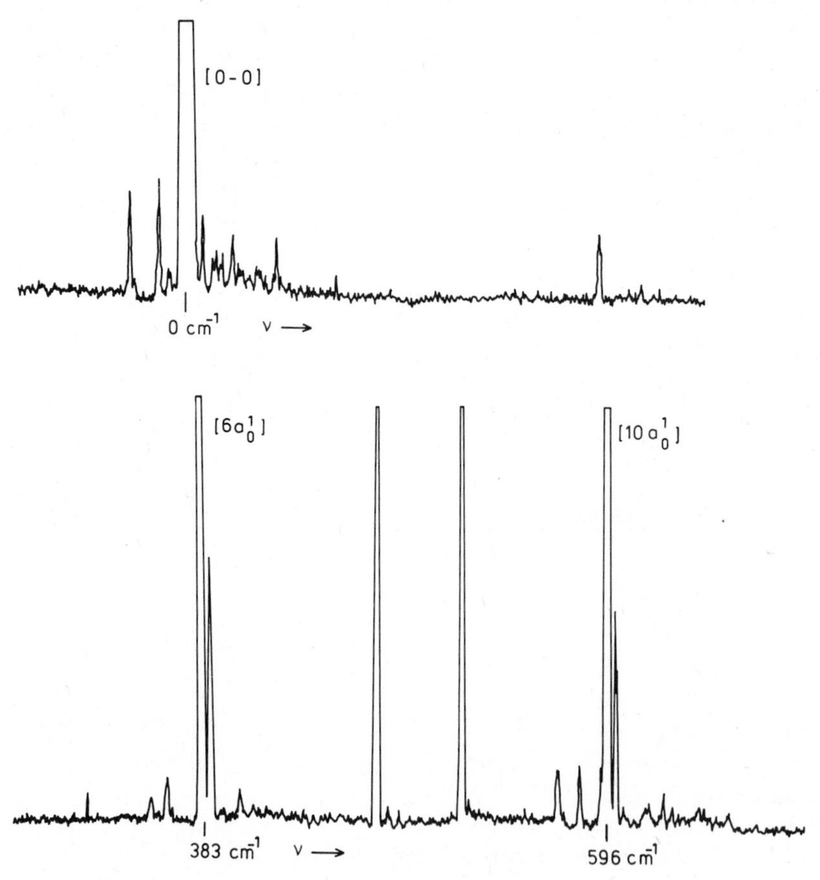


**Figure 2.** The vibronic excitation spectrum of the $^1B_{3u}$ transition in pyrazine.

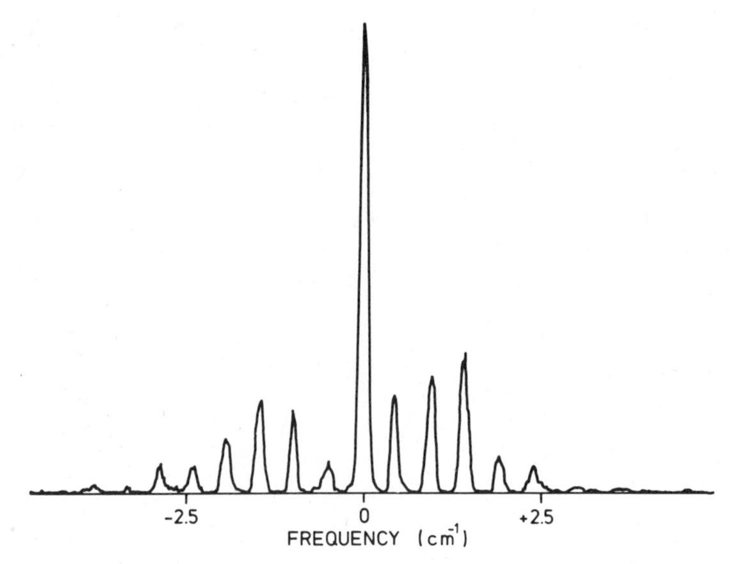


**Figure 3.** The rotational excitation spectrum of the $^1B_{3u}$ (0–0) transition of pyrazine.

By using a doubly skimmed molecular beam and a CW intracavity doubled ring laser, van der Meer et al.[6] were able to obtain the very high resolution spectrum given in Fig. 4. These absorptions occur in the region of the $P(1)$ transition and are clearly the transitions to the molecular eigenstates derived from the $J' = 0$, $K' = 0$ singlet rotational state and a set of background triplet states. Between Figs. 3 and 4 we see the transition in $\omega$ space between the large molecule and the small molecule in the spirit of our theoretical analysis. This transition comes about due to the use of a laser with a much smaller (coherence) width.

If the ME spectrum of Fig. 4 were a real absorption spectrum, it would be described by Eq. (6) and we could analyze it with that equation. The spectrum of Fig. 4 is, however, an excitation spectrum, as are most spectra taken in molecular beams. An excitation spectrum is a combination of an absorption spectrum and a quantum yield spectrum and further information is needed to disentangle the two.

Originally,[7] it was (rather blithely) assumed that the spectrum of Fig. 4 *was* an absorption spectrum, and a set of zero-order eigenstates as well as singlet–triplet coupling constants were obtained, and using the same data, more elegantly by Lawrance and Knight.[8] Later, the lifetimes of eight of the MEs of Fig. 4 were measured,[9] and it became clear that the quantum yields of the various MEs were not the same. There is clearly independent decay of the zero-order singlet and triplet states. Then Eq. (9) must be used for the analysis.

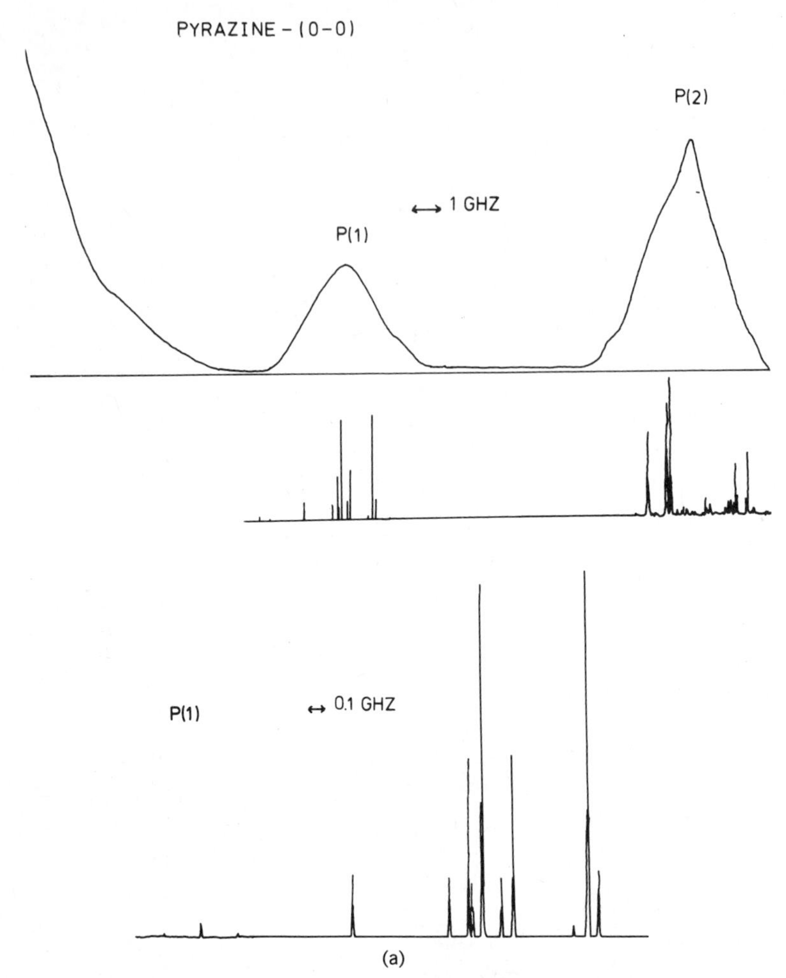


**Figure 4.** (*a*) The ME excitation spectra of *P*(1) and *P*(2). (*b*) The ME excitation spectra of *P*(1) at higher sensitivity.

Fortunately, the formalism developed by Lawrance and Knight[8] for the deconvolution into zero-order states can also be used when not only the zero-order singlet but also the zero-order triplets have a decay width, at least as long as the ME spectrum is discrete.

We then encounter a peculiar problem. In all, 36 transitions were measured in the ME spectrum. The lifetimes of only 8 could be measured, the other 28 being too weak to permit such a determination. One thus has 8 lifetimes, 36 frequencies, and 36 excitation intensities for the determination of 36 zero-

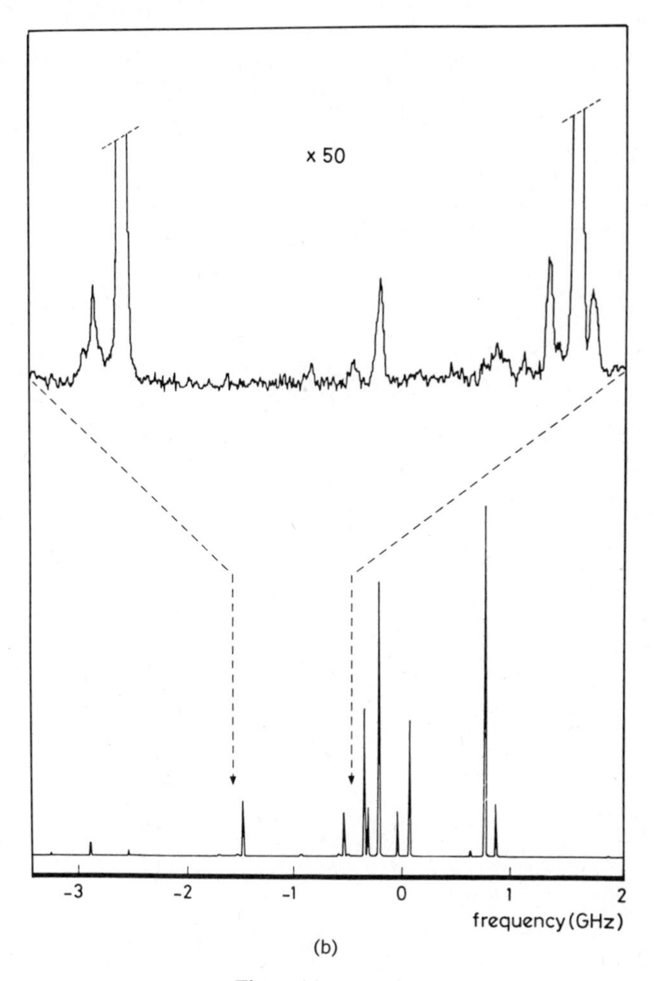


**Figure 4** (*continued*)

order states with 36 widths and 35 coupling constants. Clearly, information is lacking. On the other hand, experiment does not permit the reliable determination of the other lifetimes; therefore, we have to assume them.

We have one further piece of information, the radiative lifetime of the singlet, which has been determined from the oscillator strength of the $^1B_{3u}$ electronic transition as determined by its absorption in solution or gas phase.[10–12] The best value appears to be 290 nsec or a radiative width of 3.3 MHz.

After the deconvolution, the width of the zero-order singlet state should at

TABLE I
Energies of the Zero-Order States (Referred to the Center of Gravity of the Excitation Spectrum), Their Coupling Constants, and Their Widths for the 8-State Case and the 36-State Case[a]

| | 8 states | | | 36 states | | |
|---|---|---|---|---|---|---|
| | Energy | $v_{st}$ | $\gamma$ | Energy | $v_{st}$ | $\gamma$ |
| Singlet | | | | | | |
| | −43 | | 5.0 | −362 | | 7.3 |
| Triplets | | | | | | |
| | | | | −4705 | 270 | 2.0 |
| | | | | −4307 | 327 | 2.0 |
| | | | | −3832 | 220 | 2.0 |
| | | | | −3666 | 250 | 2.0 |
| | | | | −3216 | 267 | 1.9 |
| | | | | −2823 | 370 | 1.9 |
| | | | | −2496 | 236 | 1.9 |
| | | | | −2414 | 249 | 2.0 |
| | | | | −2319 | 265 | 2.0 |
| | | | | −1872 | 110 | 2.0 |
| | | | | −1761 | 103 | 1.9 |
| | | | | −1683 | 114 | 1.9 |
| | | | | −1533 | 39 | 1.9 |
| | | | | −1509 | 54 | 1.8 |
| | −1286 | 462 | 5.0 | −1291 | 479 | 5.6 |
| | | | | −1062 | 122 | 1.9 |
| | | | | −976 | 99 | 1.9 |
| | | | | −913 | 158 | 1.9 |
| | | | | −684 | 59 | 1.9 |
| | | | | −633 | 49 | 2.0 |
| | | | | −584 | 50 | 1.9 |
| | −502 | 119 | 1.6 | −513 | 74 | 1.0 |
| | | | | −486 | 120 | 2.6 |
| | −308 | 105 | 1.6 | −334 | 47 | 1.4 |
| | | | | −282 | 102 | 1.8 |
| | −98 | 150 | 2.7 | −89 | 160 | 2.4 |
| | 13 | 117 | 0.6 | 17 | 133 | 0.7 |
| | 463 | 457 | 3.0 | 459 | 474 | 3.9 |
| | | | | 598 | 66 | 1.8 |
| | | | | 650 | 120 | 1.5 |
| | 848 | 67 | 1.6 | 849 | 76 | 1.4 |
| | | | | 1433 | 116 | 2.0 |
| | | | | 1849 | 184 | 2.0 |
| | | | | 2702 | 245 | 2.0 |
| | | | | 2962 | 234 | 2.0 |

[a] All values are in MHz.

least be equal or exceed this value. A further constraint on the deconvolution is that all widths be of the same sign. Under these constraints the missing lifetimes must be between 450 and 700 nsec. Taking 500 nsec the system can be deconvoluted reliably, the results are given in Table I, and the composition of the ME's is given in Fig. 5 for the deconvolution of the 8 states, of which

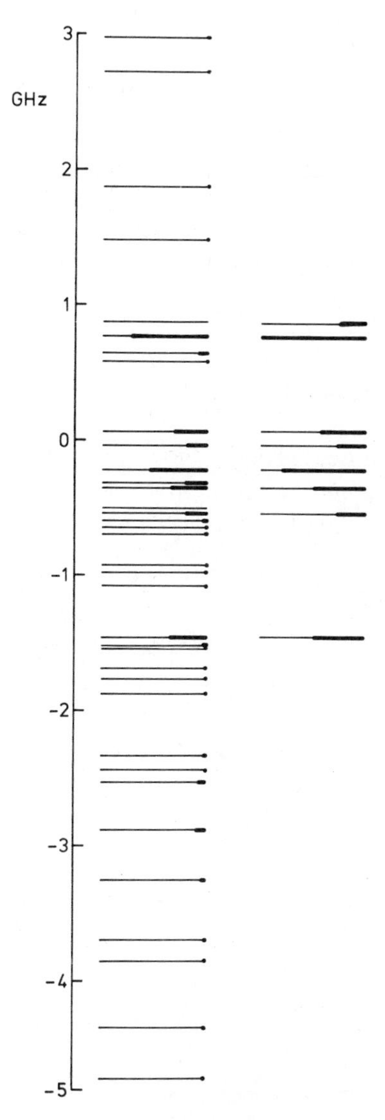


**Figure 5.** The distribution of the magnitude of the singlet amplitude over the MEs for the 36-state and the 8-state case. (A heavy full bar indicates 25% singlet.)

the lifetimes are known and for the 36 states, where 28 of the lifetimes are assumed to be 500 nsec. One notes that the inclusion of the assumed missing lifetimes leads to a minor redistribution of the singlet amplitudes.

Comparison of the results of the previous effort, where it was assumed that excitation and absorption spectrum coincided,[7] with the present results shows that there is really not much of a difference. Because of the higher sensitivity, MEs belonging to $P(1)$ are detected over a range of almost 8 GHz instead of 3.5 GHz as found previously, and in accordance with that their number has increased. We now have 35 triplet states over 7.6 GHz, which leads to a density of 140 states per $\mathrm{cm}^{-1}$. The calculated density of pyrazine triplet vibronic states around the singlet energy appears to be around 100 $\mathrm{cm}^{-1}$,[13–15] but if we want a comparison, we should take into account that they are triplets (times 3) and that nuclear symmetry permits interaction only between equal symmetry species. Table II gives the nuclear spin symmetry species of pyrazine and their statistical weight. The $J' = 0$, $K' = 0$ has the symmetry $A_g$ and therefore can only interact with $A_g$ triplet rovibronic states, which constitute $\frac{17}{48}$th of the triplet manifold. We therefore expect about $3 \times 17/48 \times 100 \approx 106$ triplet vibronic states per $\mathrm{cm}^{-1}$ to be available for the interaction, which compares favorably to the density of 140 per $\mathrm{cm}^{-1}$ as found from the density of the ME spectrum.

This also means we "see" all available triplet states without any restriction as to their vibronic nature. This must mean that 4000 $\mathrm{cm}^{-1}$ above the triplet origin all vibrations with the same nuclear spin symmetry are mixed. This may also explain the considerable variation in coupling constants observed (from 50 to 500 MHz); there is very little sign of the often assumed "democratic" distribution.

Little can be said about the nature of the zero-order states. The singlet state

TABLE II
Nuclear Statistical Weights

| $J$-States | $K_c$ States[a] | Symmetry | Weight |
|---|---|---|---|
| 0, 2, 4, ... | 0, 2, 4, ... | $A_g$ | 17 |
| | −2, −4, ... | $B_{3g}$ | 9 |
| | 1, 3, 5, ... | $B_{1g}$ | 13 |
| | −1, −3, −5, ... | $B_{2g}$ | 9 |
| 1, 3, 5, ... | 0 | $B_{3g}$ | 9 |
| | 2, 4, ... | $A_g$ | 17 |
| | −2, −4, ... | $B_{3g}$ | 9 |
| | 1, 3, ... | $B_{1g}$ | 13 |
| | −1, −3, ... | $B_{2g}$ | 9 |

[a] Axes convention: Pyrazine belongs to the molecular point group $D_{2h}$. The N–N axis is $z$, the other axis in the plane is $y$, and the axis perpendicular to the molecular plane is $x$.

is most certainly the $J' = 0$, $K' = 0$ rotational state of the $^1B_{3u}$ electronic state, but assigning the triplet vibronic states seems to be a hopeless task. Quoting D. Ramsay: "We seem to have reached the end of molecular spectroscopy here."

## IV. THE TIME DEPENDENCE OF THE FLUORESCENCE OF THE ROTATIONLESS EXCITED SINGLET STATE OF PYRAZINE

As we outlined in the theoretical section, the time dependence of the fluorescence is critically dependent on the type of exciting source used. We therefore list the experiments as a function of increasing laser width. We will first limit ourselves to the photodynamics of the $J' = 0$, $K' = 0$ rotational state of the $^1B_{3u}$ of pyrazine, since only of that state we know the ME spectrum.

Chopping their CW intracavity doubled ring laser with an electrooptical modulator van Herpen et al.[9] obtained 40 nsec pulses with a spectral width of about 25 MHz. This allowed them to study the decay of a single molecular eigenstate. Figure 6 shows a typical result of such an experiment. As might

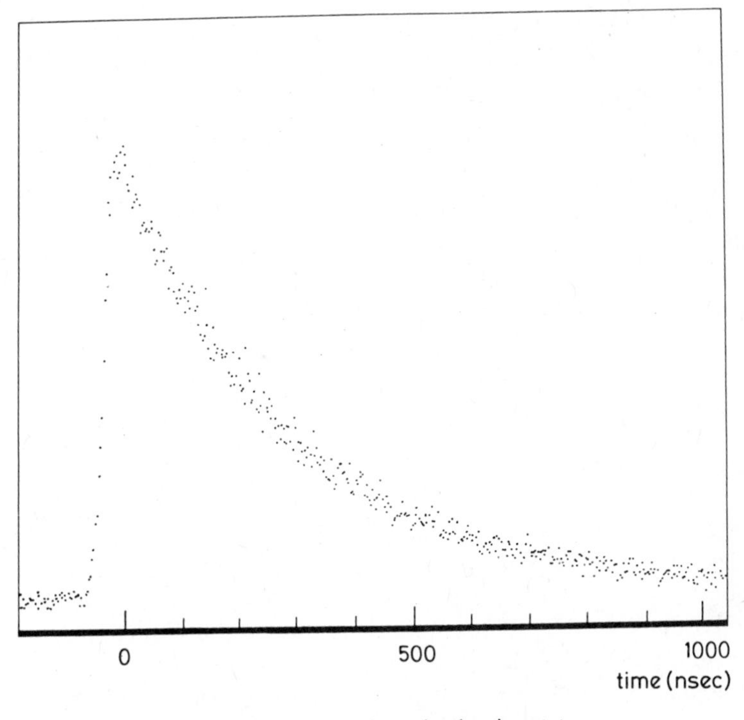


**Figure 6.** The decay of a molecular eigenstate.

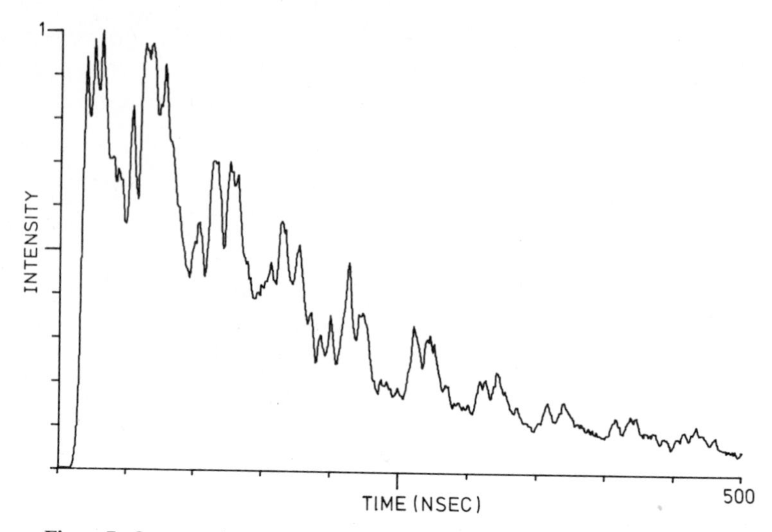


**Figure 7.** Quantum beats observed in the decay after excitation of part of $P(1)$.

be expected, pure exponential decay was observed. This was so for all MEs belonging to $P(1)$ as far as their intensity permitted such an experiment. One strong ME line in the $R(0)$ member could also be measured. Its decay showed a slight modulation with a 2-MHz quantum beat. Apparently, in that case, two transitions with a separation of 2 MHz were coherently excited.

Using a nanosecond laser with a coherence width of about 300 MHz, van der Meer et al.[16] and essentially simultaneously Okajima et al.[17] showed that quantum beats occur in the fluorescence of $P(1)$. The resulting decay is shown in Fig. 7. Of course, it should also be possible to derive this decay from the ME spectrum, using Eq. (6) and the experimental spectrum, or its deconvolution. Comparing two beating decays is hard, since small frequency mismatches lead to strong distortions of the beat pattern. Therefore, the experimental decay was Fourier transformed, the result being shown in the bottom frame of Fig. 8. Then the "theoretical" decay was obtained by placing the laser at the appropriate frequency in the ME spectrum, multiplying the square root of the intensities (the amplitudes) with the field envelope of the laser, Fourier transforming the result, and squaring it. The resulting decay was then Fourier transformed again, and the result is given in the top frame of Fig. 8. The bottom frame therefore shows $\mathrm{FT}[I(t)]$, while the top frame shows $\mathrm{FT}[\mathrm{FT}(E(\omega)\cdot I^{1/2}(\omega))]^2$. If linear response holds, the two should be equivalent, and Fig. 8 shows that this is largely true. The intensities of the peaks are not reliable, since they depend very critically on the position of the laser in the ME spectrum. Also, we should remember, as we outlined in the

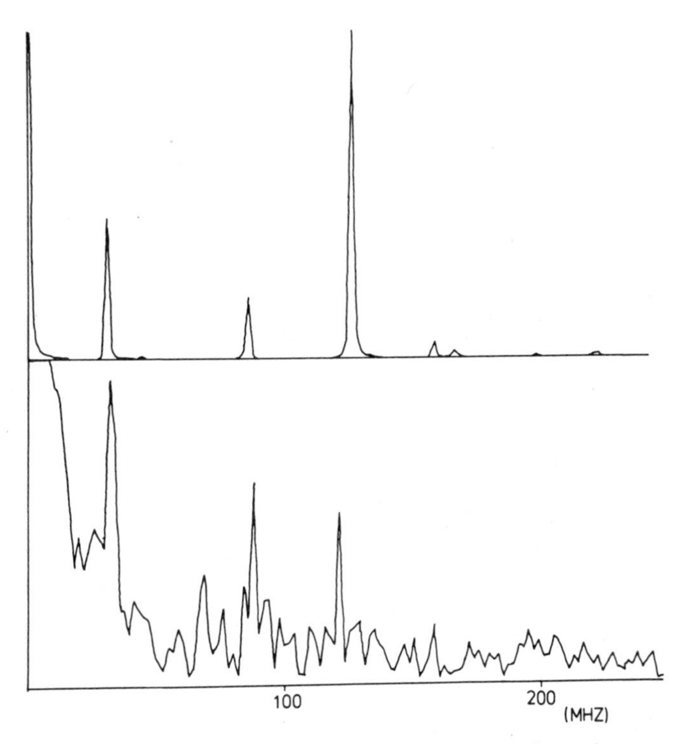


**Figure 8.** The Fourier transform of the experimental decay of $P(1)$ (bottom), compared to the Fourier transform of the decay obtained from the ME spectrum.

spectroscopy section, the excitation spectrum is close to but not equal to the absorption spectrum.

Some time later Felker et al.[13] published a result of a Fourier transformation of a beating decay obtained after picosecond excitation, that is, with a much broader laser ($\Delta\omega_c \approx 60$ GHz). Figure 9 shows a comparison of their result with the "theoretical" result, obtained as in Fig. 8. It can be concluded that again the agreement is very good, particularly if one realizes that experimentally in the decay experiment the high frequencies are somewhat attenuated.

It seems, then, that it both the spectroscopy and the photodynamics of $P(1)$ are very well understood, $\omega$ and $t$ space can be directly related to one another.

For a long time the so-called "fast component" played a confusing role. It was originally reported by Lahmani et al.[5] and gave impetus for the photodynamic study of pyrazine. It was later reported by many authors, such as Saigusa and Lim,[18] Matsumoto et al.,[19] Felker et al.,[13] and ter Horst et al.[20] Given the theory of Tramer et al.[5] it was very alluring to use the $A^+/A^-$ ratio

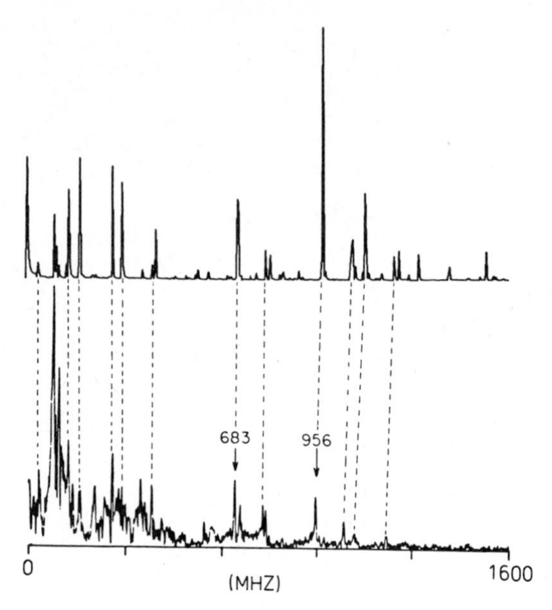


**Figure 9.** Same as Fig. 8, but now for the Fourier transform of the beats obtained with a picosecond laser (Ref. 13).

for counting the number of states coupled, and this was extensively done. However, the results from various laboratories did not seem to agree, and after the ME spectrum had become known, they certainly did not agree with the number of states as counted in that spectrum. In addition, excitation of $P(1)$ sometimes did, but sometimes did not, lead to a fast component.

These contradictory results led Jonkman et al.[21] to propose that non-resonant light scattering (NRLS) was responsible for the fast component. NRLS would yield decays that basically consisted of Raman–Rayleigh-scattered laser light together with the slower fluorescence decay. It would look like biexponential decay. Experiments where the laser was purposely detuned from the rotational line seemed to confirm their ideas.[21]

The suggestion of NRLS first led Lorincz et al.[22] and then Knee et al.[23] to study carefully the fast decay. Using picosecond lasers they could show conclusively that the fast component had a decay time of about 100 psec, which far exceeded the excitation time and therefore could not be due to NRLS. In other words, pyrazine does indeed have a real, fast component.

Very recently, the question appears to have been resolved.[24] It should be realized that the main absorptions of the MEs belonging to a rotational state occur over a very small frequency range as compared to the separations of the rotational states. Most of the spectrum is "empty." Typically, the strong

MEs occur over 3.5 MHz, spaced by 8.5 MHz. A broad laser "sees" these absorptions as a bunch, a "block," particularly for short times, when the small separations between the various lines of the block do not yet express themselves. If the laser is broad enough ($\Delta\omega_c \gg 3.5$ GHz), then Fourier transforming $c_s(\omega)$ is like Fourier transforming a block. Then $|\hat{c}_s(t)|^2$ will be like $[\sin(2\pi\Delta\nu t)/t]^2$, which for a considerable part of its range looks like an exponential. Of course, this holds for one block, but if there are many blocks, and each of them is Fourier transformed and squared, after which the results are added incoherently, the match to an exponential is even better. Many blocks are characteristic of the $Q$ branch, and since it represents the maximum of the absorption, most picosecond experiments are probably carried out there. de Lange et al.[24] followed this procedure and showed that a fast component with a decay time of 100 psec was obtained. Felker and Zewail[25] later corroborated this numerical result by using only one ($P(1)$) block. Although, as could be expected, they do obtain a fast component (with considerable recurrences at later times), their decay time looks more like 200 psec, a clear indication that the $Q$ band is responsible for most experimental results.

It should be pointed out here that the outcome of the question of the fast component is still basically Lahmani et al.'s[5] original suggestion, except that the number of states is limited, the variation of the coupling constants is considerable, and, therefore, the MEs do not show a Lorentzian distribution of the singlet amplitude. Numbers drawn from the $A^+/A^-$ amplitude may therefore not be very meaningful, also because the amplitude of the fast component will depend on the width of the laser used.

Amirav and Jortner[29] did make an attempt to redefine $N$, the number of states coupled, to $N_{\text{eff}}$, the dilution factor, which can be used when Lahmani et al.'s[5] conditions are not fulfilled. It should then also be equal to $A^+/A^-$, but their treatment takes only partially into account the effect of the coherence width of the laser. This $N_{\text{eff}}$ arose from their consideration of the quantum yield and its dependence on $J'$. In Section V we will discuss the quantum yield of the $J' = 0$, $K' = 0$ state, while in Section VI we consider its $J'$ dependence.

## V. THE QUANTUM YIELD AND THE LIFETIME OF THE ROTATIONLESS $^1B_{3u}$ STATE OF PYRAZINE

The quantum yield and the lifetime are the more common properties measured in photodynamic experiments. Both properties are in reality only defined for excitation of a single state, but, since the MEs are so closely spaced in pyrazine, it is unavoidable that experiments have been reported for quantum yields and lifetimes measured with broad lasers, encompassing more than one ME, and in this section we will pay some attention to this problem. The treatment of

TABLE III
Quantum Yields of MEs

| ME $E_i$ (in MHz) | Excitation Intensity | Lifetime (nsec) | Absorption Intensity $\|c_s^i\|^2$ | $Q = \|c_s^i\|^2 \gamma_r \tau_{ME}$ [a] |
|---|---|---|---|---|
| −1456 | 0.052 | 200 | 0.118 | 0.075 |
| −535 | 0.040 | 512 | 0.065 | 0.110 |
| −353 | 0.122 | 443 | 0.122 | 0.178 |
| −221 | 0.257 | 342 | 0.200 | 0.226 |
| −44 | 0.041 | 437 | 0.071 | 0.102 |
| +44 | 0.127 | 560 | 0.110 | 0.203 |
| +765 | 0.314 | 280 | 0.245 | 0.226 |
| +867 | 0.047 | 529 | 0.069 | 0.121 |

[a] With $\gamma_r = 3.3$ MHz.

these properties for $P(1)$ gives us a starting point for the discussion of them at higher $J'$.

We again start out with the narrow laser. The lifetimes eight MEs were measured[9] and the deconvolution led to the determination of the singlet amplitudes of the MEs. Their radiative width is then $|c_s(\omega)|^2\gamma_r^s$, where $\gamma_r^s$ is the radiative width of the singlet, which was determined to be 3.3 MHz from the absorption. The quantum yield of each ME is then determined by $Q = |c_s(\omega)|^2\gamma_r^s \cdot \tau_{ME}$, where $\tau_{ME}$ is the measured lifetime of the molecular eigenstates.

In Table III we give the position, the singlet amplitudes, and the quantum yield, when—as was discussed in Section III—only these eight states are considered in the deconvolution, which suffices to illustrate our point. Looking at Table III we see a variation of at least a factor of 3 in the quantum yields of the various strong MEs. It can be expected that the weaker MEs, of which the lieftimes are of about the same magnitude (see Section III) will have lower quantum yields, since their singlet amplitudes must be lower. The quantum yields given here for the eight measured states present an upper limit, because the singlet amplitudes of the other states were neglected, and therefore those of the eight states must be overestimated.

In any case, it is clear that one has to be careful in speaking about the quantum yield of the $J' = 0$, $K' = 0$ rotational state. When broader lasers are used, the quantum yield measured will be some intensity weighted average of those of the various MEs excited.

One could conceivably determine the quantum yield of the $J' = 0$, $K' = 0$ by ensuring that in the excitation *all* the MEs of $P(1)$ are equally excited (and no others!). Then one "prepares" the complete $J' = 0$, $K' = 0$ state at $t = 0$, and one can speak of the quantum yield of that state.

This number can be calculated from Eq. (8) by substituting in it all the zero-order energies and widths and the coupling constants as listed in Table I. $|\hat{c}_s(t)|^2$ is then found by Fourier transforming and subsequent squaring of $c_s(\omega)$. Integration of $\gamma_r^s |\hat{c}_s(t)|^2$ over all time then yields the total amount of light emitted, and dividing this number into $\gamma_r^s$ (the amount of light emitted if the state $|s\rangle$ were not coupled to any others), yields the quantum yield of $P(1)$.

Using this procedure one finds $Q(P_1) \approx 0.15$, with some uncertainty, since the lifetimes of the weaker MEs were assumed. This number, however, gives us a calibration point for the quantum yields at higher $J'$. In view of the uncertainties involved, it is in good agreement with the value of 0.22 estimated by Amirav and Jortner[29] from their experiment.

The situation for the lifetimes is simpler. Of course, again a narrow laser yields the lifetimes of the MEs directly, which are seen to vary considerably (cf. Table I). When, with a broader laser, some MEs are excited simultaneously, one obtains quantum beats, and the overall lifetime will be some intensity weighted average of the lifetimes of the MEs. If measured accurately, the widths of the beat frequencies after Fourier transformation should be the intensity weighted average of the widths of the MEs involved in the beat.

Exciting *all* MEs with a very broad laser will first give the fast ("dephasing") decay, discussed in Section IV and at longer times we obtain—in addition to a complicated beat pattern—an "exponential" representing the incoherent intensity weighted average of the various exponentials given by the MEs when they are excited individually. It seems clear that we can, from the spectroscopic and the lifetime measurements of the MEs, completely understand the behavior of the photodynamics of the rotationless ($J' = 0$, $K' = 0$) singlet $^1B_{3u}$ state of pyrazine.

A similar effort could conceivably be made for the $J' = 1$, $K' = 0$ state, since its ME spectrum can also be directly measured, because $R(0)$ also contains only one rotational transition. Unfortunately, the ME spectrum in this case consists of one strong and many weak transitions to MEs, permitting only one lifetime measurement, and therefore—as we pointed out in Section III—the spectroscopic analysis is not very meaningful at the present.

For higher $J'$ levels we have the problem of overlapping transitions—the MEs cannot be picked apart and only a more qualitative analysis of the behavior of the lifetime and the quantum yield is possible. In the next section we will make such an effort.

## VI. THE LIFETIMES AND THE QUANTUM YIELDS OF HIGHER $J'$ STATES OF THE $^1B_{3u}$ STATE OF PYRAZINE

The experimental situation for higher $J'$ states seems fairly clear. Okajima et al.[17] were the first to note that the (average) slow lifetime of the fluorescence was essentially independent of $J'$,[18] which was later corroborated by Matsumoto et al.[26] The relative quantum yield as a function of $J'$ in a room-temperature experiment was first reported by Baba et al.,[27,28] and later Amirav and Jortner[29] published a "laser-free" but higher-resolution and absolute quantum yield spectrum taken in a high-density, fairly high temperature (30 K) supersonic beam. The latter result is reproduced in Fig. 6. Any analysis shows that the quantum yield is a very strong function of $J'$, decreasing as $J'$ increases. Amirav and Jortner[29] were successful in fitting the high $J'$ behavior of the quantum yield to $(2J + 1)^{-1}$.

Very recently, de Lange et al.[30] carefully studied the intensities of the low $J'$ manifold of pyrazine with a nanosecond laser, which averages reasonably well over the quantum yields of the MEs. Taking a clue from the beautiful results obtained on benzene by Riedle et al.,[31] they fitted the intensities to a quantum yield determined by Coriolis coupling of $S_1$ to $\{S_0\}$. They assumed a Boltzmann distribution in the ground state, except for the $J'' = 0$, $K'' = 0$ state, which for nuclear symmetry reasons can only be reached by a $\Delta J'' = -2$ transition.

Furthermore, it became obvious that some $J'$-dependent quantum yield had to be introduced into the equations for the simulation of the spectrum. Coriolis coupling gave a good fit with

$$\Gamma_z = A_z K^2 \quad \text{with } A_z = (0.2 \pm 0.1)\Gamma,$$

$$\Gamma^+ = B_+(J - K)(J + K + 1), \quad B_+ = (0.25 \pm 0.05)\Gamma,$$

$$\Gamma^- = B_-(J + K)(J - K + 1), \quad B_- = (0.05 \pm 0.05)\Gamma,$$

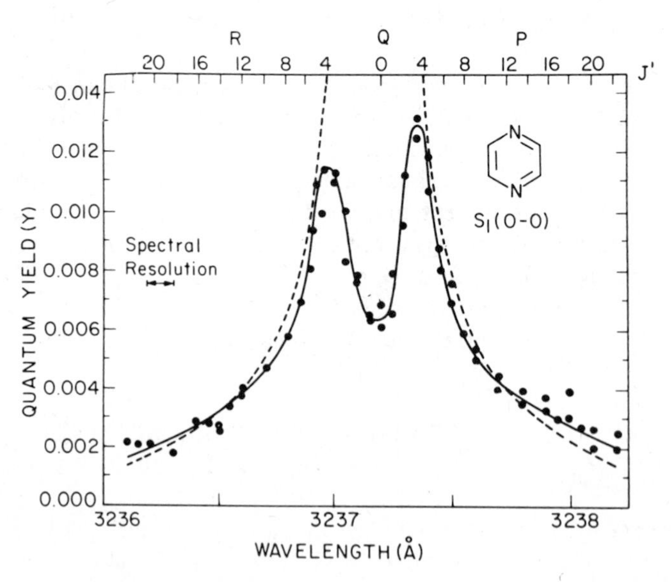


Figure 10. The dependence of the quantum yield on $J'$ as reported by Amirav and Jortner.[29]

where $\Gamma_z$, $\Gamma^+$, and $\Gamma^-$ are the radiationless rates due to the diagonal and off-diagonal Coriolis coupling; $A_z$, $B_+$, $B_-$ are the Coriolis constants, and $\Gamma$ is the inverse lifetime of the rotationless singlet state ($\Gamma \approx 5$–$10$ MHz).

With these parameters the low $J'$ rotational spectrum could very well be fitted, as is displayed in Fig. 11. Finally, the higher-temperature results of Amirav and Jortner[29] could be calculated as well as those of Baba et al.[27,28] as is shown in Figs. 12 and 13. For low $J'$ values this procedure is successful. But as can be seen from the comparison of Figs. 10 and 12, it fails for higher

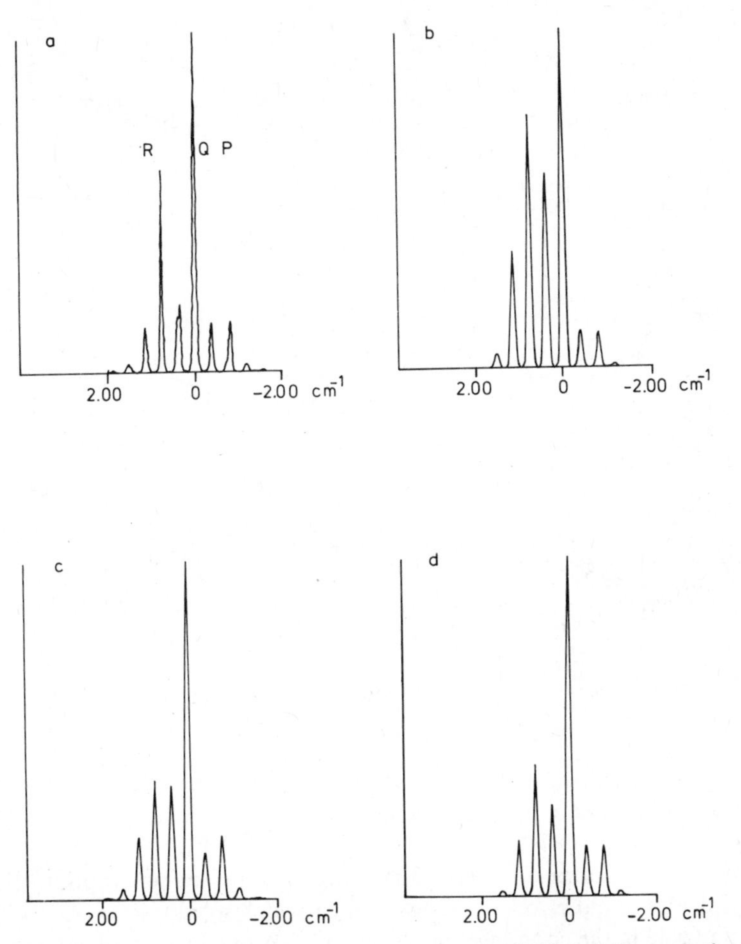


**Figure 11.** (*a*) The experimental rotational excitation spectrum. (*b*) The rotation excitation spectrum calculated with $J'$, $K'$ independent quantum yield. (*c*) The rotational excitation spectrum calculated with Coriolis coupling. (*d*) The rotational excitation spectrum calculated with Coriolis coupling *and* the $J'' = 0$, $K'' = 0$ state frozen in at 20%.

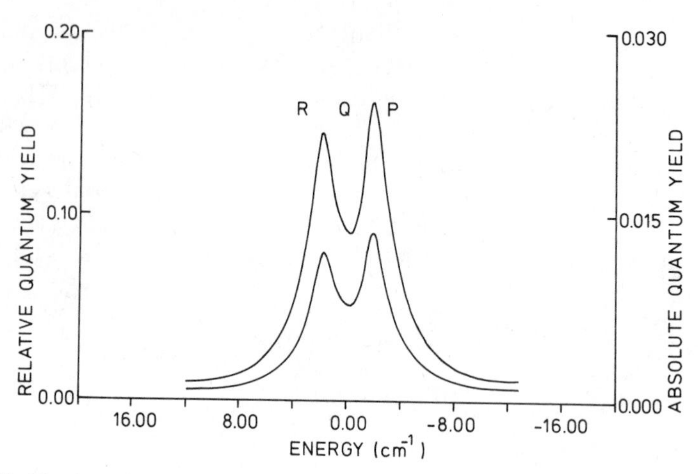


**Figure 12.** The dependence of the quantum yield on $J'$ as calculated with Coriolis coupling for high $J'$ (cf. Fig. 10) for the lowest (top) and highest (bottom) values of the Coriolis constants derived.

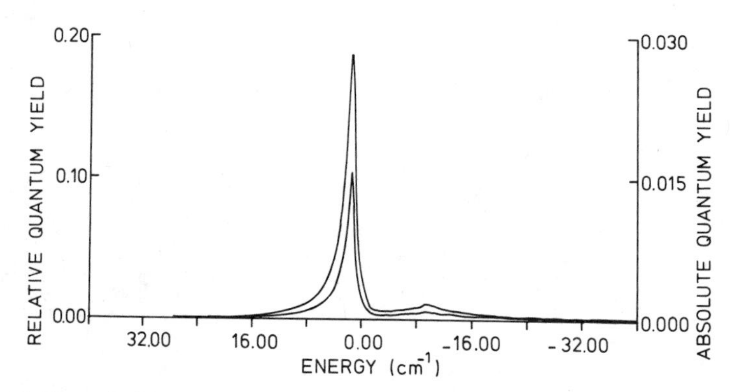


**Figure 13.** The calculated quantum yield as a function of energy for a room-temperature measurement (cf. Refs. 27 and 28).

$J'$, since Coriolis coupling gives a dependence on $(J')^{-2}$, while experiment then yields $(J')^{-1}$.

Amirav and Jortner[29] resolved this question by the assumption that in the triplet manifold $K$ mixing would occur, which would break the $\Delta K' = 0$ selection rule in the coupling and let the number of triplets coupled increases as $(2J' + 1)$. Such a suggestion was made earlier by Novak and Rice.[32] If the number of triplets increases, the singlet amplitude in the MEs must decrease, and, therefore, the radiative lifetime of the MEs decreases. Very soon the ME lifetimes are dominated by the triplet decay, and if it is independent of $J'$, the quantum yield decreases as $(2J' + 1)^{-1}$. This explanation has the advantage that it also leaves the overall lifetimes independent of $J'$[18] (they are essentially $J'$ independent triplet lifetimes),[18] which is otherwise hard to reconcile with a decreasing quantum yield.

It must be pointed out, however, that the explanation rests on two as yet unproven assumptions: the $K$ mixing in the triplet and the $J'$ independent triplet decay. It should be pointed out that for the latter there is some theoretical justification, Novak and Rice[33] having shown that Coriolis coupling between triplets and singlets is expected to be small. There is no further evidence for $K$ mixing in the triplet. At least for the lowest $J'$ levels it is absent. Counting the number of transitions in the ME spectra van der Meer et al.[7] showed that the number of triplets coupled is constant, at least up to $J' = 4$. Although the number of lines in the ME spectrum increases with $J'$, it increases as $(2J'' - 1)$ for the $P$ and as $(2J'' + 1)$ for the $R$ branch, which is just equal to the number of superposed independent transitions. Each $J'$, $K'$ state for $J' \leqq 4$ is therefore coupled to the same number of triplets.

Analysis of symmetry also leads to this conclusion. From Table II it can be noted that the lowest $J'$, $K'$ states are mostly of different nuclear symmetry and therefore cannot mix.[30] This symmetry selection rule is considerably relaxed at higher $J'$, there being many more $K'$ states of the same symmetry.

The picture evolving then seems that at low $J'$ the decay can be understood on the basis of $(S_1 - S_0)$ Coriolis coupling and that at higher $J'$, triplet decay dominates because more and more triplets are coupled to the rotational states.

This suggestion can also reconcile the behavior of the $A^+/A^-$ ratio, that is, the ratio of the amplitudes of the fast and slow decay components, as a function of $J'$. Usually it is reported that $A^+/A^-$ increases linearly with $J'$,[18,26] although very recently Terazima and Lim[34] appear to prefer a $J^2$ fit for the ${}^1B_{3u}(0\text{–}\nu_{10}^{a}{}_{0})$ transition. As we pointed out in Section IV, the theory of Lahmani et al.[5] should be taken numerically with a "grain of salt." However, it *is* true that one would expect $A^+/A^-$ to increase as the number of triplets increases. This can be understood in two ways:

1. One can use Lahmani et al.'s theory and call $A^+/A^- = N$, the number of states coupled, or use Amirav and Jortner's[29] modification of $N$ into $N_{\text{eff}}$.
2. One can also realize that the "dephasing" happens on a very short time scale, where nonradiative loss to $\{S_0\}$ has not yet occurred. The fast component must have a "quantum yield" of about 1. The slow component, however, is measured at much later times, where radiationless loss to $\{S_0\}$ has been considerable and, therefore, its "quantum yield" is much lower. One would then expect $A^+/A^-$ and the quantum yield to have the same behavior as a function of $J'$, and that is what generally appears to have been observed.

A quantitative resolution of all these questions can be derived by the use of Eq. (9), which takes into account all these effects. Unfortunately, however, this requires knowledge of the ME spectra of all $J'$, $K'$ states, which is not and probably will not become available.

It seems, then, that a qualitative interpretation of the $J'$ dependence of the various experiments should suffice, but there are two rather more unusual experiments on pyrazine, which deserve our attention.

Felker et al.[13] found that the application of a magnetic field of a fairly small magnitude ($\approx 0.01$ T) increases the $A^+/A^-$ ratio by a factor of 3. This result was later corroborated by Matsumoto et al.[19] Their explanation was fairly simple. The magnetic field breaks the selection rule $\Delta J = 0$, allows couplings with $\Delta J = \pm 1$, as well. Therefore, the number of states coupled goes up by a factor of 3 and therefore $A^+/A^-$ as well.

On the simple basis of Lahmani et al.'s original theory,[5] one would then expect the lifetime of the slow component to decrease by a factor of 3 as well, which has not been observed. If, however, the slow decay is dominated by the zero-order triplet decay, then indeed the slow lifetime is not affected, and therefore this experiment gives further evidence for the triplet dominance at higher $J'$.

Finally, an interesting experimental result was reported by Matsumoto et al.[35] At the $P(2)$ frequency they used polarized light in the excitation. Naturally, the emission was also polarized with a polarization ratio $P = [I_{\parallel}(t) - I_{\perp}(t)]/[I_{\parallel}(t) + I_{\perp}(t)]$ of about 20% at $t = 0$. During the slow decay, however, this polarization ratio decreased to zero slowly. The effect was not affected by a magnetic field and was similar for other $J'$ transitions.

Of course, collisions could be responsible for the depolarization, but the authors feel their experiments are not affected by them. It could be, they point out, that the depolarization is due to the different frequencies of the MEs that are excited, or even to the different frequencies of the $M_j$ states that are coherently prepared. Although the effects are small, it may be that further information on the high $J'$ states can be obtained from this type of experiment, particularly if circularly polarized light is used.

Summarizing this section it can be stated that, as pyrazine is excited to a higher $J'$ state, the number of levels to which the rotational state is coupled increases strongly; the "small" molecule at $J' = 0$ starts to look like a "large" molecule at high $J'$. Again, we see the transition from small to large within one system, now by exciting the rotational degree of freedom.

## VII. CONCLUSIONS

We have shown that "relatively large" is a very relative characterization. Depending on the experiments one performs; a 10-atom molecule such as pyrazine may in its spectroscopic or photodynamic behavior look small or large. If lasers were narrow enough and other sources of broadening, such as Doppler, could be eliminated, much larger molecules might also be characterized as small. In that sense, there seems to be no *a priori* boundary between large and small in the spectroscopic or photodynamic sense.

Pyrazine, on which most experiments pertaining to this problem have been performed, was a good example to illustrate the problem. Are there other molecules with similar behavior?

The closest relative of pyrazine is pyrimidine, where two CH's of benzene are replaced by N's not in the para-position as in pyrazine, but in the meta-position. A recent publication by Meerts and Majewski[36] of the ME spectrum of pyrimidine shows that this compound is already a borderline case. With a singlet–triplet gap of 2000 $cm^{-1}$, instead of 4000 $cm^{-1}$ as in pyrazine, the density of triplet vibronic states is already so low that only very occasionally is a singlet rotational state close enough to a rovibronic triplet state to yield considerable interference. Of course, this will be different at higher energies in the singlet state. A vibronic state of pyrimidine with some 2000 $cm^{-1}$ of energy should show a behavior similar to that of pyrazine.

Much significant work has been performed by Schubert et al,[37] on the so-called channel III in benzene. Apparently, however, the density of states into which the rotational states are decaying is so high that the few megahertz resolution, which they obtain in their Doppler-free experiments is not enough to show up "small"-molecule behavior in this "large" molecule.

Another case in point is the six-atom molecule glyoxal. van der Werf et al.[38] reported biexponential decay in this compound and the studies of Pebay-Peyroula et al.[39] in recent years have shown that this molecule should have properties very similar to pyrazine. Their interest, however, has been largely focused on the small-molecule behavior.

In general, it seems that if one starts with a sufficiently small molecule, at some energy it will always behave as a "relatively large" molecule. It then depends on the optical accessibility of that energy, on the nature of the lasers available at that energy, and on the magnitudes of the couplings between the various "light" and dark states, whether the properties outlined in this chapter can be observed.

In pyrazine all the conditions appear to be fulfilled, it therefore served as a useful example for the understanding of what we mean by "relatively large."

**Acknowledgments**

I could not have pursued my interest in this problem without the help of my colleagues, Leo Meerts, Harry Jonkman, and Karel Drabe, nor without the enthusiastic cooperation of my students Gerard ter Horst, Niels van der Meer, and Pieter de Lange. I gratefully acknowledge them. In addition, I benefited from extensive discussions with Joshua Jortner, André Tramer, and Stuart Rice.

## References


1. M. Bixon and J. Jortner, *J. Chem. Phys.* **48**, 715 (1968).
2. A. Tramer and R. Voltz, in *Excited States* (E. C. Lim, ed.). Academic, New York, 1978, Vol. 4, p. 281.
3. G. W. Robinson and C. A. Langhoff, *Chem. Phys.* **5**, (1974).
4. W. Rhodes, *Chem. Phys.* **22**, 95 (1977).
5. F. Lahmani, A. Tramer, and C. Tric, *J. Chem. Phys.* **60**, 4431 (1974).
6. B. J. van der Meer, H. Th. Jonkman, J. Kommandeur, W. L. Meerts, and W. A. Majewski, *Chem. Phys. Lett.* **92**, 565, (1982).
7. B. J. van der Meer, H. Th. Jonkman, and J. Kommandeur, *Laser Chem.* **2**, 77 (1983).
8. W. D. Lawrance and A. E. W. Knight, *J. Phys. Chem.* **89**, 917 (1985).
9. W. M. van Herpen, W. L. Meerts, K. E. Drabe, and J. Kommandeurr *J. Chem. Phys.* **86**, 4396 (1987).
10. W. J. Schutten (private communication, 1985).
11. K. Nakamura, *J. Am. Chem. Soc.* **93**, 3138 (1971).
12. K. K. Innes, J. P. Byrne, and I. G. Ross, *J. Mol. Spectrosc.* **12**, 125 (1967).
13. P. M. Felker, W. R. Lambert, and A. H. Zewail, *Chem. Phys. Lett.* **89**, 309 (1982).
14. A Frad, F. Lahmani, A. Tramer, and C. Tric, *J. Chem. Phys.* **60**, 4419 (1974).
15. D. B. MacDonald, G. R. Fleming, and S. A. Rice, *Chem. Phys.* **60**, 335 (1961).
16. B. J. van der Meer, H. Th. Jonkman, G. M. ter Horst, and J. Kommandeur, *J. Chem. Phys.* **76**, 2099 (1982).
17. S. Okajima, H. Saigusa, and E. C. Lim, *J. Chem. Phys.* **76**, 2096 (1982).
18. H. Saigusa and E. C. Lim, *J. Chem. Phys.* **78**, 91 (1983).
19. Y. Matsumoto, L. H. Spangler, and D. W. Pratt, *J. Chem. Phys.* **80**, 5539 (1984).
20. G. ter Horst, D. W. Pratt, and J. Kommandeur, *J. Chem. Phys.* **74**, 3616 (1981).
21. H. Th. Jonkman, K. E. Drabe, and J. Kommandeur, *Chem. Phys. Lett.* **116**, 357 (1985).
22. A. Lorincz, D. P. Smith, F. Novak, R. Kosloff, D. J. Tannor, and S. A. Rice, *J. Chem. Phys.* **82**, 1067 (1985).
23. J. L. Knee, F. E. Doany, and A. H. Zewail, *J. Chem. Phys.* **82**, 1042 (1985).
24. P. J. de Lange, K. E. Drabe, and J. Kommandeur, *J. Chem. Phys.* **84**, 538 (1986).
25. P. M. Felker and A. H. Zewail; *Chem. Phys. Lett.* **128**, 221 (1986).
26. Y. Matsumoto, L. H. Spangler, and D. W. Pratt, *Chem. Phys. Lett.* **98**, 333 (1983).
27. H. Baba, U. Fujita, and K. Kekida, *Chem. Phys. Lett.* **73**, 425 (1980).
28. H. Baba, N. Ohta, O. Sekicku, M. Jujita, and K. Kekida, *J. Chem. Phys.* **87**, 943 (1983).
29. A. Amirav and J. Jortner, *J. Chem. Phys.* **84**, 1500 (1986).
30. P. J. de Lange, B. J. van der Meer, K. E. Drabe, J. Kommandeur, W. L. Meerts, and W. A. Majewski *J. Chem,Phys.* **86**, 4004 (1987)
31. E. Riedle, H. J. Neusser, and E. W. Schlag, *Faraday Disc. Chem. Soc.* **75**, 387 (1983).
32. F. A. Novak and S. A. Rice, *J. Chem. Phys.* **71**, 4680 (1979).
33. F. A. Novak and S. A. Rice, *J. Chem. Phys.* **73**, 858 (1980).
34. M. Terazima and E. C. Lim, *Chem. Phys. Lett.* **127**, 330 (1986).
35. Y. Matsumoto, L. H. Spangler, and D. W. Pratt, *Chem. Phys. Lett.* **95**, 343 (1983).
36. W. L. Meerts and W. A. Majewski, *Laser Chem.* **6**, 339 (1986).
37. U. Schubert, E. Riedle, H. J. Neusser, and E. W. Schlag, *J. Chem. Phys.* **84**, 6182 (1986).
38. R. van der Werf, D. Zevenhuyzen, and J. Kommandeur, *Chem. Phys. Lett.* **27**, 325 (1974).
39. E. Pebay-Peyroula, R. Jost, M. Lombardi, and J. P. Pique, *Chem. Phys.* **106**, 243 (1986).

# SOLVATION EFFECTS IN FOUR-WAVE MIXING AND SPONTANEOUS RAMAN AND FLUORESCENCE LINESHAPES OF POLYATOMIC MOLECULES

SHAUL MUKAMEL

*Department of Chemistry, University of Rochester, Rochester, New York 14627*

## CONTENTS



## I. INTRODUCTION

Four-wave mixing (4WM) processes[1-6] and spontaneous Raman and fluorescence (SRF) lineshapes[7-12] provide a sensitive spectroscopic probe for polyatomic molecules in condensed phases. A 4WM process involves the interaction of three laser fields with wavevectors $\mathbf{k}_1$, $\mathbf{k}_2$, and $\mathbf{k}_3$ and frequencies $\omega_1$, $\omega_2$, and $\omega_3$, respectively, with a nonlinear medium. A coherently generated signal with wavevector $\mathbf{k}_s$ and frequency $\omega_s$ is then detected (Fig. 1), where

$$\mathbf{k}_s = \pm\mathbf{k}_1 \pm\mathbf{k}_2 \pm \mathbf{k}_3 \tag{1a}$$

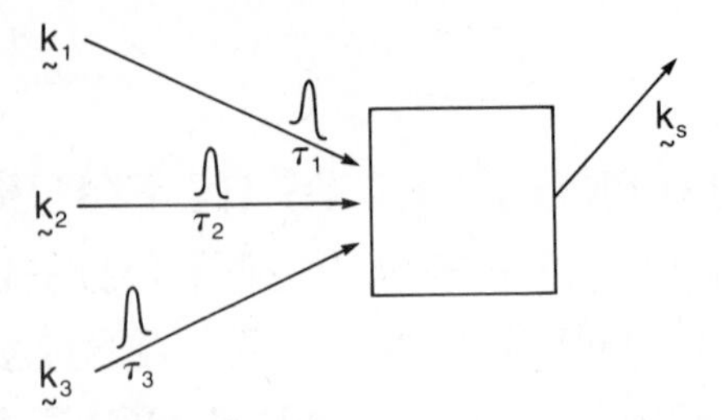


**Figure 1.** The general four-wave mixing process. The coherent $\mathbf{k}_s$ signal is generated by a nonlinear mixing of the three incoming fields $\mathbf{k}_1$, $\mathbf{k}_2$, and $\mathbf{k}_3$, at times $\tau_1$, $\tau_2$, and $\tau_3$.

and

$$\omega_s = \pm\omega_1 \pm \omega_2 \pm \omega_3. \tag{1b}$$

Equations (1) imply that $\mathbf{k}_s$ and $\omega_s$ are given by any linear combination of the incoming wavevectors and frequencies. The various types of 4WM processes differ by the particular choices of $\mathbf{k}_s$ and $\omega_s$ [i.e., the particular choice of signs in Eqs. (1)], and by the temporal characteristics of the incoming fields. A distinction is traditionally made between frequency-domain, stationary 4WM, whereby the incoming fields and the signal field are stationary,[6,13–21] and time-domain 4WM, whereby the incoming fields are infinitely short pulses. The most common time-domain 4WM techniques are photon echoes,[22–25] transient grating,[26–33] and time-resolved coherent Raman spectroscopy.[34–43] Realistic pulsed experiments involving pulses with finite duration are characterized by a finite spectral and temporal resolution and are intermediate between these two ideal frequency-domain and time-domain limits. The SRF experiment[7–12,44–56] is illustrated in Fig. 2. It involves exciting the system with one incoming field (pulsed or steady state) and monitoring the time- and frequency-resolved emission. 4WM is a phase-matched process in which a macroscopic polarization is created, and the molecules in the sample emit in phase. This results in the directionality of the signal, as given by Eq. (1a). The SRF, on the other hand, is not phase matched, and the light is emitted in all directions.

In this chapter, we develop efficient methods for the calculation of 4WM and SRF processes of large polyatomic molecules in condensed phases (e.g., solution, solid matrices, and glasses). The key quantity in the present formulation is the nonlinear response function $R(t_3, t_2, t_1)$, which contains all the microscopic information relevant for any type of 4WM and SRF.[6,11,12,19,20,57] In Section II we introduce the nonlinear response function and derive the general formal expression for 4WM. The two ideal limiting cases of time-

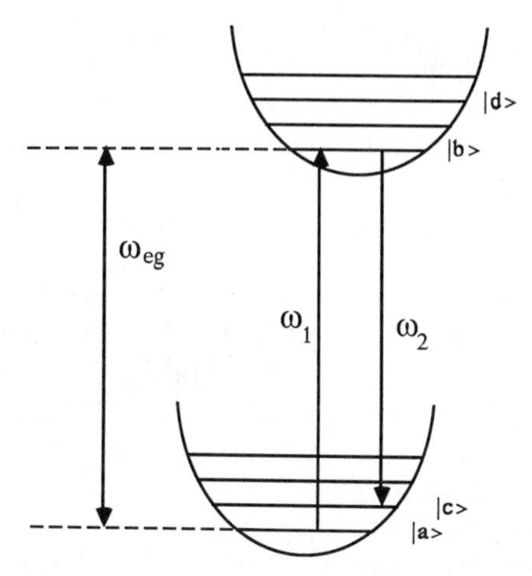


**Figure 2.** The level scheme for spontaneous Raman and fluorescence (SRF) spectra. $|a\rangle, |c\rangle, \ldots,$ denote vibronic states belonging to the ground-state manifold, and $|b\rangle, |d\rangle, \ldots,$ denote vibronic states belonging to the electronically excited manifold. In a SRF experiment, an $\omega_1$ photon is absorbed, and an $\omega_2$ photon is emitted.

domain 4WM and frequency-domain 4WM are obtained from the same unified expression. The nonlinear susceptibility $\chi^{(3)}$ that is commonly used in stationary 4WM is also expressed in terms of $R$. In Section III we calculate the response function for a general model system for molecular 4WM.[6,19,44,45] The model consists of two manifolds of vibronic levels coupled to a thermal bath, which causes dephasing. In general, the calculation involves a four-point correlation function of the dipole operator. This expression can also be written in Liouville space (i.e., in terms of the time evolution of the density matrix). The latter form is very useful for deriving approximate methods for evaluating the nonlinear response function. In Section IV we introduce the factorization approximation.[11,19] This approximation allows us to express any 4WM process in terms of generalized single-photon lineshape functions related to two-time correlation functions of the dipole operator. In Section V we define the time- and frequency-resolved spectrum in SRF. We show that these lineshapes may be expressed in terms of the same four-point correlation function of the dipole operator which appears in 4WM. Our final results are expressed further in terms of Liouville space propagators. Invoking again the factorization approximation results in a simple, easily calculable expression for SRF. The closed expressions derived in Sections III–V involve multiple summations (four) over molecular vibronic levels. Such summations restrict the usage of these results to relatively small molecules with a few relevant levels. In Section VI, we specialize to harmonic molecules. In this case, it is possible to perform two out of the four summations using Green's function techniques in the time domain.[44,45,58] This results in an expression for 4WM and SRF in terms of generalized ground-state and excited-state polarizability operators. The latter are evaluated in the time-domain using Fourier transform techniques. This result is extremely useful and allows a convenient calculation of the response functions, both in the time domain and in the frequency domain. In Section VII we apply these results toward the calculation of coherent and spontaneous Raman excitation profiles in simple model systems.[45] Detailed numerical calculations provide a comparison of these lineshapes for a stochastic model of line broadening. In Section VIII we consider the effects of vibrational relaxation on SRF lineshapes and provide a simple computational scheme for incorporating these effects using the methodology of the previous sections.[59] In Section IX, we apply the model to "intramolecular solvation" and consider the SRF spectra of ultracold anthracene in a supersonic beam.[60,61] In Section X, we carry the Green's function techniques one step further and perform all the summations over vibronic states.[45] A closed expression is obtained for the nonlinear response function in the time domain, in which all the summations over molecular states have been eliminated [Eq. (57) together with Eq. (133)]. This expression is particularly useful for ideal time-domain experiments. Finally, our results are summarized in Section XI.

## II. THE NONLINEAR RESPONSE FUNCTION FOR FOUR-WAVE MIXING

We consider a nonlinear optical medium, consisting of polyatomic molecules interacting with a classical external electromagnetic field by a dipolar interaction. The total Hamiltonian of the system is

$$H_T = H + H_{\text{int}}. \tag{2}$$

Here $H$ is the Hamiltonian for the material system in the absence of the radiation field. $H_{\text{int}}$ represents the radiation–matter interaction and is given by

$$H_{\text{int}}(t) = \sum_{\alpha} E(\mathbf{r}_\alpha, t) V_\alpha, \tag{3}$$

where $V_\alpha$ is the dipole operator of the molecule labeled $\alpha$ and located at $\mathbf{r}_\alpha$, and the summation is over all the molecules in the nonlinear medium. $E(\mathbf{r}, t)$ is the electromagnetic field, which for a general 4WM process (Fig. 1) can be decomposed into four components:

$$E(\mathbf{r},t) = \sum_{j=1}^{4} [E_j(t)\exp(i\mathbf{k}_j\cdot\mathbf{r} - i\omega_j t) + E_j^*(t)\exp(-i\mathbf{k}_j\cdot\mathbf{r} + i\omega_j t)]. \tag{4}$$

The three incoming fields $j = 1, 2, 3$ will be treated classically. For these fields, $E_j(t)$ is a classical function of time representing their temporal envelopes. The generated signal field $E_4 = E_s$ will be treated quantum mechanically, since it is generated by spontaneous emission. For that field, we have[62]

$$E_s = -i(2\pi\hbar\omega_s/\Omega)^{1/2} a_s, \tag{5a}$$

$$E_s^* = i(2\pi\hbar\omega_s/\Omega)^{1/2} a_s^\dagger. \tag{5b}$$

Here $a_s^\dagger$ ($a_s$) is the creation (annihilation) operator for the $s$th mode, and we have adopted a box normalization with $\Omega$ being the volume. In this article, we consider a simple model, in which the active nonlinear medium consists of noninteracting polyatomic molecules (absorbers) and a set of bath degrees of freedom, which do not interact with the electromagnetic field, but do interact with the absorbers. In this case, we can focus on one absorber, located at $\mathbf{r}$, and write

$$H_{\text{int}}(t) = E(\mathbf{r},t)V, \tag{6}$$

where $V$ is the dipole operator of that absorber. The nature of the bath depends on the physical system of interest; it could be the solvent, a host crystal, an amorphous medium (glass), and so forth. For isolated large molecules the bath may consist of those vibrational degrees of freedom, which are weakly coupled to the electronic transition of interest (see Section IX). Direct interactions among the absorbers lead to interesting transport phenomena, and certain 4WM techniques, such as the transient grating,[26-33] are particularly suitable to probe these intermolecular interactions. The incorporation of transport processes in 4WM will not be considered here. In order to calculate the 4WM signal, we start at $t = -\infty$ and assume that the system is in thermal equilibrium with respect to $H$ (without the radiation field)

$$\rho(-\infty) = \exp(-H/kT)/\mathrm{Tr}\exp(-H/kT). \tag{7}$$

The system then evolves in time according to the Liouville equation

$$\frac{d\rho}{dt} = -i[H,\rho] - i[H_{\text{int}},\rho]. \tag{8}$$

In Eq. (8) and in the rest of this article, we set $\hbar = 1$. The formal manipulations in this article are simplified considerably by introducing a Liouville space notation.[11,19] We define Liouville space operators $L$, $L_{int}$ and $\mathscr{V}$ by their action of an ordinary operator $A$, that is,

$$LA \equiv [H, A], \tag{9a}$$

$$L_{int} A \equiv [H_{int}, A], \tag{9b}$$

and

$$\mathscr{V} A \equiv [V, A]. \tag{9c}$$

Equation (8) then assumes the form:

$$\frac{d\rho}{dt} = -iL\rho - iL_{int}\rho. \tag{10a}$$

This notation is useful, since now Eq. (10a) is formally identical to the Schrödinger equation

$$\frac{d\psi}{dt} = -iH\psi - iH_{int}\psi. \tag{10b}$$

This allows us to carry out many formal manipulations (perturbation theory, projection operator techniques, selective averaging over bath degrees of freedom, etc.) in a straightforward way. We shall be interested in calculating the polarization $\langle P(\mathbf{r}, t)\rangle$ at position $\mathbf{r}$ at time $t$. This is given by the expectation value of the dipole operator $V$:

$$\langle P(\mathbf{r}, t)\rangle \equiv \langle\langle V|\rho(t)\rangle\rangle, \tag{11}$$

where we are using the double bracket notation[11,19] to denote an inner product of operators. For any two operators,

$$\langle\langle A|B\rangle\rangle \equiv \mathrm{Tr}(A^\dagger B). \tag{12}$$

We shall also define a Liouville space "matrix element" by

$$\langle\langle A|L|B\rangle\rangle \equiv \mathrm{Tr}(A^\dagger LB). \tag{13}$$

For a 4WM process we calculate $\rho(t)$ perturbatively to third order in $L_{int}$. We then get[6,19]

$$\langle P(\mathbf{r},t)\rangle = (-i)^3 \int_0^\infty dt_1 \int_0^\infty dt_2 \int_0^\infty dt_3 \langle\langle V| G(t_3) L_{\text{int}}(t-t_3) G(t_2)$$
$$\times L_{\text{int}}(t-t_2-t_3) G(t_1) L_{\text{int}}(t-t_1-t_2-t_3) |\rho(-\infty)\rangle\rangle. \quad (14)$$

Here the Green's function $G(\tau)$ is the formal solution of Eq. (10a) in the absence of the electromagnetic field, $L_{\text{int}} = 0$, that is,

$$G(\tau) = \exp(-iL\tau). \quad (15)$$

For subsequent manipulations, we shall also introduce the Green's function in the frequency domain

$$\hat{G}(\omega) \equiv -i \int_0^\infty d\tau \exp(i\omega\tau) G(\tau) = \frac{1}{\omega - L}. \quad (16)$$

The interpretation of Eq. (14) is as follows (Fig. 3): The system starts at $t = -\infty$ with a density matrix $\rho(-\infty)$. It then interacts three times with the electromagnetic field at times $t - t_1 - t_2 - t_3, t - t_2 - t_3$, and $t - t_3$. During the intervals between interactions ($t_1$, $t_2$, and $t_3$) it evolves in time according to $G(\tau)$. Finally, at time $t$ we calculate the polarization. We are now in a position to introduce the formal definition of the signal in 4WM. We reiterate that the only values of $\mathbf{k}_s$ giving nonvanishing contributions to the coherent nonlinear signal are those given by Eqs. (1). Hereafter we shall make a specific choice, namely,

$$\mathbf{k}_s = \mathbf{k}_1 - \mathbf{k}_2 + \mathbf{k}_3, \quad (17a)$$

and

$$\omega_s = \omega_1 - \omega_2 + \omega_3. \quad (17b)$$

Any other combination may be obtained from our final expression [Eq. (28)]

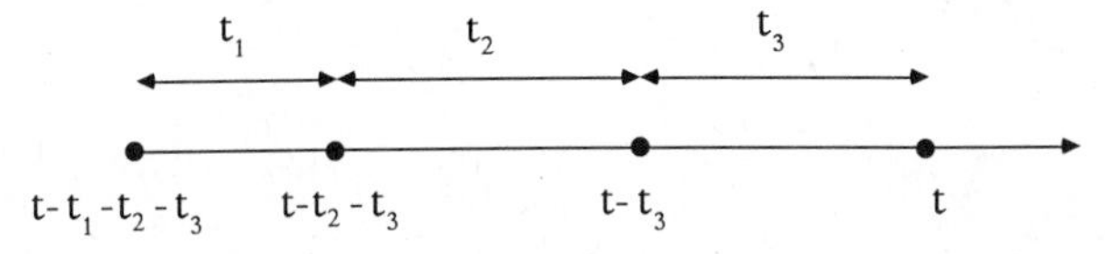


**Figure 3.** The time arguments in Eq. (14). The three interactions of the system with the radiation field take place at times $t - t_1 - t_2 - t_3$, $t - t_2 - t_3$, and $t - t_3$. The polarization is then calculated at time $t$; $t_1$, $t_2$, and $t_3$ are the intervals between the radiative interactions.

by changing one (or more) $\mathbf{k}_j$ and $\omega_j$ into $-\mathbf{k}_j$ and $-\omega_j$, and $E_j(t)$ into $E_j^*(t)$. We shall be interested in a parametric process, whereby the generated field $\omega_s$ satisfies the phase-matching condition [Eq. (17)], which implies that the signal is proportional to $\delta(\mathbf{k}_1 - \mathbf{k}_2 + \mathbf{k}_3 - \mathbf{k}_s)$. The generated coherent signal may be defined by considering the rate of emission of photons into the $\mathbf{k}_s$ mode. The operator representing this rate is

$$B = \frac{d}{dt} a_s^\dagger a_s = i[H_{\text{int}}(t), a_s^\dagger a_s], \tag{18a}$$

which in Liouville space notation assumes the form

$$|B\rangle\rangle = iL_{\text{int}}(t)|a_s^\dagger a_s\rangle\rangle. \tag{18b}$$

The photon emission rate at time $t$ into mode $s$ is the expectation value of $|B\rangle\rangle$, that is,

$$\langle\langle B|\rho(t)\rangle\rangle = \text{Tr}[B^\dagger \rho(t)], \tag{19}$$

where $\rho(t)$ is the total density matrix (system + field + bath) at time $t$. Since we shall consider only contributions to Eq. (19) that satisfy the phase-matching condition $\delta(\mathbf{k}_1 - \mathbf{k}_2 + \mathbf{k}_3 - \mathbf{k}_s)$, we may integrate the signal over all $\omega_s$, without any loss of information. Our definition for the four-wave mixing signal (rate of change of the intensity of field $s$) will, therefore, be[45,63]

$$S(\mathbf{k}_s, t) = \int d\omega_s \langle\langle B|\rho(t)\rangle\rangle, \tag{20}$$

or, using Eq. (18b),

$$S(\mathbf{k}_s, t) = -i \int d\omega_s \langle\langle a_s^\dagger a_s| L_{\text{int}}(t)|\rho(t)\rangle\rangle. \tag{21}$$

In order to proceed further, we need to expand $\rho(t)$ to the necessary order in $\mathscr{V}$. Initially, at $t \to -\infty$, we have

$$|\rho(-\infty)\rangle\rangle = |\rho_S(-\infty)\rangle\rangle |\rho_B(-\infty)\rangle\rangle |\rho_R(-\infty)\rangle\rangle, \tag{22}$$

where $\rho_S$, $\rho_B$, and $\rho_R$ stand for the density matrix of the system, of the bath, and of the radiation field, respectively. The system and bath are in thermal equilibrium, and mode $s$ of the radiation field is in the vacuum state

$$|\rho_R(-\infty)\rangle\rangle = |\text{vac}, \text{vac}\rangle\rangle. \tag{23a}$$

Equation (23a) is the Liouville space notation for

$$\rho_R(-\infty) = |\text{vac}\rangle\langle\text{vac}|. \tag{23b}$$

Note, that $\rho_R$ corresponds only to the generated mode $\mathbf{k}_s$. The incident modes $\mathbf{k}_1$, $\mathbf{k}_2$, $\mathbf{k}_3$ are treated classically and are described by the classical incident fields $E_1$, $E_2$, and $E_3$ [Eq. (4)]. It is clear that in order for the trace over the signal field $\mathbf{k}_s$ in Eq. (21) not to vanish, we need to apply the interaction [Eq. (3)] with mode $s$ twice, once from the left and once from the right. If these two interactions occur with two different particles (say $\alpha$ and $\beta$), we obtain a factor $\exp[-i\mathbf{k}_s\cdot(\mathbf{r}_\alpha - \mathbf{r}_\beta)]$ [Eq. (4)]. If $\alpha = \beta$, that is, both interactions occur with the same particle, then we lose all information about $\mathbf{k}_s$, and the process will not be phase matched. Therefore, it is clear that we need at least two different material particles in order to achieve the phase matching [Eq. (1)]. In order to satisfy the phase-matching condition, we need to expand Eq. (21) to eighth order in $V$, whereby each field $\mathbf{k}_1$, $\mathbf{k}_2$, $\mathbf{k}_3$, and $\mathbf{k}_s$ interacts once with atom $\alpha$ and once with atom $\beta$, resulting in the spatial factor $\exp[i(\mathbf{k}_1 - \mathbf{k}_2 + \mathbf{k}_3 - \mathbf{k}_s)\cdot(\mathbf{r}_\alpha - \mathbf{r}_\beta)]$. When this factor is summed over $\mathbf{r}_\alpha$, $\mathbf{r}_\beta$, it will result in the desired phase-matching condition. The reason why we need to expand our expression [Eq. (21)] to eight order in $V$, is that we are looking for a four-photon process, whose amplitude is fourth order in $V$ and intensity (amplitude square) must be eighth order in $V$. The density matrix $\rho(t)$ must therefore be expanded to seventh order in $V$ in order to obtain the signal $S$ to eighth order. We then get[63]

$$S(\mathbf{k}_s, t) = -i \int d\omega_s \langle\langle a_s^\dagger a_s | L_{\text{int}}(t) | \rho^{(7)}(t)\rangle\rangle. \tag{24}$$

After some algebraic manipulations and with the elimination of some numerical constants, the signal can be recast in the form[45,63]

$$S(\mathbf{k}_s, t) = \sum_{\alpha,\beta} \langle P(\mathbf{r}_\alpha, t) P^\dagger(\mathbf{r}_\beta, t)\rangle \exp[-i\mathbf{k}_s\cdot(\mathbf{r}_\alpha - \mathbf{r}_\beta)]. \tag{25}$$

Here $P(\mathbf{r}_\alpha, t)$, Eq. (14), is the polarization at point $\mathbf{r}_\alpha$ at time $t$ and $P^\dagger$ is the hermitian conjugate of $P$. The angular brackets denote averaging over the thermal bath with which the absorbers interact. Equation (25) reflects the fact that the 4WM signal results from a coherent parametric emission, involving many particles emitting in phase. Usually, different particles in the nonlinear medium are not correlated, so that the two-body average can be factorized in the form

$$\langle P(\mathbf{r}_\alpha, t)P^\dagger(\mathbf{r}_\beta, t)\rangle = \langle P(\mathbf{r}_\alpha, t)\rangle \langle P^\dagger(\mathbf{r}_\beta, t)\rangle. \tag{26}$$

Equations (25) and (26) result in

$$S(\mathbf{k}_s, t) = |\langle P(\mathbf{k}_s, t)\rangle|^2, \tag{27a}$$

where

$$\langle P(\mathbf{k}_s, t)\rangle = \sum_\alpha \langle P(\mathbf{r}_\alpha, t)\rangle \exp(-i\mathbf{k}_s \cdot \mathbf{r}_\alpha + i\omega_s t). \tag{27b}$$

It should be noted that in the presence of long-range spatial correlations in the sample (e.g., near critical points) the factorization [Eq. (26)] does not hold, and 4WM actually probes the two-particle correlation function. This is analogous to spontaneous light scattering. A general theory of 4WM near critical points needs to be developed. A first step in this direction was made recently.[63] In the present article, we shall not consider systems with long-range spactial correlations, and we assume that Eq. (26) holds. Using Eqs. (3), (9), and (14), we then get

$$\begin{aligned}\langle P(\mathbf{k}_s, t)\rangle = (-i)^3 \sum_{m,n,q=1,2,3} \int_0^\infty dt_3 \int_0^\infty dt_2 \int_0^\infty dt_1 R(t_3, t_2, t_1) \\ \times \exp[i(\omega_m + \omega_n + \omega_q)t_3 + i(\omega_m + \omega_n)t_2 + i\omega_m t_1] \\ \times E_m(t - t_1 - t_2 - t_3)E_n(t - t_2 - t_3)E_q(t - t_3).\end{aligned} \tag{28}$$

$R(t_3, t_2, t_1)$ is the *nonlinear response function*, which contains all relevant microscopic information for any 4WM process;

$$R(t_3, t_2, t_1) = \langle\langle V| G(t_3)\mathscr{V} G(t_2)\mathscr{V} G(t_1)\mathscr{V} |\rho(-\infty)\rangle\rangle. \tag{29}$$

The summation in Eq. (28) is over all $3! = 6$ permutations of $\omega_m$, $\omega_n$, and $\omega_q$ with $\omega_1$, $-\omega_2$, and $\omega_3$, and of $E_m$, $E_n$, and $E_q$ with $E_1$, $E_2^*$, and $E_3$. Alternatively, we may define the response function in the frequency domain by performing a triple Fourier transform of $R(t_3, t_2, t_1)$, that is,

$$\begin{aligned}\hat{R}(\omega_m + \omega_n + \omega_q, \omega_m + \omega_n, \omega_m) = (-i)^3 \int_0^\infty dt_3 \int_0^\infty dt_2 \int_0^\infty dt_1 \\ \times \exp[i(\omega_m + \omega_n + \omega_q)t_3 \\ + i(\omega_m + \omega_n)t_2 + i\omega_m t_1]R(t_3, t_2, t_1).\end{aligned} \tag{30}$$

Equation (28) may then be rearranged in the form

$$\langle P(\mathbf{k}_s, t)\rangle = \sum_{m,n,q=1,2,3} \int_{-\infty}^{\infty} d\omega_m' \int_{-\infty}^{\infty} d\omega_n' \int_{-\infty}^{\infty} d\omega_q'$$
$$\times \hat{R}(\omega_m' + \omega_n' + \omega_q', \omega_m' + \omega_n', \omega_m') J_m(\omega_m') J_n(\omega_n') J_q(\omega_q')$$
$$\times \exp[i(\omega_m + \omega_n + \omega_q - \omega_m' - \omega_n' - \omega_q')t], \tag{31}$$

where

$$\hat{R}(\omega_m + \omega_n + \omega_q, \omega_m + \omega_n, \omega_m)$$
$$= \langle\langle V | \hat{G}(\omega_m + \omega_n + \omega_q) \mathscr{V} \hat{G}(\omega_m + \omega_n) \mathscr{V} \hat{G}(\omega_m) \mathscr{V} | \rho(-\infty) \rangle\rangle \tag{32}$$

and

$$J_j(\omega_j') = (2\pi)^{-1} \int_{-\infty}^{\infty} d\tau\, E_j(\tau) \exp[i(\omega_j' - \omega_j)\tau], \quad j = m, n, q, \tag{33}$$

is the spectral density of the $j$ field.

Equations (27) and (28) or alternatively Eq. (31) provide the most general formal expression for any type of 4WM process. They show that the nonlinear response function $R(t_3, t_2, t_1)$, or its Fourier transform $\hat{R}(\omega_m + \omega_n + \omega_q, \omega_m + \omega_n, \omega_m)$, contains the complete microscopic information relevant to the calculation of any 4WM signal. As indicated earlier, the various 4WM techniques differ by the choice of $\mathbf{k}_s$ and $\omega_s$ and by the temporal characteristics of the incoming fields $E_1(t)$, $E_2(t)$, and $E_3(t)$. A detailed analysis of the response function and of the nonlinear signal will be made in the following sections for specific models. At this point we shall consider the two limiting cases of ideal time-domain and frequency-domain 4WM. In an ideal time-domain 4WM, the durations of the incoming fields are infinitely short, that is,

$$E_1(\tau) = E_1 \delta(\tau - \tau_1),$$
$$E_2(\tau) = E_2 \delta(\tau - \tau_2), \tag{34}$$
$$E_3(\tau) = E_3 \delta(\tau - \tau_3),$$

where $\tau_1 < \tau_2 < \tau_3$ (Fig. 1). We further denote $t_1 = \tau_2 - \tau_1, t_2 = \tau_3 - \tau_2$, and $t_3 = t - \tau_3$. Upon the substitution of Eq. (34) into Eq. (28), we get

$$\langle P(\mathbf{k}_s, t)\rangle = E_1 E_2^* E_3 R(t_3, t_2, t_1) \exp[i\omega_1 t_1 + i(\omega_1 - \omega_2)t_2$$
$$+ i(\omega_1 - \omega_2 + \omega_3)t_3] \tag{35}$$

and

$$S(\mathbf{k}_s, t) = |E_1 E_2 E_3|^2 |R(t_3, t_2, t_1)|^2. \tag{36}$$

The other extreme limit of 4WM is a stationary frequency-domain experiment in which the field amplitudes $E_1(\tau)$, $E_2(\tau)$, and $E_3(\tau)$ are time independent. In this case, we have

$$\begin{aligned} J_1(\omega_1') &= E_1 \delta(\omega_1' - \omega_1), \\ J_2(\omega_2') &= E_2 \delta(\omega_2' - \omega_2), \\ J_3(\omega_3') &= E_3 \delta(\omega_3' - \omega_3). \end{aligned} \tag{37}$$

Using Eqs. (31) and (37), we get

$$\langle P(\mathbf{k}_s, t) \rangle = \chi^{(3)}(-\omega_s, \omega_1, -\omega_2, \omega_3) E_1 E_2^* E_3, \tag{38}$$

where the nonlinear susceptibility $\chi^{(3)}$ is given by

$$\chi^{(3)}(-\omega_s, \omega_1, -\omega_2, \omega_3) = \sum_{m,n,q=1,2,3} \hat{R}(\omega_m + \omega_n + \omega_q, \omega_m + \omega_n, \omega_m). \tag{39}$$

The stationary signal [Eq. (27)] is given in this case by

$$S(\mathbf{k}_s) = |E_1 E_2 E_3|^2 |\chi^{(3)}(-\omega_s, \omega_1, -\omega_2, \omega_3)|^2. \tag{40}$$

## III. A MOLECULAR MICROSCOPIC MODEL FOR THE NONLINEAR RESPONSE FUNCTION

We shall now apply the results of Section II to a specific model system, commonly used in molecular 4WM. We consider a molecular-level scheme for the absorber, consisting of a manifold of vibronic levels belonging to the ground electronic state, denoted $|a\rangle, |c\rangle, \ldots$, and a manifold of vibronic levels belonging to an excited electronic state, denoted $|b\rangle$, $|d\rangle$, ..., (Fig. 4). The ground and the electronically excited states will be denoted $|g\rangle$ and $|e\rangle$, respectively. The absorber is further coupled to a thermal bath, and the combined Hamiltonian for the molecule and the bath is[6,44,57,58,64]

$$H = |g\rangle [H_g(Q_S) + h_g(Q_B)] \langle g| + |e\rangle [H_e(Q_S) + h_e(Q_B)] \langle e|. \tag{41}$$

Here the ground-state Hamiltonian is partitioned into a system part $H_g$, which

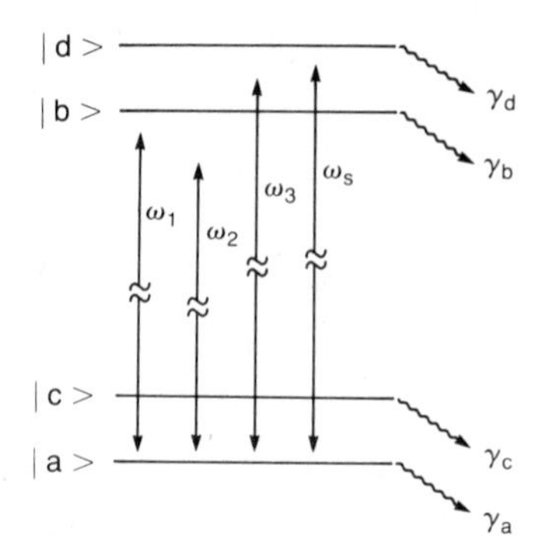


**Figure 4.** The molecular-level scheme and laser frequencies for 4WM. Levels $|a\rangle$ and $|c\rangle$ are part of the ground-state vibrational manifold, whereas levels $|b\rangle$ and $|d\rangle$ belong to an electronically excited manifold. $\gamma_\nu$ is the inverse lifetime of level $|\nu\rangle$. The electronic dipole operator [Eq. (47)] couples vibronic states belonging to different electronic states.

depends on the system coordinates $Q_S$, and a bath part $h_g$, which depends on the bath coordinates $Q_B$. Similar partitioning of the excited-state Hamiltonian into $H_e$ and $h_e$ is made. The interaction between the system and the bath is reflected in the difference between $h_g$ and $h_e$, which implies that the bath eigenstates are different, depending on the electronic state of the system. The Hamiltonian [Eq. (41)] does not couple the bath directly to the molecular vibrational degrees of freedom. The key physical assumption underlying Eq. (41) is that a polar solvent is mainly sensitive to the molecular charge distribution, which may be very different for the ground $|g\rangle$ and the electronically excited $|e\rangle$ states. The difference between $h_g$ and $h_e$ results in the solvent reorganization upon electronic excitation of the molecule.[65] The interaction of the solvent with the molecule is only weakly dependent on the nuclear coordinates and this dependence is ignored in Eq. (41). This interaction results in solvent-induced vibrational relaxation, which may play an important role in molecular spectra in condensed phases. In Section VIII we incorporate the vibrational relaxation in our theory, thus eliminating the major approximation of Eq. (41). The eigenstates of the ground-state and of the excited-state Hamiltonians will be denoted as (Fig. 4)

$$(H_g + h_g)|\nu\alpha\rangle = (\varepsilon_\nu + \varepsilon_\alpha)|\nu\alpha\rangle \qquad \nu = a, c, \ldots \tag{42a}$$

and

$$(H_e + h_e)|\nu\beta\rangle = (\varepsilon_\nu + \varepsilon_\beta)|\nu\beta\rangle \qquad \nu = b, d, \ldots . \tag{42b}$$

Here $a$, $b$, $c$, $d$, ... stand for the collection of all system quantum numbers, whereas $\alpha$ and $\beta$ represent all the bath quantum numbers. The molecule is taken to be initially at thermal equilibrium in the ground-state manifold, and its density matrix is the direct product of the system ($\sigma_g$) and the bath ($\rho_g$) components, that is,

$$\rho(-\infty) = \sigma_g(Q_S)\rho_g(Q_B) \tag{43}$$

with

$$\sigma_g = \exp(-H_g/kT)/\mathrm{Tr}\exp(-H_g/kT) = \sum_a |a\rangle P(a) \langle a|, \tag{43a}$$

$$\rho_g = \exp(-h_g/kT)/\mathrm{Tr}\exp(-h_g/kT) = \sum_\alpha |\alpha\rangle \rho_g(\alpha) \langle\alpha|. \tag{43b}$$

Here

$$P(a) = \exp(-\varepsilon_a/kT) \Big/ \sum_a \exp(-\varepsilon_a/kT), \tag{44a}$$

$$\rho_g(\alpha) = \exp(-\varepsilon_\alpha/kT) \Big/ \sum_\alpha \exp(-\varepsilon_\alpha/kT). \tag{44b}$$

In Liouville space notation, we write Eq. (43) in the form

$$|\rho(-\infty)\rangle\rangle = \sum_{a,\alpha} P(a)\rho_g(\alpha) |{}^{\alpha\alpha}_{aa}\rangle\rangle. \tag{45}$$

Here $|\nu\lambda\rangle\rangle$ is the Liouville space vector corresponding to $|\nu\rangle\langle\lambda|$. $|{}^{\alpha\alpha}_{aa}\rangle\rangle$ stands for the direct product of the system $|aa\rangle\rangle$ state and the bath $|\alpha\alpha\rangle\rangle$ state. For subsequent manipulations we further introduce the system and the bath density matrices corresponding to thermal equilibrium within the electronically excited state, that is,

$$\sigma_e = \exp(-H_e/kT)/\mathrm{Tr}\exp(-H_e/kT) = \sum_b |b\rangle P(b) \langle b|, \tag{46a}$$

$$\rho_e = \exp(-h_e/kT)/\mathrm{Tr}\exp(-h_e/kT) = \sum_\beta |\beta\rangle \rho_e(\beta) \langle\beta|, \tag{46b}$$

where the sums run over the excited-state manifold, and $P(b)$ and $\rho_e(\beta)$ are defined by Eqs. (44), by interchanging the indexes $a$ and $\alpha$ to $b$ and $\beta$, respectively. The electronic dipole operator of the absorber couples vibronic states belonging to different electronic states. We then have

$$V = \sum_{a,b,c,d} [\mu_{ab}|a\rangle\langle b| + \mu_{da}|d\rangle\langle a| + \mu_{cb}|c\rangle\langle b| + \mu_{dc}|d\rangle\langle c|], \tag{47}$$

where the summation runs over the entire manifolds of ground and electronically excited states. We are now in a position to calculate the nonlinear response function $R(t_3, t_2, t_1)$ [Eq. (29)] for the present model system. The radiative interaction $\mathscr{V}$ is a commutator that can act either from the left or from the right, and its matrix elements are[11]

$$\langle\langle \nu'\lambda'|\mathscr{V}|\nu\lambda\rangle\rangle = \mu_{\nu'\nu}\delta_{\lambda'\lambda} - \mu^*_{\lambda'\lambda}\delta_{\nu'\nu}. \tag{48}$$

The first and second terms in Eq. (48) correspond, respectively, to action of $\mathscr{V}$ from the left and from the right. Since Eq. (29) contains three factors of the radiative interaction $\mathscr{V}$, it will have $2^3 = 8$ terms corresponding to the various possible choices of the $\mathscr{V}$'s to act from the left or the right. A pictorial representation of Eq. (29) is given in Fig. 5. We start at $|\rho(-\infty)\rangle\rangle = |aa\rangle\rangle$,

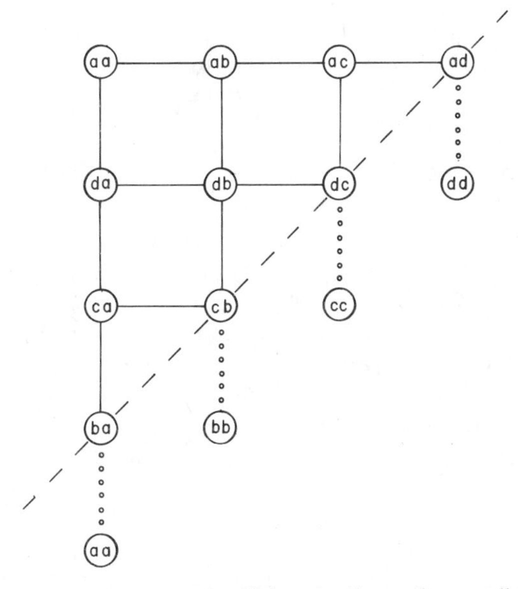


**Figure 5.** Pictoral representation of the Liouville space pathways that contribute to the nonlinear response function [Eqs. (49) and (53)]. Solid lines denote radiative coupling $V$, horizontal (vertical) lines represent action of $V$ from the right (left). Starting at *aa*, after three perturbations, the system finds itself along the dashed line. The dotted lines represent the last $V$, which acts from the left. At the end of four perturbations, the system is in a diagonal state (*aa*, *bb*, *cc*, or *dd*). The number of three-bond pathways leading to *ad*, *ba*, *dc*, and *cb* is 1, 1, 3, and 3, respectively. Altogether, there are, therefore, eight pathways, which are shown in Fig. 6. In each pathway, each of the three incoming fields acts once.

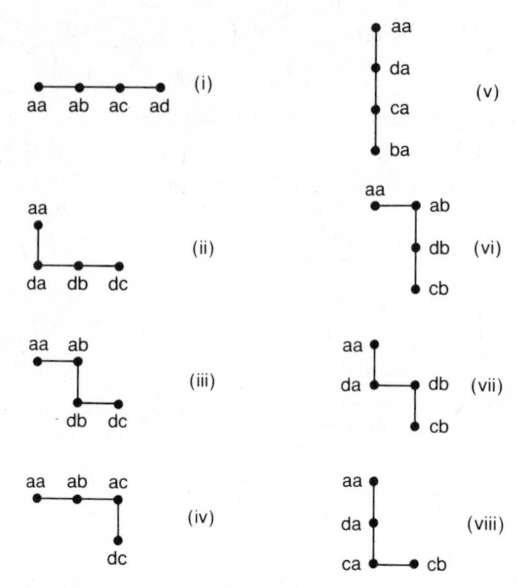


**Figure 6.** The eight Liouville space pathways that contribute to the nonlinear response function [Eq. (49) or (53)]. The eight terms in Eqs. (49), (53), (57), (60), and (63) correspond, respectively, to pathways (i)–(viii).

which is in the upper left corner. A horizontal (vertical) bond represents an interaction $\mathscr{V}$ acting from the right (left). After the first interaction (which takes place at time $t - t_1 - t_2 - t_3$) the system finds itself in either of the states $|ab\rangle\rangle$ or $|da\rangle\rangle$ (note that $b$ and $d$ are dummy indices that run over the entire excited-state manifold). The system then evolves for a period $t_1$, interacts again (at time $t - t_2 - t_3$), evolves for a period $t_2$, interacts again at time $t - t_3$, and evolves for a period $t_3$. Then, at time $t$ the polarization is calculated by acting with $V$ from the left and performing a trace. The eight pathways in Fig. 5 that contribute to $R$ are displayed in Fig. 6. Making use of Eqs. (29), (41)–(43), (47), and (48), we obtain the nonlinear response function[6,19]

$$\begin{aligned}
R(t_3, t_2, t_1) = \sum_{a,b,c,d} & P(a)\mu_{ab}\mu_{bc}\mu_{cd}\mu_{da} \\
& \times [-\langle G_{ad}(t_3)G_{ac}(t_2)G_{ab}(t_1)\rangle + \langle G_{dc}(t_3)G_{db}(t_2)G_{da}(t_1)\rangle \\
& + \langle G_{dc}(t_3)G_{db}(t_2)G_{ab}(t_1)\rangle + \langle G_{dc}(t_3)G_{ac}(t_2)G_{ab}(t_1)\rangle \\
& + \langle G_{ba}(t_3)G_{ca}(t_2)G_{da}(t_1)\rangle - \langle G_{cb}(t_3)G_{db}(t_2)G_{ab}(t_1)\rangle \\
& - \langle G_{cb}(t_3)G_{db}(t_2)G_{da}(t_1)\rangle - \langle G_{cb}(t_3)G_{ca}(t_2)G_{da}(t_1)\rangle],
\end{aligned} \quad (49)$$

where the eight terms correspond to the pathways (i)–(viii) of Fig. 6, respectively.

The following notation was introduced in Eq. (49):

$$G_{\nu\lambda}(t) \equiv \langle\langle \nu\lambda | G(t) | \nu\lambda \rangle\rangle_S = \mathrm{Tr}_S[|\lambda\rangle\langle\nu| G(t) |\nu\rangle\langle\lambda|]. \tag{50}$$

The subscript $S$ signifies that this is a matrix element in the system space only, that is, we perform a partial trace over the system degrees of freedom, and $G_{\nu\lambda}(t)$ is still a full Liouville space operator in the bath degrees of freedom. The angular brackets $\langle\cdots\rangle$ denote averaging over the bath degrees of freedom, that is,

$$\langle G \rangle = \sum_{\alpha,\alpha'} \langle\langle \alpha'\alpha' | G | \alpha\alpha \rangle\rangle \rho_g(\alpha). \tag{51}$$

This is the usual thermodynamic averaging, which amounts to averaging over initial states $|\alpha\alpha\rangle\rangle$ and summing over final states $|\alpha'\alpha'\rangle\rangle$. To clarify the notation, we shall write explicitly the first term in Eq. (49):

$$\begin{aligned}\langle G_{ad}(t_3) G_{ac}(t_2) G_{ab}(t_1) \rangle = \sum_{\alpha,\alpha_1,\alpha_2,\alpha_3,\alpha_4,\alpha'} & \langle\langle {}^{\alpha'}_{a}{}^{\alpha'}_{d} | G(t_3) |{}^{\alpha_3}_{a}{}^{\alpha_4}_{d} \rangle\rangle \langle\langle {}^{\alpha_3}_{a}{}^{\alpha_4}_{c} | G(t_2) |{}^{\alpha_1}_{a}{}^{\alpha_2}_{c} \rangle\rangle \\ & \times \langle\langle {}^{\alpha_1}_{a}{}^{\alpha_2}_{b} | G(t_1) |{}^{\alpha\alpha}_{ab} \rangle\rangle \rho_g(\alpha).\end{aligned} \tag{52}$$

Alternatively, making use of Eq. (32), we obtain the nonlinear response function in the frequency domain:

$$\begin{aligned}&\hat{R}(\omega_1 - \omega_2 + \omega_3, \omega_1 - \omega_2, \omega_1) \\ &\quad = \sum_{a,b,c,d} P(a) \mu_{ab}\mu_{bc}\mu_{cd}\mu_{da} \\ &\qquad \times [-\langle \hat{G}_{ad}(\omega_1 - \omega_2 + \omega_3) \hat{G}_{ac}(\omega_1 - \omega_2) \hat{G}_{ab}(\omega_1) \rangle \\ &\qquad + \langle \hat{G}_{dc}(\omega_1 - \omega_2 + \omega_3) \hat{G}_{db}(\omega_1 - \omega_2) \hat{G}_{da}(\omega_1) \rangle \\ &\qquad + \langle \hat{G}_{dc}(\omega_1 - \omega_2 + \omega_3) \hat{G}_{db}(\omega_1 - \omega_2) \hat{G}_{ab}(\omega_1) \rangle \\ &\qquad + \langle \hat{G}_{dc}(\omega_1 - \omega_2 + \omega_3) \hat{G}_{ac}(\omega_1 - \omega_2) \hat{G}_{ab}(\omega_1) \rangle \\ &\qquad + \langle \hat{G}_{ba}(\omega_1 - \omega_2 + \omega_3) \hat{G}_{ca}(\omega_1 - \omega_2) \hat{G}_{da}(\omega_1) \rangle \\ &\qquad - \langle \hat{G}_{cb}(\omega_1 - \omega_2 + \omega_3) \hat{G}_{db}(\omega_1 - \omega_2) \hat{G}_{ab}(\omega_1) \rangle \\ &\qquad - \langle \hat{G}_{cb}(\omega_1 - \omega_2 + \omega_3) \hat{G}_{db}(\omega_1 - \omega_2) \hat{G}_{da}(\omega_1) \rangle \\ &\qquad - \langle \hat{G}_{cb}(\omega_1 - \omega_2 + \omega_3) \hat{G}_{ca}(\omega_1 - \omega_2) \hat{G}_{da}(\omega_1) \rangle].\end{aligned} \tag{53}$$

Here again, the eight terms correspond, respectively, to the pathways (i)–(viii) of Fig. 6. $\hat{G}_{\nu\lambda}(\omega)$ and the angular brackets $\langle\cdots\rangle$ are analogous to Eq. (51).

We note that the nonlinear response function $R(t_3, t_2, t_1)$ [Eq. (49)] or $\hat{R}(\omega_1 + \omega_2 + \omega_3, \omega_1 + \omega_2, \omega_1)$ [Eq. (53)] is the fundamental quantity that contains all the relevant microscopic information for any 4WM process. Ideal time-domain 4WM probes directly $|R(t_3, t_2, t_1)|^2$ [Eq. (36)]. The response function contains eight terms that correspond to the eight distinct pathways in Liouville space (Fig. 6). Frequency-domain 4WM is described by $\chi^{(3)}$ [Eq. (40)], which has $6 \times 8 = 48$ terms corresponding to the $3! = 6$ permutations of the time ordering of the three fields that can be made for each of the eight pathways. In time-domain 4WM we have fewer terms than in frequency-domain 4WM, since in the former we can control the relative order in time of the interactions with the three fields, whereas in the latter all orderings contribute equally to the signal. The Liouville space notation used in this section is useful, since it allows us to follow the time evolution of the density matrix. It will further allow us to introduce the factorization approximation in Section IV. Before doing that, however, we note that we can express Eq. (49) in terms of ordinary (not Liouville) operators. To that end we introduce the four-point correlation function of the dipole operator,[19,57] that is,

$$\begin{aligned} F(\tau_1, \tau_2, \tau_3, \tau_4) &= \mathrm{Tr}[V(\tau_1)V(\tau_2)V(\tau_3)V(\tau_4)\rho(-\infty)] \\ &= \langle V(\tau_1)V(\tau_2)V(\tau_3)V(\tau_4)\rangle, \end{aligned} \tag{54}$$

where

$$V(\tau) = \exp(iH\tau)V\exp(-iH\tau). \tag{55}$$

Here $H$ is the molecular Hamiltonian [Eq. (41)] and $V$ is the dipole operator [Eq. (47)]. In terms of the system and the bath eigenstates [Eqs. (42)], we can recast Eq. (54) in the form

$$\begin{aligned} F(\tau_1, \tau_2, \tau_3, \tau_4) = \sum_{\substack{a,b,c,d \\ \alpha,\alpha_1,\alpha_2,\alpha_3}} &\langle a\alpha| V(\tau_1)|d\alpha_3\rangle \langle d\alpha_3| V(\tau_2)|c\alpha_2\rangle \langle c\alpha_2| V(\tau_3)|b\alpha_1\rangle \\ &\times \langle b\alpha_1| V(\tau_4)|a\alpha\rangle P(a)\rho_g(\alpha) \end{aligned} \tag{56}$$

In terms of this four-point correlation function, we have[6,19]

$$\begin{aligned} R(t_3, t_2, t_1) = &-F(0, t_1, t_1 + t_2, t_1 + t_2 + t_3) + F(t_1, t_1 + t_2, t_1 + t_2 + t_3, 0) \\ &+ F(0, t_1 + t_2, t_1 + t_2 + t_3, t_1) + F(0, t_1, t_1 + t_2 + t_3, t_1 + t_2) \\ &+ F(t_1 + t_2 + t_3, t_1 + t_2, t_1, 0) - F(0, t_1 + t_2 + t_3, t_1 + t_2, t_1) \\ &- F(t_1, t_1 + t_2 + t_3, t_1 + t_2, 0) - F(t_1 + t_2, t_1 + t_2 + t_3, t_1, 0), \end{aligned} \tag{57}$$

where the eight terms correspond, respectively, to pathways (i)–(viii) of Fig. 6. It is therefore clear that the various 4WM processes probe different features of the four-point correlation function of the dipole operator. The response function discussed in this article is closely related to that introduced by Butcher.[2] Our time arguments $t_1$, $t_2$, and $t_3$ (Fig. 3) were chosen differently. With the present choice, the relations between time-domain and frequency-domain experiments is more transparent. Since the response function is probing the four-point correlation function [Eq. (54)], it necessarily contains more information than the ordinary absorption lineshape, which is given by the two-time correlation function $\langle V(\tau)V(0)\rangle$.[11] This can be utilized, for example, to eliminate selectively inhomogeneous broadening as is done in photon echoes.[22–25,43] Specific applications to coherent Raman spectroscopy and approximate schemes for the evaluation of the nonlinear response function will be developed in the subsequent sections.

## IV. THE FACTORIZATION APPROXIMATION

Equations (49), (53), and (57) provide rigorous correlation function expressions for the nonlinear response function. In general, in order to calculate the spectral lineshapes in nonlinear optical processes, we need to solve for the correlated dynamics of the system and of the bath degrees of freedom. In this section, we shall develop an approximate expression for the nonlinear response function, which is based on the assumption of separation of time scales between the system and the bath. Under very general conditions, which will be precisely specified, it is possible to factorize the average of the product of Liouville space operators into a product of averages. This results in a considerable simplification of the final expressions and provides a useful method of modeling realistic molecules in solutions.[11,19,64] Let us examine Eq. (52) in more detail (see Fig. 6). During the time period $t_1$, the system is in an optical coherence $|a\rangle\langle b|$ between the ground and electronically excited states. The bath, which was initially in thermal equilibrium with the system in the ground electronic state, is now interacting with the system and undergoes a time evolution. Its state at the end of the $t_1$ period is $|\alpha_1\rangle\langle\alpha_2|$. Then a second radiative interaction takes place, and the system returns to the ground electronic state ($|a\rangle\langle c|$ is a part of the ground-state manifold $|g\rangle\langle g|$). During the $t_2$ period, the bath (which is now in a nonequilibrium state) changes its state to $|\alpha_3\rangle\langle\alpha_4|$. The relevant duration of the $t_2$ period is of the order of the lifetime of the vibronic states $|a\rangle$ and $|c\rangle$. This time is expected to be much longer than a typical solvent relaxation time ($10^{-13}$ sec), which restores the solvent to equilibrium with the $|g\rangle\langle g|$ system state. Therefore, after a very short time (much shorter than a typical value of $t_2$), the bath returns to the $\rho_g$ distribution. We can therefore assume that at the beginning and at the end of the $t_2$ period the bath is in thermal equilibrium and its density matrix is $\rho_g$.

Then a third radiative interaction occurs. The system returns to an optical coherence, $|a\rangle\langle d|$, and evolves with the bath for a period $t_3$. This description is appropriate for pathways (i), (iv), (v). (viii) of Eq. (49). For the other pathways [(ii), (iii), (vi), and (vii)], there is one difference. During the period $t_2$ the system is in the excited electronic state $|e\rangle\langle e|$ (rather than in $|g\rangle\langle g|$). We then expect the bath to relax rapidly to the new equilibrium $\rho_e$ [Eq. (46b)] rather than $\rho_g$. This is the solvent reorganization process.[65] During the period $t_2$ in these pathways, the state of the bath can therefore be assumed to be $\rho_e$.

These considerations suggest the introduction of the following projection operator[59,64]:

$$\hat{P} = |gg\rangle\rangle\langle\langle gg|\rho_g \mathrm{Tr}_B + |ee\rangle\rangle\langle\langle ee|\rho_e \mathrm{Tr}_B, \tag{58a}$$

or, alternatively,

$$\hat{P} = \sum_{\alpha,\alpha'} \rho_g(\alpha)\left|{}^{\alpha\alpha}_{gg}\right\rangle\rangle\langle\langle{}^{\alpha'\alpha'}_{g\ g}\right| + \sum_{\beta,\beta'} \rho_e(\beta)\left|{}^{\beta\beta}_{ee}\right\rangle\rangle\langle\langle{}^{\beta'\beta'}_{e\ e}\right|. \tag{58b}$$

The factorization approximation may now be introduced by substituting into Eq. (49):

$$G(t_2) \to \hat{P}G(t_2)\hat{P}. \tag{59}$$

This projection operator replaces the bath distribution at the end of the $t_2$ period by $\rho_g$ if the system is in the ground electronic state, and by $\rho_e$ if the system is electronically excited. Making use of Eq. (59), Eq. (49) assumes the form[19]

$$\begin{aligned}
&R(t_3, t_2, t_1)\\
&= \sum_{a,b,c,d} P(a)\mu_{ab}\mu_{bc}\mu_{cd}\mu_{da}[-I_{ad}(t_3)I_{ac}(t_2)I_{ab}(t_1)J_g^*(t_3)J_g^*(t_1)\\
&\quad + I_{dc}(t_3)I_{db}(t_2)I_{da}(t_1)J_e^*(t_3)J_g(t_1) + I_{dc}(t_3)I_{db}(t_2)I_{ab}(t_1)J_e^*(t_3)J_g^*(t_1)\\
&\quad + I_{dc}(t_3)I_{ac}(t_2)I_{ab}(t_1)J_g(t_3)J_g^*(t_1) + I_{ba}(t_3)I_{ca}(t_2)I_{da}(t_1)J_g(t_3)J_g(t_1)\\
&\quad - I_{cb}(t_3)I_{db}(t_2)I_{ab}(t_1)J_e(t_3)J_g^*(t_1) - I_{cb}(t_3)I_{db}(t_2)I_{da}(t_1)J_e(t_3)J_g(t_1)\\
&\quad - I_{cb}(t_3)I_{ca}(t_2)I_{da}(t_1)J_g^*(t_3)J_g(t_1)].
\end{aligned} \tag{60}$$

Here,

$$I_{\nu\lambda}(t) = \exp[-i\omega_{\nu\lambda}t - \Gamma_{\nu\lambda}|t|], \tag{61a}$$

where

$$\omega_{\nu\lambda} \equiv \varepsilon_\nu - \varepsilon_\lambda \tag{61b}$$

and

$$\Gamma_{\nu\lambda} \equiv \tfrac{1}{2}(\gamma_\nu + \gamma_\lambda). \tag{61c}$$

$\gamma_\nu^{-1}$ is the lifetime of level $|\nu\rangle$, which is introduced here phenomenologically. A more detailed discussion of lifetimes and relaxation is given in Sec. VIII. $I_{\nu\lambda}(t)$ represents the system contribution to the correlation function. The entire effect of the bath, in this case, is contained in the line broadening functions $J_g(t)$ and $J_e(t)$, defined as follows[64]:

$$J_g(t) = \mathrm{Tr}_B[\exp(ih_g t)\exp(-ih_e t)\rho_g] \tag{62a}$$

and

$$J_e(t) = \mathrm{Tr}_B[\exp(ih_e t)\exp(-ih_g t)\rho_e]. \tag{62b}$$

Here $\mathrm{Tr}_B$ stands for a trace over the bath degrees of freedom. $J_g(t)$ represents the dephasing effects resulting from thermal fluctuations of the bath, when the bath is equilibrated with the system in the ground electronic state, and its density matrix is $\rho_g$. $J_e(t)$ represents the same effects when the bath is equilibrated with the system in the electronically excited state $|e\rangle$ (i.e., after the solvent reorganization takes place), and its density matrix is $\rho_e$. All pathways contain $J_g(t_1)$, since the system is initially in the ground state. Pathways (i), (iv), (v), and (viii) contain $J_g(t_3)$, since during $t_2$ the bath is equilibrated with the ground-state system; whereas the other pathways contain $J_e(t_3)$, since during the $t_2$ interval the bath has relaxed to its new equilibrium with the system being excited. Upon the substitution of Eq. (60) into Eq. (30), we obtain the response function in the frequency domain:

$$\begin{aligned}
&\hat{R}(\omega_1 - \omega_2 + \omega_3, \omega_1 - \omega_2, \omega_1) \\
&\quad = \sum_{a,b,c,d} P(a)\mu_{ab}\mu_{bc}\mu_{cd}\mu_{da} \\
&\quad \times \Bigg[ -\frac{I_{da}^*(-\omega_1 + \omega_2 - \omega_3)I_{ba}^*(-\omega_1)}{\omega_1 - \omega_2 - \omega_{ac} + i\Gamma_{ac}} + \frac{\bar{I}_{dc}(\omega_1 - \omega_2 + \omega_3)I_{da}(\omega_1)}{\omega_1 - \omega_2 - \omega_{db} + i\Gamma_{db}} \\
&\quad - \frac{\bar{I}_{dc}(\omega_1 - \omega_2 + \omega_3)I_{ba}^*(-\omega_1)}{\omega_1 - \omega_2 - \omega_{db} + i\Gamma_{db}} - \frac{I_{dc}(\omega_1 - \omega_2 + \omega_3)I_{ba}^*(-\omega_1)}{\omega_1 - \omega_2 - \omega_{ac} + i\Gamma_{ac}} \\
&\quad + \frac{I_{ba}(\omega_1 - \omega_2 + \omega_3)I_{da}(\omega_1)}{\omega_1 - \omega_2 - \omega_{ca} + i\Gamma_{ac}} - \frac{\bar{I}_{bc}^*(-\omega_1 + \omega_2 - \omega_3)I_{ba}^*(-\omega_1)}{\omega_1 - \omega_2 - \omega_{db} + i\Gamma_{db}} \\
&\quad + \frac{\bar{I}_{bc}^*(-\omega_1 + \omega_2 - \omega_3)I_{da}(\omega_1)}{\omega_1 - \omega_2 - \omega_{db} + i\Gamma_{db}} + \frac{I_{bc}^*(-\omega_1 + \omega_2 - \omega_3)I_{da}(\omega_1)}{\omega_1 - \omega_2 - \omega_{ca} + i\Gamma_{ac}} \Bigg].
\end{aligned} \tag{63}$$

Here the complex lineshape functions $I$ and $\bar{I}$ are given by

$$I_{\nu\lambda}(\omega) = -i\int_0^\infty d\tau\, I_{\nu\lambda}(\tau)J_g(\tau)\exp(i\omega\tau) \tag{64a}$$

and

$$\bar{I}_{\nu\lambda}(\omega) = -i\int_0^\infty d\tau\, I_{\nu\lambda}(\tau)J_e^*(\tau)\exp(i\omega\tau). \tag{64b}$$

$I_{\nu\lambda}(\omega)$ represents an absorption lineshape broadened by $J_g(\tau)$, whereas $\bar{I}_{\nu\lambda}(\omega)$ represents an emission lineshape broadened by $J_e^*(\tau)$.

## V. CORRELATION FUNCTIONS FOR SPONTANEOUS RAMAN AND FLUORESCENCE LINESHAPES

In the previous sections, we derived general correlation function expressions for the nonlinear response function that allow us to calculate any 4WM process. The final results were recast as a product of Liouville space operators [Eqs. (49) and (53)], or in terms of the four-time correlation function of the dipole operator [Eq. (57)]. We then developed the factorization approximation [Eqs. (60) and (63)], which simplifies these expressions considerably. In this section, we shall consider the problem of spontaneous Raman and fluorescence spectroscopy. General formal expressions analogous to those obtained for 4WM will be derived. This will enable us to treat both experiments in a similar fashion and compare their information content. We shall start with the ordinary absorption lineshape. Consider our system interacting with a stationary monochromatic electromagnetic field with frequency $\omega$. The total initial density matrix is given by

$$|\rho(-\infty)\rangle\rangle = |\rho_S(-\infty)\rangle\rangle\,|\rho_B(-\infty)\rangle\rangle\,|\rho_R(-\infty)\rangle\rangle, \tag{65}$$

where the system and the bath are in thermal equilibrium, and the incident field has $n$ photons in the incoming mode, that is,

$$|\rho_R(-\infty)\rangle\rangle = |n, n\rangle\rangle. \tag{66}$$

In an absorption experiment, we measure the steady-state rate of change of the occupation number of this mode. The operator representing this quantity is

$$B = \frac{d}{dt} a_1^\dagger a_1 = iL_{\text{int}} a_1^\dagger a_1 = i[H_{\text{int}}, a_1^\dagger a_1]. \tag{67}$$

The absorption lineshape $\sigma(\omega)$ is given by the linear response to the stationary external field. To that end, we need to evaluate the density matrix to first order in $L_{\text{int}}$. We then have

$$\sigma(\omega) = \langle\langle B | \rho^{(1)} \rangle\rangle. \tag{68}$$

Making use of Eqs. (67) and (68), and eliminating some numerical constants, we obtain

$$\sigma(\omega) = (-i)^2 \int_0^\infty dt \langle\langle a_1^\dagger a_1 | \mathscr{V} G(t) \mathscr{V} | \rho(-\infty) \rangle\rangle \exp(i\omega t) \tag{69a}$$

or

$$\sigma(\omega) = -i \langle\langle a_1^\dagger a_1 | \mathscr{V} \hat{G}(\omega) \mathscr{V} | \rho(-\infty) \rangle\rangle. \tag{69b}$$

Equation (69b) can be recast in the form

$$\sigma(\omega) = -2 \,\text{Im} \sum_{a,b} P(a) |\mu_{ab}|^2 \langle \hat{G}_{ba}(\omega) \rangle. \tag{70}$$

We next turn to the spontaneous Raman and fluorescence lineshapes. In an SRF experiment, we have a single incident classical field ($\omega_1$) and a single scattered mode ($\omega_2$). We shall use the Hamiltonian [Eq. (2)] with the only difference that the sum in Eq. (4) runs over $j = 1, 2$, with $E_1$ being the classical incident field and $E_2$ being the scattered field, which will be treated quantum mechanically. In an SRF experiment, we monitor the scattered field with both time and frequency resolution. The operator representing the rate of emission of $\omega_2$ photons is

$$B = \frac{d}{dt} a_2^\dagger a_2 = iL_{\text{int}} a_2^\dagger a_2 = i[H_{\text{int}}, a_2^\dagger a_2]. \tag{71}$$

The observed rate of photon emission is

$$S'(\omega_1, \omega_2, t) = \langle\langle B | \rho(t) \rangle\rangle. \tag{72}$$

$\rho(t)$ will now be evaluated using third-order perturbation theory, resulting in

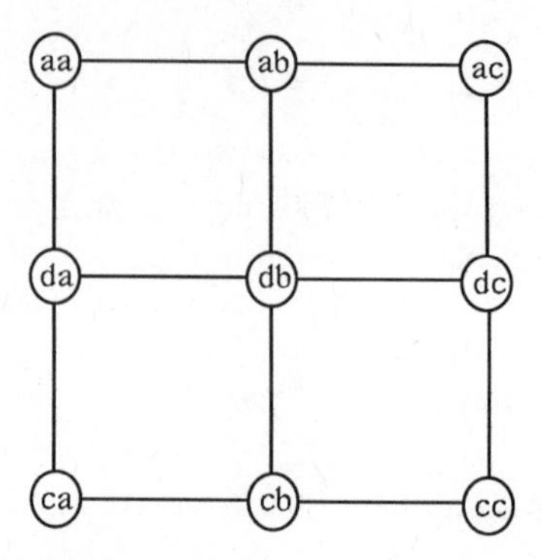


**Figure 7.** Pictorial representation of the pathways in Liouville space that contribute to SRF spectra. Solid lines denote radiative coupling $V$. Horizontal (vertical) lines represent action of $V$ from the right (left). The SRF process is obtained by all pathways that start at $aa$ and end at $cc$ in fourth order (four bonds). There are six pathways that contribute. However, owing to symmetry, we need consider only the three pathways shown in Fig. 8. The other three are obtained by a complex conjugation and permutation of $b$ and $d$.

$$S'(\omega_1,\omega_2,t) = \int_0^\infty dt_1 \int_0^\infty dt_2 \int_0^\infty dt_3 \langle\langle a_2^\dagger a_2 | L_{\text{int}}(t) G(t_3) L_{\text{int}}(t-t_3) \times G(t_2) L_{\text{int}}(t-t_2-t_3) G(t_1) L_{\text{int}}(t-t_1-t_2-t_3) | \rho(-\infty)\rangle\rangle. \tag{73}$$

When Eq. (73) is explicitly evaluated, we find that within the rotating wave approximation we need to calculate six pathways (Fig. 7). The contribution of three of them is the complex conjugate of the others. We therefore need to consider only the three pathways shown in Fig. 8. We then get[11,12,57,64]

$$\begin{aligned} S'(\omega_1,\omega_2,t) = 2\,\text{Re} \sum_{a,b,c,d} P(a)\mu_{ab}\mu_{bc}\mu_{cd}\mu_{da} \int_0^\infty dt_1 \int_0^\infty dt_2 \int_0^\infty dt_3 \\ \times \{\langle G_{cb}(t_3) G_{db}(t_2) G_{ab}(t_1)\rangle \exp[-i\omega_2 t_3 - i\omega_1 t_1] \\ \times E^*(t-t_1-t_2-t_3) E(t-t_2-t_3) \\ + \langle G_{dc}(t_3) G_{db}(t_2) G_{ab}(t_1)\rangle \exp[i\omega_2 t_3 - i\omega_1 t_1] \\ \times E^*(t-t_1-t_2-t_3) E(t-t_2-t_3) \\ + \langle G_{dc}(t_3) G_{ac}(t_2) G_{ab}(t_1)\rangle \exp[i\omega_2 t_3 - i(\omega_1-\omega_2)t_2 - i\omega_1 t_1] \\ \times E^*(t-t_1-t_2-t_3) E(t-t_3)\}, \end{aligned} \tag{74}$$

where the notation is identical to that of Eq. (49). The three terms in Eq. (74) correspond, respectively, to pathways (i)–(iii) of Fig. 8. In a steady-state

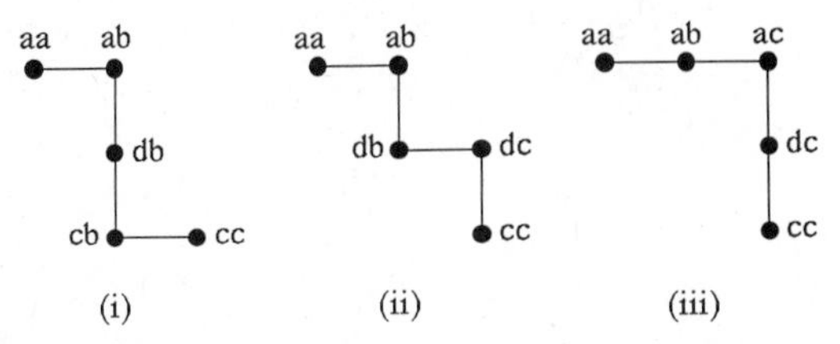


**Figure 8.** The three Liouville space pathways that contribute to SRF spectra. The three terms in each of Eqs. (74)–(78) correspond, respectively, to pathways (i), (ii), and (iii).

experiment, $E(\tau)$ is independent of $\tau$. Setting $E(\tau) = E$ in Eq. (74) results in the following expression for steady-state SRF[64]:

$$S_{\mathrm{SRF}}(\omega_1, \omega_2) = 2\,\mathrm{Im} \sum_{a,b,c,d} P(a)\mu_{ab}\mu_{bc}\mu_{cd}\mu_{da}[\langle \hat{G}_{cb}(-\omega_2)\hat{G}_{db}(0)\hat{G}_{ab}(-\omega_1)\rangle + \langle \hat{G}_{dc}(\omega_2)\hat{G}_{db}(0)\hat{G}_{ab}(-\omega_1)\rangle + \langle \hat{G}_{dc}(\omega_2)\hat{G}_{ac}(\omega_2 - \omega_1)\hat{G}_{ab}(-\omega_1)\rangle]. \tag{75}$$

In pathways (i) and (ii), the system first interacts twice with the $\omega_1$ field at times $t - t_1 - t_2 - t_3$ and $t - t_2 - t_3$ and then twice with the $\omega_2$ mode at times $t - t_3$ and $t$. In pathway (iii), the system first interacts with $\omega_1$ at time $t - t_1 - t_2 - t_3$, then, at time $t - t_2 - t_3$, it interacts with $\omega_2$; then, at time $t - t_3$, it interacts again with $\omega_1$, and finally, at time $t$, it interacts with $\omega_2$. It also should be noted that the steady-state rate [Eq. (75)] is closely related to $\mathrm{Im}\,\chi^{(3)}(-\omega_2, \omega_2, -\omega_1, \omega_1)$. All the terms appearing in Eq. (75) appear also in $\chi^{(3)}$. However, $\chi^{(3)}$ contains more terms, which do not contribute to spontaneous emission, due to the quantum nature of $\omega_2$. In analogy to Section IV, we can further recast Eq. (75) using the four-point correlation function of the dipole operator [Eq. (54)] resulting in[11,12,64]

$$\begin{aligned} S'(\omega_1, \omega_2, t) = 2\,\mathrm{Re} \int_0^\infty dt_1 \int_0^\infty dt_2 \int_0^\infty dt_3 \{ & F(0, t_1 + t_2 + t_3, t_1 + t_2, t_1) \\ & \times \exp[-i\omega_2 t_3 - i\omega_1 t_1] E^*(t - t_1 - t_2 - t_3) E(t - t_2 - t_3) \\ & + F(0, t_1 + t_2, t_1 + t_2 + t_3, t_1) \\ & \times \exp[i\omega_2 t_3 - i\omega_1 t_1] E^*(t - t_1 - t_2 - t_3) E(t - t_2 - t_3) \\ & + F(0, t_1, t_1 + t_2 + t_3, t_1 + t_2) \\ & \times \exp[i\omega_2 t_3 - i(\omega_1 - \omega_2) t_2 - i\omega_1 t_1] \\ & \times E^*(t - t_1 - t_2 - t_3) E(t - t_3) \}. \end{aligned} \tag{76}$$

Here the three terms correspond, respectively, to the three terms of Eq. (75). Finally, let us consider the factorization approximation. In pathways (i) and (ii), the system is in the excited state during the $t_2$ interval, whereas in pathways (iii), it is in the gound state. Introducing the projection operator [Eq. (59)] in Eq. (74) results in[64]

$$
\begin{aligned}
S'(\omega_1, \omega_2, t) = 2\,\mathrm{Re} \sum_{a,b,c,d} P(a)\mu_{ab}\mu_{bc}\mu_{cd}\mu_{da} \int_0^\infty dt_1 \int_0^\infty dt_2 \int_0^\infty dt_3 \\
&\{I_{cb}(t_3)I_{db}(t_2)I_{ab}(t_1)J_e(t_3)J_g^*(t_1)\exp(-i\omega_2 t_3 - i\omega_1 t_1) \\
&\times E^*(t - t_1 - t_2 - t_3)E(t - t_2 - t_3) \\
&+ I_{dc}(t_3)I_{db}(t_2)I_{ab}(t_1)J_e^*(t_3)J_g^*(t_1)\exp[i\omega_2 t_3 - i\omega_1 t_1] \\
&\times E^*(t - t_1 - t_2 - t_3)E(t - t_2 - t_3) \\
&+ I_{dc}(t_3)I_{ac}(t_2)I_{ab}(t_1)J_g(t_3)J_g^*(t_1) \\
&\times \exp[i\omega_2 t_3 - i(\omega_1 - \omega_2)t_2 - i\omega_1 t_1] \\
&\times E^*(t - t_1 - t_2 - t_3)E(t - t_3)\}. \qquad (77)
\end{aligned}
$$

For steady-state experiments we take $E(t)$ in Eq. (77) to be independent of $t$, and we get

$$
\begin{aligned}
&S_{\mathrm{SRF}}(\omega_1, \omega_2) \\
&= 2\,\mathrm{Im} \sum_{a,b,c,d} P(a)\mu_{ab}\mu_{bc}\mu_{cd}\mu_{da} \\
&\times \left( \frac{\bar{I}_{bc}^*(\omega_2)I_{ba}^*(\omega_1)}{\omega_{bd} + i\Gamma_{bd}} - \frac{\bar{I}_{dc}(\omega_2)I_{ba}^*(\omega_1)}{\omega_{bd} + i\Gamma_{bd}} - \frac{I_{dc}(\omega_2)I_{ba}^*(\omega_1)}{\omega_2 - \omega_1 - \omega_{ac} + i\Gamma_{ac}} \right). \qquad (78)
\end{aligned}
$$

In conclusion, in this section we presented the formal expressions for the absorption lineshape [Eq. (70)] and for spontaneous Raman and fluorescence spectroscopy. For the latter, we derived Liouville space expressions in the time and the frequency domain [Eqs. (74) and (75)], an ordinary correlation function expression [Eq. (76)], and, finally, the factorization approximation resulted in Eqs. (77) and (78). The factorization approximation is expected to hold in many cases for steady-state experiments and for time-resolved experiments with low temporal resolution. It is possible to observe a time-dependent shift of spontaneous emission lineshapes using picosecond excitation and detection [66–68]. This shift arises from the reorganization process of the solvent and also from vibrational relaxation that occurs during the $t_2$ time interval. A proper treatment of these effects requires going beyond the present factorization approximation, which assumes an instantaneous reorganization. Our formalism is ideally suited for developing microscopic models for the solvent reorganization and incorporating them properly in $G(t_2)$. We have recently developed such a theory by introducing a solvation coordinate. The theory is based on a factorization approximation for all the solvent degrees of freedom, except for the solvation coordinate whose dynamics is related to the dielectric properties of the solvent.[68] Some calculations of time-dependent fluorescence lineshapes will be presented in Sec. VIII, following the introduction of the vibrational relaxation processes, which constitute another factor contributing to these lineshapes.

## VI. NONLINEAR OPTICS OF POLYATOMIC HARMONIC MOLECULES IN CONDENSED PHASES—"EIGENSTATE-FREE" SPECTROSCOPY

The expressions developed in Sections IV and V for the nonlinear response function and for the SRF lineshape include four summations over molecular states $(a, b, c, d)$. These summations can be carried out easily for small molecules with a few relevant levels. However, for large polyatomic molecules, they become intractable. In this section, we recast the previous expressions in terms of generalized molecular polarizabilities. This allows us to formally eliminate two of the four summations. For a simple, but realistic, model of harmonic molecules, it is possible to evaluate these polarizabilities using Green's function Fourier-transform techniques.[12,44,45,58] We thus end up with a closed expression most suitable for numerical computations, both in the time domain $[R(t_3, t_2, t_1)]$ and in the frequency domain $[\hat{R}(\omega_1 - \omega_2 + \omega_3, \omega_1 - \omega_2, \omega_1)]$. The time-domain formula can be simplified even further for this model by carrying out all four summations, resulting in an explicit expression for $R(t_3, t_2, t_1)$ in which all the summations have been eliminated. This formula,[45] which is particularly useful for time-domain 4WM, will be developed in Section X [Eq. (133)]. The transformation to the frequency domain is, however, more complex in this case, and the calculation of $\hat{R}(\omega_1 - \omega_2 + \omega_3, \omega_1 - \omega_2, \omega_1)$ via Eq. (30) requires the performing of a triple Fourier transform (rather than the single Fourier transforms required in this section). For simplicity, we hereafter assume that the lifetimes of electronically excited vibronic states are the same, that is, $\gamma_b = \gamma_d = \gamma$, and that for the ground electronic state $\gamma_a = \gamma_c = \gamma'$. We shall now introduce the following operators[64]:

$$T(t) = V \exp(-iH_e t - \gamma' t/2) V J_g(t), \tag{79a}$$

$$T(t) = V \exp(-iH_g t - \gamma t/2) V J_e(t), \tag{79b}$$

$$\tilde{T}(t) = V \sigma_g \exp(-iH_g t - \gamma t/2) V J_g^*(t), \tag{79c}$$

We further introduce the ground-state partition function

$$Z = \mathrm{Tr}_S \exp(-H_g/kT). \tag{79d}$$

Equation (60) can then be written in the form[45]

$$\begin{aligned}
R(t_3, t_2, t_1) = \sum_{a,c} P(a)\{ & -T^*_{ac}(t_3) I_{ac}(t_2) T^*_{ca}(t_1) \exp[-i\varepsilon_a t_3 - i\varepsilon_a t_1] \\
& + T_{ca}(t_3) I_{ac}(t_2) T^*_{ca}(t_1) \exp[i\varepsilon_c t_3 - i\varepsilon_a t_1] \\
& + T_{ac}(t_3) I_{ca}(t_2) T_{ca}(t_1) \exp[i\varepsilon_a t_3 + i\varepsilon_a t_1] \\
& - T^*_{ca}(t_3) I_{ca}(t_2) T_{ca}(t_1) \exp[-i\varepsilon_c t_3 + i\varepsilon_a t_1]\} \\
& + \sum_{b,d} \{\bar{T}^*_{db}(t_3) I_{db}(t_2) \tilde{T}^*_{bd}(t_1) \exp[-i\varepsilon_d t_3 - i\varepsilon_d t_1] \\
& + \bar{T}^*_{db}(t_3) I_{db}(t_2) \tilde{T}_{db}(t_1) \exp[-i\varepsilon_d t_3 + i\varepsilon_b t_1] \\
& - \bar{T}_{bd}(t_3) I_{db}(t_2) \tilde{T}_{db}(t_1) \exp[i\varepsilon_b t_3 + i\varepsilon_b t_1] \\
& - \bar{T}_{bd}(t_3) I_{db}(t_2) \tilde{T}^*_{bd}(t_1) \exp[i\varepsilon_b t_3 - i\varepsilon_d t_1]\}.
\end{aligned} \tag{80}$$

Upon the substitution of Eq. (80) into Eq. (30), we obtain the response function in the frequency domain. To that end, we introduce the following definitions:

$$K(\omega) = -i \int_0^\infty d\tau \exp(i\omega\tau) T(\tau), \tag{81a}$$

$$\bar{K}(\omega) = -i \int_0^\infty d\tau \exp(i\omega\tau) \bar{T}(\tau), \tag{81b}$$

$$\tilde{K}(\omega) = -i \int_0^\infty d\tau \exp(i\omega\tau) \tilde{T}(\tau), \tag{81c}$$

We then have

$$\begin{aligned}
& \hat{R}(\omega_1 - \omega_2 + \omega_3, \omega_1 - \omega_2, \omega_1) \\
& = \sum_{a,c} P(a) \Bigg( -\frac{K^*_{ac}(\varepsilon_a - \omega_1 + \omega_2 - \omega_3) K^*_{ca}(\varepsilon_a - \omega_1)}{\omega_1 - \omega_2 - \omega_{ac} + i\Gamma_{ac}} \\
& - \frac{K_{ca}(\varepsilon_c + \omega_1 - \omega_2 + \omega_3) K^*_{ca}(\varepsilon_a - \omega_1)}{\omega_1 - \omega_2 - \omega_{ac} + i\Gamma_{ac}} \\
& + \frac{K_{ac}(\varepsilon_a + \omega_1 - \omega_2 + \omega_3) K_{ca}(\varepsilon_a + \omega_1)}{\omega_1 - \omega_2 - \omega_{ca} + i\Gamma_{ca}}
\end{aligned}$$

$$
\begin{aligned}
&+ \frac{K^*_{ca}(\varepsilon_c - \omega_1 + \omega_2 - \omega_3) K_{ca}(\varepsilon_a + \omega_1)}{\omega_1 - \omega_2 - \omega_{ca} + i\Gamma_{ca}}\Bigg) \\
&+ \sum_{b,d} \left( \frac{\bar{K}^*_{db}(\varepsilon_d - \omega_1 + \omega_2 - \omega_3) \tilde{K}^*_{bd}(\varepsilon_d - \omega_1)}{\omega_1 - \omega_2 - \omega_{db} + i\Gamma_{db}} \right. \\
&- \frac{\bar{K}^*_{db}(\varepsilon_d - \omega_1 + \omega_2 - \omega_3) \tilde{K}^*_{db}(\varepsilon_b - \omega_1)}{\omega_1 - \omega_2 - \omega_{db} + i\Gamma_{db}} \\
&- \frac{\bar{K}_{bd}(\varepsilon_b + \omega_1 - \omega_2 + \omega_3) \tilde{K}_{db}(\varepsilon_b + \omega_1)}{\omega_1 - \omega_2 - \omega_{db} + i\Gamma_{db}} \\
&\left. + \frac{\bar{K}_{bd}(\varepsilon_b + \omega_1 - \omega_2 + \omega_3) \tilde{K}^*_{bd}(\varepsilon_d - \omega_1)}{\omega_1 - \omega_2 - \omega_{db} + i\Gamma_{db}} \right).
\end{aligned} \tag{82}
$$

Similarly, the absorption lineshape [Eq. (70)] can be recast in the form

$$
\sigma(\omega) = -2\,\mathrm{Im} \sum_a P(a) K_{aa}(\varepsilon_a + \omega). \tag{83}
$$

For the SRF lineshapes [Eq. (78)], we have

$$
\begin{aligned}
S_{\mathrm{SRF}}(\omega_1, \omega_2) = 2\,\mathrm{Im} &\left( \sum_{b,d} \frac{[\bar{K}_{bd}(\varepsilon_b - \omega_2) - \bar{K}^*_{db}(\varepsilon_d - \omega_2)] \tilde{K}_{db}(\varepsilon_b - \omega_1)}{\omega_{bd} + i\Gamma_{bd}} \right. \\
&\left. - \sum_{a,c} P(a) \frac{K_{ca}(\varepsilon_c + \omega_2) K^*_{ca}(\varepsilon_a + \omega_1)}{\omega_2 - \omega_1 + \omega_{ca} + i\Gamma_{ca}} \right).
\end{aligned} \tag{84}
$$

The expressions derived in Sections IV and V explicitly contain four summations over molecular states (over $a$, $b$, $c$, and $d$). In Eqs. (80), (82), and (84), these summations have been rearranged, so that only two summations appear explicitly (either over $a$, $c$ or over $b$, $d$). The other summations are "buried" in the definitions of $T(t)$, $\bar{T}(t)$, and $\tilde{T}(t)$ [Eq. (79)]. A significant reduction in computational effort may therefore be achieved if we can find an "eigenstate-free" procedure for the evaluation of the matrix elements of $T(t)$, $\bar{T}(t)$, and $\tilde{T}(t)$ without performing any summations. This may be done for a specific model of harmonic molecules, which will now be introduced. Consider a polyatomic harmonic molecule with $N$ vibrational modes. Its Hamiltonian is

$$
H = |g\rangle H_g \langle g| + |e\rangle [\omega_{eg} + H_e - i\gamma/2] \langle e|, \tag{85a}
$$

with

$$H_g = \frac{1}{2}\sum_{j=1}^{N} \omega_j''(p_j''^2 + q_j''^2 - 1), \tag{85b}$$

$$H_e = \frac{1}{2}\sum_{j=1}^{N} \omega_j'(p_j'^2 + q_j'^2 - 1) \tag{85c}$$

and

$$p_j' = \left(\frac{1}{m_j\omega_j'\hbar}\right)^{1/2} P_j', \tag{86}$$

$$q_j' = \left(\frac{m_j\omega_j'}{\hbar}\right)^{1/2} Q_j'. \tag{87}$$

We adopt here the common spectroscopic notation and use a superscript double prime and prime to denote quantities belonging to the ground and to the electronically excited states, respectively. $P_j'$ and $Q_j'$ are the conjugate momentum and the normal coordinate of the $j$th mode of the excited state. The transformation [Eqs. (86) and (87)] defines $p_j'$ and $q_j'$, which are the dimensionless momentum and normal coordinate of the $j$th mode. A similar transformation between $p_j''$, $q_j''$, and $P_j''$, $Q_j''$ is defined by changing all prime indexes in Eqs. (86) and (87) to double primes. $\omega_j'$ $(\omega_j'')$ and $m_j$ are the frequency and the mass of the $j$th mode. The present Hamiltonian [Eq. (85)] is a special case of the general two-manifold Hamiltonian [Eq. (41)]. We shall now introduce a vector notation and define the $N$ component vectors $\mathbf{q}'$ and $\mathbf{q}''$, whose components are $q_j'$ *and* $q_j''$, $j = 1, \ldots, N$, respectively. The normal modes $\mathbf{q}'$ and $\mathbf{q}''$ are not necessarily the same, and, most generally, they may be related by the transformation

$$\mathbf{q}' = \mathbf{S}\mathbf{q}'' + \mathbf{D}, \tag{88}$$

where $\mathbf{S}$ is the Dushinsky transformation matrix. $\mathbf{D}$ is an $N$-component vector, whose components $D_j$ denote the dimensionless displacements of the equilibrium positions between the two electronic states. In this section, we shall introduce an additional simplifying assumption, the Condon approximation, which implies that the electronic dipole operator is weakly dependent on the nuclear coordinates, so that $\mu_{ij}$ may be simply replaced by the Franck–Condon factors. Setting the electronic dipole matrix element to be unity, we get

$$\mu_{ij} = \langle \mathrm{i} | j \rangle. \tag{89}$$

The molecular eigenstates will be denoted in this section by $|\mathbf{n}\rangle, |\mathbf{m}\rangle, \ldots,$ where

$$|\mathbf{n}\rangle = \prod_{j=1}^{N} |n_j\rangle, \tag{90a}$$

$$|\mathbf{m}\rangle = \prod_{j=1}^{N} |m_j\rangle. \tag{90b}$$

In this case, we have:

$$T_{\mathbf{m}''\mathbf{n}''}(\tau) = G_{\mathbf{m}''\mathbf{n}''}(\tau) J_g(\tau), \tag{91a}$$

$$\bar{T}_{\mathbf{m}'\mathbf{n}'}(\tau) = \tilde{G}_{\mathbf{m}'\mathbf{n}'}(\tau) J_e(\tau), \tag{91b}$$

$$\tilde{T}_{\mathbf{m}'\mathbf{n}'}(\tau) = Z^{-1} \tilde{G}_{\mathbf{m}'\mathbf{n}'}(\tau - i\beta) J_g^*(\tau), \tag{91c}$$

where $Z$ is given by Eq. (79d) and the Green's function is defined by

$$G_{\mathbf{m}''\mathbf{n}''}(\tau) = \langle \mathbf{m}''| \exp(-iH_e \tau) |\mathbf{n}''\rangle, \tag{92a}$$

$$\tilde{G}_{\mathbf{m}'\mathbf{n}'}(\tau) = \langle \mathbf{m}'| \exp(-iH_g \tau) |\mathbf{n}'\rangle. \tag{92b}$$

We shall be interested in calculating the matrix elements of $G$ and $\tilde{G}$ [Eq. (92)]. Their calculation is formally very similar, and the expressions for $G_{\mathbf{m}'\mathbf{n}''}(t)$ may be applied to calculate the matrix element $\tilde{G}_{\mathbf{m}'\mathbf{n}'}(t)$ by simply changing $\mathbf{D}$ to $-\mathbf{S}^{-1}\mathbf{D}$, $\mathbf{S}$ to $\mathbf{S}^{-1}$, and exchanging $\omega_j'$ and $\omega_j''$. It will therefore be sufficient to focus on the evaluation of $G_{\mathbf{m}''\mathbf{n}''}(t)$. By using the generating function of Hermite polynominals and performing a Gaussian integration, it is possible to derive a general expression for $G_{\mathbf{m}''\mathbf{n}''}(\tau)$ for the present model. The final result is considerably simplified when the normal modes in the ground and in the excited electronic states are identical, that is, when the Dushinsky transformation matrix is diagonal

$$S_{ij} = (\omega_i'/\omega_i'')^{1/2} \delta_{ij}. \tag{93}$$

In this case we have (where for brevity we drop the "indexes in $\mathbf{m}''$ and $\mathbf{n}''$"),

$$G_{\mathbf{mn}}(t) = \prod_{j=1}^{N} G_{m_j n_j}(t). \tag{94}$$

To simplify the notation, we shall hereafter consider a single mode and omit all the $j$ subscripts from $\omega_j'$, $\omega_j''$, $D_j$, $n_j$, $m_j$. The total Green's function can then be calculated using Eq. (94). For a single mode, we have[58]

$$G_{mn}(t) = G_{00}(t) W_{mn}(t) \tag{95}$$

where

$$G_{00}(t) = |\psi(t)|^{-1/2} \exp[D^2 f(t)], \tag{96a}$$

$$\psi(t) = \frac{(\omega_+)^2}{4\omega'\omega''}[1 - (\omega_-/\omega_+)^2 \exp(-2i\omega' t)], \tag{96b}$$

and

$$f(t) = -\frac{\omega''[1 - \exp(-i\omega' t)]}{\omega_+ - \omega_- \exp(-i\omega' t)}. \tag{96c}$$

$W_{mn}(t)$ is given by

$$W_{mn}(t) = (m!n!2^{m+n})^{-1/2} \sum_{p=0}^{m} \sum_{q=0}^{n} (-1)^q \binom{m}{p}\binom{n}{q} [\alpha(t)]^{m+n-p-q} [\tilde{\alpha}(t)]^{p+q}$$
$$\cdot H_{m+n-p-q}[\lambda f(t) D/\alpha(t)] H_{p+q}(0), \tag{96d}$$

with

$$\omega_\pm = \omega' \pm \omega'', \tag{97a}$$

$$\lambda = (\omega'/\omega'')^{1/2}, \tag{97b}$$

$$\alpha(t) = \left(\frac{1}{2}\frac{\omega_- - \omega_+ \exp(-i\omega' t)}{\omega_+ - \omega_- \exp(-i\omega' t)}\right)^{1/2}, \tag{97c}$$

$$\tilde{\alpha}(t) = \left(\frac{1}{2}\frac{\omega_- + \omega_+ \exp(-i\omega' t)}{\omega_+ + \omega_- \exp(-i\omega' t)}\right)^{1/2}. \tag{97d}$$

Here $H_n(q)$ are the Hermite polynomials. We further have

$$H_n(0) = \begin{cases} 0, & n \text{ odd}, \\ (-1)^{n/2} n!/(n/2)!, & n \text{ even}. \end{cases} \tag{97e}$$

In addition, the ground-state partition function [Eq. (79d)] is given by

$$Z = \prod_{j=1}^{N} [1 - \exp(-\hbar\omega_j''/kT)]^{-1}. \tag{97f}$$

$\tilde{G}_{\mathbf{m}'\mathbf{n}'}(\tau)$ can be obtained simply by using the substitutions $\omega_j'' \to \omega_j'$, $\omega_j' \to \omega_j''$,

$D_j \to -(\omega_j''/\omega_j')^{1/2} D_j$ in the expression for $G_{\mathbf{m}''\mathbf{n}''}(\tau)$. The more general expression of $G_{\mathbf{m}''\mathbf{n}''}(\tau)$, which includes the Dushinsky rotation, was derived elsewhere.[58] Equations (91) and (95), together with Eqs. (80)–(84), provide a general, efficient algorithm for calculating any 4WM and SRF lineshape from large polyatomic molecules.

In concluding this section, we present an expression for the absorption lineshape [Eq. (83)]. Using the above definitions, it assumes the form

$$\sigma(\omega) = -2\,\mathrm{Im} \int_0^\infty dt\, \sigma(t) J_g(t) \exp(i\omega t), \tag{98a}$$

where

$$\sigma(t) = \sum_{\mathbf{n}} P(\mathbf{n}) G_{\mathbf{nn}}(t). \tag{98b}$$

The absorption correlation function $\sigma(t)$ at temperature $T$, is given by[58]

$$\sigma(t) = [\psi_T(t)]^{-1/2} \exp[D^2 f_T(t)], \tag{99a}$$

with

$$\psi_T(t) = \frac{1}{4\omega'\omega''}(\omega'' C_+ A_- + \omega' C_- A_+)(\omega' C_+ A_- + \omega'' C_- A_+) \tag{99b}$$

and

$$f_T(t) = -\frac{\omega'' C_- A_-}{\omega'' C_+ A_- + \omega' C_- A_+}, \tag{99c}$$

where

$$C_\pm = 1 \pm \exp(-i\omega' t), \tag{99d}$$

$$A_\pm = (\bar{n} + 1) \pm \bar{n} \exp(i\omega'' t), \tag{99e}$$

$$\bar{n} = [\exp(\hbar\omega''/kT) - 1]^{-1}. \tag{99f}$$

When $T = 0$ K, $\bar{n} = 0$, and $A_\pm = 1$, we have $\sigma(t) = G_{00}(t)$ [Eq. (96a)]. When $\omega = \omega' = \omega''$, $\psi_T(t) = 1$, $f_T(t) = -\frac{1}{2} C_- A_-$ and Eq. (99a) reduces to,

$$\sigma(t) = \exp\{\tfrac{1}{2} D^2 (\bar{n} + 1)[\exp(-i\omega t) - 1] + \tfrac{1}{2} D^2 \bar{n} [\exp(i\omega t) - 1]\}. \tag{100}$$

## VII. COHERENT VERSUS SPONTANEOUS RAMAN SPECTROSCOPY

There is currently a considerable interest in extracting useful structural and dynamical information on complex molecular systems from spontaneous and coherent Raman lineshapes. Some of the systems extensively studied are conjugated polyenes, aromatic molecules, and molecules of biological interest, such as porphyrins and $\beta$-carotenes in solution.[7–12,34–56] Raman lines are narrow, two-photon resonances, which occur both in spontaneous emission and in 4WM (coherent Raman) spectroscopy. In this section, we make a specific application of our general results to Raman lineshapes. We shall first present the expressions for these two types of Raman experiments and then compare them in detail. We start with spontaneous Raman spectroscopy. Typically, it is possible to distinguish between two types of contributions to the spontaneous emission spectra of polyatomic molecules in condensed phases. These are denoted as Raman and fluorescence, respectively.[11,69–75] The fluorescence component is sometimes referred to as hot luminescence or redistribution. The Raman components are relatively narrow emission lines occurring at $\omega_1 - \omega_2 = \omega_{ca}$, where $|a\rangle$ and $|c\rangle$ are any pair of ground-state vibronic levels, and their widths are narrower than a few $\text{cm}^{-1}$. The fluorescence lines in solution are much broader (typical width of a few hundred $\text{cm}^{-1}$), and their maxima occur at $\omega_2 = \omega_{bc}$, where levels $|b\rangle$ and $|c\rangle$ belong to the excited and the ground electronic states, respectively (Fig. 2). A clear distinction between these two components may be made by tuning $\omega_1$ across the absorption spectrum. The Raman lines, which occur at a fixed $\omega_1 - \omega_2$, will follow the tuning of $\omega_1$, whereas the fluorescence components will remain roughly in the same positions. The origin of these components can be understood by a close examination of Eq. (84). Consider the third term [pathway (iii)] in Eq. (84). Let us further assume that $\Gamma_{ca} = 0$, which implies an infinite lifetime for vibronic states belonging to the ground electronic state. We then have

$$\frac{1}{\omega_2 - \omega_1 + \omega_{ca}} = -i\pi\delta(\omega_2 - \omega_1 + \omega_{ca}) + \text{PP}\left(\frac{1}{\omega_2 - \omega_1 + \omega_{ca}}\right), \quad (101)$$

where PP stands for the principal part. The $\delta(\omega_2 - \omega_1 + \omega_{ca})$ term will contribute to the Raman component, whereas all the rest of the terms in Eq. (84) will not contain any such resonance and will constitute the "fluorescence." We thus have[64]

$$S_{\text{SRF}}(\omega_1, \omega_2) = S_{\text{RAMAN}}(\omega_1, \omega_2) + S_{\text{FL}}(\omega_1, \omega_2), \quad (102a)$$

$$S_{\text{RAMAN}}(\omega_1, \omega_2) = 2\pi \sum_{a,c} P(a)|K_{ca}(\varepsilon_a + \omega_1)|^2 \delta(\omega_2 - \omega_1 + \omega_{ca}), \quad (102b)$$

and

$$S_{FL}(\omega_1, \omega_2) = 2\,\mathrm{Im}\left[ \sum_{b,d} \frac{[\bar{K}_{bd}(\varepsilon_b - \omega_2) - \bar{K}^*_{db}(\varepsilon_d - \omega_2)]\tilde{K}_{db}(\varepsilon_b - \omega_1)}{\omega_{bd} + i\Gamma_{bd}} + \mathrm{PP}\left( \sum_{a,c} P(a) \frac{K_{ca}(\varepsilon_c + \omega_2) K^*_{ca}(\varepsilon_a + \omega_1)}{\omega_2 - \omega_1 - \omega_{ca}} \right) \right]. \tag{102c}$$

For isolated molecules, in the absence of a bath, an interesting destructive interference takes place, whereby all the contributions to the fluorescence cancel exactly. We then have $S_{FL} = 0$, and the total emission is of the Raman type. Equations (102) then reduce to the Kramer–Heisenberg formula,[76] which can be derived using second-order perturbation theory. The situation is more complicated, however, for spontaneous emission spectra in condensed phases (solutions, matrices, molecular crystals, glasses, chromophores on proteins, etc.). In these cases, the molecule is subject to a random force, resulting from its interaction with a macroscopic number of external degrees of freedom, which constitute a thermal bath. The random force results in a significant line broadening and a loss of much of the structure in these spectra. It also "redistributes" the emitted intensity by reducing the Raman and building the fluorescence component instead. There are three types of experimental observables in a molecular Raman experiment: (1) the absorption spectrum, $\sigma(\omega_1)$ [Eq. (83)]; (2) the excitation profiles $|K_{ca}(\varepsilon_a + \omega_1)|^2$ [Eq. (102b)], obtained by tuning $\omega_1$ and detecting the narrow Raman component at $\omega_2 - \omega_1 = \omega_{ac}$ each pair of ground vibronic states $|a\rangle$ and $|c\rangle$ has its distinct excitation profile; and (3) the fluorescence spectrum, $S_{FL}(\omega_1, \omega_2)$ [Eq. (102c)]. These will be discussed following the introduction of coherent Raman lineshapes.

We shall now consider coherent Raman measurements (CARS and CSRS),[3,4,14,34–43] which are the coherent analogue of the spontaneous Raman spectra. Coherent Raman experiments are a special type of 4WM. They involve two incoming fields (i.e., $\mathbf{k}_3 = \mathbf{k}_1$), and the signal mode is

$$\mathbf{k}_s = 2\mathbf{k}_1 - \mathbf{k}_2, \tag{103a}$$

$$\omega_s = 2\omega_1 - \omega_2. \tag{103b}$$

Since two fields are identical, there are only three permutations of the frequencies (and not six). Writing the frequency permutations explicitly, we get

$$\begin{aligned}\chi^{(3)}(-\omega_s, \omega_1, -\omega_2, \omega_1) = {} & \hat{R}(2\omega_1 - \omega_2, \omega_1 - \omega_2, \omega_1) \\ & + \hat{R}(2\omega_1 - \omega_2, \omega_1 - \omega_2, -\omega_2) \\ & + \hat{R}(2\omega_1 - \omega_2, 2\omega_1, \omega_1),\end{aligned} \tag{104}$$

where $\hat{R}$ is given in Eq. (63). $\chi^{(3)}$ thus contains $3 \times 8 = 24$ terms. In coherent Raman spectroscopy we look for two-photon resonances in the signal, which occur when $\omega_1 - \omega_2$ equals an energy difference between two ground-state or excited-state vibrational states. For our level scheme (Fig. 4) such resonances occur for $\omega_1 - \omega_2 = \pm\omega_{ca}$ or for $\omega_1 - \omega_2 = \pm\omega_{db}$. The names CARS and CSRS refer to the cases where $\omega_1 > \omega_2$ or $\omega_1 < \omega_2$, respectively. Since there are no fundamental differences between the theoretical treatments of the two, we shall focus on the CARS resonances,

$$\omega_1 - \omega_2 = \omega_{ca} \tag{105a}$$

and

$$\omega_1 - \omega_2 = \omega_{db}. \tag{105b}$$

Equation (105a) represents ground-state CARS and Eq. (105b) represents excited-state CARS. It is clear that the third term in Eq. (104), $\hat{R}(2\omega_1 - \omega_2, 2\omega_1, \omega_1)$, cannot contribute to CARS since its two-photon frequency is $2\omega_1$ and not $\omega_1 - \omega_2$. We shall therefore ignore this term. Let us denote the eight pathways of the first term in Eq. (104) by (i)–(viii), and the corresponding ones for the second term in Eq. (104) by (i)′–(viii)′. (See Fig. 6.) Starting with ground-state CARS, we note that there are eight terms containing an $\omega_1 - \omega_2 \pm \omega_{ca}$ denominator. These correspond to diagrams (i), (iv), (v), (viii), and (i)′, (iv)′, (v)′, and (viii)′, that is,

$$\chi_{ca}^{(3)} = \sum_{b,d} \mu_{ab}\mu_{bc}\mu_{cd}\mu_{da}[I_{ba}(2\omega_1 - \omega_2) + I_{bc}^*(\omega_2 - 2\omega_1)] \times \frac{P(a)I_{da}(\omega_1) + P(a)I_{da}(-\omega_2) - P(c)I_{dc}^*(-\omega_1) - P(c)I_{dc}^*(\omega_2)}{\omega_1 - \omega_2 - \omega_{ca} + i\Gamma_{ac}} \tag{106a}$$

where the subscript $ca$ denotes that these are ground-state ($\omega_1 - \omega_2 = \omega_{ca}$) resonances. In the terms corresponding to pathways (i), (i)′, (iv), (iv)′, we have interchanged the dummy summation indexes $a$ and $c$, and $b$ and $d$, in order to recast these expressions into a more compact form. We shall now invoke the rotating wave approximation (RWA), in which we retain only resonant terms in which all denominators contain a difference of a field frequency and a molecular optical frequency, and neglect all terms where at least one denominator is antiresonant. For the ground-state CARS, the only surviving terms are (iv)′ and (v), that is,

$$\chi_{ca}^{(3)} = \sum_{b,d} \mu_{ab}\mu_{bc}\mu_{cd}\mu_{da} \frac{I_{ba}(2\omega_1 - \omega_2)[P(a)I_{da}(\omega_1) - P(c)I_{dc}^*(\omega_2)]}{\omega_1 - \omega_2 - \omega_{ca} + i\Gamma_{ac}}. \tag{106b}$$

In addition, there are eight terms in Eq. (104), which contain an $\omega_1 - \omega_2 - \omega_{db}$ denominator. These correspond to pathways (ii), (iii), (vi), (vii) and (ii)′, (iii)′, (vi)′, and (vii)′, that is,

$$\chi_{db}^{(3)} = \sum_{a,c} P(a)\mu_{ab}\mu_{bc}\mu_{cd}\mu_{da}[\bar{I}_{dc}(2\omega_1 - \omega_2) + \bar{I}_{bc}^*(\omega_2 - 2\omega_1)] \times \frac{I_{da}(\omega_1) + I_{da}(-\omega_2) - I_{ba}^*(-\omega_1) - I_{ba}^*(\omega_2)}{\omega_1 - \omega_2 - \omega_{db} + i\Gamma_{db}} \tag{107a}$$

Within the RWA, only two terms contribute to $\chi_{db}^{(3)}$, which come from pathways (ii) and (iii)′, that is,

$$\chi_{db}^{(3)} = \sum_{a,c} P(a)\mu_{ab}\mu_{bc}\mu_{cd}\mu_{da}\frac{\bar{I}_{dc}(2\omega_1 - \omega_2)[I_{da}(\omega_1) - I_{ba}^*(\omega_2)]}{\omega_1 - \omega_2 - \omega_{db} + i\Gamma_{db}}. \tag{107b}$$

Making use of Eqs. (106b) and (107b), we have finally

$$\begin{aligned}\chi_{\text{CARS}}^{(3)} &= \sum_{a,c} \chi_{ac}^{(3)} + \sum_{b,d} \chi_{db}^{(3)} \\ &= \sum_{a,b,c,d} P(a)\mu_{ab}\mu_{bc}\mu_{cd}\mu_{da} \\ &\quad \times \left(\frac{I_{ba}(2\omega_1 - \omega_2)[I_{da}(\omega_1) - \exp(-\omega_{ca}/kT)I_{dc}^*(\omega_2)]}{\omega_1 - \omega_2 - \omega_{ca} + i\Gamma_{ca}}\right. \\ &\quad \left. + \frac{\bar{I}_{dc}(2\omega_1 - \omega_2)[I_{da}(\omega_1) - I^*{}_{ba}(\omega_2)]}{\omega_1 - \omega_2 - \omega_{db} + i\Gamma_{db}}\right).\end{aligned} \tag{108}$$

Making use of Eqs. (81), we may recast the CARS signal generated at $\mathbf{k}_s = 2\mathbf{k}_1 - \mathbf{k}_2$ into the form

$$S_{\text{CARS}}(\omega_1, \omega_2) = |\chi_{\text{CARS}}^{(3)}(-\omega_s, \omega_1, -\omega_2, \omega_1)|^2, \tag{109a}$$

where

$$\begin{aligned}&\chi_{\text{CARS}}^{(3)}(-\omega_s, \omega_1, -\omega_2, \omega_1) \\ &\quad = \sum_{a,c} \frac{P(a)K_{ac}(\varepsilon_a + 2\omega_1 - \omega_2)[K_{ca}(\varepsilon_a + \omega_1) - \exp(-\omega_{ca}/kT)K_{ac}^*(\varepsilon_c + \omega_2)]}{\omega_1 - \omega_2 - \omega_{ca} + i\Gamma_{ca}} \\ &\quad + \sum_{b,d} \frac{\tilde{K}_{db}^*(\varepsilon_d - 2\omega_1 + \omega_2)[\tilde{K}_{bd}^*(\varepsilon_d - \omega_1) - \tilde{K}_{db}(\varepsilon_b - \omega_2)]}{\omega_1 - \omega_2 - \omega_{db} + i\Gamma_{db}}.\end{aligned} \tag{109b}$$

For comparison, the spontaneous Raman signal [Eq. (102b)] is given by

$$S_{\mathrm{RAMAN}}(\omega_1, \omega_2) = -2\,\mathrm{Im}\,\chi^{(3)}_{\mathrm{RAMAN}}(-\omega_2, \omega_2, -\omega_1, \omega_1), \tag{110a}$$

where

$$\chi^{(3)}_{\mathrm{RAMAN}}(-\omega_2, \omega_2, -\omega_1, \omega_1) \equiv \sum_{a,c} P(a) \frac{K_{ca}(\varepsilon_c + \omega_2) K^*_{ca}(\varepsilon_a + \omega_1)}{\omega_2 - \omega_1 + \omega_{ca} + i\Gamma_{ca}}. \tag{110b}$$

For molecules in solution, $\Gamma_{ac}$ and $\Gamma_{bd}$ are usually much smaller than a characteristic line broadening, which is typically a few hundred $\mathrm{cm}^{-1}$. A Lorentzian with a width of $\Gamma_{ac}$ or $\Gamma_{bd}$ may therefore be approximated by a delta function. We then write

$$\left| \frac{1}{\omega_1 - \omega_2 - \omega_{ca} + i\Gamma_{ca}} \right|^2 \cong \frac{\pi\delta(\omega_1 - \omega_2 - \omega_{ca})}{\Gamma_{ca}}, \tag{111a}$$

$$\left| \frac{1}{\omega_1 - \omega_2 - \omega_{db} + i\Gamma_{db}} \right|^2 \cong \frac{\pi\delta(\omega_1 - \omega_2 - \omega_{db})}{\Gamma_{db}}, \tag{111b}$$

and

$$\mathrm{Im} \frac{1}{\omega_2 - \omega_1 + \omega_{ca}} = -\pi\delta(\omega_2 - \omega_1 + \omega_{ca}). \tag{111c}$$

Making use of Eqs. (109)–(111) and ignoring background terms, we get finally

$$S_{\mathrm{CARS}} = \sum_{a,c} Q^g_{ac}(\omega_1)\delta(\omega_1 - \omega_2 - \omega_{ca}) + \sum_{b,d} Q^e_{bd}(\omega_1)\delta(\omega_1 - \omega_2 - \omega_{db}) \tag{112a}$$

and

$$S_{\mathrm{RAMAN}} = \sum_{a,c} Q_{ac}(\omega_1)\delta(\omega_1 - \omega_2 - \omega_{ca}), \tag{112b}$$

where

$$Q^g_{ac}(\omega_1) = \frac{\pi}{\Gamma_{ca}} |P(a) K_{ac}(\varepsilon_c + \omega_1)|^2 |K_{ca}(\varepsilon_a + \omega_1) - \exp(-\omega_{ca}/kT) K^*_{ac}(\varepsilon_a + \omega_1)|^2, \tag{112c}$$

$$Q^e_{bd}(\omega_1) = \frac{\pi}{\Gamma_{db}} |\bar{K}_{db}(\varepsilon_b - \omega_1)|^2 |\tilde{K}^*_{bd}(\varepsilon_d - \omega_1) - \tilde{K}_{db}(\varepsilon_d - \omega_1)|^2, \tag{112d}$$

$$Q_{ac}(\omega_1) = 2\pi P(a) |K_{ca}(\varepsilon_a + \omega_1)|^2. \tag{112e}$$

$Q_{ac}^{g}(\omega_1)$ is the coherent Raman excitation profile corresponding to a ground-state resonance, and $Q_{bd}^{e}(\omega_1)$ is the coherent Raman excitation profile for an excited-state resonance. $Q_{ac}(\omega_1)$ is the spontaneous Raman excitation profile. These profiles are easily measured by probing the intensity of a particular Raman line ($\omega_1 - \omega_2$ fixed) as a function of $\omega_1$. It should be noted that the excited-state resonances $Q_{db}^{e}$ are induced by the interaction with the bath. In the absence of a bath, an interference causes these resonances to vanish.[6,16,17,21] The excited-state cars resonances are therefore induced by dephasing.

The present formulation provides rigorous microscopic relations of 4WM and SRF lineshapes to standard correlation functions of the medium that can be evaluated by a variety of methods. Such methods include density expansions for pressure broadening in the gas phase and in liquids,[11,77–80] cumulant expansions for phonon broadening,[81,82] and semiclassical and molecular dynamics simulations.[83–86] For polar solvents, it is possible to use a continuum dielectric model for the solvent and relate the necessary correlation functions to the frequency-dependent dielectric constant $\varepsilon(\omega)$.[68,87–92] For the sake of illustration, we now present some calculations performed using a simplified stochastic model, whereby the effect of the bath is modeled as a stochastic Gaussian modulation of the electronic energy gap.[93–95] We assume that due to the random force exerted on the molecule by the bath, the electronic energy gap $\omega_{eg}$ becomes a stochastic function of time (Figs. 2 and 9) and the Hamiltonian [Eq. (85)] assumes the form[44,45,64]

$$H = |g\rangle H_g(Q_s)\langle g| + |e\rangle[\omega_{eg} + \delta\omega_{eg}(t) + H_e(Q_s) - i\gamma/2]\langle e|, \quad (113a)$$

where $\delta\omega_{eg}(t)$ is assumed to be a Gaussian Markov process with

$$\langle \delta\omega_{eg}(t)\rangle = 0, \quad (113b)$$

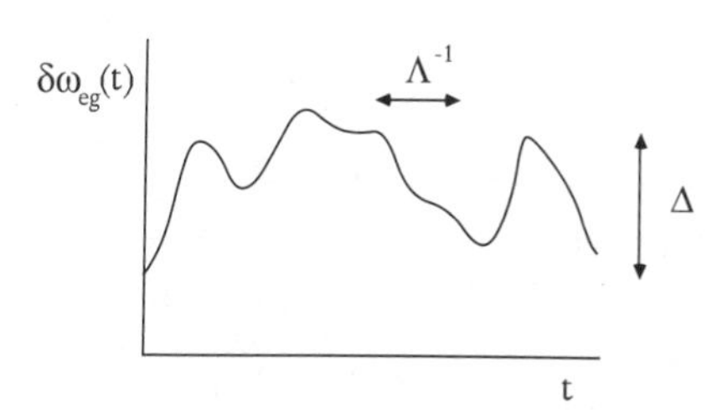


**Figure 9.** The stochastic process $\delta\omega_{eg}(t)$ representing the time-dependent molecular electronic energy gap in a solution [Eq. (113)]. $\Delta$ represents the magnitude of the fluctuations, and $\Lambda^{-1}$ represents their time scale.

and

$$\langle \delta\omega_{eg}(t)\delta\omega_{eg}(0)\rangle = \Delta^2 \exp(-\Lambda t). \tag{113c}$$

Here, the angular brackets $\langle \cdots \rangle$ denote averaging over the stochastic process, $\Delta$ is the root-mean-squared amplitude of the fluctuations, and $\Lambda^{-1}$ is the correlation time of the bath fluctuations. Equations (113) can be obtained from our microscopic model [Eq. (41)] by taking the bath degrees of freedom to be classical and assuming that the bath is sufficiently large that its motions are independent of the absorber. By going to the interaction picture (with respect to the bath Hamiltonian), we may then recast Eq. (41) in the form of Eqs. (113). Within this model, the bath is affecting the system, but the system is not affecting the bath. The Gaussian nature of $\delta\omega_{eg}(t)$ may usually be justified using the central limit theorem. This choice of a stochastic Hamiltonian is based on the assumption that the bath couples mainly to the electronic degrees of freedom, so that the ground-state and the excited-state vibronic manifolds are being stochastically modulated with respect to each other, but no modulation occurs for frequencies of levels belonging to the same electronic manifold, that is, vibrational dephasing is neglected. This is often a realistic assumption. The nonlinear response function can be evaluated also for a more general class of stochastic models.[19] However, for the sake of simplicity we shall restrict the present discussion to this special case. For this model, we have[93–95]

$$J_e(t) = J_g(t) = \exp\{-(\Delta/\Lambda)^2[\exp(-\Lambda t) - 1 + \Lambda t]\}. \tag{114}$$

This simple model provides a good insight on the effect of the bath. Generalization to more complicated situations, whereby $J_e(t) \neq J_g(t)$, and relating $\Delta$ and $\Lambda$ to more microscopic properties of the solvent can be made without a major difficulty.[68] The nature of the lineshape function [Eqs. (64)] depends on the dimensionless parameter

$$\kappa = \Lambda/\Delta. \tag{115}$$

The *fast modulation* (homogeneous) limit is obtained when the correlation time of the bath fluctuations is very fast compared with their magnitude, that is, $\kappa \gg 1$. In this case, the $\exp(-\Lambda\tau)$ on the right-hand side of Eq. (114) vanishes very rapidly and may be ignored. We then get

$$J_e(t) = J_g(t) = \exp(-\hat{\Gamma}t), \tag{116a}$$

where

$$\hat{\Gamma} = \Delta^2/\Lambda. \tag{116b}$$

The absorption lineshape function [Eq. (64)] assumes a Lorentzian form corresponding to homogeneous dephasing, that is,

$$I_{\nu\lambda}(\omega) = \frac{1}{\omega - \omega_{\nu\lambda} + i\Gamma}, \tag{116c}$$

where $\Gamma = \hat{\Gamma} + \gamma/2$. The *slow modulation* (static, inhomogeneous) limit is obtained when $\Lambda \ll \Delta$, that is, $\kappa \ll 1$. In this case, we can make a Taylor expansion of Eq. (114) (short-time approximation) resulting in

$$J_e(t) = J_g(t) = \exp[-\Delta^2 t^2/2]. \tag{116d}$$

The absorption lineshape [the imaginary part of Eq. (64)] assumes, in this case, the Voigt profile (a convolution of a Gaussian and a Lorentzian), and, when $\gamma \ll \Delta$, it becomes a Gaussian. As $\kappa$ is increased, the absorption lineshape thus changes continuously from a Gaussian to a Lorentzian. The full width at half maximum $\Gamma_0$ of the absorption lineshape [the imaginary part of $I_{\nu\lambda}(\omega)$] is displayed in Fig. 10 versus $\kappa$ and may be adequately approximated by the Padé approximant[44]

$$\frac{\Gamma_0}{\Delta} = \frac{2.355 + 1.76\kappa}{1 + 0.85\kappa + 0.88\kappa^2}. \tag{117}$$

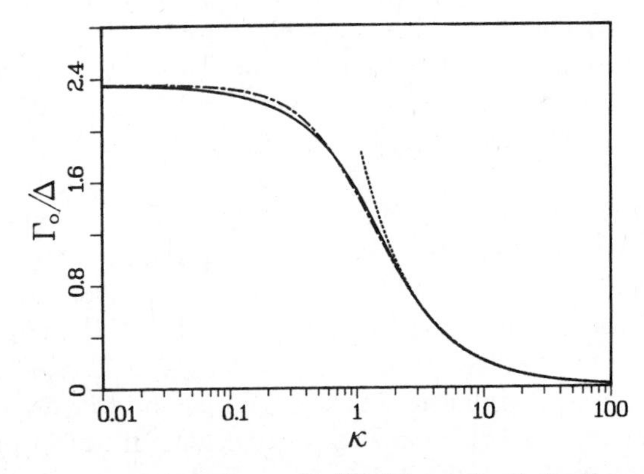


**Figure 10.** The full width at half maximum ($\Gamma_0/\Delta$) of the stochastic lineshape function $\sigma(\omega)$ for a two-level system versus $\kappa$ [Eqs. (98a) and (114) with $\sigma(t) = 1$)], $\gamma = 0$. Solid line is the exact curve (calculated numerically); dot-dashed line is the Padé approximant, Eq. (117); short dashed line is $\Gamma_0/\Delta = 2/\kappa$, which holds in the fast modulation limit when $\kappa \gg 1$ [Eq. (116a)].

The model Hamiltonian defined by Eqs. (85) and (113) is exactly solvable, and we have derived explicit closed-form expressions for the nonlinear response function [Eq. (49)] and for the SRF lineshape [Eq. (75)].[44,45,64] The exact result is in the form of an infinite series, whose first term corresponds to the factorization approximation [Eqs. (82) and (84)]. The factorization approximation was shown to be exact in the fast modulation limit, $\kappa \gg 1$. In addition, under quite general conditions ($\Lambda \gg \gamma$), the "rapid fluctuation" limit, the factorization approximation holds for an arbitrary value of $\kappa$ and provides an excellent approximation over a wide range of broadening parameters. We shall now present numerical calculations of coherent and spontaneous Raman spectroscopy using Eqs. (112) and (113).[45] We first consider a diatomic molecule, with a single vibrational mode, $\omega'' = 1000\ \mathrm{cm}^{-1}$, $\omega' = 875\ \mathrm{cm}^{-1}$, and the dimensionless displacement $D = 1.5$. The temperature is assumed to be low compared with $\omega''$ ($\hbar\omega'' \ll kT$), so that at equilibrium the molecule is in the ground vibronic state. We have calculated the absorption lineshape [Eq. (98)], the dispersed fluorescence $S_{\mathrm{SRF}}(\omega_1, \omega_2)$ [Eq. (102)], and the CARS signal $S_{\mathrm{CARS}}(\omega_1, \omega_2)$ [Eq. (109)]. $K(\omega)$, $\bar{K}(\omega)$, and $\tilde{K}(\omega)$ were calculated using Eqs. (81) and a standard fast Fourier routine. We used 8192 points of integration with a time step of $\Delta t = 10^{-15}$ sec. In Fig. 11 we display the absorption $\sigma(\omega)$, the dispersed fluorescence $S_{\mathrm{SRF}}(\omega_1, \omega_2)$ and the CARS signal $S_{\mathrm{CARS}}(\omega_1, \omega_2)$ for our model. $S_{\mathrm{SRF}}(\omega_1, \omega_2)$ and $S_{\mathrm{CARS}}(\omega_1, \omega_2)$ are displayed as a function of $\omega_2$, with $\omega_1$ fixed at the origin 0–0 transition frequency. The absorption spectrum shows a progression of the excited-state frequency ($\omega' = 1000\ \mathrm{cm}^{-1}$). The dispersed fluorescence consists of a series of broad bands ("fluorescence") and narrow lines ("Raman"), which are peaked at $\omega_1 - \omega_2 = n\omega''$. Similarly, the CARS signal $S_{\mathrm{CARS}}$ contains a series of ground-state narrow resonances at $\omega_1 - \omega_2 = n\omega''$ and a series of excited-state narrow resonances peaked at $\omega_1 - \omega_2 = n\omega'$. $S_{\mathrm{CARS}}$ thus reveals both the ground-state and the excited-state frequencies and is more informative than $S_{\mathrm{SRF}}$.[16,17,21] We shall now consider the Raman excitation profiles for our model $Q^g_{ac}(\omega_1)$, $Q^e_{bd}(\omega_1)$, and $Q_{ac}(\omega_1)$. In Fig. 12, we compare the spontaneous and the coherent Raman excitation profiles of $\beta$ carotene, calculated by using a harmonic model with three Raman active modes. Eqs. (112) were modified to take the degeneracy in $\omega_{ac}$ and $\omega_{bd}$ into account. The broadening parameters were obtained by a fit to experimental spontaneous Raman profiles, and the coherent Raman profiles are predictions using our theory. The details of the calculations are presented in Ref. 96. We have further used the present stochastic model for harmonic molecules to calculate the absorption and the Raman excitation profiles of the 700 nm ($S_0 - S_1$) transition of azulene.[44] The available experimental data[46] consist of the absorption; the Raman fundamentals of the 1400, 1200, and the 900 $\mathrm{cm}^{-1}$ modes; and the first Raman overtone of the 825 $\mathrm{cm}^{-1}$ mode. The measurements were made at room

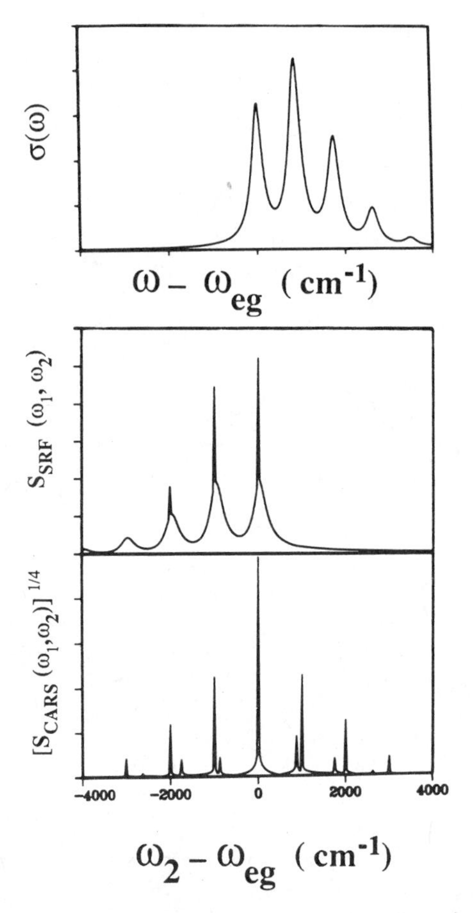


**Figure 11.** Absorption, SRF, and coherent Raman emission for a diatomic molecule with a single vibrational mode $\omega'' = 1000\ \text{cm}^{-1}$, $\omega' = 875\ \text{cm}^{-1}$, and the dimensionless displacement $D = 1.5$. The calculation was done in the fast modulation limit [Eq. (116a)] with $\hat{\Gamma} = 1000\ \text{cm}^{-1}$ and $\gamma = 0$. The absorption spectrum $\sigma(\omega)$ was calculated using Eq. (98a). The SRF spectrum $S_{\text{SRF}}(\omega_1, \omega_2)$ [Eq. (102)] and the coherent Raman signal $S_{\text{CARS}}(\omega_1, \omega_2)$ [Eq. (109)] are plotted versus $\omega_2$. For clarity we have plotted the fourth root of the coherent Raman signal $[S_{\text{CARS}}(\omega_1, \omega_2)]^{1/4}$. $\omega_1$ was taken to be on resonance with the fundamental (0–0) transition, $\omega_1 = \omega_{eg}$.[45]

temperature using $CS_2$ as a solvent. All calculations were made using Eqs. (112). $G_{\mathbf{mn}}(t)$ were first calculated using Eq. (95), and $K_{\mathbf{mn}}(\omega)$ [Eqs. (81)] were then computed using a standard fast Fourier transform routine. We used 8192 time points with a step size of $\Delta t = 6.4 \times 10^{-15}$ sec. The ground-state and the excited-state seven-mode frequencies $\omega'$ and $\omega''$ were taken from Ref. 46, and the displacements $D_j$ and the broadening parameters ($\Delta$ and $\Lambda$) were adjusted

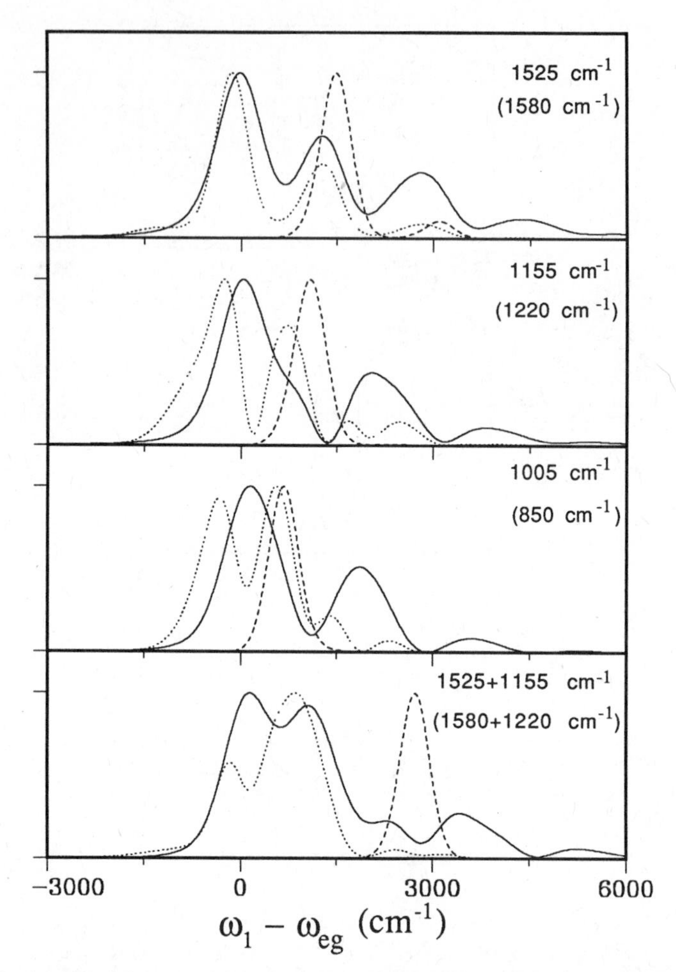


**Figure 12.** Calculated Raman excitation profiles of $\beta$ carotene. A harmonic three-mode model was used with ground state frequencies $\omega_1'' = 1525$, $\omega_2'' = 1155$, and $\omega_3'' = 1005\ \mathrm{cm}^{-1}$ and excited state frequencies $\omega_1' = 1580$, $\omega_2' = 1220$, and $\omega_3' = 850\ \mathrm{cm}^{-1}$. The displacements are $D_1 = 1.12$, $D_2 = .95$, and $D_3 = .65$. The broadening parameters are $\Delta = 362\ \mathrm{cm}^{-1}$ and $\Lambda = 109\ \mathrm{cm}^{-1}$ corresponding to $\kappa = 0.3$ and $\Gamma_0 = 760\ \mathrm{cm}^{-1}$. In each panel we show the spontaneous Raman $Q_{ac}(\omega_1)$ (solid line), the Raman excitation profile corresponding to ground-state resonance $Q_{ac}^{g}(\omega_1)$ (dotted line), and the corresponding excited-state resonance $Q_{bd}^{e}(\omega_1)$ (dashed line). The Raman frequency $\omega_{ca}$ is indicated in each panel; the corresponding excited-state frequency $\omega_{db}$ is indicated in parentheses.[96]

TABLE I
Ground-State and Excited-State Frequencies and the Dimensionless Displacements Used in our Best-Fit Seven-Mode Calculation of Azulene[44–46]

| Mode | $\omega_j''$ ($cm^{-1}$) | $\omega_j'$ ($cm^{-1}$) | $D_j$ |
|---|---|---|---|
| 1 | 406 | 384 | 0.61 |
| 2 | 679 | 662 | 0.76 |
| 3 | 825 | 857 | 1.23 |
| 4 | 900 | 900 | 0.55 |
| 5 | 1260 | 1193 | 0.77 |
| 6 | 1400 | 1388 | 1.09 |
| 7 | 1562 | 1550 | 1.12 |

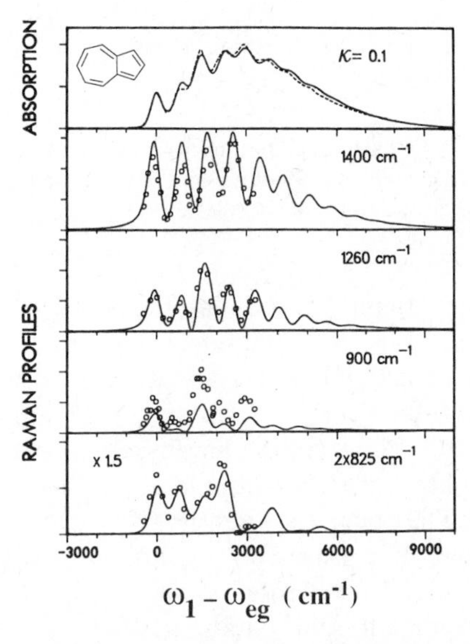


**Figure 13.** Absorption and resonance Raman profiles for azulene in $CS_2$ at 300 K.[44] The solid lines are theoretical curves computed using a seven-mode stochastic harmonic model without Dushinsky rotation (Table I). The absorption curve (upper panel) is $\omega\sigma(\omega)$, where $\sigma(\omega)$ is given in Eq. (98a). The dashed line represents the experimental data.[46] The Raman profiles (lower four panels) were calculated using Eq. (112e). Shown are the experimental data[46] (circles) and the calculated profiles $Q_{ac}(\omega_1)$ for four different Raman transitions, as indicated in each panel. The broadening parameters are $\Delta = 180$ $cm^{-1}$, $\Lambda = 18$ $cm^{-1}$, and $\Gamma_0 = 408$ $cm^{-1}$. $\gamma = 0$ and $\omega_{eg} = 14{,}286$ $cm^{-1}$.

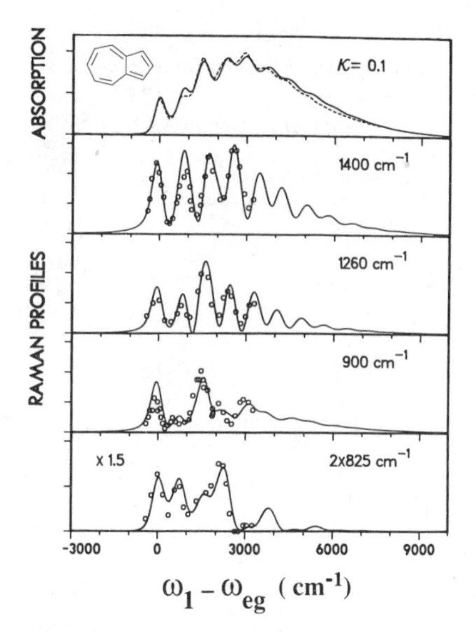


**Figure 14.** The same as Fig. 13, but with a Dushinsky rotation included. We have used a 2 × 2 Dushinsky rotation matrix, coupling the 900 $cm^{-1}$ and the 1400 $cm^{-1}$ modes. The off-diagonal element $S_{4,6}$ between these modes is $-0.20$ [Eq. (88)].

to fit the data. We first attempted to obtain the best fit without a Dushinsky rotation, making use of Eqs. (93)–(95) for $G_{\mathbf{mn}}(t)$. The structural parameters of our best fit are given in Table I. The solvent line-broadening parameters of our fit are $\Delta = 180$ $cm^{-1}$, $\Lambda = 18$ $cm^{-1}$ corresponding to $\kappa = 0.1$, and $\Gamma_0 = 408$ $cm^{-1}$ (Fig. 10). The resulting lineshapes are shown in Fig. 13. It is clear that the overall fit for the absorption and the 1400, 1260, and 825 $cm^{-1}$ modes is satisfactory. The calculated intensity of the 900 $cm^{-1}$ vibration (mode 4) is too weak, and that of the 1400 $cm^{-1}$ vibration (mode 6) is too strong. We therefore introduced a Dushinsky rotation parameter to couple the 1400 and 900 $cm^{-1}$ modes, keeping all other parameters fixed, and we were then able to obtain a good simultaneous fit of the absorption and all the Raman profiles.[44,58] The resulting Dushinsky rotation parameter ($S_{4,6} = -0.20$) is the same as that obtained in Ref. 46. Our best fit is displayed in Fig. 14. In Figs. 15 and 16 we demonstrate the dependence of our lineshapes on $\kappa$. In these figures we kept the same structural parameters of Fig. 14 and varied $\kappa$ ($\kappa = 1, \infty$) keeping the full width at half maximum of the absorption lineshape $\Gamma_0$ [Eq. (117)] fixed. The absorption lineshape for these new parameters is virtually identical to that obtained from the parameters of Figs. 13 and 14. The excitation profiles, however, are significantly different. These calculations demonstrate the advantages of the present theory. As the number of modes

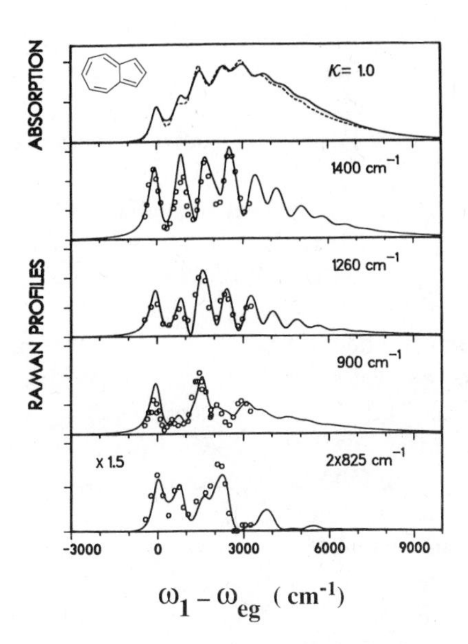


**Figure 15.** The same as Fig. 14 with different broadening parameters, $\Delta = 266$ cm$^{-1}$, $\Lambda = 266$ cm$^{-1}$ corresponding to $\kappa = 1$ and with the same value of $\Gamma_0$, that is, $\Gamma_0 = 408$ cm$^{-1}$.[44]

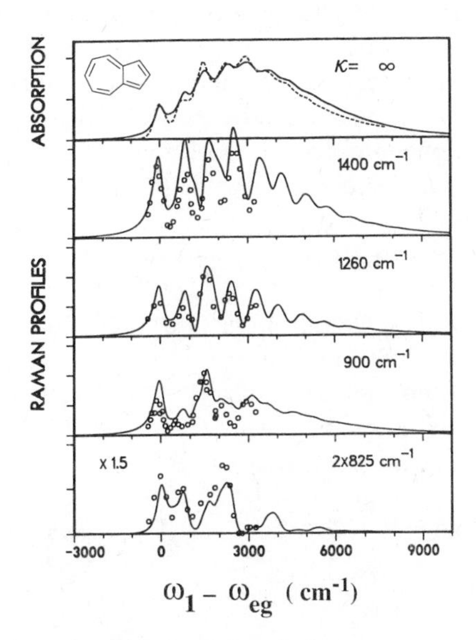


**Figure 16.** The same as Fig. 14 with different broadening parameters corresponding to the fast modulation limit, $\kappa = \infty$, with the same value of $\Gamma_0$, that is, $\Gamma_0 = 2\Delta^2/\Lambda = 408$ cm$^{-1}$.[44]

increases, the conventional expressions, based on summations over eigenstates, become intractable. The computational effort involved in the present Fourier transform method does not increase substantially as the molecular size increases, since a single Fourier transform is required in any case.

## VIII. THE ROLE OF VIBRATIONAL RELAXATION IN SPONTANEOUS RAMAN AND FLUORESCENCE SPECTROSCOPY

So far we considered only dephasing effects of the thermal bath, that is, the interaction between the system and the bath was taken to be diagonal in the system states [Eq. (41)]. This interaction results in line broadening and also induces new components in the spectra (e.g., the fluorescence and the excited-state resonances in CARS). An important process, which takes place in polyatomic molecules in solution, is vibrational relaxation (VR).[59] The vibrational state of the system following the optical transition is determined by the incident light frequency and by the Franck–Condon factors, and it is usually a nonequilibrium state. The interactions with the solvent will result in vibrational relaxation, which will tend to equilibrate the optically excited molecule with the bath temperature. A microscopic treatment of these processes requires introducing a specific model for the interaction between the system and the bath into the Hamiltonian, Eq. (41). However, it is convenient and often satisfactory to treat these processes phenomenologically via the Pauli master equation.[97] This is done by adding a relaxation matrix to the Liouville equation. The procedure is similar for 4WM and for SRF lineshapes. In this section we shall focus explicitly on SRF. The extension to 4WM is straightforward. The phenomenological VR matrix is introduced by modifying the Liouville operator [Eq. (9a)]. We define

$$L\rho = [H, \rho] - i\mathbf{\Gamma}\rho, \tag{118}$$

where $H$ is the molecular Hamiltonian [Eq. (41)] and $\mathbf{\Gamma}$ is the relaxation matrix, which includes vibrational and radiative relaxation. For simplicity we consider only VR within the electronically excited state. Including ground-state VR can be made in a similar way. The Liouville space operator $\mathbf{\Gamma}$ is defined by its matrix elements, that is,

$$\langle\langle b'b'|\mathbf{\Gamma}|bb\rangle\rangle \equiv -\gamma_{b'b}, \qquad b' \neq b, \tag{119a}$$

$$\langle\langle bb|\mathbf{\Gamma}|bb\rangle\rangle = \sum_{b' \neq b} \gamma_{b'b} + \gamma_r \equiv \gamma_b. \tag{119b}$$

$\mathbf{\Gamma}$ further satisfies the detailed balance condition

$$\gamma_{b'b} = \gamma_{bb'} \exp(-\omega_{b'b}/kT). \tag{119c}$$

In addition, we have the decay rate of coherences

$$\langle\langle bb'|\Gamma|bb'\rangle\rangle = -\tfrac{1}{2}(\gamma_b + \gamma_{b'}) \equiv \Gamma_{bb'}. \tag{119d}$$

Here, $\gamma_{b'b}$ is the rate of VR from state $|b\rangle$ to state $|b'\rangle$. $\gamma_b$ is the total decay rate of level $|b\rangle$, consisting of a VR part and a radiative part ($\gamma_r$). Equation (119c) is the detailed balance condition, which guarantees that the system will evolve to thermal equilibrium at long times. Finally, Eq. (119d) accounts for the decay of coherences due to the relaxation matrix. The relaxation matrix [Eqs. (119)] corresponds to the following equations of motion:

$$\frac{d\rho_{bb}}{dt} = -\gamma_b \rho_{bb} + \sum_{b' \neq b} \gamma_{bb'} \rho_{b'b'}, \tag{120a}$$

which can be written in a matrix form

$$\frac{d\boldsymbol{\rho}}{dt} = -\hat{\gamma}\boldsymbol{\rho}, \tag{120b}$$

where $\boldsymbol{\rho}$ is a vector, whose components are $\rho_{bb}$. The equation of motion for the coherences, corresponding to Eq. (119d), is

$$\frac{d\rho_{bb'}}{dt} = -\Gamma_{bb'}\rho_{bb'}, \qquad b \neq b'. \tag{120c}$$

Making use of Eqs. (118) and (119), we can solve for the SRF lineshapes by repeating the procedure of Sections V and VI. The final result is[59,96]

$$S_{\mathrm{SRF}}(\omega_1, \omega_2) = S_{\mathrm{RAMAN}}(\omega_1, \omega_2) + S_{\mathrm{FL}}(\omega_1, \omega_2), \tag{121a}$$

where

$$S_{\mathrm{RAMAN}}(\omega_1, \omega_2) = 2\pi \sum_{a,c} P(a)|K_{ca}(\varepsilon_a + \omega_1)|^2 \delta(\omega_1 - \omega_2 - \omega_{ca}), \tag{121b}$$

$$S_{\mathrm{FL}}(\omega_1, \omega_2) = S_{\mathrm{FL}}^{\mathrm{I}}(\omega_1, \omega_2) + S_{\mathrm{FL}}^{\mathrm{II}}(\omega_1, \omega_2) + S_{\mathrm{FL}}^{\mathrm{III}}(\omega_1, \omega_2), \tag{121c}$$

$$S_{\mathrm{FL}}^{\mathrm{I}}(\omega_1, \omega_2) = 4 \sum_{b',b} [\mathrm{Im}\, \bar{K}_{b'b'}(\varepsilon_{b'} - \omega_2)](\hat{\gamma}^{-1})_{b'b}[\mathrm{Im}\, \tilde{K}_{bb}(\varepsilon_b - \omega_1)], \tag{121d}$$

$$S_{\mathrm{FL}}^{\mathrm{II}}(\omega_1,\omega_2) = 2\,\mathrm{Im} \sum_{b \neq d} \frac{[\bar{K}_{bd}(\varepsilon_b - \omega_2) - \bar{K}_{db}^{*}(\varepsilon_d - \omega_2)]\tilde{K}_{db}(\varepsilon_b - \omega_1)}{\omega_{db} + i\Gamma_{db}}, \tag{121e}$$

$$S_{\mathrm{FL}}^{\mathrm{III}}(\omega_1,\omega_2) = 2\,\mathrm{Im} \sum_{a,c} P(a)\,\mathrm{PP}\frac{K_{ca}(\varepsilon_c + \omega_2)K_{ca}^{*}(\varepsilon_a + \omega_1)}{\omega_1 - \omega_2 - \omega_{ca}}. \tag{121f}$$

Here $K$, $\bar{K}$, and $\tilde{K}$ are given by Eqs. (81). The definition of $T(t)$ now need to be changed. Previously, all levels had the same lifetime. Now we have

$$T_{ca}(t) = \sum_b V_{cb} \exp(-i\varepsilon_b t - \tfrac{1}{2}\gamma_b t) V_{ba} J_g(t). \tag{122}$$

In the absence of VR, $\gamma_b = \gamma_r$, and Eq. (122) reduces to Eq. (79) with $\gamma = \gamma_r$. The off-diagonal terms ($b' \neq b$) in $S_{\mathrm{FL}}^{\mathrm{I}}$ are induced by the VR. The nature of this term is determined by the relative magnitude of the VR rates $\gamma_{b'b}$ and the radiative lifetime $\gamma_r$.[59,96] In the slow VR limit $\gamma_r \gg \gamma_{b'b}$, we have

$$(\hat{\gamma}^{-1})_{b'b} = \gamma_r^{-1}\delta_{bb'}. \tag{123}$$

In this case, Eq. (121) reduce to Eq. (84). In the opposite (fast VR) limit, $\gamma_{b'b} \gg \gamma_r$, and we can analyze Eq. (121) as follows. Since $\hat{\gamma}$ is non-Hermitian its left ($\langle\lambda_i|$) and right ($|\lambda_i\rangle$) eigenvectors are not the same. The left and right eigenvalues $\lambda_i$ are, however, identical. We then have

$$\hat{\gamma}|\lambda_i\rangle = \lambda_i|\lambda_i\rangle, \tag{124a}$$

$$\langle\lambda_i|\hat{\gamma} = \lambda_i\langle\lambda_i|, \tag{124b}$$

and

$$\hat{\gamma}^{-1} = \sum_i \frac{|\lambda_i\rangle\langle\lambda_i|}{\lambda_i}. \tag{124c}$$

Let us now consider the $\hat{\gamma}$ matrix in the absence of radiative lifetime ($\gamma_r = 0$). In that case, it has an equilibrium vector with eigenvalue $\lambda_0 = 0$, that is,

$$|\lambda_0\rangle = \sum_b \sigma_e(b)|b\rangle, \tag{125a}$$

$$\langle\lambda_0| = \sum_b \langle b|, \tag{125b}$$

where $\sigma_e(b)$ is the equilibrium distribution of $|b\rangle$ in the electronically excited state [Eq. (46a)], that is,

$$\sigma_e(b) = \exp(-\varepsilon_b/kT) \Big/ \sum_b \exp(-\varepsilon_b/kT). \tag{126}$$

All other eigenvectors $|\lambda_i\rangle$ will have eigenvalues $O(\gamma_{b'b})$. When $\gamma_{b'b} \gg \gamma_r$, we can treat $\gamma_r$ perturbatively. Using first-order perturbation theory, we have

$$\lambda_0 \cong \gamma_r, \tag{127a}$$

$$\lambda_i \cong \gamma_{b'b} + O(\gamma_r), \qquad i = 1, 2, \ldots. \tag{127b}$$

We thus have separation of time scales and $\lambda_0 \ll \lambda_i$, $i \neq 0$. $\lambda_0$ will then be dominant in Eq. (124c), resulting in

$$\hat{\gamma}^{-1} \cong \frac{|\lambda_0\rangle\langle\lambda_0|}{\lambda_0}, \tag{128a}$$

that is,

$$\hat{\gamma}^{-1} = \sum_{b,b'} \frac{\sigma_e(b')|b'\rangle\langle b|}{\gamma_r}. \tag{128b}$$

$S_{\rm FL}^{\rm I}$ thus assumes the form

$$S_{\rm FL}^{\rm I}(\omega_1, \omega_2) = \frac{4}{\gamma_r} \sum_{b'} \sigma_e(b')[\operatorname{Im} \bar{K}_{b'b'}(\varepsilon_{b'} - \omega_2)] \sum_b [\operatorname{Im} \tilde{K}_{bb}(\varepsilon_b - \omega_1)]. \tag{129}$$

This corresponds to "fully relaxed" emission coming from the equilibrated distribution of vibronic states. Equation (121) thus interpolates between the previous expression, Eq. (84), which contains no VR, and the fast relaxation limit, Eq. (129), whereby the emission is fully relaxed. The key parameter controlling this behavior is $\gamma_{b'b}/\gamma_r$.

The role of vibrational relaxation and solvation dynamics can be probed most effectively by fluorescence experiments, which are both time- and frequency-resolved,[66-68] as indicated at the end of Sec. V. We have recently developed a theory for fluorescence of polar molecules in polar solvents.[68] The solvaion dynamics is related to the solvent dielectric function $\varepsilon(\omega)$ by introducing a solvation coordinate. When $\varepsilon(\omega)$ has a Lorentzian dependence on frequency (the Debye model), the broadening is described by the stochastic model [Eqs. (113)], where the parameters $\Delta$ and $\Lambda$ may be related to molecular

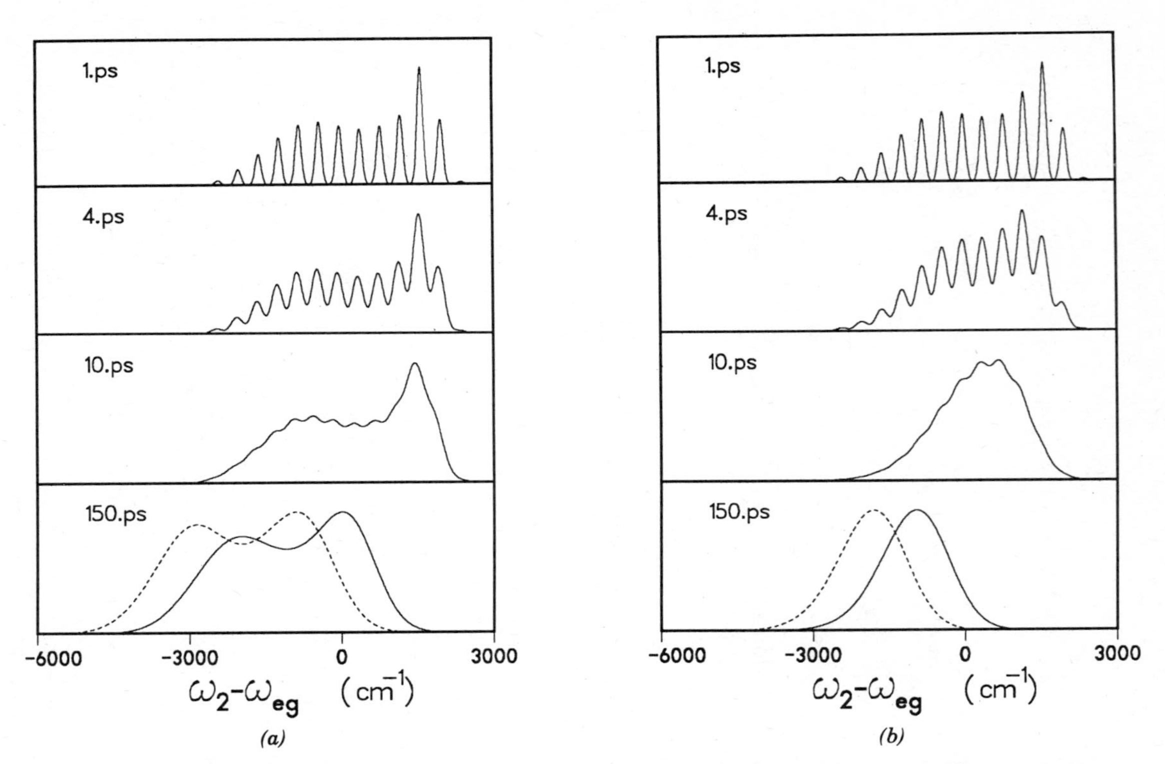


**Figure 17.** Time-resolved fluorescence spectra of a solute with one vibrational mode in ethanol at 247 K.[68] The various frames show the fluorescence spectrum measured at successively later times after the application of a 1 ps excitation pulse. Each spectrum is labeled with the observation time. The steady-state fluorescence spectrum is given by the dashed curve in the bottom frame. In the electronic ground state, the solute vibrational frequency is 400 $\text{cm}^{-1}$, and in the excited state, the frequency is 380 $\text{cm}^{-1}$. The dimensionless displacement is 1.4. The permanent dipole moment changes by 10 Debye upon electronic excitation. The Onsager radius is 3Å. The longitudinal dielectric relaxation time, $\tau_L$, is 150 ps. $\omega_1 - \omega_{eg} = 2000$ $\text{cm}^{-1}$. (a) Vibrational relaxation is not included. (b) Finite vibrational relaxation rate of $\gamma = 25/\tau_L = 0.167$ $\text{psec}^{-1}$ is included.

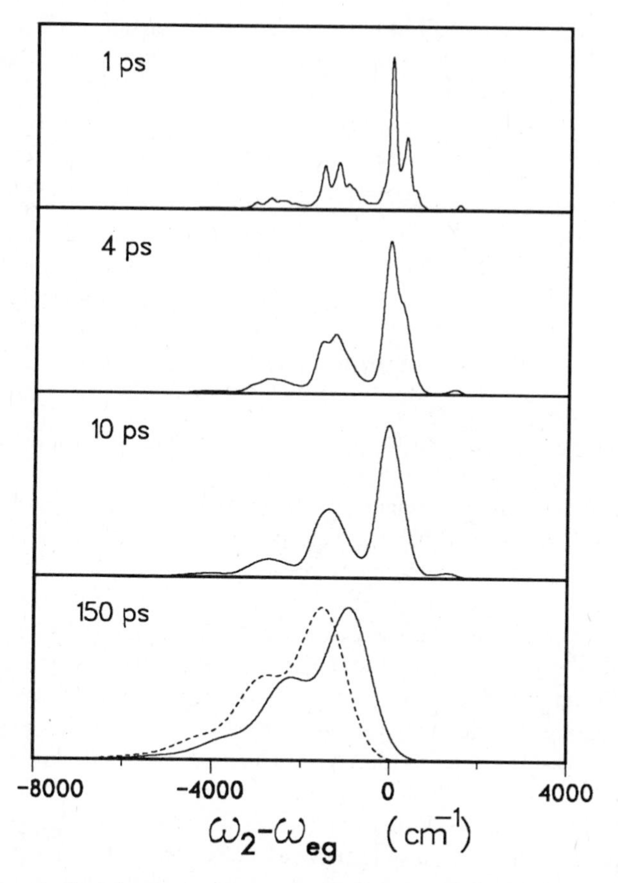


**Figure 18.** Time-resolved fluorescence spectra of a polyatomic solute in ethanol at 247 K following a 1 ps excitation pulse.[68] The model solute has the 29 Raman active vibrational modes of the retinal chromophore in bacteriorhodopsin and undergoes rapid vibrational relaxation. $\omega_1 - \omega_{eg} = 1528 \text{ cm}^{-1}$. The solvent and the broadening parameters are the same as used in Fig. 17.

and solvent properties. In addition, the model exhibits a time-dependent solvent shift which reflects the solvent reorganization. We have also incorporated vibrational relaxation using the master equation introduced in this section. In Fig. 17, we display the time-resolved fluorescence spectrum of a model solute with one harmonic vibrational mode, calculated by using Eq. (76). Fig. 17a is calculated in the absence of vibrational relaxation and reflects the solvation dynamics. In Fig. 17b, a finite vibrational relaxation rate is included. In Fig. 18, we show a similar calculation for the retinal chromophore in bacteriorhodopsin with 29 vibrational modes. Fast vibrational relaxation was assumed in this calculation, which used our eigenstate-free procedure. These spectra are narrow at short times due to the capability of the excitation process to select a subgroup of solvent configurations. As time evolves, a broadening and red shift take place, which reflect the solvent reorganization process. Time- and frequency-resolved fluorescence lineshapes thus provide a unique probe for solvation dynamics on microscopic length scales and time scales.

## IX. INTRAMOLECULAR VIBRATIONAL REDISTRIBUTION (IVR) IN ULTRACOLD MOLECULES IN SUPERSONIC BEAMS

The calculation of molecular fluorescence and Raman spectra in large anharmonic molecules is one of the fundamental problems in molecular dynamics and spectroscopy. Recent experiments, particularly involving ultracold molecules in supersonic beams,[60,98–102] are yielding accurate and detailed information (both time-resolved and frequency-resolved). In the previous sections we demonstrated how correlation function techniques, which are based on a *reduced description*, may be used to calculate SRF and 4WM lineshapes in macroscropic systems. The reduced correlation function formulation allows us to calculate any lineshape function without attempting to calculate the exact eigenstates of the macroscopic system. Calculating the eigenstates of a macroscopic system is extremely difficult owing to the large number of degrees of freedom involved. Moreover, the experimental broad line-shapes contain highly averaged information and do not reveal properties of individual eigenstates. The calculation of individual eigenstates of macroscopic systems is therefore neither feasible nor desirable. The analysis of spectra of isolated molecules is, on the other hand, traditionally made in terms of properties of individual molecular eigenstates (level positions and dipole matrix elements).[103] Such an approach is appropriate for small or intermediate size molecules, but for large molecules (10 atoms or more), it is impractical. The spectra show intramolecular line broadening in which information on individual eigenstates is highly averaged. This situation is reminiscent of the behavior of macroscopic systems, and it seems natural to adopt the techniques developed in the previous sections toward the treatment of intramolecular line broadening in large isolated polyatomic molecules.

In order to develop a reduced description for the spectroscopy of ultracold, isolated molecules in supersonic beams, we partition the molecular vibrational degrees of freedom into a few "system" modes, which are strongly coupled to the electronic transition of interest and are characterized by large values of the displacements $D_j$, and the rest of the vibrational degrees of freedom are treated as an intramolecular bath.[59,61,102] Once we have identified the "system" modes and the "bath" modes, the calculation of SRF lineshapes of ultracold molecules may be carried out using the formulation developed in Sections V and VIII. There are several points that make this calculation quite different from SRF in solution. First, the incident frequency is usually tuned to a narrow, isolated vibronic level in the electronically excited state. This state serves as a *doorway state*,[102] and its contribution is dominant in the emission. An excellent approximation will therefore be to eliminate the summation over intermediate states $|b\rangle$ and $|d\rangle$ in Eq. (78) and consider a single doorway state $|b\rangle$. This simplifies the calculation considerably. Another fundamental difference exists between the intramolecular bath and the solvent treated in previous sections. The solvent has an infinite number of degrees of freedom, and it is normally at a finite temperature. It is therefore reasonable to assume that its motions are very weakly correlated with the state of the system. This enabled us to use the projection operator Eq. (58) and the factorization approximation. For isolated polyatomic molecules (say anthracene or di-fluorobenzene), the situation is very different. The bath is finite and is initially cold ($T = 0^o K$). As the vibrational relaxation proceeds, the excess vibrational energy is released into the bath. *The state of the bath is therefore strongly correlated with the vibronic state of the system.* A reduced description of the emission can then be obtained by introducing the following projection operator[59,61]

$$\hat{P} = |gg\rangle\rangle\langle\langle gg|\rho_g \mathrm{Tr}_B + \sum_b |bb\rangle\rangle\langle\langle bb|\rho_e^b \mathrm{Tr}_B. \tag{130}$$

The ground-state projection $|gg\rangle\rangle\langle\langle gg|$ is identical to Eq. (58), and $\rho_g$ is, in this case, the zero-temperature density matrix of the bath. For the excited state, we define $\rho_e^b$ as the bath density matrix, when the system is in the $|b\rangle$ state. As the vibrational relaxation proceeds, the bath energy increases, and $\rho_e^b$ changes. A simple way to model $\rho_e^b$ is by assuming a harmonic bath with a microcanonical distribution. Note that the previous projection operator [Eq. (58)] is a special case of Eq. (130) obtained by taking $\rho_e^b$ to be independent of $|b\rangle$. This is the case for an infinite bath that is weakly correlated with the state of the system. Our procedure for incorporating the effects of IVR processes on the emission lineshapes of supercooled molecules is based on introducing an IVR rate $\gamma_{b'b}$, whereby the doorway state $|b\rangle$ relaxes to state $|b'\rangle$ (see Section VIII). Once the system is in the $|b'\rangle$ state, the bath modes become hot, since they absorb the excess vibrational energy $\omega_{bb'}$. As a result, the emission from the $|b'\rangle$ state can be represented by our stochastic model, whereby the parameters $\Delta$ and $\Lambda$ [Eq. (113)] now depend on the amount of

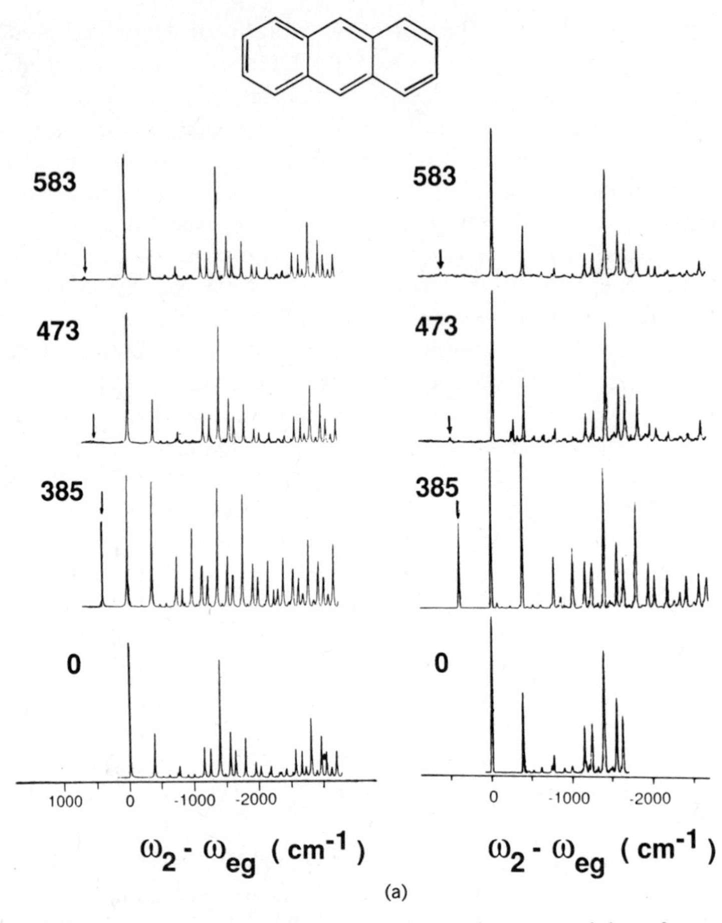


**Figure 19.** Dispersed fluorescence of ultracold anthracene in a supersonic beam for several values of the excess vibrational energy, which is indicated (in $cm^{-1}$) on each plot. The right column is experimental,[60] and the left column is our 17-mode harmonic calculation (Table II).[61] The agreement is excellent at low excitation energies ($<766\ cm^{-1}$), but fails at higher energies owing to the onset of IVR processes.

vibrational energy in the bath. For the doorway state $|b\rangle$, $\Delta = 0$, and there is no broadening. We expect that the stochastic fluctuation amplitude $\Delta$ for emission from a given $|b'\rangle$ state will increase as the available vibrational energy of the bath $\omega_{bb'}$ increases. The total emission thus consists of a progression of narrow lines originating from the doorway state and a series of broad emission lines corresponding to the redistributed emission. Our expression for the emission, in this case, is[61]

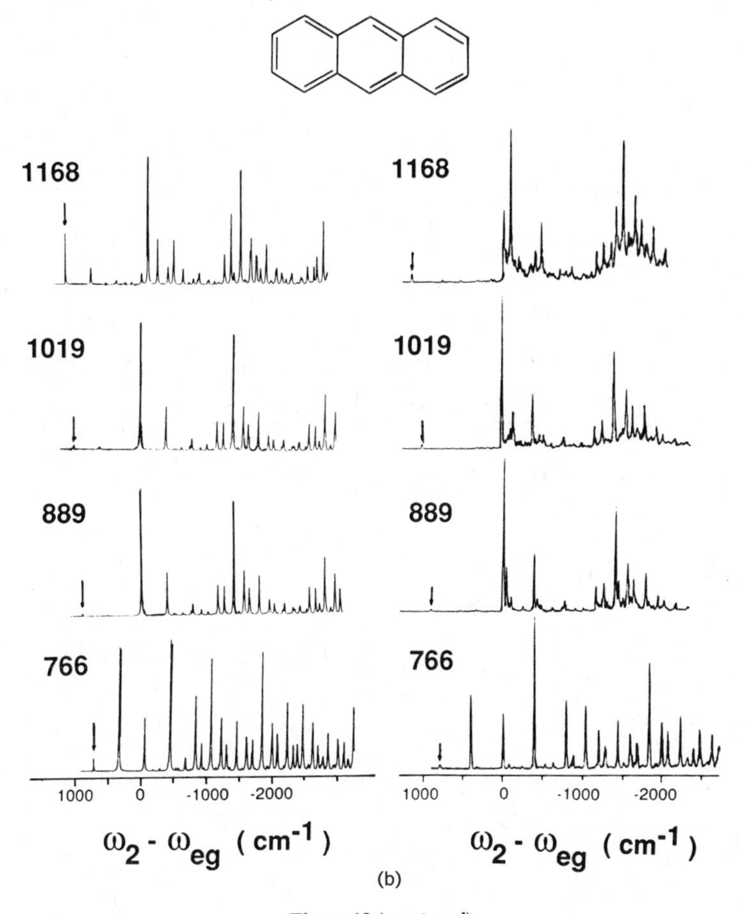


**Figure 19** (*continued*)

$$S_{\mathrm{SRF}}(\omega_1, \omega_2) = \frac{-2\gamma_b |\mu_{ba}|^2}{(\omega_1 - \omega_{ba})^2 + (\gamma_b/2)^2} \sum_{b'} (\gamma^{-1})_{b'b} [\mathrm{Im}\, \bar{K}_{b'b'}(\varepsilon_{b'} - \omega_2)], \quad (131)$$

where $\bar{K}_{b'b}$ is given by Eqs. (79) and (81) together with Eq. (114), and the bath parameters $\Delta$ and $\Lambda$ are taken to be dependent on $|b'\rangle$. Equation (131) is a special case of $S_{\mathrm{FL}}^{\mathrm{I}}(\omega_1, \omega_2)$ [Eq. (121d)]. We have applied Eq. (131) toward the calculation of the emission spectra of ultracold anthracene. A comprehensive supersonic beam study of this molecule was conducted recently by Zewail and co-workers.[60] Our results are displayed in Figs. 19–21. Anthracene has 17 system modes that are strongly coupled to the electronic transition. The parameters of these modes used on our calculation are summarized in Table II.

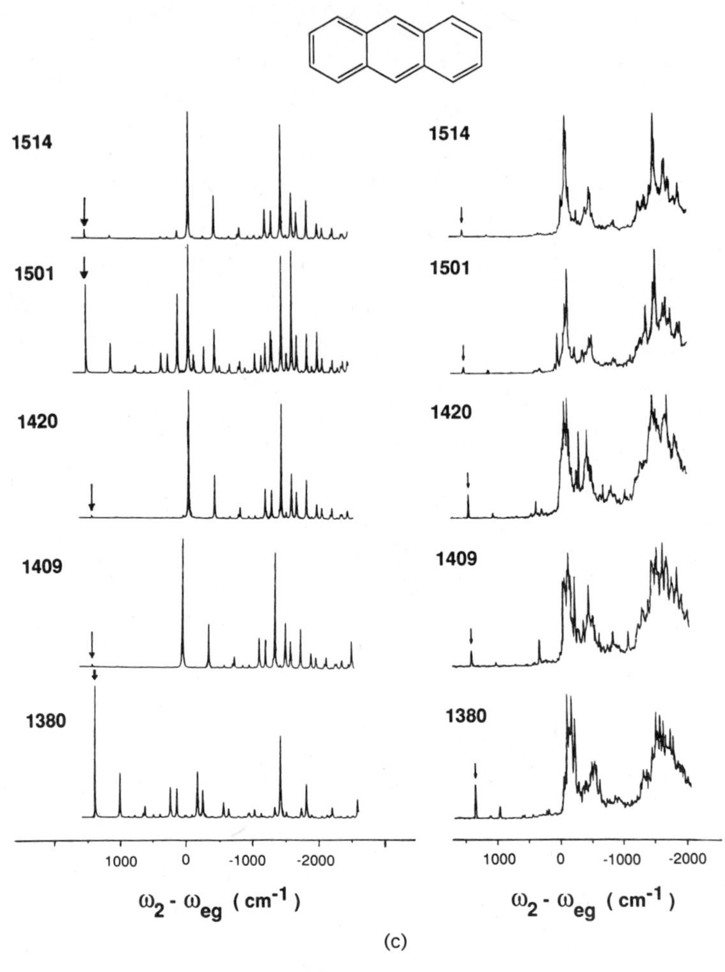


(c)

**Figure 19** (*continued*)

In Fig. 19, we compare the calculated emission spectra in the harmonic limit (no vibrational relaxation) with experiment. At low excess vibrational energy ($<766$ cm$^{-1}$), the agreement is excellent. At higher energies. the harmonic calculations fail to reproduce the broad redistributed emission, which becomes increasingly dominant with increasing excess energy. The redistributed emission can be quantitatively accounted for by incorporating the IVR processes via Eq. (131). Figure 20 focuses on excess vibrational energy of 1792 cm$^{-1}$ in which IVR processes play an important role in the emission, since the purely harmonic emission (bottom figure) clearly fails to reproduce

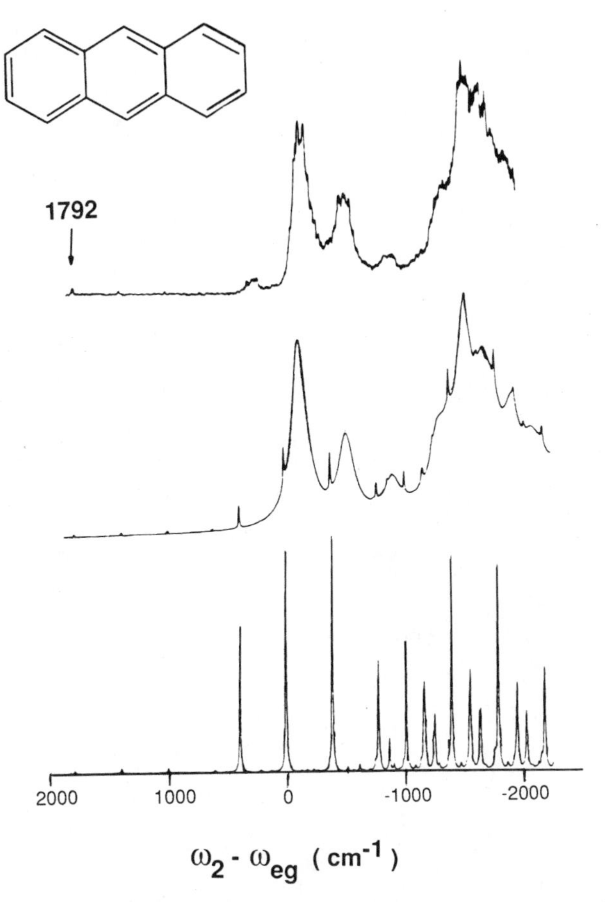


**Figure 20.** The $\bar{5}^1_0 12^1_0$ dispersed fluorescence of ultracold anthracene in a supersonic beam. The available vibrational energy is 1792 $cm^{-1}$. The parameters of the optically active modes are given in Table II. The top figure is the experimental spectrum.[60] The bottom figure is the emission in the harmonic approximation ($\gamma_{b'b} = 0$). The calculation clearly fails to reproduce the broad redistributed emission. The middle figure was calculated with IVR [Eqs. (131)]. Only one $|b'\rangle$ state (the ground vibrational state $|b'\rangle = |0\rangle$) was used. $\gamma_{b'b}/\gamma_r = 40$. The relaxed emission was calculated in the fast modulation limit [Eq. (116a)], with $\Gamma_0 = 2\hat{\Gamma} = 75$ $cm^{-1}$.[61]

the broad experimental spectrum (top figure). The middle figure, which introduced IVR via our stochastic model [Eq. (131)], adequately reproduces the experimental spectrum. This effect becomes more dramatic as the excess vibrational energy is increased. Figure 20 was calculated in the fast modulation (Lorentzian) limit [Eq. (116a)]. Finally, Fig. 21 shows the calculated emission spectrum of vibrationally hot anthracene at three temperatures. This calcula-

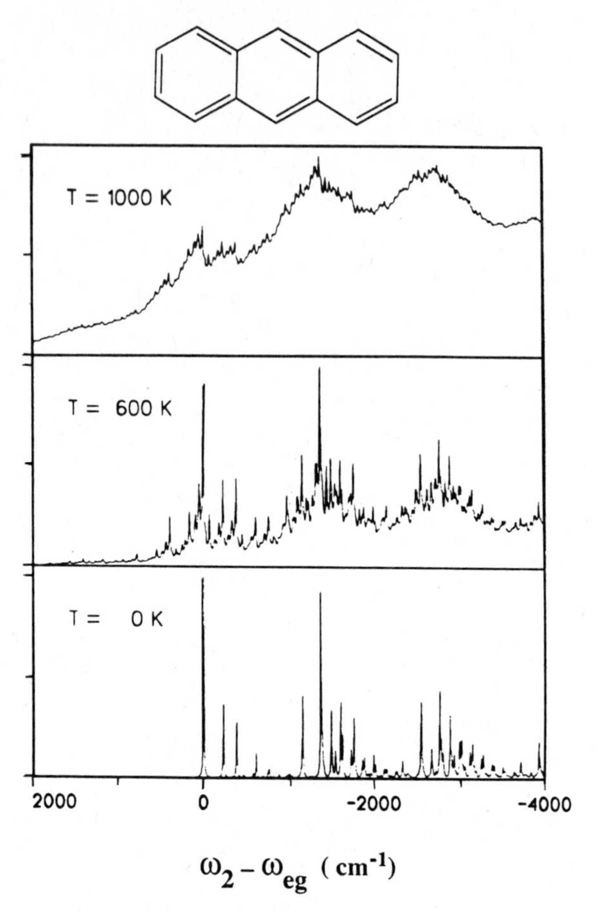


**Figure 21.** The total emission spectrum of anthracene at finite temperatures,[61] taking the electronically excited state to be in thermal equilibrium at three different temperatures [Eq. (46a)]. The calculation was made using Eq. (131), with a linewidth $\Gamma_0 = 2\Gamma = 10\ \mathrm{cm}^{-1}$.

tion was made in the harmonic limit in the absence of IVR processes. The present calculations demonstrate the simplicity in which this model can be used to interpret the spectra of large polyatomic molecules and molecular clusters.[99,104] The amount of computational effort does not increase dramatically as the molecular size increases. This is in sharp contrast to conventional expressions, based on summations over eigenstates of the system and the bath, which are intractable for large molecules. Methods for controlling IVR processes in SRF lineshapes were discussed recently.[105] The effects of

TABLE II
Frequencies and Dimensionless Displacements of Anthracene[60,61]

| Mode | $\omega_j''$ ($cm^{-1}$) | $\omega_j'$ ($cm^{-1}$) | $D_j$ |
|---|---|---|---|
| 12 | 390 | 385 | 0.79 |
| 11 | 624 | 583 | 0.21 |
| 10 | 753 | 755 | 0.24 |
| 9 | 1012 | 1019 | 0.20 |
| 8 | 1165 | 1163 | 0.67 |
| 7 | 1263 | 1168 | 0.69 |
| 6 | 1408 | 1380 | 1.35 |
| 5 | 1486 | 1420 | 0.20 |
| 4 | 1566 | 1501 | 0.80 |
| $\overline{11}$ | 391 | 232 | 0.33 |
| $\overline{10}$ | 524 | 473 | 0.15 |
| $\overline{9}$ | 912 | 889 | 0.20 |
| $\overline{8}$ | 1100 | 1046 | 0.15 |
| $\overline{7}$ | 1184 | 1183 | 0.24 |
| $\overline{5}$ | 1382 | 1409 | 0.20 |
| $\overline{4}$ | 1576 | 1514 | 0.35 |
| $\overline{3}$ | 1643 | 1635 | 0.63 |

incoherence in the radiation field on SRF[106] and 4WM[63] lineshapes may also be analyzed using the present formalism.

## X. AN EIGENSTATE-FREE EXPRESSION FOR THE NONLINEAR RESPONSE FUNCTION

In the previous sections, we have utilized Green's function techniques to eliminate some of the summations involved in the calculations of nonlinear susceptibilities. The general expression for $R(t_3, t_2, t_1)$ [Eq. (49) or (60)], involves four summations over molecular states *a*, *b*, *c*, *d*. In Eq. (80) we carried out two of these summations for harmonic molecules. It should be noted that for this particular model it is possible to carry out formally all the summations involved, resulting in a closed time-domain expression for $R(t_3, t_2, t_1)$. This expression, however, cannot be written in terms of simple products of functions of $t_1$, $t_2$, and $t_3$. Therefore, calculating the frequency-domain response function $\hat{R}$ via Eq. (30) requires the performing of a triple Fourier transform (rather than three one-dimensional transforms). This formula is, therefore, useful for extremely short pulses when a time-domain expression is needed. Otherwise, it is more convenient to use the expressions of Section VI, whereby only two of the four summations were carried out, but the transformation to the frequency domain is much simpler. We have shown that the response function contains eight terms (Fig. 6). We can express each term using the four-point correlation function of the dipole operator, $F(t_1, t_2, t_3, t_4)$, Eq. (54). We have recently evaluated $F$ for our harmonic model [Eq. (85)] in the absence of Dushinsky rotation.[64] In this case, the four-point correlation function $F$ can be factorized as a product of terms corresponding to the various modes:

$$F(t_1, t_2, t_3, t_4) = \prod_{j=1}^{N} F_j(t_1, t_2, t_3, t_4), \tag{132}$$

making use of the Condon approximation [Eq. (89)], and assuming that the ground and the excited state frequencies are the same $\omega_j' = \omega_j'' \equiv \omega_j$, we get

$$F_j(t_1, t_2, t_3, t_4) = \exp[(D_j^2/2)(f_j + \bar{f}_j - \bar{n}_j f_j f_j^*)], \tag{133a}$$

where

$$f_j = \exp[-i\omega_j(t_1 - t_4)] - \exp[-i\omega_j(t_2 - t_4)] - \exp[-i\omega_j(t_3 - t_4)] - 1, \tag{133b}$$

$$\bar{f}_j = \exp[-i\omega_j(t_1 - t_2)] - \exp[-i\omega_j(t_1 - t_3)] - \exp[-i\omega_j(t_2 - t_3)] - 1, \tag{133c}$$

and

$$\bar{n}_j = [\exp(\hbar\omega_j/kT) - 1]^{-1}. \tag{133d}$$

$F(t_1, t_2, t_3, t_4)$ can be calculated for the general harmonic model [Eq. (85)] with $\omega_j' \neq \omega_j''$ by using the methodology developed in Ref. 58. For anharmonic systems, it may be possible to calculate $F(t_1, t_2, t_3, t_4)$ approximately by making use of a semiclassical propagation scheme in Liouville space.[107]

## XI. CONCLUDING REMARKS

In this chapter, we developed a general theory of 4WM processes in terms of the nonlinear response function of the nonlinear medium $R(t_3, t_2, t_1)$. The response function is an intrinsic molecular property that contains all the microscopic information relevant to any type of 4WM process. The details of a particular 4WM experiment are contained in the external fields $E_1(t)$, $E_2(t)$, $E_3(t)$, and in the particular choice of the observable mode $\mathbf{k}_s$. The generated signal is calculated by convolving the response function with the external fields and choosing $\mathbf{k}_s$ [Eqs. (27), (28), and (31)]. It is only at this stage that the distinction among the various 4WM techniques (photon echo, transient grating, CARS, CSRS, etc.) is made. We have further shown that SRF lineshapes may be expressed in terms of the same four-point correlation function of the dipole operator that is required for the description of 4WM processes. Microscopic and stochastic models and approximation schemes for the explicit evaluation of these lineshapes for polyatomic molecules in condensed phases and in supersonic beams were developed and analyzed.

## Acknowledgments

The support of the National Science Foundation, the Office of Naval Research, the U.S. Army Research Office, the Petroleum Research Fund, administered by the American Chemical Society, and the Camille and Henry Dreyfus Foundation is gratefully acknowledged. I wish to thank Z. Deng, R. F. Loring, K. Shan, J. Sue, and Y. J. Yan for stimulating discussions. The critical comments of A. H. Zewail are most appreciated. Special thanks go to Irmhild O. Barnett for the careful typing.

## References

1. N. Bloembergen, *Nonlinear Optics*. Benjamin, New York, 1965.
2. P. N. Butcher, *Nonlinear Optical Phenomena*. Ohio University Press, Athens, Ohio, 1965.
3. S. A. J. Druet and J. P. E. Taran, *Progr. Quantum Electron.* **7**, 1 (1981).
4. Y. R. Shen, *The Principles of Nonlinear Optics*. Wiley, New York, 1984.
5. G. R. Fleming, *Chemical Applications of Ultrafast Spectroscopy*. Oxford, London, 1986.
6. S. Mukamel and R. F. Loring, *J. Opt. Soc. Am. B* **3**, 595 (1986).
7. J. Tang and A. C. Albrecht, in *Raman Spectroscopy* (H. A. Szymanski, ed.). Plenum, New York, 1970, Vol 2, p. 33; P. M. Champion and A. C. Albrecht, *Annu. Rev. Phys. Chem.* **333**, 353 (1982).
8. B. Johnson and W. Peticolas, *Annu. Rev. Phys. Chem.* **27**, 465 (1976); L. Chinsky, A. Laigle, W. L. Peticolas, and P. Y. Turpin, *J. Chem. Phys.* **76**, 1 (1982).
9. T. G. Spiro, *Adv. Protein Chem.* **37**, 111 (1985).
10. A. B. Myers and R. A. Mathies, in *Biological Applications of Raman Spectroscopy*, Vol. 2 (T. G. Spiro, ed.). Wiley, N.Y. 1987, p. 1.
11. S. Mukamel, *Phys. Reports* **93**, 1 (1982).
12. S. Mukamel, *J. Phys. Chem.* **88**, 3185 (1984).
13. N. Bloembergen, H. Lotem, and R. T. Lynch, *Indian J. Pure Appl. Phys.* **16**, 151 (1978).
14. T. K. Yee and T. K. Gustafson, *Phys. Rev. A* **18**, 1597 (1978).
15. J. L. Oudar and Y. R. Shen, *Phys. Rev. A* **22**, 1141 (1980).
16. Y. Prior, A. R. Bogdan, M. Dagenais, and N. Bloembergen, *Phys. Rev. Lett.* **46**, 111 (1981); A. R. Bogdan, M. W. Downer, and N. Bloembergen, *Phys. Rev. A* **24**, 623 (1981); L. J. Rothberg and N. Bloembergen, *Phys. Rev. A* **30**, 820 (1984).
17. J. R. Andrews and R. M. Hochstrasser, *Chem. Phys. Lett.* **82**, 381 (1981); J. R. Andrews, R. M. Hochstrasser, and H. P. Trommsdorff, *Chem. Phys.* **62**, 87 (1981).
18. T. Yajima and H. Souma, *Phys. Rev. A* **17**, 309 (1978); T. Yajima, H. Souma, and Y. Ishida, *Phys. Rev. A* **17**, 324 (1978).
19. S. Mukamel, *Phys. Rev. A* **28**, 3480 (1983).
20. R. W. Boyd and S. Mukamel, *Phys. Rev. A* **29**, 1973 (1984).
21. V. Mizrahi, Y. Prior, and S. Mukamel, *Opt. Lett.* **8**, 145 (1983).
22. I. D. Abella, N. A. Kurnit, and S. R. Hartmann, *Phys. Rev.* **141**, 391 (1966); S. R. Hartmann,

*IEEE J. Quantum Electron.* **4**, 802 (1968); T. W. Mossberg, R. Kachru, A. M. Flusberg, and S. R. Hartmann, *Phys. Rev. A* **20**, 1976 (1979).
23. H. W. Hesselink and D. A. Wiersma, in *Modern Problems in Condensed Matter Sciences*, (V. M. Agranovich and A. A. Maradudin, eds.). North-Holland, Amsterdam, 1983, Vol. 4, p. 249.
24. R. W. Olson, F. G. Patterson, H. W. H. Lee, and M. D. Fayer, *Chem. Phys. Lett.* **78**, 403 (1981).
25. R. F. Loring and S. Mukamel, *Chem. Phys. Lett.* **114**, 426 (1985).
26. J. R. Salcedo, A. E. Siegman, D. D. Dlott, and M. D. Fayer, *Phys. Rev. Lett.* **41**, 131 (1978).
27. M. D. Fayer, *Ann. Rev. Phys. Chem.* **33**, 63 (1982).
28. P. F. Liao, L. M. Humphrey, and D. M. Bloom, *Phys. Rev. B* **10**, 4145 (1979).
29. J. K. Tyminski, R. C. Powell, and W. K. Zwicker, *Phys. Rev. B* **29**, 6074 (1984).
30. H. J. Eichler, *Optica Acta* **24**, 631 (1977).
31. A. von Jena and H. E. Lessing, *Opt. Quant. Electron.* **11**, 419 (1979).
32. D. K. Garrity and J. L. Skinner, *J. Chem. Phys.* **82**, 260 (1985).
33. R. F. Loring and S. Mukamel, *J. Chem. Phys.* **83**, 4353 (1985); **84**, 1228 (1985); **85**, 1950 (1986).
34. N. Bloembergen, *Am. J. Phys.* **35**, 989 (1967).
35. A. Laubereau and W. Kaiser, *Rev. Mod. Phys.* **50**, 607 (1978); W. Zinth, H. J. Polland, A. Laubereau, and W. Kaiser, *Appl. Phys. B* **26**, 77 (1981).
36. S. M. George, A. L. Harris, M. Berg, and C. B. Harris, *J. Chem. Phys.* **80**, 83 (1984).
37. I. I. Abram, R. M. Hochstrasser, J. E. Kohl, M. G. Semack, and D. White, *J. Chem. Phys.* **71**, 153 (1979); F. Ho, W. S. Tsay, J. Trout, S. Velsko, and R. M. Hochstrasser, *Chem. Phys. Lett.* **97**, 141 (1983).
38. E. L. Chronister and D. D. Dlott, *J. Chem. Phys.* **79**, 5286 (1984); C. L. Schosser and D. D. Dlott, *J. Chem. Phys.* **80**, 1394 (1984).
39. B. H. Hesp and D. A. Wiersma, *Chem. Phys. Lett.* **75**, 423 (1980); D. P. Weitekamp, K. Duppen, and D. A. Wiersma, *Phys. Rev. A* **27**, 3089 (1983).
40. A. M. Weiner, S. DeSilvestri, and E. P. Ippen, *J. Opt. Soc. Am. B* **2**, 654 (1985).
41. B. S. Hudson, W. H. Hetherington, S. P. Cramer, I. Chabay, and G. K. Klauminzer, *Proc. Nat'l. Acad. Sci.* (*USA*) **73**, 3798 (1976).
42. L. A. Carreira, T. C. Maguire, and T. B. Malloy, *J. Chem. Phys.* **66**, 2621 (1977).
43. R. F. Loring and S. Mukamel, *J. Chem. Phys.* **83**, 2116 (1985).
44. S. Mukamel, *J. Chem. Phys.* **82**, 5398 (1985); J. Sue, Y. J. Yan, and S. Mukamel, *J. Chem. Phys.* **85**, 462 (1986).
45. Z. Deng and S. Mukamel, *J. Chem. Phys.* **85**, 1738 (1986).
46. O. Brafman, C. K. Chan, B. Khodadoost, J. B. Page, and C. T. Walker, *J. Chem. Phys.* **80**, 5406 (1984); C. K. Chan, J. B. Page, D. L. Tonks, O. Brafman, B. Khodadoost, and C. T. Walker, *J. Chem. Phys.* **82**, 4813 (1985).
47. F. Inagaki, M. Tasumi, and T. Miyazawa, *J. Mol. Spectrosc.* **50**, 286 (1974); S. Saito, M. Tasumi, and C. H. Eugster, *J. Raman Spectrosc.* **14**, 299 (1983); S. Saito and M. Tasumi, *J. Raman Spectrosc.* **14**, 310 (1983).
48. J. B. Page and D. L. Tonks, *J. Chem. Phys.* **75**, 5694 (1981); C. K. Chan and J. B. Page, *J. Chem. Phys.* **79**, 5234 (1983); *Chem. Phys. Lett.* **104**, 609 (1984).
49. T. Kitagawa, M. Abe, and H. Ogoshi, *J. Chem. Phys.* **69**, 4516 (1978); M. Abe, T. Kitagawa, and Y. Kyogoku, *J. Chem. Phys.* **69**, 4526 (1978).
50. B. E. Kohler, T. A. Spiglanin, R. J. Hemley, and M. Karplus, *J. Chem. Phys.* **80**, 23 (1984); J. R. Ackerman, S. A. Forman, H. Hossain, and B. E. Kohler, *J. Chem. Phys.* **80**, 39 (1984).
51. A. Warshel, *Annu. Rev. Biophys. Bioeng.* **6**, 273 (1977); A. Warshel and P. Dauber, *J. Chem. Phys.* **66**, 5477 (1977).

52. B. S. Hudson, B. E. Kohler, and K. Schulten, in *Excited States*, Vol. 6 (E. C. Lim, ed.). Academic, New York, 1982, p. 1.
53. B. R. Stallard, P. M. Champion, P. R. Collis, and A. C. Albrecht, *J. Chem. Phys.* **78**, 712 (1983); D. Lee, B. R. Stallard, P. M. Champion, and A. C. Albrecht, *J. Phys. Chem.* **88**, 6693 (1984).
54. Z. Z. Ho, R. C. Hanson, and S. H. Lin, *J. Chem. Phys.* **77**, 3414 (1982); Z. Z. Ho, T. A. Moore, S. H. Lin, and R. C. Hanson, *J. Chem. Phys.* **74**, 873 (1981); Z. Z. Ho, R. C. Hanson, and S. H. Lin, *J. Phys. Chem.* **89**, 1014 (1985).
55. W. Siebrand and M. Z. Zgierski, *J. Chem. Phys.* **71**, 3561 (1979); in *Excited States*, (E. C. Lim, ed.). Academic, New York, 1979, Vol. 4, p. 1; W. Siebrand and M. Z. Zgierski, *J. Chem. Phys.* **71**, 3561 (1979).
56. A. B. Myers, R. A. Harris, and R. A. Mathies, *J. Chem. Phys.* **79**, 603 (1983).
57. S. Mukamel, *J. Chem. Phys.* **71**, 2884 (1979).
58. S. Mukamel, S. Abe, Y. J. Yan, and R. Islampour, *J. Phys. Chem.* **89**, 201 (1985); Y. J. Yan and S. Mukamel, *J. Chen Phys.* **85**, 5908 (1986).
59. S. Mukamel and R. E. Smalley, *J. Chem. Phys.* **73**, 4156 (1980).
60. W. R. Lambert, P. M. Felker, and A. H. Zewail, *J. Chem. Phys.* **81**, 2209 (1984); **81**, 2195 (1984); **81**, 2217 (1984); W. R. Lambert and A. H. Zewail, *J. Chem. Phys.* **82**, 3003 (1985); P. M. Felker and A. H. Zewail, *J. Chem. Phys.* **82**, 2961 (1985); **82**, 2975 (1985); **82**, 2994 (1985).
61. S. Mukamel, K. Shan, and Y. J. Yan, in *Polycyclic Aromatic Hydrocarbons and Astrophysics*, (A. Leger and L. d'Hendercourt, and N. Boccara, NATO ASI Series, Vol. C **191**, pp. 129). Reidel, Dordrecht, (1986); K. Shan, Y. J. Yan, and S. Mukamel *J. Chem. Phys.* **87**, 2021 (1987).
62. W. H. Louisell, *Quantum Statistical Properties of Radiation.* Wiley, New York, 1973.
63. S. Mukamel and E. Hanamura, *Phys. Rev. A* **33**, 1099 (1986); E. Hanamura and S. Mukamel, *J. Opt. Soc. Am. B* **3**, 1124 (1986).
64. Y. J. Yan and S. Mukamel *J. Chem. Phys.* **86**, 6085 (1987).
65. R. A. Marcus, *J. Chem. Phys.* **24**, 979 (1956).
66. Yu. T. Mazurenko and V. S. Udaltsov, *Opt. Spec.* **44**, 417 (1977); L. A. Hallidy and M. R. Topp, *J. Phys. Chem.* **82**, 2415 (1978); E. W. Castner, M. Maroncelli, and G. R. Fleming, *J. Chem. Phys.* **86**, 1090 (1987); M. Maroncelli, E. W. Castner, S. P. Webb, and G. R. Fleming, in *Ultrafast Phenomena V* (G. R. Fleming and A. E. Siegman, eds.). Springer-Verlag, Berlin, 1986.
67. Yu. T. Mazurenko and N. G. Bakshiev, *Opt. Spec.* **28**, 490 (1970); B. Bagchi, D. W. Oxtoby, and G. R. Fleming, *Chem. Phys.* **86**, 257 (1984); G. Van der Zwan and J. T. Hynes, *J. Phys. Chem.* **89**, 4181 (1985).
68. R. F. Loring, Y. J. Yan, and S. Mukamel, *Chem. Phys. Lett.* **135**, 23 (1987); *J. Phys. Chem.* **91**, 1302 (1987); *J. Chem. Phys.* **87**, (1987) (in press).
69. D. L. Huber, *Phys. Rev.* **170**, 418 (1968); **178**, 93 (1969); **187**, 392 (1969); A. Omont, E. W. Smith, and J. Cooper, *Astrophys. J.* **175**, 185 (1972).
70. P. F. Liao, J. E. Bjorkholm, and P. R. Berman, *Phys. Rev. A* **20**, 1489 (1977); J. L. Carlsten, A. Szoke, and M. G. Raymer, *Phys. Rev. A* **15**, 1029 (1977).
71. R. M. Hochstrasser and C. A. Nyi, *J. Chem. Phys.* **70**, 1112 (1979); J. M. Friedman and R. M. Hochstrasser, *Chem. Phys. Lett.* **33**, 225 (1975); *Chem. Phys.* **6**, 155 (1974).
72. Y. R. Shen, *Phys. Rev. B* **9**, 622 (1974).
73. K. Burnett, *Phys. Rep.* **118**, 339 (1985).
74. Y. Toyozawa, *J. Phys. Soc. Jpn.* **41**, 400 (1976); A. Kotani and Y. Toyozawa, *J. Phys. Soc. Jpn.* **41**, 1699 (1976).
75. V. Hizhnyakov and I. Tehver, *Phys. Stat. Solidi* **21**, 755 (1967); *Opt. Comm.* **32**, 419 (1980).
76. H. A. Kramers and W. Heisenberg, *Z. Phys.* **31**, 681 (1925).

77. R. G. Breene, *Theories of Spectral Lineshapes.* Wiley, New York, 1981.
78. D. Grimbert and S. Mukamel, *J. Chem. Phys.* **75**, 1958 (1981); **76**, 834 (1982).
79. S. Mukamel, *Phys. Rev. A* **26**, 617 (1982).
80. S. Mukamel and D. Grimbert, *Optics Comm.* **40**, 421 (1982).
81. J. J. Markham, *Rev. Mod. Phys.* **31**, 956 (1959).
82. R. Kubo and Y. Toyozawa, *Prog. Theo. Phys.* **13**, 160 (1955).
83. S. Mukamel, *J. Chem. Phys.* **77**, 173 (1982); A. Warshel, P. S. Stern, and S. Mukamel, *J. Chem. Phys.* **78**, 7498 (1983).
84. S. Mukamel, *J. Phys. Chem.* **88**, 3185 (1984).
85. J. Chesnoy and J. J. Weis, *J. Chem. Phys.* **84**, 5378 (1986); M. J. Saxton and J. M. Deutch, *J. Chem. Phys.* **60**, 2800 (1974).
86. D. Thirumalai, E. J. Bruskin, and B. J. Berne, *J. Chem. Phys.* **83**, 230 (1985).
87. L. Onsager, *J. Am. Chem. Soc.* **58**, 1486 (1936).
88. N. S. Baylis, *J. Chem. Phys.* **18**, 292 (1950).
89. C. J. F. Bottcher, *Theory of Electric Polarization.* Elsevier, Amsterdam, 1973, Vols. I, II.
90. J. Jortner and C. A. Coulson, *Mol. Phys.* **4**, 451 (1961); J. Jortner, *Mol. Phys.* **5**, 257 (1962).
91. D. Chandler, K. S. Schweizer, and P. G. Wolynes, *Phys. Rev. Lett.* **49**, 1100 (1982); K. S. Schweizer and D. Chandler, *J. Chem. Phys.* **78**, 4118 (1983).
92. R. W. Hall and P. G. Wolynes, *J. Chem. Phys.* **83**, 3214 (1985).
93. N. Bloembergen, E. M. Purcell, and R. V. Pound, *Phys. Rev.* **73**, 679 (1948).
94. P. W. Anderson and P. R. Weiss, *Rev. Mod. Phys.* **25**, 269 (1953).
95. R. Kubo, in *Fluctuations, Relaxation and Resonance in Magnetic Systems* (D. ter Haar, ed.). Plenum, New York, 1962, p. 23.
96. J. Sue and S. Mukamel (unpublished)
97. N. G. Van Kampen, *Stochastic Processes in Physics and Chemistry.* North-Holland, Amsterdam, 1981.
98. See papers in *Farad. Disc. Chem. Soc.* **75** (1983).
99. D. H. Levy, *Ann. Rev. Phys. Chem.* **31**, 197 (1980).
100. C. S. Parmenter, *J. Phys. Chem.* **86**, 1735 (1982); *Farad. Disc. Chem. Soc.* **75**, 7 (1983); D. A. Dolson, K. W. Holtzclaw, S. H. Lee, S. Munchak, C. S. Parmenter, and B. M. Stone, *Laser Chem.* **2**, 271 (1983); K. W. Holtzclaw and C. S. Parmenter, *J. Chem. Phys.* **84**, 1099 (1986); D. A. Dolson, K. W. Holtzclaw, D. B. Moss, and C. S. Parmenter, *J. Chem. Phys.* **84**, 1119 (1986).
101. J. B. Hopkins, D. E. Powers, S. Mukamel, and R. E. Smalley, *J. Chem. Phys.* **72**, 5049 (1980).
102. S. Mukamel, *J. Chem. Phys.* **82**, 2867 (1985).
103. G. Herzberg, *Molecular Spectra and Molecular Structure.* Van Nostrand, New York, 1966.
104. R. Islampour and S. Mukamel, *J. Chem. Phys.* **80**, 5487 (1984); *Chem. Phys. Lett.* **107**, 239 (1984).
105. S. Mukamel and K. Shan, *J. Phys. Chem.* **89**, 2447 (1985); *Chem. Phys. Lett.* **117**, 489 (1985).
106. J. Sue and S. Mukamel, *Chem. Phys. Lett.* **107**, 398 (1984).
107. J. Grad, Y. J. Yan, and S. Mukamel, *Chem. Phys. Lett.* **134**, 291 (1987); J. Grad, Y. J. Yan, A. Haque, and S. Mukamel, *J. Chem. Phys.* **86**, 3441 (1987).

# ON THE STATISTICAL THEORY OF UNIMOLECULAR PROCESSES

DAVID M. WARDLAW*

*Department of Chemistry, Queen's University, Kingston, Ontario K7L 3N6, Canada*

R. A. MARCUS

*Arthur Amos Noyes Laboratory of Chemical Physics†, California Institute of Technology, Pasadena, California 91125*

## CONTENTS



* NSERC of Canada University Research Fellow.
† Contribution No. 7473.

# I. INTRODUCTION

In recent years there has been an increasing use of laser spectroscopic and other techniques to investigate unimolecular dissociations of molecules, both to initiate a dissociation and to measure the formation of the individual quantum states of the immediate reaction products. This chapter is concerned with a description of statistical theories used to calculate the rates of such dissociations, for example, of a molecule $AB$,

$$AB \rightarrow A + B, \tag{1.1}$$

and to calculate the distribution of the quantum states of the fragments $A$ and $B$. Typically, reaction rates in the literature have been calculated using RRKM theory, while distributions over various quantum states of products have been calculated mostly with phase-space theory and other models described subsequently.

Upon excitation of $AB$ with a laser pulse, the molecule may be excited to a high vibrational state of the lowest electronic state of $AB$ or to a vibrational state of an excited electronic state $AB^*$, depending on the method and on the molecule. More specifically, a wave packet of vibrational states is typically formed in each case. When the excitation is to $AB^*$, the subsequent dissociation may proceed either directly from that electronic state or, after an internal conversion, from the ground electronic state. We concentrate on the dissociation itself, and so, in the case of an internal conversion, on the behavior subsequent to the conversion.

One topic of interest is the comparison of the measured rates with those calculated using statistical theory. Commonly, the initial vibrational excitation is a nonstatistical one, as for example in the excitation of a high CH overtone of the molecule (a wave packet), which is then followed by a tendency to energy redistribution as well as by reaction. Infrared multiphoton excitation and chemical activation can be expected to be nonstatistical. If the redistribution is sufficiently rapid, the observed time decay of $AB$ in Eq. (1.1) will be a single exponential, according to statistical theory, when the energetic $AB$ has been formed with a sufficiently narrow range of energy $E$ and of angular momentum $J$, both of which affect the dissociation rate. When the redistribution is not sufficiently rapid, a more complicated time evolution for the formation of $A$ and $B$ is expected. Indeed, the assumption of a kinetic model[1] for the redistribution (involving a subdivision of the phase space into coupled subspaces) and for the reaction leads to the result that under certain conditions the long-time behavior can still be described by statistical (RRKM) theory.[2]

The study of unimolecular reactions was originally confined to thermal studies as a function of pressure and, later, owing particularly to the pioneering work of B. S. Rabinovitch, to chemical activation studies.[3,4] In the past two or so decades a wide variety of other methods have also been introduced,[5] including the study of translational and vibrational energy distribution of products of chemical activation reactions in molecular beams, energy distributions of products of infrared multiphoton dissociation in bulk and in beams, use of high CH or OH overtone excitation to initiate unimolecular reactions, study of the angular distribution of reaction products of unimolecular reactions in molecular beams, use of picosecond techniques to study unimolecular reaction rates and quantum states of products, and the investigation of molecular ion dissociations using a variety of techniques, including coincidence measurements.

We describe several statistical approaches for calculating rates and product quantum state distributions in the following section. Illustrative applications are given in Section III, and dynamical aspects and statistical behavior are considered in a concluding Section IV.

## II. OUTLINE OF STATISTICAL THEORIES

### A. RRKM Theory

The expression for the rate constant $k_{EJ}$ for dissociation of a molecule as a function of its energy $E$ and of any other constants of the motion, such as $J$, the total angular momentum quantum number, is given in RRKM theory as[6]

$$k_{EJ} = \frac{N_{EJ}^{\ddagger}}{h\rho_{EJ}}, \tag{2.1}$$

where $N_{EJ}^{\ddagger}$ is the number of accessible quantum states of the transition state for the dissociation and $\rho_{EJ}$ is the density of states of the parent molecule at the same $E$ and $J$.

In treating isomerizations or dissociations, two limiting forms of the transition state were identified—loose and tight, or rigid, as it was termed then.[6a] In the former the separating fragments rotate freely in the transition state, while in the latter their motion is more or less similar to that in the parent molecule. In the case of reactions with a marked maximum of the potential energy along a reaction coordinate $R$, the position of the transition state $R^{\ddagger}$ was taken to occur at that $R$. In some dissociations there is little or no such maximum, and another criterion is needed for the determination of $R^{\ddagger}$. When the reactants are assumed to rotate freely at $R^{\ddagger}$, that is, when the transition state is loose, the reaction coordinate $R$ has been chosen to be the distance between the centers of mass of the separating fragments, and $R^{\ddagger}$ has then been equated, approximately, to the separation distance in a van der Waals' complex of the fragments, for example, as in the Gorin model[7] described in the next section. When the orbital angular momentum (quantum number $l$) is the dominant contribution (as it frequently is) to the total angular momentum (quantum number $J$), that is, when $l \simeq J$, the orbital contribution to $N_{EJ}^{\ddagger}$ is simply $2l + 1$, that is, $2J + 1$.[4] The remaining contribution to $N_{EJ}^{\ddagger}$ is also simple, when the various remaining coordinates, apart from $R$, can be approximated by free rotations and vibrations.

The choice of the location of the transition state $R^{\ddagger}$ has been generalized by choosing it to correspond to the position of the minimum of the reactive flux along $R$ (now often called microcanonical variational transition state theory[10]). Combining this choice with a method for calculating $N_{EJ}^{\ddagger}$ for any transition state (loose, tight, or in between), Eq. (2.1) has recently been implemented for such more general systems.[11]

In the case of a loose transition state, the distribution function for the quantum states of the separated products is the same as that in the transition state, since there is no coupling between $R$ and the internal coordinates. Thus, if the number of states in the transition state that have a given quantum number $n_i$ for the $i$th coordinate is $N_{EJn_i}^{\ddagger}$ the probability $P_{EJn_i}$ that the fragments have the quantum number $n_i$ for this coordinate is

$$P_{EJn_i} = N_{EJn_i}^{\ddagger}/N_{EJ}^{\ddagger}. \tag{2.2}$$

Ideas analogous to this, together with a model for the distribution of $J$'s due to the energetic dissociating molecule being formed in a bimolecular collision, have been used to treat translational energy distributions[8] and other properties[8b] of products of bimolecular reactions involving molecular complexes as intermediates. In further work[9] we extended this type of calculation, with some added approximations, to tight transition states in the exit channel.

In general, except when the transition state is loose, the calculation of the distribution function $P_{EJn_i}$ of excess energy among the quantum states of the reaction products involves an additional approximation over and above those needed to calculate $k_{EJ}$. Examples are given in Section III.

### B. Loose Transition State and Angular Momentum Conservation (PST)

In a loose transition state there is no interaction of the radial coordinate with the internal degrees of freedom, and the reaction coordinate $R$ is chosen to be that coordinate. The effective potential $V_{\text{eff}}$ for motion along $R$ is then given (in units of $\hbar = 1$) as the sum of the actual potential and a centrifugal potential,

$$V_{\text{eff}}(R) = V(R) + \frac{l(l+1)}{2\mu R^2}, \tag{2.3}$$

where $V(R)$ is the $R$-dependent part of the potential energy and $\mu$ is the reduced mass of the separating fragments. The position of the transition state is assumed to be at the maximum of $V_{\text{eff}}(R)$,

$$\frac{\partial V_{\text{eff}}(R)}{\partial R} = 0 \quad \text{at } R = R^{\ddagger}, \tag{2.4}$$

and so this $R^{\ddagger}$ depends on $l$. Equations (2.3) and (2.4) were used in an early calculation of Gorin[7] for the recombination of methyl radicals.

If $N^{\ddagger}_{EJl}$ is the number of states for which, for the given $l$, the maximum of $V_{\text{eff}}$ is less than or equal to $E$, the $N^{\ddagger}_{EJ}$ in Eq. (2.1) equals

$$N^{\ddagger}_{EJ} = \sum_{l} N^{\ddagger}_{EJl}. \tag{2.5}$$

If the approximation $l \simeq J$ is introduced, the sum in (2.5) consists only of one term.

We consider next phase-space theory (PST), which was designed principally to calculate energy distributions of the reaction products.[12] In PST a loose transition state is assumed and, rather than using the approximation of $l \simeq J$, $l$ is chosen so as to satisfy angular momentum conservation, namely, the triangle inequality,

$$|J - k| \le l \le J + k, \tag{2.6}$$

where $k$ is an integer, a quantum number, corresponding to the vector sum of the rotational angular momenta of the two fragments (quantum numbers $j_1$ and $j_2$); $k$, itself, satisfies the triangle inequality

$$|j_1 - j_2| \le k \le j_1 + j_2. \tag{2.7}$$

The various states contributing to $N^{\ddagger}_{EJ}$ contain as quantum numbers $J, l, j_1, j_2, k, \kappa_1, \kappa_2$, together with the vibrational quantum numbers of each fragment. Here, $\kappa_i$ is the quantum number which, in addition to $j_i$ $(i = 1, 2)$, specifies the rotational energy of fragment $i$. (There are $2j_i + 1$ values of $\kappa_i$, for any $j_i$. If fragment $i$ is a symmetric top, $\kappa_i$ is the component of $j_i$ along the symmetry axis.) The absence of coupling terms between the radial motion $R$ and the remaining coordinates in Eq. (2.3) reflects the fact that the separating fragments rotate freely in a loose transition state. The calculation of each $N^{\ddagger}_{EJl}$ is relatively straightforward albeit numerical because of the constraints (2.6) and (2.7). The quantity of major interest in applications of PST is the distribution function $P_{EJn_i}$, which is given by Eq. (2.2).

The assumption of a loose transition state provides only an upper bound to the dissociation rate constant, apart from possible tunneling effects, which are usually quite minor for dissociations into polyatomic fragments. Frequently, the $R$-motion in the exit channel is coupled to the other coordinates, for example to the developing rotations of the fragments, and the latter are coupled to each other, resulting in a hindered rather than free rotational motion of the separating fragments in the transition state. The resulting quantum states contributing to $N_{EJ}^{\ddagger}$ in Eq. (2.1) are then more widely spaced than when the fragments rotate freely, and so the states are fewer in number. The actual $k_{EJ}$ is then less than the PST value, a result well known for many reactions.[13]

Again, in the case of a dissociation reaction in which the transition state was not loose (e.g., the reverse reaction of methyl radical addition to an olefin had a steric factor), PST did not adequately describe the translational energy distribution of the products, and resort was made to a more general treatment that made some allowance for hindered motions.[9]

We turn in the next section to a recent treatment that allows for these coupled hindered rotational motions.[11]

### C. General Transition State and Angular Momentum Conservation (RRKM)

We consider here the implementation of Eq. (2.1) for more general transition states. When the transition state is loose, it occurs at the maximum in $V_{\text{eff}}$, as in Eq. (2.4). However, when the rotational motion of the separating fragments is hindered, owing to their coupled behavior, a more general criterion for $R^{\ddagger}$ is needed. Often in dissociations into free radicals there is no appreciable potential energy maximum to suggest an $R^{\ddagger}$, and the criterion (2.4) is also clearly inadequate for the more general case. An appropriate criterion arises from Wigner's and Keck's classical version of variational transition state theory,[14] where the transition state, a hypersurface in phase space now, is chosen so as to have the minimum flux from reactants to products passing through it. In this case, there are the fewest recrossings of the transition state hypersurface by classical trajectories and the transition state calculated rate becomes an increasingly better bound to the actual rate.[14b]

When the motion along the reaction coordinate is treated classically and the remaining coordinates quantum mechanically, it is convenient to define $N_{EJ}(R)$, the number of states with energy equal to or less than $E$ for the given $J$ and for that $R$. The flux along $R$ at any point $R$ is proportional to $N_{EJ}(R)$, as can be seen, for example, from Eq. (2.1), since $k_{EJ}$ is proportional to the flux. Thus, the location $R^{\ddagger}$ of the transition state along $R$ (loose, tight, or in between) is determined as the $R$ that minimizes $N_{EJ}(R)$:[10]

$$\frac{\partial N_{EJ}(R)}{\partial R} = 0 \quad \text{at} \quad R = R^{\ddagger}. \tag{2.8}$$

Recently, we described a method for implementing Eq. (2.1) using (2.8) in a way in which $N_{EJ}(R)$ is calculated so as to satisfy the angular momentum triangle inequalities (2.6) and (2.7) for any prescribed potential energy function in the exit channel.[11] There is at least one price to pay for this generalization: While the expression for $N_{EJ}^{\ddagger}$ in the case of a tight transition state, or in the case of a loose transition state with $l \simeq J$, was evaluated analytically,[6] the expression in this more general implementation of Eq. (2.1) is evaluated numerically. In applications to dissociations thus far, we have chosen the reaction coordinate $R$ to be the distance between the centers of mass of the separating fragments.

The complexity of evaluating $N_{EJ}(R)$ and, by its minimization $N_{EJ}^{\ddagger}$, in the case of transition states that are neither loose nor tight, arises because of the coupled bending and rotational motion of the parent molecule, motions that become free rotations and orbital motion of the separated fragments.[11] A quantum calculation of the energy eigenvalues of these typically six or so coordinates (termed "transitional coordinates") at each $R$, for the given potential energy surface, would permit the determination of all such states contributing to $N_{EJ}(R)$, for the given $E$ and $J$. When these states of the transitional coordinates are convoluted with $N_V$, the number of states from the more simply treated remaining coordinates, the total number of states $N_{EJ}(R)$ would then be obtained.

We have adopted the following procedure.[11] $N_{EJ}(R)$ is written as a convolution for each $R$,

$$N_{EJ}(R) = \int_0^E N_V(E - \varepsilon)\Omega_J(\varepsilon)\, d\varepsilon, \tag{2.9}$$

where $\Omega_J(\varepsilon)$ is the density of states of the "transitional" (coupled bending vibrational–rotational) degrees of freedom in the energy interval $(\varepsilon, \varepsilon + d\varepsilon)$, and $N_V(E - \varepsilon)$ is the number of quantum states for the coordinates not present in $\Omega_J$, when their energy is equal to or less than $E - \varepsilon$. The two types of coordinates are taken to be uncoupled from each other, apart from their dependence on $R$. The value of $N_V$ in Eq. (2.9) is obtained by the usual quantum count. The main approximation is in the evaluation of $\Omega_J(\varepsilon)$.

We have found it convenient to calculate $\Omega_J(\varepsilon)$ classically and to express the coordinates contributing to $\Omega_J(\varepsilon)$ in terms of action-angle variables. (Approximate quantum corrections can be made.[11]) The actions are the classical counterparts of the quantum numbers $j_1$, $j_2$, $\kappa_1$, $\kappa_2$, $k$, $l$, and $J$ previously mentioned, and we shall simply denote the actions by the same symbols to simplify the notation. When the $i$th fragment is an atom the numbers $j_i$, $\kappa_i$ and $k$ are absent, while if one of the fragments is a linear molecule, its $\kappa_i$ is absent. For two nonlinear fragments we have, in units of $\hbar = 1$,[11] for a particular electronic state, and apart from any electronic multiplicity factor,

$$N_{EJ}(R) = (2J + 1)(2\pi)^{-6}\sigma^{-1} \times \int \cdots \int N_V(E - H_{cl})\Delta(J, k, l)\Delta(k, j_1, j_2)\, dj_1\, dj_2\, d\kappa_1\, d\kappa_2\, dk\, dl\, d\alpha, \tag{2.10}$$

where $d\alpha$ is the volume element for the angle variables conjugate to the six action integration variables in Eq. (2.10). The $\sigma$ denotes the symmetry number for transformations involving these six angle coordinates.[6] Each angle is integrated over the interval 0 to $2\pi$, within the limits imposed by energy conservation. The action variables are restricted both by energy conservation and by the triangle inequalities (2.6) and (2.7), each $\Delta$ being unity when the relevant inequality is fulfilled and zero otherwise. Each $\kappa_i$ is restricted by the condition $|\kappa_i| \leq j_i$. (In practice, a symmetric top approximation will be typically used for fragment $i$, and then $\kappa_i$ refers to the component of $j_i$ along the symmetry axis.) $H_{cl}$ is the part of the Hamiltonian that refers to the coordinates contributing to $\Omega_J(\varepsilon)$, for example, as in Eq. (II.2) of Ref. 11*c*. Equations (2.1) and (2.8)–(2.10) are used for the calculation of the rate constants $k_{EJ}$, the integral in (2.10) being evaluated by a Monte Carlo method. The calculation of the thermally averaged value of $k_{EJ}$ is given later.

A calculation of the distribution function $P_{EJn_i}$ for the quantum states of the reaction products requires an additional approximation over and above that needed for $k_{EJ}$. One approximation is to assume an adiabaticity in the exit channel, wherein the lowest states in $N_{EJ}^{\ddagger}$ correlate adiabatically with the lowest states in the products.[2] The relative probability $P_{EJn_i}$ that the quantum state of the $i$th coordinate is $n_i$ is then given by[2]

$$P_{EJn_i} = N_{E'-E_iJ}^{\infty}/N_{EJ}^{\ddagger}, \tag{2.11}$$

where $E'$ is defined by the condition

$$N_{EJ}^{\ddagger} = N_{E'J}^{\infty}, \tag{2.12}$$

$N_{E'J}^{\infty}$ being the number of quantum states of the products at $R = \infty$ with energy less than or equal to $E'$, $N_{E'-E_iJ}^{\infty}$ is the corresponding number at $R = \infty$ when there is an energy $E_i$ in the $i$th coordinate (or set of coordinates) of the products and the quantum number for this coordinate (or totality of quantum numbers) is $n_i$. When states of different symmetry are involved along the entire reaction coordinate, a suitable classification of the states in (2.11) by symmetry is first made. A variant of (2.11) assumes that $l$ is a constant in the exit channel; this variant reduces to PST in the limit that the hindered rotations at $R^{\ddagger}$ become free.[2] However, when $l \simeq J$, $l$ becomes essentially a constant of the motion automatically, and so even Eqs. (2.11)–(2.12) reduce approximately to PST in the limit that the rotations at $R^{\ddagger}$ are free. Otherwise, Eqs. (2.11)–(2.12) reduce to PST only on the average in the free rotation limit.

### D. Vibrationally Adiabatic Transition State and SACM

Adiabatic, or as it has been termed vibrationally adiabatic,[15] transition state theory has its origin in a paragraph in an article by Hirschfelder and Wigner.[16] The treatment was developed further by a number of authors.[17] In this type of transition state theory the eigenvalues of the system at each $R$, which are the vibrationally adiabatic eigenvalues, are plotted versus $R$. The $N_{EJ}^{\ddagger}$ in Eq. (2.1) then becomes the number of such states whose maximum energy on this plot does not exceed $E$, that is, $N_{EJ}^{\ddagger}$ now denotes the sum of all open adiabatic reaction channels.

In principle, this $N_{EJ}^{\ddagger}$ can be determined if the potential energy surface is known: For the given $J$ value and for each value of the reaction coordinate $R$ (e.g., the bond length of the dissociating bond or the distance between the centers of mass of the two separating fragments), the vibrational–rotational quantum eigenvalues $E_v(R)$ are obtained. The result is a set of nonintersecting adiabatic channel curves that smoothly connect, and hence correlate, the reactant (parent molecule) and product (separated fragments) vibrational–rotational energy levels. Each curve $E_v(R)$ has a maximum $E_{\max}$ at some $R$ in the exit channel. $N_{EJ}^{\ddagger}$ is then the sum of all open adiabatic channel curves, that is, those curves for which $E_{\max} < E$. In practice, the full potential energy functions are usually unavailable and, even if they were, obtaining the quantum eigenvalues (perhaps by matrix diagonalization) at all relevant values of $R$ is simply not practical at present for most systems. The intermediate region between reactants ($R = R_e$) and products ($R = \infty$) presents the greatest computational difficulties for both the potential energy function and the eigenvalues, and this region often determines the reaction rate of the bond fission process.

The model of Quack and Troe,[18] the statistical adiabatic channel model (SACM), is an approximate prescription for calculating the number of open adiabatic channels. A universal function $g(R)$ is assumed for interpolating between all eigenvalues of reactants and products,[19]

$$g(R) = \exp[-\alpha(R - R_e)], \tag{2.13}$$

with $\alpha$ being treated as an adjustable parameter. In fitting SACM to thermal rate data, typically $\alpha \sim 1$ Å$^{-1}$.[20] Even with this approximation, as indicated in the original presentation[18] of SACM, direct numerical count for $N_{EJ}^{\ddagger}$

becomes exceedingly time consuming for large molecules, owing to the large number of reactant and product energy levels to be correlated and the correspondingly large number of channel maxima to be located (usually numerically). Various additional approximations have been postulated to simplify the calculation.[21]

The distribution (relative probability) of products having a quantum number $n_i$ for mode $i$ is again given by Eq. (2.2), where $N^{\ddagger}_{EJn_i}$ now denotes the number of open adiabatic channels leading to products in state $n_i$ and $N^{\ddagger}_{EJ}$ is the total number of open adiabatic channels at the given $E$ and $J$.[22]

## III. APPLICATIONS

In this section several recent applications of the methods previously described are summarized for some selected bond fissions or for the reverse reactions. No attempt is made to provide a comprehensive review of all such applications. Applications to ion–molecule reactions[23,24] have also been omitted. Some specialized statistical models have been developed[23] for this topic.

### A. $CH_3$ Recombination

The recombination of methyl radicals to form ethane

$$2CH_3 \xrightarrow{k_r} C_2H_6 \tag{3.1}$$

has been widely studied experimentally under various conditions, the recombination rate constant $k_r$ being defined by

$$-\frac{d[CH_3]}{dt} = 2k_r[CH_3]^2. \tag{3.2}$$

Many statistical models have been applied to reaction (3.1), and it might be considered a test case for theoretical treatments of the rate constant. The process inverse to (3.1), the dissociation of ethane, has also been extensively studied experimentally[25,26] and theoretically.[11b,22,27] The theoretical predictions for the rate of dissociation are, of course, quite sensitive to the value of the bond dissociation energy. On the other hand, recombination rates depend only weakly on that quantity. In the present review, attention is focused on the prediction of the recombination rate using the transition state theory outlined in Section II C. First, the high-pressure limit of $k_r$, denoted by $k_\infty$, is considered, particularly its temperature dependence. This is followed by a brief description of some results for the pressure dependence of $k_r$ and for the dissociation of a vibrationally excited $C_2H_6$ molecule.

### 1. *High-Pressure Rate*

In the following the salient features of the general transition state theory of Sections II A and II C, as applied to methyl radical recombination, are summarized. Details, where omitted, are given in Ref. 11*c*.

The high-pressure recombination rate constant $k_\infty$ is given as a function of the temperature $T$ by[28]

$$k_\infty(T) = \frac{g_e}{hQ_r(T)} \int_0^\infty dE \sum_{J=0}^{\infty} (2J+1) N_{EJ}(R^\ddagger) e^{-E/kT}, \tag{3.3}$$

where $E$ is the total energy in the center-of-mass frame, $J$ and $N_{EJ}(R)$ were defined earlier, and $R^\ddagger$ was defined by Eq. (2.8); $Q_r$ is the partition function, in the center-of-mass frame, for the pair of reactants at infinite separation and is evaluated in a standard fashion[29]; $g_e$ is the ratio $g^\ddagger/g_1 g_2$ of electronic partition functions for the transition state and separated radicals and is taken to be $\frac{1}{4}$, corresponding to a common situation that only systems initially on the singlet potential energy surface lead to recombination products.

The main ideas in the calculation of $N_{EJ}$ were described earlier (Section II C). For the specific case of reaction (3.1), Eq. (2.10) is used for $N_{EJ}(R)$. The classical Hamiltonian $H_{cl}$ in (2.10) is written as

$$H_{cl} = \sum_{i=1,2} E_{ri} + \frac{l^2}{2\mu R^2} + V_t(r_{mn}, \theta_1, \theta_2), \tag{3.4}$$

where $E_{ri}$ is the rotational energy obtained from a rigid oblate symmetric top model of methyl radical $i (i = 1, 2)$ with $R$-dependent moments of inertia; $\mu$ is the reduced mass for relative motion of the two fragments; $V_t$ is the potential energy function for the transitional modes (specifically, for variation in the separation distance and in the orientation of the $CH_3$ groups); $r_{mn}$ denotes a $4 \times 4$ set of interfragment distances between atom $n (n = 1$ to $4)$ of fragment 1 and atom $m (m = 1$ to $4)$ of fragment 2; $\theta_i$ is the angle between the carbon–carbon displacement vector $\mathbf{r}_{CC} \equiv \mathbf{r}_{44}$ and the symmetry axis of methyl radical $i (i = 1, 2)$. The internal coordinates $r_{mn}, \theta_1, \theta_2$ that determine $V_t$ in Eq. (3.4) are completely specified (using a transformation given in Ref. 11*c*) by $R$, $J$ and the 12 variables of integration in Eq. (2.10). The $l$ in Eq. (3.4) denotes an action, the orbital angular momentum $l_{cl}$, and is related semiclassically to the quantum number $l$ in Eq. (2.3) by $l_{cl} \approx (l + \frac{1}{2})$ (in units of $\hbar = 1$) and so the $l^2$ in Eq. (3.4) denotes $l_{cl}^2$, that is, $(l + \frac{1}{2})^2$, which is the semiclassical (Langer) approximation to the $l(l+1)$ in Eq. (2.3). The potential $V_t$ is intended to be physically reasonable, and can be constructed from *ab initio* points when they become available. It is to be emphasized that the present method can be used regardless of the explicit form of the potential energy surface for the transitional modes (or for the conserved modes) and regardless of whether or not $l$ is a constant of the motion.

The transitional mode potential was assumed to arise from nonbonded and bonded interactions

$$V_t = V_{NB} + V_B, \tag{3.5}$$

where $V_{NB}$ is a sum of Lennard-Jones potentials between all nonbonded C···H and H···H pairs and $V_B$ is a Morse function modified by an orientational factor to include in an approximate way to the now bent C···C bond. The explicit forms of $V_{NB}$ and $V_B$ are given in Ref. 11*c*.

In the absence of a current precise knowledge of the potential energy surface, interpolations were used to obtain approximate normal mode frequencies for the conserved modes and methyl radical moments of inertia at intermediate $R$ values, as described in Ref. 11*c*, using the interpolation $g(R)$ specified by Eq. (2.13). The number of quantum states for the conserved modes [$N_V$ in Eq. (2.10)] was obtained at each value of $R$ by a direct count of the approximately harmonic levels. Calculations were made for two values of $\alpha$ in Ref. 11*c*, as reported below. The potential energy and structural parameters that determine $V_t$, the conserved mode frequencies, and the moments of inertia are given in tables in Ref. 11*c*.

Three related approaches to the calculation of the recombination rate constant are described subsequently, the second two being approximations to the first. In each case the integral over $E$ and the sum over $J$ in the numerator of Eq. (3.3) are approximated by $N$-point Laguerre and $2M$-point extended Simpson's rule quadratures,[30] respectively, yielding

$$k_\infty^I(T) = \frac{g_e kT}{hQ_r(T)} \frac{\Delta J}{3} \sum_{i=1}^{N} \sum_{j=0}^{2M} w_i w_j N_{E_iJ_j}(R_I^\ddagger), \tag{3.6}$$

where $E_i = kTx_i$, $w_i$ and $x_i$ are Laguerre weights and points,[30] $\Delta J$ is the constant step size for the $J$-quadature, $J_j = j\Delta J$, with $j$ being an integer, not to be confused with the $j_i$'s in Eq. (2.10), and $w_j = 1$ for $j = 0$ and $2M$, $w_j = 2$ for $j$ even $w_j = 4$ for $j$ odd.

The transition state location $R_I^\ddagger$ was determined for each $(E_i, J_j)$ pair in Eq. (3.6) by minimizing $N_{E_iJ_j}$ with respect to $R$ on a 0.1 Å grid over an appropriate range of $R$ values. The variation in $R_I^\ddagger$ with energy is depicted in Fig. 1 for several of the $J$ values employed in the evaluation of Eq. (3.6). The integral appearing in the expression for $N_{EJ}(R)$ in Eq. (2.10) was evaluated by the Monte Carlo method described in Ref. 11*b* and is not discussed here. The Monte Carlo approximation to $N_{EJ}(R)$ was denoted by $N_{EJ} \pm \sigma^{MC}$, where $\sigma^{MC}$

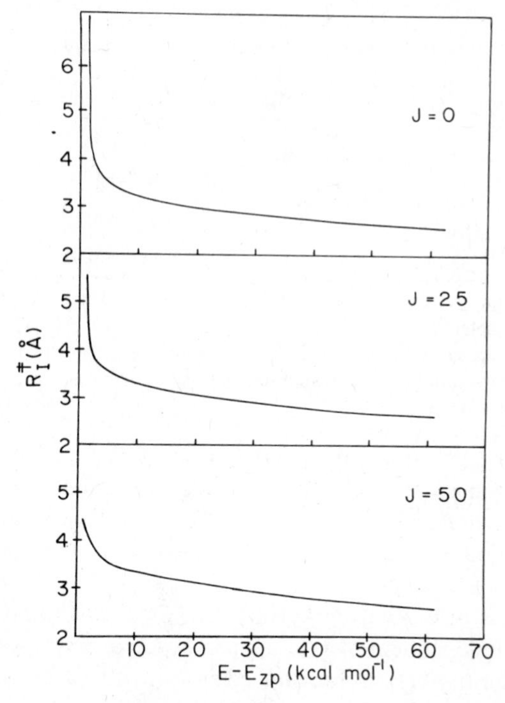


**Figure 1.** Plot of the transition state location versus the energy in excess of the total zero-point energy of separated methyl radicals for several values of the total angular momentum $J$. The solid curves result from smoothly connecting $R_I^\ddagger$ values for discrete values of the energy $E$. Each $R_I^\ddagger$ was obtained for a particular $(E, J)$ pair by minimizing $N_{E,J}$ with respect to $R$ on a 0.1 Å grid over an appropriate range of $R$ values.

is one standard deviation. Estimates of the upper and lower bounds on $k_\infty^I(T)$ were obtained by using $(N_{E_iJ_j} + \sigma^{MC})$ and $(N_{E_iJ_j} - \sigma^{MC})$, respectively, in Eq. (3.6). Convergence of the quadratures in Eq. (3.6) was ascertained by the agreement (within upper and lower bounds) of $k_\infty^I$ values obtained with different values of $N$ and $2M$.[11c] Since the model potential $V_t$ for the transitional modes admits, for sufficiently small values of $R$ and $E$, disjoint regions of energetically accessible classical phase space, an *ad hoc* restriction in the Monte Carlo sampling procedure was introduced[11c] to account for this feature.

The symmetry factor $\sigma$ in Eq. (2.10) deserves special consideration. A loose transition state will be characterized by large values of $R_I^\ddagger$ and, hence, will have essentially planar methyl radicals. In this case each of the identical fragments has both $C_{2v}$ and $C_{3v}$ axes, resulting in a $\sigma = (3 \times 2)^2 \times 2 = 72$. The results given in Table I were obtained using this value of $\sigma$. Results obtained by an

TABLE I
High-Pressure Recombination Rate Constants[a]

| | Rate Constants ($10^{13}$ cm$^3$ mol$^{-1}$ sec$^{-1}$) | | |
|---|---|---|---|
| Temperature (K) | $k_\infty^I \pm \delta^I$ | $k_\infty^{II} + \delta^{II}$ | $k_\infty^{III} \pm \delta^{III}$ |
| 300 | 4.33 ± 0.12 | 4.28 ± 0.10 | 5.08 ± 0.13 |
| 500 | 3.67 ± 0.08 | 3.58 ± 0.08 | 4.38 ± 0.06 |
| 1000 | 2.37 ± 0.06 | 2.29 ± 0.06 | 2.79 ± 0.06 |
| 2000 | 1.09 ± 0.04 | 1.08 ± 0.03 | 1.32 ± 0.03 |

[a] All rate constants calculated with a value of 1.0 Å$^{-1}$ for the potential energy surface interpolation parameter $\alpha$; $k_\infty^I = \frac{1}{2}[(k_\infty^I)_{\max} + (k_\infty^I)_{\min}]$ and $\delta = \frac{1}{2}[(k_\infty^I)_{\max} - (k_\infty^I)_{\min}]$, where $(k_\infty^I)_{\max}$ and $(k_\infty^I)_{\min}$ are estimated upper and lower bounds obtained by using $N_{E_iJ_j} + \sigma^{MC}$ and $N_{E_iJ_j} - \sigma^{MC}$, respectively, in Eq. (3.6); analogous expressions define $(k_\infty^{II}, \delta^{II})$ and $(k_\infty^{III}, \delta^{III})$.

approximate interpolation taking into account the umbrella shape of the methyl radicals at smaller $R$ values are also reported in Ref. 11*c* and are discussed briefly here. Motivation for use of an $R$-dependent symmetry correction is provided by Fig. 1 in which the decrease in $R_I^\ddagger$ with increasing $E$ is evident.

In the process of evaluating $k_\infty^I$ in Eq. (3.6) it was found[11c] that $R_I^\ddagger$ was approximately independent of $J$ for a given $E$. Accordingly, as an approximation to $k_\infty^I$, a quantity $k_\infty^{II}$ was introduced as follows: A $J$-averaged $N_{EJ}(R)$ was first defined, for use in (3.3),

$$N_E(R) = \int_0^{J_{\max}} dJ\, N_{EJ}(R), \tag{3.7}$$

where $J_{\max}$ is the maximum $J$ for a rigid symmetric top model of $C_2H_6$ when all of the available energy appears as overall rotation of the $C_2H_6$. The integral over $E$ in Eq. (3.3) was then replaced by an $N$-point Laguerre quadrature, as for $k_\infty^I$, and the resulting approximation to Eq. (3.3) became

$$k_\infty^{II}(T) = \frac{g_e kT}{hQ_r(T)} \sum_{i=1}^{N} w_i N_{E_i}(R_{II}^\ddagger). \tag{3.8}$$

In constrast with $k_\infty^I$ the transition state location $R_{II}^\ddagger$ is now determined for each $E_i$ in Eq. (3.8) by minimizing $N_{E_i}$ with respect to $R$ on a 0.1 Å grid. The integral for $N_E(R)$ in Eq. (3.7) was evaluated by extending the Monte Carlo method described in Ref. 11*b* to include the additional integration over $J$. $k_\infty^{II}(T)$ will equal $k_\infty^I(T)$ when $R_I^\ddagger$ is independent of $J$. The advantage of using Eq. (3.8) instead of Eq. (3.6) lies in the elimination of individual Monte Carlo calculations for each of the discrete $J_j$'s in Eq. (3.6) and a concomittant 3–5-fold reduction in the total number of required Monte Carlo points; the precise reduction factor varied with temperature. Agreement of the $k_\infty^I$ and $k_\infty^{II}$ values was found[11c] for the particular model of methyl radical recombination under consideration.

In a third approach, a canonical rate constant, denoted by $k_\infty^{III}$, was obtained in order to assess the error introduced by neglecting the dependence of the transition state location on $E$ and $J$. This rate constant $k_\infty^{III}$ is given by

$$k_\infty^{III}(T) = \frac{g_e kT}{hQ_r(T)} \sum_{i=1}^{N} w_i N_{E_i}, \tag{3.9}$$

where the sum is evaluated at the transition state location $R_{III}^\ddagger$, which is determined by minimizing that sum with respect to $R$ on a 0.2 Å grid; this value of $R_{III}^\ddagger$ is thereby independent of $E_i$. The evaluation of $N_{E_i}$ in Eq. (3.9) is as described above. $k_\infty^{III}$ equals $k_\infty^I$ when $R_I^\ddagger$ is independent of both $E$ and $J$. The procedure of using $k_\infty^{III}$ [though not using Eqs. (2.10) and (3.7) specifically] has often been employed in some of the earlier literature on reaction rates, where the procedure is referred to as a maximization of free energy of activation and, more recently, as canonical variational transition state theory.

Results for $k_\infty^I$, $k_\infty^{II}$, and $k_\infty^{III}$ are given in Table I for the temperatures 300, 500, 1000, and 2000 K, which approximately span the temperature range studied in the collective experimental work on the high-pressure limit of the methyl radical recombination rate. All results in Table I were obtained with $\alpha = 1$ Å$^{-1}$ [Eq. (2.13)]. The $k_\infty^I$ and $k_\infty^{II}$ values are seen to agree, within the numerical uncertainty (denoted by $\delta$ in Table I), for the four temperatures studied. The canonical rate constant $k_\infty^{III}$ is seen to be greater than $k_\infty^I$ or $k_\infty^{II}$ by about 20% at each temperature. Since at each $R$, $N_{EJ}(R) \geq N_{EJ}(R_I^\ddagger)$, $k_\infty^{III}$ should indeed be an upper bound to $k_\infty^I$. For the Monte Carlo method in Ref. 11$c$ the computational effort used to obtain $k_\infty^{III}$ via the energy quadrature in Eq. (3.9) was approximately the same as that for the preferred $k_\infty^{II}$ [Eq. (3.8)]. (An attempt to include the energy integral in the most straightforward way in the Monte Carlo procedure in Ref. 11$c$ substantially increased the required computer time. However, an improved weighting factor for the Monte Carlo sampling may be available.)

Of particular interest both theoretically and experimentally is the temperature dependence of $k_\infty$. Much of the available experimental data on $k_\infty$ is displayed in Fig. 2 of Ref. 11$c$. These data span the temperature range 250–2500 K and were published over a period from 1951 to 1985. A comprehensive review of pre-1980 experimental work on methyl radical recombination has been published by Baulch and Duxbury.[26] The large scatter in the data is evident in Fig. 2 of Ref. 11$c$ and some of the experimental uncertainties (not shown in that figure) are large. Nevertheless, it appears that $k_\infty$ may exhibit

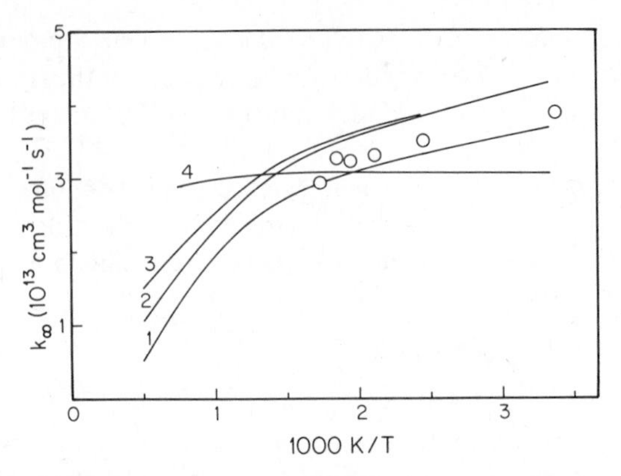


**Figure 2.** Plot of the high-pressure rate constant for $CH_3$ recombination versus reciprocal temperature. Theoretical results are presented as solid curves 1 through 4 and are discussed in the text: (1) generalized RRKM with $\alpha = 0.8$ Å$^{-1}$; (2) generalized RRKM with $\alpha = 1$ Å$^{-1}$; (3) generalized RRKM with $\alpha = 1$ Å$^{-1}$ and an approximate symmetry correction; (4) a simplified SACM treatment. The open circles are the experimental results of Ref. 33.

a negative temperature dependence, that is, a decrease in $k_\infty$ with increasing temperature. In their survey Olson and Gardiner[25] judged that "the consensus value of the recombination rate constant near 1000 K is seen to be a factor of 3 lower than the room-temperature consensus value." The present theoretical calculation yielded a similar result.[11c] However, accurate experimental determinations of $k_\infty$ require an extrapolation of results at various pressures, yielding some uncertainty in the quantitative temperature dependence, particularly at high temperatures, where a large extrapolation was needed.

Some of the present flexible transition state theory results for methyl radical recombination are given in Fig. 2, a plot of $k_\infty$ versus 1/T. Curves 1 and 2 depict $k_\infty^{II}$ for $\alpha = 0.8$ and 1 Å$^{-1}$, respectively,[31] curve 3 depicts the results, denoted by $k_\infty^{II'}$, of a $k_\infty^{II}$ calculation with $\alpha = 1$ Å$^{-1}$ in which the anticipated deviation of the symmetry number from its loose value was included in an approximate way.[11c] Each curve was obtained by smoothly connecting the calculated values of $k_\infty^r$ at $T = 300$, 500, 1000, and 2000 K. Plots of $k_\infty^{II}$, rather than the correct $k_\infty^I$, were presented because the $k_\infty^{II}$ calculations involved significantly less computation time and yielded results in close numerical agreement with those of the corresponding $k_\infty^I$ calculations. As might have been expected, the rate constant decreased when $\alpha$ was decreased from 1 to 0.8 Å$^{-1}$, in this case the decrease being about 10% at all four temperatures studied. At $T = 300$ K, $k_\infty^{II'}$ (curve 3) agrees with $k_\infty^{II}$ (curve 2) but does not decrease as rapidly with increasing $T$ and is about 40% greater than $k_\infty^{II}$ at $T \sim 2000$ K. In terms of the present model, the trend of decreasing $k_\infty$ with increasing $T$ is attributable to the decrease in $R^{\ddagger}$ with the increasing available energy $E$ associated with increasing temperature (Fig. 1). As the transition state becomes tighter deviations from a loose symmetry number become more pronounced.

A noteworthy feature of all three curves in Fig. 2 is the distinct negative temperature dependence. A number of other statistical models have been used to obtain $k_{\infty}$ values for this system. Many of these results are displayed in Fig. 2 of Ref. 11*c* and are discussed in the text accompanying that figure. Only two other models displaying a negative temperature dependence have been reported: a simplified SACM treatment[18] displayed a very small negative temperature dependence (curve 4 in Fig. 2). For a Gorin model loose transition state, modified by the *ad hoc* inclusion of geometrically derived steric factors for each radical (not given in Fig. 2), qualitative considerations indicated a rough estimate of $T^{-1/6}$ for the temperature dependence of $k_{\infty}$.[32]

Comparison with experiment in the present Fig. 2 is given for the 1985 results of Macpherson et al.[33] whose data appear as open circles. There is agreement with the general transition state theory calculations over the experimentally studied temperature range 296–577 K for curve 1. A feature of the data in Ref. 33 is the accurate determination of the absorption cross section $\sigma_a$ for the $CH_3$ absorption at 216.36 nm. This determination was important since most experimental studies of the $CH_3$ recombination rate accurately measure the ratio $k_{\infty}/\sigma_a$. In many earlier determinations of $k_{\infty}$, one source of error appears to lie in the absorption cross section, a case in point being an earlier work of Macpherson et al.[34]

From a theoretical perspective, it is clear that a knowledge of the full potential energy surface is needed for the improved application of a detailed statistical theory to this system, particularly at high temperatures, thereby avoiding the use of any interpolation parameter $\alpha$.

### 2. *Pressure-Dependent Rates*

The strengths and problems of studying the pressure-dependent recombination rates using a weak colliding gas is evident.[35] A brief preliminary report is given here of some calculations of Wagner et al.[35] on methyl radical recombination. Experimental rates for a discrete set of temperatures ranging from 295 to 1200 K were available as functions of Ar buffer gas concentration. These rates are plotted in Fig. 3 (with associated error bars, if available) versus the [Ar], which is seen to span one to three orders of magnitude, depending on the temperature studied. The theoretical analysis of the rates in Ref. 35 followed that in an earlier analysis of a different reaction,[36] except that a standard RRKM treatment of the transition state was replaced by the general transition state treatment for $k_{EJ}$ described in the previous section. Specifically, the $N_{EJ}(R)$ as given by Eq. (2.10) was evaluated by the method mentioned

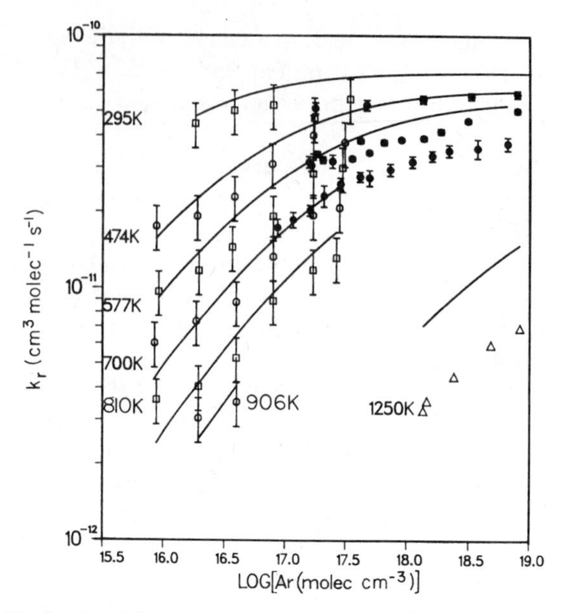


**Figure 3.** Logarithmic plot of the rate constant for $CH_3$ recombination versus argon buffer gas concentration at various temperatures. The solid curves are theoretical results obtained using generalized RRKM theory with $\alpha = 1$ Å$^{-1}$ and assuming an average energy transfer of $-190$ cm$^{-1}$ between buffer gas and metastable $C_2H_6$ per up and down collision. Experimental rates are indicated by various sysmbols with, in most cases, accompanying error bars; different symbols indicate rates obtained independently by different workers. Details are given in Ref. 35.

after Eq. (3.6) and employed for the sum of states at the transition state. A parameter $\langle \Delta E \rangle$ was varied from $\sim -175$ to $-300$ cm$^{-1}$ to obtain best agreement, $\langle \Delta E \rangle$ being the average energy transferred between buffer gas and metastable $C_2H_6$ per up and down collision. (The quantity $\langle \Delta E \rangle$ is related to an efficiency factor chosen so as to empirically mimic a master equation solution.[37]) The stabilization rate constant was taken to equal the product of the efficiency factor and a Lennard-Jones gas kinetic rate constant. For the value $\alpha = 1$ Å$^{-1}$ used earlier (Fig. 2), a value of $\langle \Delta E \rangle \simeq -190$ cm$^{-1}$ provided good agreement. The solid lines in Fig. 3 were computed with this value. Comparing theory and experiment on pressure effects with a weak collider, such as Ar, does introduce a new aspect, the unknown collision cross sections for the various deactivation and activation steps.

### 3. *Dissociation of Vibrationally Excited* $C_2H_6$

Growcock et al. determined the decomposition rate constant for vibrationally excited ethane produced by chemical activation, $C_2H_6^* \rightarrow 2CH_3$.[38] The reported[38] excitation energy of $114.9 \pm 2$ kcal mol$^{-1}$ with respect to the zero-point energy of ethane corresponds, for the molecular parameters used in Ref. 11*c*, to an excess energy of $27.3 \pm 2$ kcal mol$^{-1}$ with respect to the zero-point energy of separated methyl radicals. Here, it was assumed that this excitation energy was deposited in internal degrees of freedom and the overall rotational energy of ethane was approximated by $E_J = J(J+1)/2I_{A,r}$ (in units of $\hbar = 1$) with a symmetric top moment of inertia $I_{A,r}$ defined in Ref. 11*b*. For $J$-dependent total energies $E - E_{zp} = 27.3 + E_J$ and various $J$'s, the $N_{EJ}(R_I^\ddagger)$'s were determined. In each case $R_I^\ddagger$ was found to be $2.9 \pm 0.1$ Å. The calculated rate constants given by Eq. (2.1) were found to be $k_{EJ}$ (in $10^9$ sec$^{-1}$) = 4.6, 4.2, 4.4, 6.9, 14, and 33 for $J = 0$–125 in increments of 25.[11c] (At the lower $J$'s, $E_J$ at this $E$ is presumably too small to affect $k_{EJ}$.) These calculated $k_{EJ}$'s are consistent, for $J < 75$, with the reported[38] experimental value of $(4.6 \pm 1.2) \times 10^9$ sec$^{-1}$. The experimental $J$ distribution is not known. If $J$ had its thermally averaged value at room temperature (the temperature in the photolysis experiment[38] was not specified however), then the average $J$ would be about 25. In any event a $J$ distribution in which the dominant $J$'s are less than 75 is likely.

### B. $H_2O_2$ Dissociation

The unimolecular dissociation of isolated hydrogen peroxide in its ground electronic state

$$\mathrm{HOOH} \rightarrow 2\mathrm{OH} \tag{3.10}$$

has recently been studied experimentally[39–41] in a detailed manner. Statistical[39–41] theories have also been applied and classical trajectory[42] calculations have been made. In recent experiments the unimolecular decomposition of HOOH has been initiated by excitation of different OH stretching overtones (e.g., $5\nu_{\mathrm{OH}}$ and $6\nu_{\mathrm{OH}}$), and limited results for the partially deuterated species HOOD have also been obtained.[39–41] The yields of the different quantum states of the OH fragments were probed by time-resolved laser-induced fluorescence.

A vibrational–torsional potential model, in which the high-frequency OH vibration and the low-frequency torsional motion are adiabatically separated, was fit to the observed excitation spectra,[40] a model later modified to include the excitation of the O–O stretching vibration.[41] Peaks in the excitation spectrum were then assigned on the basis of this model (a main band involving pure OH stretching excitation, combination bands involving OH stretching excitation and torsional excitation, and hot bands for OH stretching excitation, in which the initial O–O vibration or torsional vibration was in an excited state).

In the work described subsequently, comparisons of experimental product internal energy distributions to each other or to theoretical models, are given. Complementary information is provided by the rate of product formation, a quantity whose measurement is precluded in the experiments described herein by the 10 nsec time resolution. The study of the direct time-resolved picosecond measurements of HOOH decomposition and the rates of energy relaxation or reaction has recently been done by Zewail and co-workers.[43] Furthermore, experiments have been reported using Doppler spectroscopy and the laser-induced fluorescence technique to measure the translational energy distribution and the angular distribution of OH products.[44]

Results have been obtained by Rizzo et al.[39] for HOOH excited to the region of the fifth overtone ($6\nu_{OH}$) and by Ticich et al.[41a] for HOOH excited to the region of the fourth overtone ($5\nu_{OH}$). With six quanta in OH stretching, there is a minimum of 4.6 kcal mol$^{-1}$ of energy available for disposal in the products, when the molecule is in its lowest vibrational state initially. On the other hand, five quanta of OH stretching fall 12.9 kcal mol$^{-1}$ short of the O–O bond dissociation energy, so that only reactant molecules that initially have at least this much thermal energy can dissociate. Consequently, the resulting product distributions are expected to reflect the influence of threshold effects on the decay dynamics. For both the $6\nu_{OH}$ and $5\nu_{OH}$ cases, all excess energy is observed to appear in translation and rotation of the products; there is insufficient energy in the light quantum itself to produce a vibrationally excited OH. Measured product rotational distributions resulting from HOOH initially prepared in the main band ($6\nu_{OH}$) and in a combination band ($6\nu_{OH} + \nu_x$), where one quantum of torsional motion, $\nu_x$, is excited are compared in Fig. 4 for both the $R_1$ and $Q_1$ branches of the OH spectrum, $N$ denoting the rotational quantum number of an OH.

Statistical and experimental product distributions for excitation wavelengths in the region of HOOH ($5\nu_{OH}$) main and combination bands and in the region of the HOOH ($6\nu_{OH}$) main and combination bands are given in Fig. 5. The statistical product distributions calculated in Ref. 41 are based on two of the statistical models described in Section II: PST and the original SACM. (PST corresponds to the limiting SACM case of $\alpha \to \infty$, and in Ref. 41 the PST results were essentially identical to the SACM product distributions when $\alpha = 1.5$ Å$^{-1}$.) In the application of PST to this system it was found that the rather strong long-range attraction between OH fragments yielded centrifugal barriers at large separations and hence there were only small centrifugal barrier heights for physically reasonable values of the orbital angular momentum. The PST calculation was therefore found to be insensitive to the exact form of the long-range radial potential. The SACM and PST calculations included the initial distribution of the unexcited $H_2O_2$ and the effect of photoexcitation.[41a]

The experimental product distributions arising from decomposition of HOOH excited to the region of $5\nu_{OH}$ (Figs. 5*a* and 5*b*) are seen to be qualita-

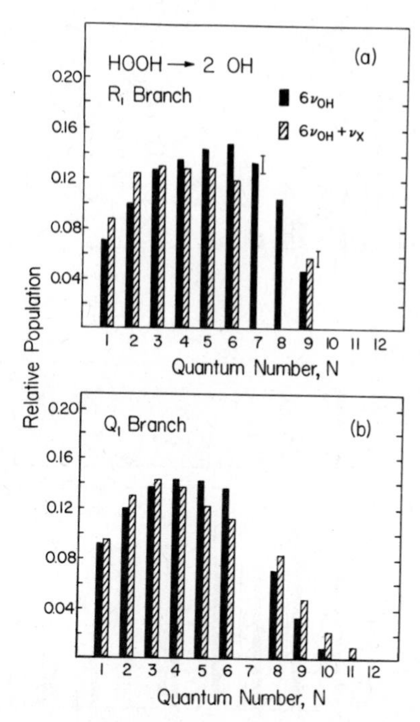


**Figure 4.** Relative rotational state distributions of OH products from overtone-vibration-induced unimolecular decomposition of HOOH. The solid bars are populations for excitation of the main local mode transition ($6v_{OH}$) and hatched bars are populations for excitation of the combination transition ($6v_{OH} + v_x$). The quantum number $N$ denotes the rotational OH angular momentum. Figures 4*a* and 4*b* show results obtained probing the $Q_1$ and $R_1$ branches, respectively, of OH. The error bars in Fig. 4(*a*) show the maximum range of values obtained and are typical of the uncertainties for all states. (Reproduced with permission from Ref. 39.)

tively different from the corresponding $6v_{OH}$ distributions (Figs. 5*c* and 5*d*). The $5v_{OH}$ distributions arise from decomposition of reactant states excited from the high-energy tail of the initial Boltzmann distribution, and many of these states therefore have energy near threshold. This is not the case for the $6v_{OH}$ decompositions. In parallel with new picosecond measurements of rates,[43] PST calculations are under way,[45] and calculations with the general transition state theory in Section II C are planned.

## C. NCNO Dissociation

Another unimolecular reaction that has been studied in great detail experimentally[46–49] is the collision-free decomposition of nitrosyl cyanide:

$$NCNO \rightarrow CN + NO. \qquad (3.11)$$

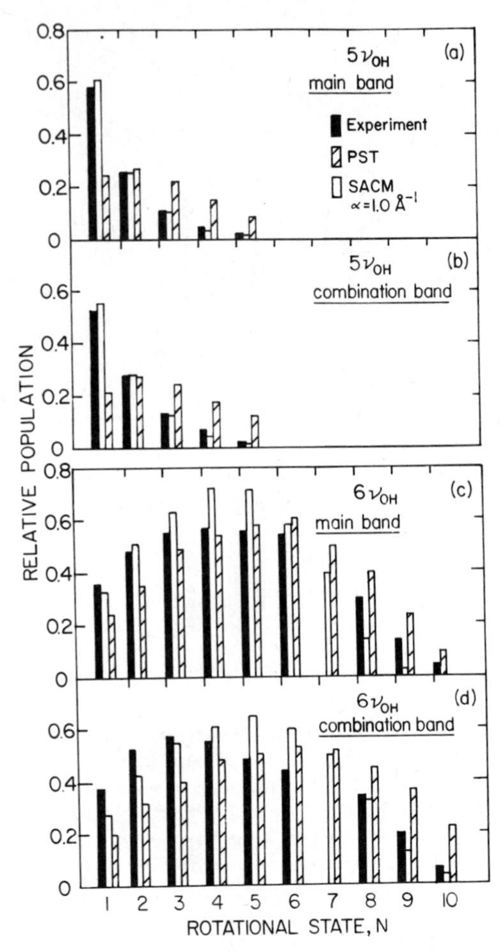


**Figure 5.** Comparison of the observed product rotational state distributions (solid bars) and the results of two statistical models: the statistical adiabatic channel model (SACM, open bars) and phase space theory (PST, hatched bars). The distributions are an average of those observed, or those calculated, at several excitation wavelengths in the region of (*a*) the $5\nu_{OH}$ main band, (*b*) the $5\nu_{OH}$ combination band, (*c*) the $6\nu_{OH}$ main band, and (*d*) the $6\nu_{OH}$ combination band of HOOH. (Reproduced with permission from Ref. 41.)

Using a pulsed free jet expansion the vibrational modes of the NCNO were in their ground states and, with He as a carrier gas, the rotational temperature of NCNO was about 2 K, corresponding to the symmetric top quantum states $J = 0$ to 6 and $K = 0$ to 1, where $J$ is the total rotational angular momentum and $K$ is its projection on the symmetry axis. The excitation of NCNO was to a low-lying electronic state, which underwent internal conversion to the ground electronic state. The subsequent dissociation to CN and NO was slow on a rotational time scale.[49] The yields of the different rotational–vibrational quantum states of CN were monitored by laser-induced fluorescence. An accurate estimate of the dissociation energy $D_0$ of the NC–NO bond was obtained. Observation of the maximum value of $N''$, the rotational angular momentum of CN, permitted by having all available product energy, $E - D_0$, in CN rotation, provided evidence that the dissociation occurs on $S_0$ with no significant potential energy maximum in the exit channel.

Both CN and NO distributions were obtained, although the NO populations are considerably more difficult to measure.[49] In contrast to the $H_2O_2$ experiments in the previous section, it was possible to pump NCNO with sufficiently high-energy photons to obtain vibrationally excited CN or NO products, although the majority of experiments were performed with excitation energies below the thresholds for CN ($v'' = 1$) or NO ($v'' = 1$). Measured rotational distributions for CN and NO, as a function of their respective quantum numbers $N''$ and $J''$, are given in Fig. 6 (open circles) for three values of the photolysis wavelength, $\lambda_p$. The peaks in the photodissociation spectrum corresponding to the selected $\lambda_p$'s are indicated in the center panel of the figure. None of the three wavelengths was sufficiently short to result in vibrationally excited products. The spin-orbit splitting (123 $cm^{-1}$)[50] of the $X^2\Pi$ electronic state of NO resulted in separate rotational distributions for the $^2\Pi_{3/2}$ and $^2\Pi_{1/2}$ states at each excitation wavelength. There were no abrupt changes in the CN distributions with increasing $E - D_0$.[48]

PST is compared with the data in Fig. 6. When the energy available to products $E - D_0$ was below the product vibration excitational threshold, 1876 $cm^{-1}$ [NO at ($v'' = 1$)], the agreement of PST (solid circles) with the experimental rotational distributions in Fig. 6 was excellent. When $E - D_0$ was increased beyond 1876 $cm^{-1}$, the PST description of the rotational distributions became somewhat poorer, although by some standards still very good (e.g., Fig. 7). Wittig and co-workers obtain improved agreement by introducing a modification (denoted by SSE) in which the vibrational distribution was assumed to arise not from a sharing of energy among all coordinates (PST), but from a sharing of energy among a slightly more restricted number, namely, the vibrations of the products plus four one-dimensional translations in the case of two diatoms (or plus five one-dimensional translations in the case of a diatomic and a polyatomic product).[50] For any specified vibrational state of CN, the rotational state distribution of CN for each energetically accessible NO vibrational state was then calculated using PST, weighting each such distribution by the distribution of vibrational states of NO, calculated as above. In SSE, as expected, the vibrational excitation is now enhanced slightly and so the rotational excitation is decreased compared with PST. Below the

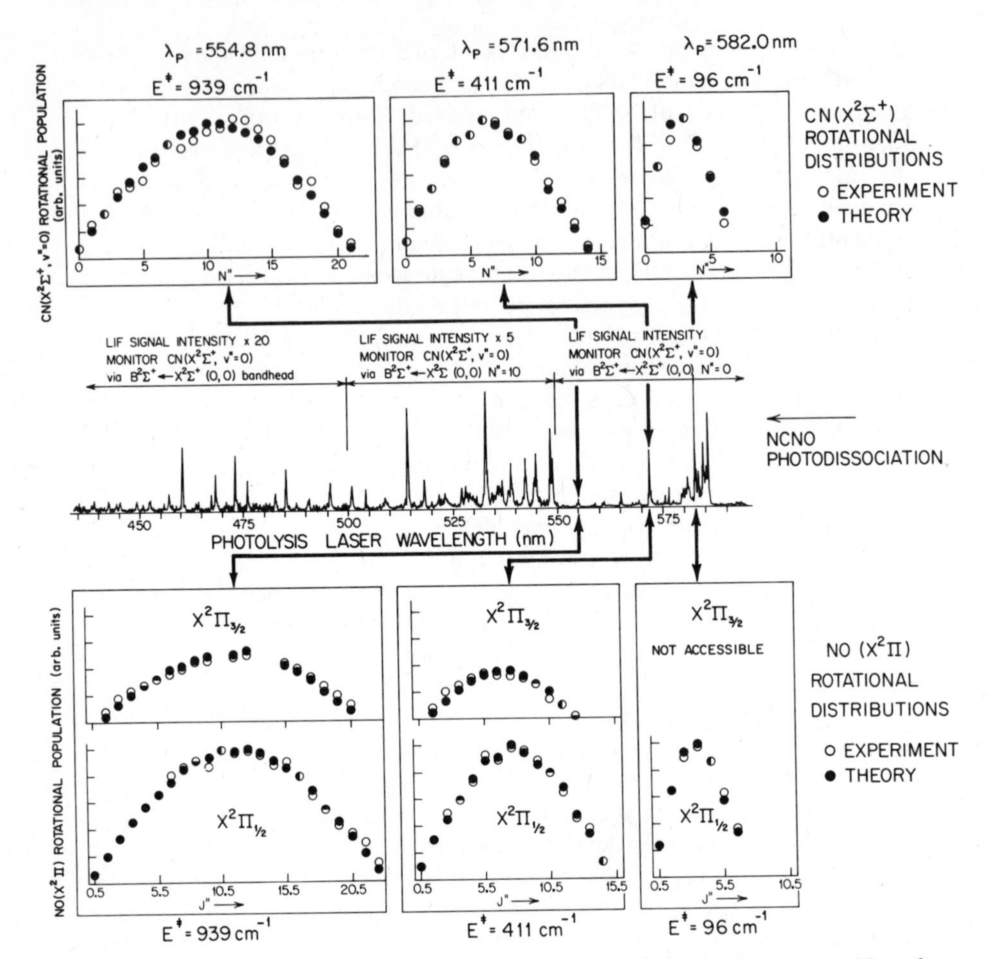


**Figure 6.** A composite view of the NO and CN data for the unimolecular decomposition of NCNO. The central trace shows the photodissociation spectrum of jet-cooled NCNO. The NO and CN rotational populations result from dissociation at the indicated absorption maxima. Experimental distributions for the CN rotational quantum number $N''$ and the NO rotational quantum number $J''$ are given as open circles in the upper and lower panels, respectively. The theoretical results are from PST. For a discussion of the calculated versus experimental NO spin-orbit population ratio see Ref. 49. (Reproduced with permission from Ref. 49.)

threshold for vibrational excitation, SSE and PST are identical. The results are given in Fig. 7, where PST and SSE are compared to the CN ($v'' = 0$) rotational distribution at $E - D_0 = 2348\ \text{cm}^{-1}$. Figure 8 gives a comparison of SSE and the data for both spin-orbit states of NO ($v'' = 1$) at the same excess energy. In order to compare theory with experiment it is necessary, in

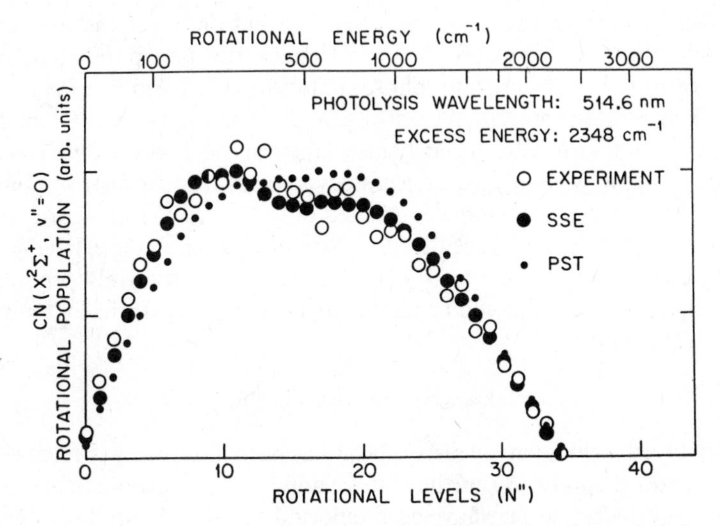


**Figure 7.** Comparison between experimental rotational distributions (open circles) and theoretical PST (small solid circles) and SSE (large solid circles) distributions for CN ($v'' = 0$) at $\lambda_p = 514.6$ nm and $E - D_0 = 2348$ cm$^{-1}$. (Reproduced with permission from Ref. 48.)

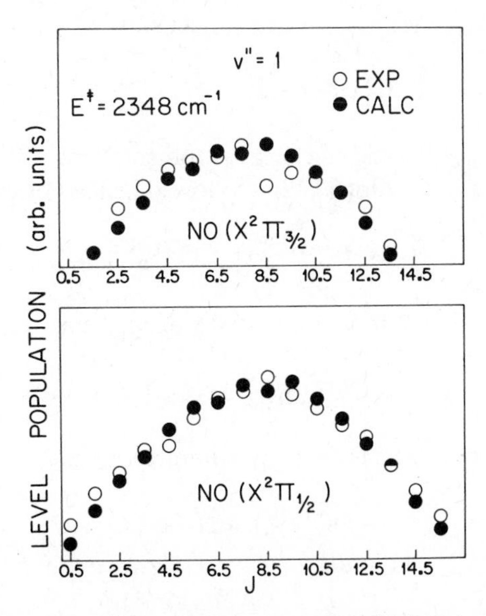


**Figure 8.** Experimental (open circles) and theoretical SSE (solid circles) rotational distributions for both spin-orbit states of NO ($v'' = 1$) with $E - D_0 = 2348$ cm$^{-1}$. See comment concerning the NO spin-orbit ratio in caption of Fig. 6. (Reproduced with permission from Ref. 49.)

principle, to average the PST and SSE distributions over the appropriate $P(E, J)$ distribution. Here only a few rotational states were populated initially ($T_{rot} \sim 2K; 0 \leq J \leq 6$) and, after photoexcitation, the ensemble of parent molecules had a rather narrow spread of $E$ and $J$ values. As the PST and SSE calculations were found to be fairly insensitive to the $J$ values over the range (0, 10), averaging was deemed unnecessary and so product distributions and product energy averages were determined only for $J = 5$. Further discussion of this system was given recently in a pair of articles.[50b] The RRKM plus adiabatic treatment in Section II C is expected to qualitatively produce the same effect as SSE, and it will be interesting to make a quantitative comparison with the data.

### D. $H_2CO$ Dissociation

Laser-induced formaldehyde dissociation has been intensively studied experimentally and theoretically in the last decade.[51–54] Detailed calculations of the potential energy surface have been reported and various studies have been undertaken regarding the possibility of mode selective decay. The low density of vibrational states at the typical energies studied make $H_2CO$ an interesting candidate for possible non-RRKM effects, a point to which we return later.

Experiments on the laser photochemistry of formaldehyde[51] indicate that the probable mechanism involves the excitation of the ground state $S_0$ to a ro-vibrational state of the first excited singlet state $S_1$:

$$H_2CO\ (S_0) + h\nu \rightarrow H_2CO^*\ (S_1). \tag{3.12}$$

At the usual energies the excess energy is too small for dissociation to occur on the $S_1$ surface, and reaction (3.12) is followed instead by either fluorescence

$$H_2CO^*\ (S_1) \rightarrow H_2CO\ (S_0) + h\nu' \tag{3.13}$$

or by internal conversion to a vibrationally excited state of $S_0$

$$H_2CO^*\ (S_1) \rightarrow H_2CO^\dagger\ (S_0), \tag{3.14}$$

followed by the unimolecular decomposition reactions

$$H_2CO^\dagger\ (S_0) \rightarrow H_2 + CO \tag{3.15}$$

$$\rightarrow H + HCO. \tag{3.16}$$

The relative rates of processes (3.13) and (3.14) depend on the proximity of the ro-vibrational state on the $S_1$ electronic surface to a highly vibrationally excited state on $S_0$. The radiationless decay rate [Eq. (3.14)] will be maximized when two vibronic states of $S_0$ and $S_1$ are in resonance, a condition exploited in recent Stark tuning experiments.[53,54]

The threshold for decay to radical products [Eq. (3.16)] is higher than that for molecular products [Eq. (3.15)]. At excitation energies sufficiently close to the origin of the $S_0 \rightarrow S_1$ transition only molecular products are observed and Eq. (3.16) becomes more important with increasing excitation energy.[55] Whereas there is presumably no barrier maximum for the dissociation in the simple bond fission process [Eq. (3.16)], there is one for the molecular elimination channel [Eq. (3.15)]. *Ab initio* calculations[56] and a fit to experimental data[54] put the barrier height (including zero-point energies) for reaction (3.15) in the ~80–90 kcal mol$^{-1}$ range.

The density of vibrational states in the $S_0$ state formed in the internal conversion (3.14) is low at the usual photon energies used in (3.12), because of the smallness of the molecule and the relatively high values of the vibration frequencies, half the atoms being H's and there being a C=O bond. The sparseness of these vibrationally excited levels of the $S_0$ state is evident in recent work[53,54] in which the lifetime of a $J$, $K$, $|M|$-resolved $S_1$ state was considerably altered by Stark tuning of the vibronic levels of the two states in and out of resonance. The differences in lifetime in Ref. 53 were found not to be due to symmetry or $J$ effects in their case.[53] The tuning was possible because of the difference in dipole moments of the $S_0$ and $S_1$ states. For $D_2CO(S_1)$ excited in the vicinity of the $4^1$ vibrational state, decay rates ranged irregularly from $2 \times 10^7$ to $5 \times 10^8$ sec$^{-1}$, when the applied voltage responsible for the Stark field was varied from 0 to 20 kV.

A variety of theoretical calculations have been performed, including classical trajectory studies of chaotic versus quasiperiodic motion.[57] Approximate tunneling corrections within the framework of RRKM (i.e., microcanonical transition state) theory have been used, in which the one-dimensional barrier along the reaction coordinate was represented by the generalized Eckart potential and applied to the reaction (3.15) of $J = 0$ formaldehyde by Miller[52a] and Gray et al.[52b] In Ref. 52*a* *ab initio* transition state vibrational frequencies were used and the *ab initio* potential energy function along the reaction path was found to be fitted well by an Eckart function for energies up to 8 kcal mol$^{-1}$ below the barrier maximum. The computed rates were sensitive to the assumed barrier height and fairly insensitive to the transition state frequencies as obtained from various levels of *ab initio* calculations. The RRKM tunneling rates were found to agree with rates obtained from a "reaction path Hamiltonian model," to within 20% at energies for which the rate constant was greater than $10^5$ sec$^{-1}$. Best estimates of the statistical microcanonical rate in Ref. 52*b* yielded a rate constant of $5.9 \times 10^6$ sec$^{-1}$ at 6.4 kcal mol$^{-1}$ below threshold, faster than the radiative decay rate (~$2 \times 10^5$ sec$^{-1}$). Possible symmetry effects for states symmetrical and antisymmetrical with respect to the $H_2CO$ plane were also examined, and an RRKM plus tunneling model[58a] was applied to each irreducible representation to obtain $J = 0$ microcanonical rate constants for formaldehyde decomposition. A factor of $\sim 20$ difference between them at $J = 0$ was predicted in the tunneling region, which became a factor of 2 at 5–6 kcal mol$^{-1}$ above threshold, and unity as $E \rightarrow \infty$. Results for nonzero total angular momentum revealed a diminished symmetry specificity for $J > 0$.[58b]

Troe has calculated $k_{EJ}$ for the molecular elimination channel (3.15),[59] using an RRKM plus tunneling model, namely, a model analogous to that in Ref. 57 but with slightly different molecular parameters and including approximate anharmonicity correction factors based on a coupled Morse oscillator model. Using a simplified SACM for (3.16), the calculations led him to suggest a channel switching of the two mechanisms at $J \sim 35$, channel (3.15) being more important than (3.16) when $J < 35$, this being reversed for $J > 35$.

## IV. DYNAMICAL ASPECTS AND STATISTICAL BEHAVIOR

In the preceding sections several statistical approaches for calculating reaction rates and distributions of quantum states were described and illustrated using recent studies. We conclude with some remarks on the dynamical aspects leading to and creating deviations from statistical theories.

A dynamical basis for statistical theories is a matter of much current interest.[60] We consider first the case where the vibrational or really vibrational–rotational quantum states of the molecule are those of a bound system, that is, where there is no dissociation. The Hamiltonian $H$ can be regarded as a perturbation (parameter $\lambda$) from some integrable Hamiltonian $H_0(\lambda = 0)$, for example, from a collection of normal modes or Morse oscillator modes to, say, $\lambda = 1$. The eigenvalues in $H_0$ occur in sequences, sequences which can have spacings slowly varying with the quantum numbers (as in a collection of Morse oscillators, for example).[60a] If the spacings of the eigenvalues are sufficiently large, the perturbation from $H_0$ to $H$ produces no avoided crossings of the eigenvalues in the vicinity of $\lambda = 1$ and so the wavefunctions of $H$ at $\lambda = 1$, like those of $H_0$, can be expected to be "mode selective,"[60a,61] and the eigenvalues can be expected to occur in sets of regular sequences.[60a] If, on the other hand, the levels are sufficiently closely spaced, they may approach each other closely at $\lambda = 1$, and one obtains then a "mixing" of states. If a particular such level participates in many such avoided crossings simultaneously (overlapping avoided crossings[62]), the wavefunction takes on a statistical character (highly irregular nodal patterns), and its behavior will differ only randomly from that of nearby states.[63] Such then are two limiting dynamical pictures of the vibrational eigenstates of $H$.

If now the states of $H$ are coupled to a continuum via a nuclear tunneling through a potential energy barrier, the same two limiting pictures as given previously, can be expected to occur, but now each state of $H$ is broadened by the coupling to the continuum. A case in point may be $H_2CO$, whose states in the tunneling regime showed, as noted earlier, considerable differences.

The next case in increasing complexity is where the predissociative states of $H$ do not involve a nuclear tunneling. Here, a suitable choice for an $H_0$ may not be evident.[64] (One possibility is for $H_0$ to consist of a Morse oscillator for the dissociating mode and normal modes for each fragment, but this would need to be supplemented by coupled $R$-dependent hindered rotations of the fragments.) Indeed, the difficulty in finding such an $H_0$ may be consistent with a considerable "mixing" of the states as a whole in such systems. As before, the more widely spaced the states, the less the extent of "mixing" and the greater the opportunity for observing a mode-selective behavior.

Regardless of whether the $H$ has "mode-selective" or "highly mixed" eigenstates, the excitation process can involve, when the eigenstates are sufficiently closely spaced, the formation of a wavepacket of states: Many states whose energies lie within the width of the laser pulse may be optically active. An example of mixed and mode-selective behavior is seen in recent work on infrared fluorescence after excitation by a single infrared photon.[65] Here, depending on the number of coordinates, the infrared excitation of a molecule by a single vibrational quantum yielded infrared fluorescence from many parts of the molecule, when the molecule was sufficiently large. Examples of "mode selectivity" in vibrational states, that is, nonstatistical behavior, include the observation of vibrational quantum beats in some organic molecules at low energies,[66] the behavior of some van der Waals' complexes,[67] and perhaps the studies in $H_2CO$ mentioned earlier.[53,54] Experiments on "mode selectivity" in the tunneling and nontunneling regimes, for molecules of different complexity is clearly of considerable interest in defining some limitations of statistical theoreies.

In a concluding comment we recall some ideas on a topic that has frequently arisen in discussions of dynamical and statistical behavior, namely, the relation between what was previously termed "highly mixed quantum states" and "chaos." As noted elsewhere,[68] classical chaos is expected to imply some corresponding behavior in the quantum case, such as "highly mixed states," but only when the classical chaos is, on a phase-space scale, large compared with $h$.[69] A more detailed discussion, based on Chirikov's idea for the origin of classical chaos (overlapping resonances) and the connection between classical resonances and quantum-mechanical avoided crossings, has been given elsewhere.[62,69]

It may also be recalled that there are two types of "global states": In particular, one can have states that, owing to the presence of an isolated quantum-mechanical resonance, may have a delocalized character. Such states will, when projected onto a set of localized basis set states, have many components. Nevertheless, the state may still be regular, that is, not participate in overlapping avoided crossings and be a member of a set of states which has a high regularity in its eigenvalue sequences. It was shown elsewhere[70] that the "global states" of Nordholm and Rice, for example, were of this type.[71] There is also another type of global state, namely, a state that is delocalized by virtue of being involved in many overlapping avoided crossings. These two types of global states will exhibit some differences in properties, as well as some similarities. Detailed experimental pump-probe measurements are only now just beginning to unravel the nature of the vibrational and rotational states of vibrationally excited molecules.

## Acknowledgment

It is a pleasure to acknowledge the support of this research by the Natural Sciences and Engineering Research Council of Canada (D.M.W.) and by the National Science Foundation (R.A.M.).

## References

1. R. A. Marcus, W. L. Hase, and K. N. Swamy, *J. Phys. Chem.* **88**, 6717 (1984).
2. R. A. Marcus, *J. Chem. Phys.* **85**, 5035 (1986).
3. For example, P. J. Robinson and K. A. Holbrook, *Unimolecular Reactions.* Wiley, New York, 1972. W. Forst, *Theory of Unimolecular Reactions.* Academic, New York, 1973. A. B. Callear, in *Comprehensive Chemical Kinetics* (C. H. Bamford and C. F. H. Tipper, eds.) Elsevier, New York, 1983, Vol. 24, Chap. 4.
4. For example, I. Oref and B. S. Rabinovitch, *Acc. Chem. Res* **12**, 166 (1979), I. W. M. Smith, *Kinetics and Dynamics of Elementary Gas Reaction Rates.* Pergamon Oxford, 1966, Chap. 3.
5. There is a large body of literature on these newer developments, for example, W. B. Miller, S. A. Safron, and D. R. Herschbach, *J. Chem. Phys.* **56**, 3581 (1972); J. M. Farrar and Y. T. Lee, *ACS Symp. Ser.* **66**, 191 (1978); S. Stolte, A. E. Proctor, W. M. Pope, and R. B. Bernstein, *J. Chem. Phys.* **66**, 3468 (1977); R. B. Bernstein, *Chemical Dynamics via Molecular Beam and Laser Techniques.* Oxford University Presss, 1982; K. V. Reddy and M. J. Berry, *Chem. Phys. Lett.* **66**, 223 (1979); J. W. Hudgens and J. D. McDonald, *J. Chem. Phys.* **76**, 173 (1982), F. F. Crim, *Annu. Rev. Phys. Chem.* **35**, 657 (1984); K. Rynefors, P. A. Elofson, and L. Holmlid, *Chem. Phys.* **90**, 347 (1984), N. Scherer, F. E. Doany, A. H. Zewail, and J. Perry, *J. Chem. Phys.* **84**, 1932 (1986), and references cited in these articles. Many more recent examples include those in various articles in *J. Phys. Chem.* **90**, No. 16 (1986), such as in R. B. Bernstein and A. H. Zewail, p. 3467; D. L. Snavely, R. N. Zare, J. A. Miller, and D. W. Chandler, p. 3544; A. M. Wodtke, E. J. Hintsa, and Y. T. Lee, p. 3549; S. Olesik, T. Baer, and J. C. Morrow, p. 3563; H.-S. Kim, M. F. Jarrold, and M. T. Bowers, p. 3584.
6. (*a*) R. A. Marcus, *J. Chem. Phys.* **20**, 359 (1952); R. A. Marcus and O. K. Rice, *J. Phys. Colloid. Chem.* **55**, 894 (1951); (*b*) R. A. Marcus, *J. Chem. Phys.* **43**, 2658 (1965); **52**, 1018 (1970).
7. E. Gorin, *Acta Physicockim. U.R.S.S.* **9**, 691 (1938).
8. (*a*) S. A. Safron, N. D. Weinstein, D. R. Herschbach, and J. C. Tully, *Chem. Phys. Lett.* **12**, 564 (1972); (*b*) D. M. Wardlaw, Ph.D. dissertation, University of Toronto, 1982.
9. R. A. Marcus, *J. Chem. Phys.* **62**, 1372 (1975); G. Worry and R. A. Marcus, *J. Chem. Phys.* **67**, 1636 (1977).

10. Cf. R. A. Marcus, *J. Chem. Phys.* **45**, 2630 (1966). This paper contains this criterion (p. 2635), but mistakenly ascribes it to Bunker, who actually uses, instead, a minimized density of states criterion [D. L. Bunker and M. Pattengill, *J. Chem. Phys.* **48**, 772 (1968)]. This minimum number of states criterion has been used by various authors, for example, W. L. Hase, *J. Chem. Phys.* **57**, 730 (1972); **64**, 2442 (1976); M. Quack and J. Troe (Ref. 21); B. C. Garrett and D. G. Truhlar, *J. Chem. Phys.* **70**, 1593 (1979). The transition state theory utilizing it is now frequently termed microcanonical variational transition state theory ($\mu$VTST). A recent review of $\mu$VTST and of canonical VTST is given in D. G. Truhlar and B. C. Garrett, *Ann. Rev. Phys. Chem.* **35**, 159 (1984).
11. D. M. Wardlaw and R. A. Marcus, (*a*) *Chem. Phys. Lett.* **110**, 230 (1984), (*b*) *J. Chem. Phys.* **83**, 3462 (1985), (*c*) *J. Phys. Chem.* **90**, 5383 (1986). The notation in the present paper is that employed in (*c*). There are several minor errors in (*c*): the factor $(2J + 1)$ should be deleted from Eq. (I.3); in Table V, the heading for the fourth column should be $N_{EJ}(R^{\dagger})\sigma_l/(2J + 1)$, with $\sigma_l$ as defined in the text; in Table IX, the headings for the second and third columns should be $[N_E \pm \sigma^{MC}]\sigma_l/10^{18}$ and $[N_E^A \pm \sigma_A^{MC}]\sigma_l/10^{18}$, respectively; in Eqs. (IV.1), (IV.3), and (IV.4), the quantity $g_e$ should be in the numerator rather than in the denominator.
12. P. Pechukas and J. C. Light, *J. Chem. Phys.* **42**, 3281 (1965); P. Pechukas, J. C. Light, and C. Rankin, *J. Chem. Phys.* **44**, 794 (1966).
13. For example, C. J. Cobos and J. Troe, *J. Chem. Phys.* **83**, 1010 (1985); S. W. Benson, *Can. J. Chem.* **61**, 881 (1983); W. L. Hase and R. J. Duchovic, *J. Chem. Phys.* **83**, 3448 (1985), and references cited therein.
14. (*a*) E. Wigner, *J. Chem. Phys.* **5**, 720 (1937); (*b*) *Trans. Faraday Soc.* **34**, 29 (1938); (*c*) J. C. Keck, *Adv. Chem. Phys.* **13**, 85 (1967).
15. R. A. Marcus, *J. Chem. Phys.* **43**, 1598 (1965).
16. J. O. Hirschfelder and E. Wigner, *J. Chem. Phys.* **7**, 616 (1939).
17. For example, M. A. Eliason and J. O. Hirschfelder, *J. Chem. Phys.* **30**, 1426 (1959); L. Hofacker, *Z. Naturforsch.* **18a**, 607 (1963); R. A. Marcus. Ref. 15; R. A. Marcus, *J. Chem. Phys.* **45**, 4493, (1966); **45**, 4500 (1966). Additional references are given in M. M. Kreevoy and D. G. Truhlar, in *Investigations of Rates and Mechanisms of Reactions.* (C. F. Bernasconi, ed.). Wiley, New York, 1986, Vol 6, Part 1, p. 13.
18. M. Quack and J. Troe, *Ber. Bunsenges. Phys. Chem.* **78**, 240 (1974).
19. (*a*) Another interpolation function, $\frac{1}{3}(2g + g^2)$, has been reported to reproduce satisfactorily *R*-dependent eigenvalues for a particular 1*D*-hindered rotor [M. Quack, *J. Phys Chem.* **83**, 150 (1979)]. The $k_{EJ}$ in one study changed by a factor of 2, compared with the use of (2.13).[11b] (*b*) Various views on the use of an interpolation function to describe a potential energy surface are given in R. J. Duchovic, W. L. Hase, and H. B. Schlegel, *J. Phys. Chem.* **88**, 1339 (1984); F. B. Brown and D. G. Truhlar, *Chem. Phys. Lett.* **113**, 441 (1985); S. Peyerimhoff, M. Lewerenz, and M. Quack, *Chem. Phys. Lett.* **109**, 563 (1984).
20. J. Troe, *J. Phys. Chem.* **88**, 4375 (1984); C. J. Cobos and J. Troe, Ref. 13.
21. M. Quack and J. Troe, *Ber. Bunsenges. Phys. Chem.* **81**, 329 (1977); J. Troe, *J. Chem. Phys.* **79**, 6017 (1983), and references cited therein.
22. M. Quack and J. Troe, *Ber. Bunsenges. Phys. Chem.* **79**, 469 (1975).
23. W. J. Chesnavich, L. Bass, T. Su, and M. T. Bowers, *J. Chem. Phys.* **74**, 2228 (1981); M. F. Jarrold, L. M. Bass, P. R. Kemper, P. A. M. van Koppen, and M. T. Bowers, *J. Chem. Phys.* **78**, 3756 (1983); M. T. Bowers, M. F. Jarrold, W. Wagner-Redeker, P. R. Kemper, and L. M. Bass, *Faraday Disscuss. Chem. Soc.* **75**, 57 (1983); S. Olesik, T. Baer, and J. C. Morrow, Ref. 5; H.-S. Kim, M. F. Jarrold, and M. T. Bowers, Ref. 5
24. J. A. Dodd, D. M. Golden, and J. I. Brauman, *J. Chem. Phys.* **80**, 1894 (1984); D. G. Truhlar, *J. Chem. Phys.* **82**, 2166 (1985); W. J. Chesnavich and M. T. Bowers, *J. Chem. Phys.* **82**, 2168 (1985); K. N. Swamy and W. L. Hase, *J. Chem. Phys.* **77**, 3011 (1982); S. L. Mondro, S. Vande Linde, and W. L. Hase, *J. Chem. Phys.* **84**, 3783 (1986).

25. D. B. Olson and W. C. Gardiner, *J. Phys. Chem.* **83**, 922 (1979), and references cited therein.
26. D. L. Baulch and J. Duxbury, *J. Combust. Flame* **37**, 313 (1980), and references cited therein.
27. W. L. Hase, *J. Chem. Phys.* **64**, 2442 (1976).
28. For example, R. A. Marcus, *J. Chem. Phys.* **43**, 2658 (1965); Eq. (3.3) is a straightforward adaptation of the unimolecular case treated in this reference.
29. For example, see R. E. Weston and H. A. Schwarz, *Chemical Kinetics*. Prentice-Hall, Englewood Cliffs, NJ, 1972.
30. M. Abramowitz and I. Stegun, *Handbook of Mathematical Functions*. Dover, New York, 1965; the Laguerre quadrature is given on p. 890.
31. Calculations with $\alpha = 0.8$ Å$^{-1}$ (curve 1) were performed, in addition to $\alpha = 1$ Å$^{-1}$, because the choice for the potential energy surface interpolation parameter although reasonable is somewhat arbitrary. Hase[27] used $\alpha = 0.82$ Å$^{-1}$, based on a fit of ethane decomposition rates to the experimental data.
32. S. W. Benson, in Ref. 13.
33. M. T. Macpherson, M. J. Pilling, and M. J. C. Smith, *J. Phys. Chem.* **89**, 2268 (1985).
34. M. T. Macpherson, M. J. Pilling, and M. J. C. Smith, *Chem. Phys. Lett.* **94**, 430 (1983).
35. A. F. Wagner and D. M. Wardlaw, submitted to *J. Phys. Chem.*
36. L. B. Harding and A. F. Wagner, *XXI International Symposium on Combustion* (to be published).
37. J. Troe, *J. Phys. Chem.* **87**, 1800 (1983).
38. F. B. Growcock, W. L. Hase, and J. W. Simons, *Int. J. Chem. Kinet.* **5**, 77 (1973).
39. T. R. Rizzo, C. C. Hayden, and F. F. Crim, *J. Chem. Phys.* **81**, 4501 (1984).
40. H.-R. Dubal and F. F. Crim, *J. Chem. Phys.* **83**, 3863 (1985).
41. (*a*) T. M. Ticih, T. R. Rizzo, H.-R. Dubal, and F. F. Crim, *J. Chem. Phys.* **84**, 1508 (1986). (*b*) See also L. J. Butler, T. M. Ticich, M. D. Likar, and F. Crim, *J. Chem. Phys.* **85**, 2331 (1986).
42. T. Uzer, J. T. Hynes, and W. P. Reinhardt, *Chem. Phys. Lett.* **117**, 600 (1985); *J. Chem. Phys.* **85**, 5791 (1986); R. Bersohn and M. Shapiro, *J. Chem. Phys.* **85**, 1396 (1986).
43. N. F. Scherer, F. E. Doany, A. H. Zewail, and J. W. Perry, in Ref. 5.
44. S. Klee, K.-H. Gericke, and F. J. Comes, *J. Chem. Phys.* **85**, 40 (1986).
45. N. F. Scherer and A. H. Zewail (private communication).
46. I. Nadler, H. Reisler, M. Noble, and C. Wittig, *Chem. Phys. Lett.* **108**, 115 (1984).
47. M. Noble, I. Nadler, H. Reisler, and C. Wittig, *J. Chem. Phys.* **81**, 4333 (1984).
48. I. Nadler, M. Noble, H. Reisler, and C. Wittig, *J. Chem. Phys.* **82**, 2608 (1985).
49. C. X. W. Qian, M. Noble, I. Nadler, H. Reisler, and C. Wittig, *J. Chem. Phys.* **83**, 5573 (1985).
50. (*a*) C. Wittig, I. Nadler, H. Reisler, M. Noble, J. Catanzarite, and G. Radhakrishnan, *J. Chem. Phys.* **83**, 5581 (1985); (*b*) C. Wittig, I. Nadler, H. Reisler, M. Noble, J. Catanzarite, and G. Radhakrishnan *J. Chem. Phys.* **85**, 1710 (1986); J. Troe, *J. Chem. Phys.* **85**, 1708 (1986).
51. C. B. Moore and J. C. Weisshaar, *Annu. Rev. Phys. Chem.* **34**, 525 (1983), and references cited therein.
52. (*a*) W. H. Miller, *J. Am. Chem. Soc.* **101**, 6810 (1979); (*b*) S. K. Gray, W. H. Miller, Y. Yamaguchi, and H. F. Schaefer, *J. Am. Chem. Soc.* **103**, 1900 (1981).
53. H.-L. Dai, R. W. Field, and J. L. Kinsey, *J. Chem. Phys.* **82**, 1606 (1985).
54. D. R. Guyer, W. F. Polik, and C. B. Moore, *J. Chem. Phys.* **84**, 6519, (1986).
55. Th. Just, *Symp (Int.) Combust.* [Proc.], 17th, 584 (1979); J. H. Clark, C. B. Moore and N. S. Nogar, *J. Chem. Phys.* **70**, 5135 (1979).
56. J. D. Goddard, Y. Yamaguchi, and H. F. Schaefer, *J. Chem. Phys.* **75**, 3459 (1981); M. J. Frisch, R. Krishnan, and J. A. Pople, *J. Phys. Chem.* **85**, 1467 (1981); M. Dupuis, W. A. Lester, B. H. Lengsfield, and B. Lui, *J. Chem. Phys.* **79**, 6167 (1983).
57. K. N. Swamy and W. L. Hase, *Chem. Phys. Lett.* **92**, 371 (1982); S. K. Gray and M. S. Child, *Mol. Phys.* **53**, 961 (1984).

58. (*a*). W. H. Miller, *J. Am. Chem. Soc.* **105**, 216 (1983); R. A. Marcus, *J. Chem. Phys.* **45**, 2138 (1966); (*b*) W. H. Miller, *J. Phys. Chem.* **87**, 2731 (1983); see also B. A. Waite, S. K. Gray, and W. H. Miller, *J. Chem. Phys.* **78**, 259 (1983).
59. J. Troe, *J. Phys. Chem.* **88**, 4375 (1984).
60. (*a*) D. W. Noid, M. L. Koszykowski, and R. A. Marcus, *Annu. Rev. Phys. Chem.* **32**, 267 (1981); (*b*) F. H. Mies, *J. Chem. Phys.* **51**, 787 (1969); **51**, 798 (1969); K. G. Kay, *J. Chem. Phys.* **65**, 3813 (1976); recent articles in *J. Phys. Chem.* **90**, No. 16 (1986), such as S. K. Gray, S. A. Rice, and M. J. Davis, p. 3470; J. M. Bowman, p. 3492; R. S. Dumont and P. Brumer, p. 3509; K. N. Swamy, W. L. Hase, B. C. Garrett, C. W. McCurdy, and J. F. McNutt, p. 3517; R. C. Brown and R. E. Wyatt, p. 3590; M. S. Child, p. 3595; P. Pechukas, p. 3603; E. Pollak, p. 3619.
61. Examples of mode selectivity in calculations appear in Ref. 58 and in several of the articles in Ref. 60.
62. R. A. Marcus, in *Horizons in Quantum Chemistry*. (K. Fukui and B. Pullman, eds.). Reidel, Dordrecht, 1980, p. 107; *Ann. N. Y. Acad. Sci.* **357**, 169 (1980); *Faraday Discuss. Chem. Soc.* **75**, 103 (1983).
63. For example, R. A. Marcus, Extended Abstracts, 27th Annual Conference on Mass Spectrometry and Allied Topics, Seattle, WA (Am. Soc. Mass Spectroscopy, 1979), p. 568; V. Buch, R. B. Gerber, and M. A. Ratner, *J. Chem. Phys.* **81**, 3393 (1984); M. Feingold, N. Moiseyev, and A. Peres, *Chem. Phys. Lett.* **117**, 344 (1985).
64. Recent examples of choices of a Hamiltonian for dissociation, isomerization, or intramolecular energy transfer include those in Z. Bacic, R. B. Gerber, and M. A Ratner, *J. Phys. Chem.* **90**, 3606 (1986); in S. K. Gray, W. H. Miller, Y. Yamaguchi, and H. F. Schaefer, *J. Chem. Phys.* **73**, 2733 (1980); and in R. D. Levine and J. L. Kinsey, *J. Phys. Chem.* **90**, 3653 (1986).
65. G. M. Stewart and J. D. McDonald, *J. Chem. Phys.* **78**, 3907 (1983); G. M. Stewart, M. D. Ensminger, T. J. Kulp, R. S. Ruoff, and J. D. McDonald, *J. Chem. Phys.* **79**, 3190 (1983); T. J. Kulp, H. L. Kim, and J. D. McDonald, *J. Chem. Phys.* **85**, 211 (1986).
66. W. R. Lambert, P. M. Felker, and A. H. Zewail, *J. Chem. Phys.* **75**, 5958 (1981); **81**, 2217 (1984).
67. D. H. Levy, *Annu. Rev. Phys. Chem.* **31**, 197 (1980); K. W. Butz, D. L. Catlett, G. E. Ewing, D. Krajnovich, and C. S. Parmenter, *J. Phys. Chem.* **90**, 3533 (1986).
68. A brief summary and references are given in R. A. Marcus, in *Chaotic Behavior in Quantum Systems* (G. Casati, ed.). Plenum, New York, 1985, p. 293.
69. An example where a system showed classical chaos but regular eigenvalue sequences is given in D. W. Noid, M. L. Koszykowski, M. Tabor, and R. A. Marcus, *J. Chem. Phys.* **73**, 391 (1980).
70. D. W. Noid and R. A. Marcus, *J. Chem. Phys.* **67**, 559 (1977). The regularity of the eigenvalue sequences was demonstrated in Ref. 69.
71. K. S. J. Nordholm and S. A. Rice, *J. Chem. Phys.* **61**, 203 (1974); **61**, 768 (1974); **62**, 157 (1975).

# PICOSECOND TIME-RESOLVED DYNAMICS OF VIBRATIONAL-ENERGY REDISTRIBUTION AND COHERENCE IN BEAM-ISOLATED MOLECULES

PETER M. FELKER* AND AHMED H. ZEWAIL

*Arthur Amos Noyes Laboratory of Chemical Physics,[†] California Institute of Technology, Pasadena, California, 91125*

## CONTENTS



* *Current address*: Department of Chemistry, University of California, Los Angeles, CA 90024.

[†] Contribution No. 7504.

# I. INTRODUCTION AND PERSPECTIVES

Among the most fundamental of current problems in chemical physics is the characterization of dynamical processes in individual molecules effectively isolated from external perturbations. Such intramolecular processes, which arise following the excitation of a gaseous molecule, may be expected to be markedly dependent on the nature of the initially excited state. In large part, it is this expected dependence of dynamics on initial state that provides the justification for detailed studies of these processes, for it implies that one may be able to intimately control the physics and/or chemistry of a molecule simply by varying the excitation parameters. Thus, for the past 20 years or so, experimentalists and theorists have been interested in developing propensity rules by which one can predict intramolecular dynamics given knowledge of molecular parameters and the initially prepared state.

Without question, and perhaps not surprisingly, the greatest success in this effort has been at the coarsest level of molecular-level structure—that is, at the level of the electronic state. The reasons for this success are many, but for the most part are related to (1) the relatively small number of relevant electronic states, (2) the success that theories of electronic structure have had in characterizing these states and their couplings to other electronic states, and (3) the fact that the rates of the relevant processes can be directly measured by experiment (e.g., by studying fluorescence, phosphorescence, or photochemical processes).

For the next coarsest regime of molecular-level structure, the *vibrational regime*, the characterization of intramolecular processes is not nearly as complete. This relative lack of knowledge of vibrational effects compared to the knowledge of electronic effects on intramolecular dynamics is due primarily to the larger number of relevant states (often millions of states per $cm^{-1}$), less precise knowledge of relevant couplings, and experimental difficulties in measuring vibrational state-specific rates. But it is precisely at the vibrational level where one might expect initial state-dependent processes to yield the greatest measure and variety of control over subsequent dynamics. For chemical processes this expectation derives from the nature of vibrational motions as distortions of chemical bonds; different vibrational motions might lend different reactivities to a molecule. For photophysical processes, Franck–Condon factors and promoting vibrational modes are expected to be intimately involved in determining rates. Numerous processes such as multiphoton dissociation, laser-selective chemistry, radiationless transitions, and unimolecular reactions may therefore be strongly dependent on the vibrational state initially excited.[1,2]

At the heart of the question of vibrational state specificity in intramolecular dynamics is the matter of intramolecular vibrational-energy redistribution

(IVR). This is the process by which energy that is initially localized in a particular vibrational motion redistributes in time such that different vibrational motions gain energy at the expense of the initial motion. One can readily see that the extent to which this redistribution occurs and the time scale on which it happens has direct bearing on vibrational state-specific processes. For, if IVR spreads energy over a large number of vibrations in a time short compared to other intramolecular decay processes, then memory of the initial state will be lost and with it the possibility of state-specific dynamics. Thus, insofar as vibrational character influences molecular dynamics, the characterization of IVR is essential to the understanding of the dynamics.

In large part, one can accurately characterize early experimental approaches to the study of IVR as involving either the measurement of unimolecular reaction rates or the measurement of steady-state emission spectra of isolated molecules.[3] In the earliest variant of the former approach (as pioneered by Rabinovitch's group[4]) information concerning IVR is obtained by comparison of the observed rates of gas-phase reactions with rates calculated assuming instantaneous and complete vibrational-energy randomization subsequent to activation of the molecule. In many cases, the predictions of such "statistical" theories (e.g., RRKM theory) closely match (in an average sense) observed behavior,[5] which agreement implies rapid and extensive IVR. In other cases, "nonstatistical" behavior was taken to indicate vibrational-energy redistribution times on the order of picoseconds.[6] In another variant of the measurement of rates as a means to characterize IVR, Lim, Schlag, Fischer, Rice, and others[7] have measured decay rates (and quantum yields) of total fluorescence from excited molecules in the gas phase as a function of the total energy in the excited molecules. The shape of such rate versus energy curves can be used as an indication of whether statistical IVR is occurring in the molecules.

Estimates of IVR rates have also been inferred from spectral studies. The principal spectral approach, the measurement of the emission spectra of isolated molecules (for a review, see, for example, Ref. 3), relies on the fact that the spectral characteristics of emission from a molecule will depend intimately on the vibrational character of the excited molecular state. If one prepares a gaseous molecule in a well-defined vibrational state and subsequently observes emission bands that would not be expected to arise from this initially prepared state, then some IVR process can be inferred. Moreover, a "rate" can be calculated for the process by comparing the intensities of expected emission bands (vibrationally unredistributed emission) and unexpected emission bands (vibrationally redistributed emission). "Redistributed" and "unredistributed" emission often are loosely termed "relaxed" and "unrelaxed." Since energy is conserved in isolated molecules, there is no real energy relaxation, as occurs in solution or in solids.

Until the development of seeded supersonic jets for molecular spectroscopy both of the previously mentioned general methods of studying IVR processes were seriously handicapped by the requirement that the gaseous samples be at temperatures on the order of, or greater than, room temperature just to achieve vapor densities large enough to render experiments possible. For reasonably large molecules (e.g., naphthalene and anthracene) at such temperatures, the thermal distribution of ro-vibrational levels is composed of a large number of states with significant population. The situation makes it all but impossible to excite well-defined initial states, even with narrow band light sources. Thus, experiments done on large molecules in gas bulbs generally suffer from this thermal congestion problem.

The use of ultracold gaseous samples generated by free-jet expansion[8] can eliminate much of the ambiguity associated with thermal effects while still providing a sample in which the molecules of interest are effectively isolated. In the past decade or so, these features of jet-cooled samples have been exploited extensively in the study of IVR by emission spectroscopy,[9–14] unimolecular rate measurements,[15–17] and quantum yield and fluorescence decay measurements.[18] More recently, linewidth measurements (excitation spectra) by Neusser, Schlag, and others[19] and by Amirav and Jortner[20] have provided relaxation times (by the use of the uncertainty principle). The results of these studies have been very fruitful in contributing to the characterization of IVR processes. Nevertheless, it has been apparent that there is a definite need to circumvent problems arising because the measurements made do not directly monitor the IVR process in time, but instead measure quantities that can only be related to IVR indirectly. Alternative methods for probing IVR dynamics have been developed by the groups of Parmenter (chemical timing of molecules in bulbs), Hochstrasser (picosecond fluorescence gating in bulbs), and this group (picosecond techniques in molecular beams), all of which will be discussed here.

In an effort to study IVR processes of large molecules in the time domain directly, since 1980[21] we have used picosecond laser techniques to make measurements on molecules cooled by free-jet expansion (the picosecond-beam technique[22]). In this manner we have hoped to eliminate both the ambiguity associated with thermal congestion (although it must be noted that even in the jet at very low temperatures, significant rotational congestion persists for large molecules) and the deficiencies of time-integrated or low-time-resolution experiments. The key features, therefore, of this experimental approach are the relative absence of thermal congestion in the molecular samples (and the concomitant ability to prepare well-defined initial vibronic states) and the capability of observing dynamical processes on a picosecond time scale.

In this chapter we focus on the real-time dynamics of IVR, with particular emphasis on the results of our application of picosecond-beam techniques in probing the dynamics of IVR and ro-vibrational coherence in a number of systems. In so doing, we will not be able to provide a literature survey of all work pertaining to IVR, particularly that work which does not directly time-resolve the dynamics. Recent reviews by Parmenter,[23] Smalley,[24] and Bondybey[25] consider in detail much of the work in this area. We also will not consider in detail the implications of IVR measurements as they pertain to theoretical interests in the area of chaos and quasiperiodic behavior in molecules. A number of reviews by Rice,[26] Noid et al.,[27] Hase,[28] and others are available on this subject. Our concern here regarding the theory of IVR is focused on its manifestations in time-domain experiments. For this, we borrow from the formalism of radiationless transition theory[29] to relate observables to the nature of the vibrational coupling in real molecules.[30]

This chapter is divided into two major sections. The first (Section II) considers the information available from other studies of IVR. The section has essentially two purposes: (1) to point out some of the capabilities and limitations that (the very popular) time-integrated spectroscopic studies have in application to IVR and (2) to provide an overview of results obtained on dynamics in two laboratories other than our own.

The second major section (Section III), comprising the bulk of the chapter, pertains to the studies of IVR from this laboratory, studies utilizing either time- and frequency-resolved fluorescence or picosecond pump-probe methods. Specifically, the interest is to review (1) the theoretical picture of IVR as a quantum coherence effect that can be manifest in time-resolved fluorescence as quantum beat modulated decays, (2) the principal picosecond-beam experimental results on IVR and how they fit (or do not fit) the theoretical picture, (3) conclusions that emerge from the experimental results pertaining to the characteristics of IVR (e.g., time scales, coupling matrix elements, coupling selectivity), in a number of systems, and (4) experimental and theoretical work on the influence of molecular rotations in time-resolved studies of IVR. Finally, in Section IV we provide some concluding remarks.

## II. EARLIER ADVANCES AND OTHER RECENT APPROACHES

### A. Steady-State Spectroscopic Studies of IVR

The exploitation of the link between IVR and spectral broadening by several groups utilizing jet specroscopic techniques has provided a good deal of information pertinent to IVR and has catalyzed tremendous activity in the field. Nevertheless, there are limitations to time-integrated techniques in revealing the full details of IVR processes. The limitations pertain particularly to the determination of (1) IVR "rates," (2) the temporal characteristics of IVR, (3) the coupling matrix elements involved, and (4) the extent of the process. In what follows, we discuss the method used to obtain information about IVR from (emission) spectral parameters and the bases for some of the limitations of this method.

Emission spectra as a function of the excess vibrational energy in a large number of medium to large-sized molecules undergo an evolution from being characterized by resolvable vibrational structure and having bands assignable in terms of transitions between single vibronic levels, to having unresolvable vibrational bands contributing to broad, only grossly structured, features.[23,24] The evolution typically occurs over a range of vibrational energy from 0 to 1000–3000 $cm^{-1}$ depending on the size, and other characteristics, of the molecule, and is present even though the absorption (or fluorescence excitation) spectrum corresponding to this range of energy in the molecule does not exhibit analogous broadening in the high-energy regime. Such behavior, if due solely to IVR, can be interpreted as being indicative of a lack of IVR in the low-vibrational-energy regime, with IVR becoming increasingly important as the vibrational energy increases. One considers the IVR process as being the result of coupling between an optically active state $|a\rangle$ and a set of optically inactive states $|b\rangle$. The excitation process prepares a state having only the $|a\rangle$ state excited. We shall denote this state as $|a^*b\rangle$. As a result of the coupling between $|a\rangle$ and $|b\rangle$, this prepared state evolves to contain some excited $b$-type modes, as well—states denoted as $|ab^*\rangle$. This evolution is the process of IVR. Now, the emission from $|a^*b\rangle$ will give rise to a structured, relatively uncongested spectrum, the form of which is determined by the Franck–Condon overlap of a single vibronic level with other single vibronic levels. In contrast, an IVR-evolved state, consisting of $|a^*b\rangle$ and the manifold of coupled $|ab^*\rangle$ states, will produce a considerably more congested emission spectrum, corresponding to the superposition of spectra arising from a number of vibronic levels. Thus, the observed changes in emission spectra as a function of vibrational energy are explained in terms of IVR.

The simple kinetic model usually employed to relate spectral characteristics of emission to rates of IVR is as follows. As in the previous discussion, one assumes that the species is excited to some optically active state $|a^*b\rangle$. Subsequently, an *irreversible* IVR process with rate constant $k_{IVR}$ occurs such that the state of the species eventually evolves to $|ab^*\rangle$. With this model, application of kinetics gives

$$[a^*b]_t = [a^*b]_0 e^{-(k_f + k_{IVR})t}, \tag{2.1}$$

where $[a^*b]_0$ is the initial population of excited molecules in the $|a^*b\rangle$ state and $k_f$ is the fluorescence rate constant. Likewise,

$$[ab^*]_t = [a^*b]_0 (e^{-k_f t} - e^{-(k_f + k_{IVR})t}). \tag{2.2}$$

Now consider two types of experiments, two cases that underscore the need for *temporal as well as spectral resolution* in revealing the details of IVR.

*Case 1. Total fluorescence detection:* Consider an experiment in which one does not resolve the separate emission bands from $|a^*b\rangle$ and the $|ab^*\rangle$ states, but in which one does temporally resolve the fluorescence. In such a case one observes fluorescence versus time $[I(t)]$ proportional to $[a^*b]_t + [ab^*]_t$, which from the preceding equations goes as $[a^*b]_0 e^{-k_f t}$. That is, without spectrally resolving the vibrational structure of the emission, all information about the IVR dynamics is lost.

*Case 2. Time-integrated fluorescence spectra:* Now consider experiments in which one spectrally resolves time-integrated fluorescence so as to relate spectral features to IVR rates. In such a case one is interested in $\overline{[a^*b]}/\overline{[ab^*]}$, where the bars denote integration over time. This ratio gives the ratio of fluorescence intensity from the $|ab^*\rangle$ states relative to that from the $|a^*b\rangle$ states, and can be obtained by integrating Eqs. (2.1) and (2.2) over time:

$$\frac{I_{ab^*}}{I_{a^*b}} = \frac{\overline{[ab^*]}}{\overline{[a^*b]}} = \frac{k_{\mathrm{IVR}}}{k_f}. \tag{2.3}$$

Equation (2.3) is the key result for the use of spectral information to obtain quantitative information concerning IVR. By determining experimentally the ratio of vibrationally redistributed ($I_{ab^*}$) to vibrationally unredistributed ($I_{a^*b}$) emission, together with knowledge of $k_f$, one can obtain $k_{\mathrm{IVR}}$. $k_{\mathrm{IVR}}$, in turn, can provide other information about the dynamics (within the constraints of the kinetic model). For instance, average vibrational coupling matrix elements can be obtained from a Fermi golden rule expression for the rate.

*There are, however, a number of assumptions and approximations in the treatment leading to Eq. (2.3) that may be invalid in certain cases or that are invalid much of the time.* First, the model assumes that IVR can be described by a kinetic equation. As shown by Lahmani et al.[31] and Mukamel and co-workers,[32] and as discussed in Ref. 30, there are limits of correspondence between kinetic behavior and quantum-mechanical descriptions. Nevertheless, a kinetic treatment precludes the consideration of recurrences (quantum beats) in the vibrational-energy distribution. As discussed at length in Section III, such recurrences often are a major feature of IVR. Second, one assumes in the preceding picture that the excitation and decay processes are separable and that the $|a^*b\rangle$ state is truly the initially prepared state. These are tantamount to assuming that the coherence width of the excitation source is large enough and the nature of the molecule is such that one can prepare the zero-order $|a^*b\rangle$ state exclusively. Third, to apply Eq. (2.3), one must be able to resolve and assign $I_{ab^*}$ and $I_{a^*b}$ emissions. In practice, this is very often not an easy task. Finally, related to the second and third points above, the simple model assumes that a vibrationally and rotationally pure state is prepared via the excitation process. This need not always be the case. For example, Fermi resonance interactions with nearby $|a'^{*}b\rangle$ states outside the laser bandwidth can contaminate the $|a^{*}b\rangle$ state. The contaminated $|a^{*}b\rangle$ state, upon preparation by the laser, will produce congested emission, even when not coupled to any state besides $|a'^{*}b\rangle$. Thus, the spectrum may indicate the presence of dynamics, even though there is none. Spectral congestion in emission can also be a product of rotational inhomogeneous broadening arising from the thermal distribution of initial rotational levels in the sample. Clearly, such congestion is not dynamical in origin, yet it can easily interfere with the characterization of spectral features that do have their origins in dynamics.[33]

### B. Chemical Timing Studies of IVR in Bulbs

In 1980, Coveleski et al.[34a] reported on a method of study of IVR known as "chemical timing." Since this first report, the Parmenter group has published a number of detailed papers on the theory of chemical timing[34b] and its application to studies of IVR in *p*-difluorobenzene[34] and *p*-fluorotoluene.[35]

The essence of chemical timing is to measure the fluorescence spectra of a vibronically excited gas-phase molecule in the presence of a buffer gas (typically $O_2$) that is an efficient quencher of the electronic excitation of the molecule. The quenching process affords one an internal timing mechanism that can be varied by changing the concentration of the buffer gas; the emission observed in the experiment is only that emission that can occur during a time interval (after the excitation process) on the order of the electronic quenching time. Thus, one is able to monitor the changes in fluorescence spectra as the time interval over which fluorescence is collected varies from $\sim 10$ to thousands of picoseconds. By monitoring the ratio of vibrationally relaxed versus unrelaxed emission (see Section II A) as a function of this time interval and by applying a kinetic model to the analysis of such results, one is able to obtain values for the rate and the extent of IVR as a function of the initially prepared state of the molecule. This affords the opportunity to calculate other parameters associated with the dynamics, for example, average coupling matrix elements.

Some characteristics associated with IVR in *p*-difluorobenzene that have been deduced via the application of chemical timing[34] are characteristics that substantially match those of IVR in other molecules as revealed by picosecond-beam studies (Section III). First, the extent of IVR in the $S_1$ state of the molecule changes qualitatively as its excess vibrational energy increases. Second, the time scales associated with IVR in *p*-difluorobenzene range from about 15 to hundreds of picoseconds. Third, these time scales are dependent on the initially excited vibrational level, with one mode in particular ($\nu'_{30}$)

seeming to act as an accelerating mode in the dynamics.[34] Fourth, average coupling matrix elements associated with IVR dynamics are found to be in the range from 0.02 to 0.1 $cm^{-1}$ (0.6 to 3.0 GHz). Finally, manifestations of the involvement of molecular rotations in IVR appear in the chemical timing results.

Clearly, from the previous paragraph, the chemical timing method is an important means by which to study IVR. What is particularly attractive about the technique is that one need not generate picosecond excitation pulses or use fast detection electronics; convenient CW or nanosecond-pulsed light sources, together with standard fluorescence detection techniques, are sufficient to do the experiments. A major drawback to chemical timing lies in the fact that the timing imposed by collisional quenching is somewhat "fuzzy" in the sense that at any given concentration of the quenching gas there is a distribution of time intervals over which excited molecules may emit. A second disadvantage is that one observes fluorescence integrated over an interval from $t = 0$ (the time of excitation) to $t = \tau$ (with $\tau$ changing as a function of quencher concentration) rather than observing fluorescence in sequential "slices" of time. These two characteristics of the method preclude the direct observation of oscillatory ("restricted") IVR (although one can distinguish between IVR in the intermediate case and that in the statistical limit). This has been noted by the Parmenter group, who have also noted the importance of (1) a proper kinetic analysis, (2) due consideration of the presence of spectral congestion in the bulb, and (3) due consideration of collisional effects in analyzing the results of chemical timing so as to obtain parameters relating to IVR dynamics by this clever method.

### C. Picosecond-Gated Fluorescence in Bulbs

Hochstrasser and co-workers[36] have applied frequency down-conversion of fluorescence, together with detection employing an optical multichannel analyzer, to obtain picosecond-gated fluorescence spectra of gaseous *p*-difluorobenzene. In the experiments, the sample molecules were excited to the $3^1 30^3$ vibrational level of the $S_1$ state by the fourth harmonic of a picosecond $Nd^{3+}$: phosphate glass laser. Fluorescence was collected and mixed in a KDP crystal with the variably delayed fundamental (1054 nm) pulse of the laser to produce frequency difference spectra, which were then measured by using a spectrograph/OMA detection system. This scheme provides a means by which to obtain time-gated fluorescence spectra with temporal resolution limited only by the pulsewidth of the laser (8 psec). Utilizing the scheme, evolution of the $3^1 30^3$ fluorescence spectrum in the bulb was observed to occur on a time scale of approximately 100 psec. Discussion and comparison of this result on *p*-difluorobenzene to chemical timing results on the molecule appear in Ref. 34.

## III. PICOSECOND-MOLECULAR-BEAM STUDIES OF IVR

### A. The Technique and Its Application to IVR

As stated in Section I, the approach of our group at Caltech toward the elucidation of the characteristics of IVR in large molecules has employed variants of what we call the picosecond-beam technique.[22] The technique, as the name suggests, involves the combination of picosecond spectroscopy with supersonic jet-cooled gaseous samples. In a generic experimental scheme, a picosecond pulse of light impinges on the ultracold, isolated molecules of a seeded supersonic free jet expansion (or skimmed beam). The pulse interacts with sample molecules in such a way to excite them to a vibrational level of an excited electronic state. (Individual rotational transitions are generally not resolved.) The dynamics of these vibronically excited molecules is then monitored, in time, by measuring the temporal behavior of some observable, the observable in most of our IVR studies being spectrally resolved fluorescence. In some experiments ion currents, generated by variably delayed *probe* picosecond pulses, have been used as observables in pump-probe studies of IVR. These techniques have been applied[22] to studies of (1) photoisomerization, (2) charge transfer, (3) photodissociation, (4) intramolecular hydrogen bonding, (5) interstate electronic couplings, and (6) IVR and coherence. Here, we focus only on the studies of IVR and coherence.

The desirable characteristics of the picosecond-beam scheme with respect to its application to IVR studies (and studies of other vibronic state-specific processes) are several. First, the conditions of the experiments allow for the preparation of vibronically well-defined states, the dynamics of which are then monitored. This selectivity in excitation is possible because (1) the samples, being at low vibrational temperatures, have a large fraction of molecules in the lowest vibrational level of the molecular ground electronic state, (2) most molecules have relatively few excited vibronic levels that can combine in absorption with the vibrationless ground state level, and (3) the frequency bandwidth of the picosecond pulses used to excite the molecules is typically several wavenumbers, less than the average spacings between those vibronic transitions that have significant intensity. The ability to excite to well-defined vibronic states, an ability that is generally lacking in "bulb" samples, owing to thermal spectroscopic congestion, is extremely important in the study of IVR, since different vibrational levels may be expected to exhibit quantitatively and qualitatively different IVR dynamics.

A second desirable characteristic is that the molecules in free jet expansions can be probed in spatial regions where they are effectively noninteracting. (A free jet sample is similar to a sparse bulb sample in this regard.) This fact permits exclusively *intra*molecular energy relaxation processes to be probed by picosecond-beam spectroscopy.

Finally, the technique is one that permits the direct characterization of dynamical processes on a picosecond time scale. This ability to ascertain the temporal details of IVR is important in that it is necessary to the observation of vibrational coherence and to a complete understanding of the nature of the IVR process.

The first application of picosecond spectroscopy to jet-cooled samples was communicated[21] in 1981. The letter reported our first observation of quantum beats in the decay of spectrally resolved fluorescence from anthracene excited to its $S_1 + 1380\ \mathrm{cm}^{-1}$ vibronic level. The observation, which was made with temporal resolution of several hundreds of picoseconds, was surprising and, in a number of ways, a novelty. First, anthracene is much larger than any other molecule[37] that up to that time had exhibited beat-modulated decays due to electronic-state mixing. Second, the beats were found to be very sensitive to experimental conditions associated with the spectral characteristics of the fluorescence detected (only spectrally resolved fluorescence at a particular wavelength showed beat-modulated decays). Third, the beats were found to be insensitive to applied magnetic and electric fields. These novel characteristics suggested that the quantum beats in anthracene could not be attributed to experimentally proven sources of quantum coherence effects (e.g., singlet–triplet,[37] hyperfine, and fine structure coupling, and Zeeman level splittings[38]). On the other hand, there remained the possibility that the beats *were* associated with IVR, as suggested in Ref. 21.

During the two years following our initial observation of beats, extensive picosecond-beam studies[39] (again with resolution of several hundreds of picoseconds) of anthracene fluorescence were performed. These studies revealed an apparent scarcity, if not total lack, of quantum beats in the spectrally resolved fluorescence of the molecule. All decays measured, except those corresponding to excitation of anthracene at $\bar{\nu}_x = S_1 + 1380\ \mathrm{cm}^{-1}$ and detection of the fluorescence band at $\bar{\nu}_x - 1125\ \mathrm{cm}^{-1}$, were found to be unmodulated and single exponential. This situation not only frustrated the effort to ascertain the source of the beats, but also seemed to indicate that even if that source were determined, the beats were a manifestation of some fortuitous coupling, a coupling having only minor importance in the scheme of molecular, and, indeed, anthracene's dynamics.

The situation changed, however, with two advances. The first advance was the discovery that in the $S_1 \leftarrow S_0$ spectrum of jet-cooled anthracene a second band exists (at $S_1 + 1420\ \mathrm{cm}^{-1}$), the excitation of which gives rise to quantum beat-modulated fluorescence decays.[40] Besides indicating a somewhat more global importance to the beat phenomenon in anthracene, the characteristics of these new beats provided very strong evidence that they arose as a manifestation of IVR. In particular, the beats were shown to have *phases and modulation depths dependent on the fluorescence band detected*. Such behavior, which we had previously predicted should occur, is a signature for coherence effects arising as a result of IVR between a small number of vibrational levels (so-called "restricted" IVR).[40]

The second advance was an increase in the temporal resolution of our fluorescence detection system to $\leq 80$ psec. This roughly fivefold increase in resolution, by allowing the observation of beat modulations having frequencies of up to 12 GHz, made it possible to expose the general presence of vibrational quantum beats in anthracene fluorescence decays,[30a,41–43] to use the characteristics of the beats to obtain semiquantitative information about vibrational coupling parameters,[42] to trace the changing nature of IVR in anthracene as a function of vibrational energy,[42] and to assess the effects of rotational level structure on that IVR which arises via anharmonic coupling.[43]

Subsequent to the anthracene studies, picosecond-beam measurements of IVR in a number of other molecules have been made. These molecules include deuterated anthracenes,[44] *t*-stilbene,[45] and some alkyl anilines.[46] One of the most significant results of these studies is that they have indicated that vibrational coherence[30,40] (phase-shifted quantum beats) is a general phenomenon in molecules. Thus, it appears that an accurate understanding of IVR must rest firmly on an accurate understanding of vibrational coherence.

Recent application of the picosecond-beam technique to problems associated with IVR has pertained to the role of *ro*-vibrational coherence in the IVR process.[47–50] Stimulated by the results of picosecond pump-probe photoionization experiments,[51,52] which revealed the presence of polarization-dependent early time transients in the decays of jet-cooled and bulb samples of large molecules, and by theoretical studies,[53] which have indicated that polarization measurements of time-integrated fluorescence may serve as a probe of rovibrational energy flow, we have performed time-resolved fluorescence polarization experiments on jet-cooled large molecules. The results have shown that the manifestations of *purely rotational* coherence can be observed.[47–50] While the phenomenon is unrelated to IVR, its manifestations can mimic those of IVR, and may interfere with any "clean" observation of ro-vibrational energy flow in time-resolved fluorescence. Therefore, purely rotational coherence, aside from being intrinsically interesting (we have made use of the phenomenon as a tool for sub-Doppler spectroscopic measurements of excited state rotational constants[47,48]), is of interest in its relation to measurements of IVR.

The remainder of this section is a detailed review and summary of results from the application of picosecond-beam spectroscopy to the study of IVR.

### B. Theoretical Description of Vibrational Coherence and IVR

If there is one central concept that has been strengthened by picosecond-beam studies of IVR, it is that IVR and quantum coherence are inextricably linked—

IVR is equivalent to the evolution of a particular form of quantum superposition state (for example, Refs. 30 and 54–57). In retrospect, the fact that this is so may not be particularly surprising, given that the general formalism of radiationless transition theory in isolated molecules identifies time-dependent behavior with coherence effects.[29,31] On the other hand, prior to the picosecond studies, the process of IVR was typically described in the language of kinetics. While such kinetic descriptions can be consistent with radiationless transition theory,[31,32] the application of kinetics presupposes that certain limits of the theory pertain to the IVR problem. Notably, these limits preclude from consideration the details of vibrational energy flow between a small number of levels, cases that give rise to vibrational quantum beats. Furthermore, the picosecond-beam studies show that the effect of rotations on IVR is not as complicated as previously thought; in particular, rotations in large molecules do not necessarily wash out coherence effects.

The observation of "novel" quantum beats in the spectrally resolved fluorescence of anthracene[21] forced one to consider, within the context of radiationless transition theory, the details of how IVR might be manifested in beat-modulated fluorescence decays. This work led to the concepts of "phase-shifted" quantum beats and "restricted" IVR,[30a,40] and to a general set of results[30b] pertaining to the decays of spectrally resolved fluorescence in situations where an arbitrary number of vibrational levels, coupled by anharmonic coupling, participate in IVR. Moreover, three regimes of IVR have been identified: no IVR, restricted (or coherent) IVR, and dissipative IVR.[42]

The purpose of this section is to review these results and their derivation. We begin with a special case, that of vibrational energy flow between two levels.[40] This case exhibits many of the concepts of the general case and gives one an intuitive feel for the description of vibrational energy flow in terms of quantum coherence. Then, we review more general results,[30b] corresponding to IVR between an arbitrary number ($N$) of levels.

### 1. *Two-Level IVR*

In this subsection we consider the fluorescence signal that arises in the situation depicted in Fig. 1 and relate that signal to IVR. The figure corresponds to a molecule in which two, and only two, zero-order $S_1$ vibrational states ($|a\rangle$ and $|b\rangle$), separated by zero-order energy $E_{ab}$, are coupled by anharmonic coupling (matrix element $V_{ab}$). By virtue of this coupling, $|a\rangle$ and $|b\rangle$ are not eigenstates of the molecular Hamiltonian. Instead, two eigenstates, $|1\rangle$ and $|2\rangle$ (having energies $E_1$ and $E_2$, respectively), which are linear combinations of $|a\rangle$ and $|b\rangle$, arise from the coupling:[58]

$$|1\rangle = \alpha_{1a}|a\rangle + \alpha_{1b}|b\rangle, \tag{3.1a}$$

$$|2\rangle = \alpha_{2a}|a\rangle + \alpha_{2b}|b\rangle, \tag{3.1b}$$

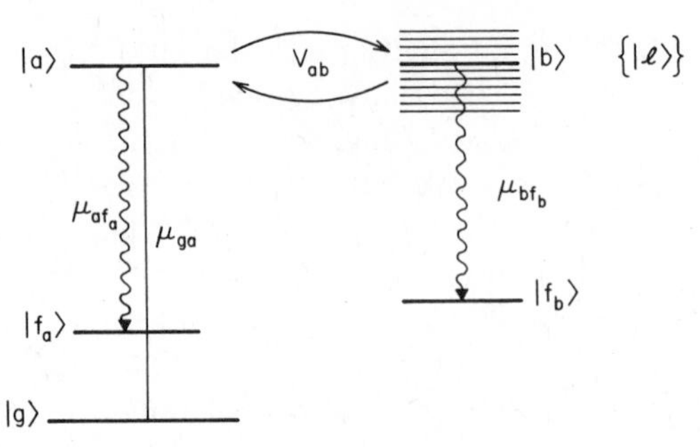


**Figure 1.** A schematic level diagram pertaining to an experiment involving two zero-order excited state vibrational levels, $|a\rangle$ and $|b\rangle$, coupled by an interaction $V_{ab}$. $|a\rangle$ is active in absorption from $|g\rangle$ and in emission to $|f_a\rangle$, while $|b\rangle$ is not active in absorption from $|g\rangle$ but is active in emission to $|f_b\rangle$.

where $\alpha_{1a} = -\alpha_{2b} \equiv \alpha$ and $\alpha_{2a} = \alpha_{1b} \equiv \beta$, $\alpha$ and $\beta$ are real, and $\alpha^2 + \beta^2 = 1$. $\alpha$, $\beta$, and $|E_1 - E_2|$ depend on the coupling between $|a\rangle$ and $|b\rangle$ (i.e., on $E_{ab}$ and $V_{ab}$).

Suppose, now, that the molecule, initially in the ground-state vibrational level $|g\rangle$, is excited by a delta-function pulse of light to the excited state manifold, whereupon time- and frequency-resolved fluorescence to the ground-state vibrational level $|f\rangle$ is detected. The signal[38] in such an experiment is

$$I(t) = K \sum_{I,J=1}^{2} (\mu_{Ig})(\mu_{gJ})(\mu_{Jf})(\mu_{fI}) e^{-(iE_{IJ}/\hbar + \Gamma)t}, \tag{3.2}$$

where $K$ is a constant, $\mu_{nm} \equiv \langle n|\mu|m\rangle$ is the transition electric dipole moment matrix element between $|n\rangle$ and $|m\rangle$, $|1\rangle$ and $|2\rangle$ have been assumed to have the same decay rate $\Gamma$, and $E_{IJ} \equiv E_I - E_J$. Note that in obtaining Eq. (3.2) we have assumed that we can completely neglect the rotational level structure of the molecule and any polarization of the exciting light or the fluorescence. We shall consider in Section III D some of the consequences when this approximation is not made.

To proceed further, we now assume that, because of Franck–Condon factors, only $|a\rangle$ can combine in absorption from the initial state $|g\rangle$. That is, $|a\rangle$ is an absorption "doorway" state.[29] (The assumption that only a few out of many $S_1$ vibrational levels have any appreciable dipole-induced transition probability to or from any given $S_0$ vibrational level is a good approximation for a large number of molecules.) This, with the definition of the $\mu_{nm}$ and Eqs. (3.1), gives $\mu_{1g} = \alpha\mu_{ag}$ and $\mu_{2g} = \beta\mu_{ag}$. We also assume, for similar reasons, that either $|a\rangle$ or $|b\rangle$ combines in emission with the final state $|f\rangle$ (i.e., $|a\rangle$ or $|b\rangle$ is a doorway state in emission to $|f\rangle$). If $|a\rangle$ is the emission doorway state, giving rise to what we shall call an *a-type fluorescence band*, then $\mu_{1f} = \alpha\mu_{af}$ and $\mu_{2f} = \beta\mu_{af}$. If $|b\rangle$ is the emission doorway state, giving rise to a *b-type band*, then $\mu_{1f} = \beta\mu_{bf}$ and $\mu_{2f} = -\alpha\mu_{bf}$.

Using the $\mu$'s for an $a$-type band, Eq. (3.2) simplifies to the following:[40]

$$I_a(t) \sim e^{-\Gamma t}(1 - 2\alpha^2\beta^2 + 2\alpha^2\beta^2 \cos \omega_{12} t), \tag{3.3}$$

where $\omega_{12} \equiv E_{12}/\hbar$. Similarly, for a $b$-type band,

$$I_b(t) \sim e^{-\Gamma t}\alpha^2\beta^2(1 - \cos \omega_{12} t). \tag{3.4}$$

One notes immediately that $I_a(t)$ and $I_b(t)$ are both beat-modulated at an angular frequency of $\omega_{12}$. Yet, one also notes two aspects in which the decays are qualitatively different. First, the $a$-type decay is modulated by a cosine term with positive coefficient. In contrast, the $b$-type decay has a cosine term with negative coefficient. *The beats in the b-type decay are phase-shifted 180° from those in the a-type decay.* Second, the $a$-type decay is not, in general, 100% modulated $[2\alpha^2\beta^2 \leq (1 - 2\alpha^2\beta^2)]$. The $b$-type decay, however, is always 100% modulated. *The modulation depths of the beat-modulated decays of the different bands are different.*

Now, the preceding treatment predicts specifics about fluorescence decays that arise from two coupled vibrational levels within the same electronic-state manifold. One might ask what these decays have to do with IVR. It turns out, in fact, and this is the particular utility of time-resolved fluorescence in the study of IVR, that *the decay of a given fluorescence band is a direct picture of the energy flow in to and out of the zero-order state that gives the band its emission intensity.* To see this, consider first the excited state that exists instantaneously after the delta-function excitation of the two-level system:

$$\Psi(0^+) \sim \alpha|1\rangle + \beta|2\rangle = |a\rangle. \tag{3.5}$$

The temporal evolution of $\Psi$ is given by

$$\Psi(t) \sim (\alpha|1\rangle e^{-iE_1t/\hbar} + \beta|2\rangle e^{-iE_2t/\hbar})e^{-\Gamma t/2}. \tag{3.6}$$

From Eq. (3.6) and Eqs. (3.1), one can determine the temporal evolution of the contribution of $|a\rangle$ to $\Psi(t)$:

$$|\langle a|\Psi(t)\rangle|^2 \sim e^{-\Gamma t}(1 - 2\alpha^2\beta^2 + 2\alpha^2\beta^2 \cos \omega_{12} t). \tag{3.7}$$

This function, which can be interpreted as that portion of the vibrational energy of the molecule that resides in the $|a\rangle$ vibration, *is proportional to $I_a(t)$*, the decay of the $a$-type fluorescence band [Eq. (3.3)]. One can also determine the contribution of $|b\rangle$ to $\Psi(t)$:

$$|\langle b|\Psi(t)\rangle|^2 \sim e^{-\Gamma t}[2\alpha^2\beta^2(1 - \cos\omega_{12}t)]. \tag{3.8}$$

Comparing this with Eq. (3.4), one sees that *the contribution of* $|b\rangle$ *to* $\Psi(t)$ *is proportional to* $I_b(t)$.

The fact that time- and frequency-resolved fluorescence gives one a direct picture of the vibrational-energy flow in two-level IVR gives added meaning to the characteristics of the fluorescence decays $I_a(t)$ and $I_b(t)$. The quantum beats in the decays reflect the oscillatory flow of vibrational energy between $|a\rangle$ and $|b\rangle$.[40] The fact that the beats in $I_a(t)$ are out of phase with those in $I_b(t)$ is a reflection of the conservation of energy (or, more accurately, probability)—the vibrational energy that leaves $|a\rangle$ must end up in $|b\rangle$ (and vice versa) in this two-level system. The modulation depths of the different types of decays also have meaning with respect to IVR. The less than 100% modulation of $I_a(t)$ is a manifestation of the fact that vibrational-energy flow out of the initially prepared state $\Psi(0^+) \sim |a\rangle$, is incomplete. (Only when $|\alpha| = |\beta|$, or equivalently, when $E_{ab} = 0$, $V_{ab} \neq 0$, is there complete transfer of vibrational energy back and forth between $|a\rangle$ and $|b\rangle$.) The 100% modulation of $I_b(t)$ is a manifestation of the fact that $\Psi(0^+) \sim |a\rangle$. That is, because of this fact, the energy in the $|b\rangle$ vibrational level is necessarily zero at $t = 0^+$, which means that $I_b(0^+) = 0$ and that, therefore, $I_b(t)$ must be completely modulated.

Finally, before considering $N$-level IVR, it is pertinent to make three more points about the two-level case. First, one notes that beat-modulated decays arising from situations of two-level (and multilevel) IVR have characteristics that are qualitatively different from modulated decays that arise from other sources of coupling within molecules (e.g., $S_1 - T_1$[37,59] and $S_1 - S_0$[60] coupling). In particular, the changes in beat phase and modulation depth that occur with changes in fluorescence detection wavelength in the IVR case do not occur in the other cases. This is so because in these other cases the coupling is between electronic state manifolds, and one of the manifolds is effectively "dark" with respect to emission to the ground electronic state. On the other hand, the vibrational coupling we consider occurs within a single electronic-state manifold. Although any given vibrational level involved in the coupling will not emit to all ground-state levels, there are some levels in the ground-state manifold with which it *can* combine in emission. Thus, in IVR none of the coupled levels is dark.

Second, it is notable that by measurement of $\omega_{12}$ and the modulation depth of $I_a(t)$, it is possible[30,40] to obtain quantitative information on $|E_{ab}|$ and $|V_{ab}|$, the parameters describing the vibrational coupling in the two-level system. This is possible because the two experimental parameters depend solely on the two coupling parameters in well-known ways.

Finally, it is useful to point out the analogy between the oscillatory "energy flow" among quantum oscillators and the similar energy flow that can occur between classical oscillators. A classical case in point is that of two pendula weakly coupled to each other (by a weak spring for instance). If one of these pendula ($a$) is initially set in motion, the other (b) being initially at rest, then as time progresses, pendulum $b$ will gain energy at the expense of $a$. As time progresses further, the energy in $b$ will reach a maximum, after which point net energy flow will be toward $a$, and so on in an oscillatory fashion. This energy flow occurs because of the coupling between the oscillators. Clearly, this classical case of energy flow between coupled oscillators is closely analogous to the vibrational energy flow in two-level IVR, as manifested in time-resolved fluorescence.

### 2. *IVR between N Levels*

While useful as a simple, model case with which one can demonstrate many of the principles involved in vibrational coherence, IVR between two vibrational levels is clearly not a very general situation. In view of the expectation that vibrational coupling in molecules may involve any number of levels and that many such cases of multilevel IVR are amenable to study by picosecond spectroscopy, it is useful to have theoretical guidelines[30b] pertaining to the manifestations that a system of $N$ coupled vibrational levels might exhibit in time-resolved fluorescence. Hence, we consider now the situation depicted in Fig. 2.

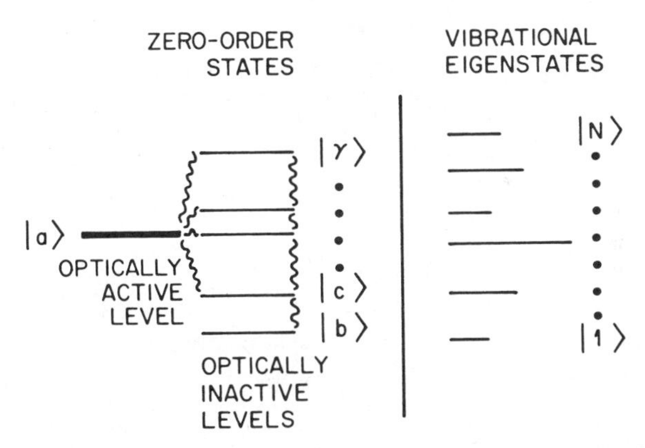


**Figure 2.** Level diagram illustrating the $N$-level situation to be treated herein. The zero-order harmonic vibrational states on the left are coupled by anharmonic interactions. Only one of these states, the optically active $|a\rangle$ level, can be reached from the ground state via the excitation process. All of the other levels are optically inactive (dark). On the right are pictured the vibrational eigenstates that result from the interactions among the zero-order states. The different lengths of the lines representing these levels are meant to indicate that each eigenstate has a different contribution from $|a\rangle$, and hence has a different absorption strength from the ground state.

*a. Expression for the Decays of Fluorescence Bands.* Instead of two coupled $S_1$ vibrational levels, we are now concerned with $N$ such levels. (Again, we neglect the presence of any rotational level structure, an approximation that is valid when the vibrational coupling is exclusively anharmonic in nature and when rotational coherence effects can be neglected.) These zero-order states are denoted as $|a\rangle$, $|b\rangle$, $|c\rangle$, etc., and $|\gamma\rangle$ is used to denote a generic zero-order state. Diagonalization of the molecular Hamiltonian $H$ in this zero-order basis set gives rise to $N$ vibrational eigenstates $|1\rangle$, $|2\rangle$, etc., of the form

$$|I\rangle = \sum_{\gamma=a,b,\ldots} \alpha_{I\gamma}|\gamma\rangle, \qquad I = 1, \ldots, N, \tag{3.9}$$

where the matrix of the $\alpha$ coefficients

$$C = \begin{bmatrix} \alpha_{1a} & \alpha_{2a} & \cdots & \alpha_{Na} \\ \alpha_{1b} & \alpha_{2b} & \cdots & \alpha_{Nb} \\ \vdots & \vdots & \ddots & \vdots \end{bmatrix} \tag{3.10}$$

is an $N \times N$, real, orthonormal matrix. The energies of the eigenstates are denoted as $E_1$, $E_2$, etc.

We now consider an experiment in which molecules in the ground-state vibrational level $|g\rangle$ are excited by a delta-function light pulse to the coupled $N$-level system, whereupon the fluorescence from the excited state thereby prepared to some final ground-state level $|f\rangle$ is monitored as a function of time. (Note that this implies some degree of spectral resolution of the fluorescence.) The fluorescence signal in such an experiment is a generalization of Eq. (3.2):

$$I_\gamma(t) = K \sum_{I,J=1}^{N} (\mu_{Ig})(\mu_{gJ})(\mu_{Jf})(\mu_{fI})e^{-(i\omega_{IJ}+\Gamma)t}, \tag{3.11}$$

where the factors in the equation have the same meaning as in Eq. (3.2).

Now we assume, as in the two-level case, that because of Franck–Condon factors only one state, $|a\rangle$, has any appreciable absorption strength from $|g\rangle$ and that only one state, $|\gamma\rangle$, has any appreciable emission strength to $|f\rangle$. (Note, however, that as $|f\rangle$ changes, $|\gamma\rangle$ can change.) This gives for the $\mu$ factors in Eq. (3.11) $\mu_{Ig} = \alpha_{Ia}\mu_{ag}$, $\mu_{gJ} = \alpha_{Ja}\mu_{ag}$, $\mu_{Jf} = \alpha_{J\gamma}\mu_{\gamma f}$, and $\mu_{fI} = \alpha_{I\gamma}\mu_{\gamma f}$. The intensity versus time for the fluorescence band, which we call a "$\gamma$-type band" in reference to the zero-order state that provides its emission strength, is denoted $I_\gamma(t)$ and is found from Eq. (3.11) to be

$$I_\gamma(t) = K'\left(\sum_{I=1}^{N} \alpha_{Ia}^2\alpha_{I\gamma}^2 + 2\sum_{I>J}^{N} \alpha_{Ia}\alpha_{Ja}\alpha_{I\gamma}\alpha_{J\gamma}\cos\omega_{IJ}t\right)e^{-\Gamma t}. \tag{3.12}$$

*b. $I_\gamma(t)$ as a Direct View of IVR.* Before considering in detail the implications of Eq. (3.12), consider how it relates to IVR. The connection can be established, just as it was in the two-level case, by finding an expression for $|\langle\gamma|\Psi(t)\rangle|^2$, the contribution that $|\gamma\rangle$ makes, in time, to the excited superposition state created by the laser pulse. Under the assumptions of delta-function excitation and $|a\rangle$ being the only zero-order state with any absorption strength,

$$|\Psi(t)\rangle = \sum_{I=1}^{N} \alpha_{Ia}|I\rangle e^{-(i\omega_I+\Gamma/2)t} = \sum_{I=1}^{N} \sum_{\gamma=a,b,\ldots} \alpha_{Ia}\alpha_{I\gamma}|\gamma\rangle e^{-(i\omega_I+\Gamma/2)t}, \quad (3.13)$$

where $\omega_I = E_I/\hbar$. Therefore,

$$|\langle\gamma|\Psi(t)\rangle|^2 = \left|\sum_{I=1}^{N} \alpha_{Ia}\alpha_{I\gamma}e^{-(i\omega_I+\Gamma/2)t}\right|^2$$

$$= \sum_{I,J=1}^{N} \alpha_{Ia}\alpha_{Ja}\alpha_{I\gamma}\alpha_{J\gamma}e^{-(i\omega_{IJ}+\Gamma)t}. \quad (3.14)$$

Comparison of Eqs. (3.14) and (3.12) reveals that $|\langle\gamma|\Psi(t)\rangle|^2 \sim I_\gamma(t)$. In other words, $I_\gamma(t)$ has the same temporal behavior as the contribution of $|\gamma\rangle$ to $|\Psi(t)\rangle$. Therefore, *the fluorescence decay of a $\gamma$-type band is a chronicle of the energy flow in to and out of the vibrational motion associated with the zero-order $|\gamma\rangle$ state.* This fact, which is analogous to the two-level result, should be borne in mind as the implications of Eq. (3.12) are considered below.

*c. General Characteristics of $I_\gamma(t)$.* Returning to Eq. (3.12), one may note several points concerning the general characteristics of $I_\gamma(t)$. First, it is clear that $I_\gamma(t)$ is beat-modulated. Second, the angular frequencies ($\omega_{IJ}$) associated with the beats are just equal to $1/\hbar$ times the energy differences between distinct pairs of eigenstates in the $N$-level system. This means that $N(N-1)/2$ cosine beat terms enter into the fluorescence decay. Third, since the $\alpha$'s are real, the Fourier amplitude of any given cosine term in $I_\gamma(t)$ is either positive or negative (i.e., is either *in-phase or out-of-phase*). Moreover, these amplitudes (magnitude and phase) depend on the vibrational coupling parameters, which determine the values of the $\alpha$'s. Fourth, one notes that

$$\omega_{IJ} + \omega_{JK} = \omega_{IK}. \quad (3.15)$$

Thus, there is a condition relating triplets of beat frequencies (i.e., $\omega_{IJ}$, $\omega_{JK}$, $\omega_{IK}$) in Eq. (3.12). One can show that each beat frequency in $I_\gamma(t)$ is a member of $(N-2)$ such triplets, and that present in $I_\gamma(t)$ are $N(N-1)(N-2)/6$ distinct triplets obeying Eq. (3.15). Finally, one notes that $\gamma$ can take on any one of $N$ values, corresponding to each of the zero-order vibrational levels.

[In an experiment one changes the value of $\gamma$ by detecting different fluorescence bands. This changes the pertinent final states, $|f\rangle$, and thereby can change the zero-order state ($|\gamma\rangle$) that is the one active in emission.] Now, the number and values of the beat frequencies in Eq. (3.12) do not depend on $\gamma$. But, it is clear that the Fourier amplitudes (magnitude and sign) of these beat frequencies *do* depend on $\gamma$. Therefore, the $N$ different band-types arising from a coupled $N$-level system are modulated with identical beat frequencies, but are different from each other with respect to the modulation depths and phases of these frequencies.

*d. Quantum Beat Modulation Depths.* From Eq. (3.12) one can see that the modulation depth $M_\gamma(\omega_{IJ})$ of beat frequency $\omega_{IJ}$ is given by

$$M_\gamma(\omega_{IJ}) \equiv \frac{2\alpha_{Ia}\alpha_{Ja}\alpha_{I\gamma}\alpha_{J\gamma}}{\sum_{I=1}^{N} \alpha_{Ia}^2\alpha_{I\gamma}^2}. \tag{3.16}$$

Several useful relations, derived entirely from the orthonormality of the $\alpha$'s, can be obtained for these modulation depths. In doing so, it is convenient to make a distinction between band types, since the general characteristics of the $M_\gamma(\omega_{IJ})$ are different for $\gamma = a$ (identical absorption and emission doorway states) as opposed to $\gamma \neq a$ (different absorption and emission doorway states).

For $a$-type bands, Eq. (3.16) becomes

$$M_a(\omega_{IJ}) = \frac{2\alpha_{Ia}^2\alpha_{Ja}^2}{\sum_{I=1}^{N} \alpha_{Ia}^4}. \tag{3.17}$$

From this expression (and the orthonormality of the $\alpha$'s) it is possible to show that there are limits on the values of individual $M_a(\omega_{IJ})$:

$$0 \leq M_a(\omega_{IJ}) \leq 1 \qquad \text{for all } \omega_{IJ}. \tag{3.18}$$

Moreover, one can also place limits on the *sum* of all modulation depths in an $a$-type decay:

$$\sum_{I>J=1}^{N} M_a(\omega_{IJ}) \leq N - 1. \tag{3.19}$$

Non-$a$-type modulation depths have limits analogous to Eqs. (3.18) and (3.19). For individual $M_\gamma(\omega_{IJ})$

$$0 \leq |M_\gamma(\omega_{IJ})| \leq 1 \qquad \text{for all } \omega_{IJ},\ \gamma \neq a. \tag{3.20}$$

Note that this equation allows for negative $M_\gamma(\omega_{IJ})$, in contrast to the $M_a(\omega_{IJ})$, all of which must be positive. The sum of modulation depths in a non-$a$-type decay obeys the following:

$$\sum_{I>J=1}^{N} M_\gamma(\omega_{IJ}) = -1. \tag{3.21}$$

This result is a generalization of the 100% modulation of $b$-type decays in the two-level case. Just as in the two-level case, the result ensures that $I_\gamma(0^+) = 0$ when $\gamma \neq a$. This, in turn, is a reflection of the fact that only $|a\rangle$ contributes to the initially prepared superposition state, $|\Psi(t)\rangle$.

*e. Quantum Beat Phase Distributions.* As we have stated above, one signature of IVR as manifested in time- and frequency-resolved fluorescence is that quantum beat phases change upon detection of different fluorescence band types. The phase $s_\gamma(\omega_{IJ})$ of beat component $\omega_{IJ}$ in decay $I_\gamma(t)$ is defined as

$$s_\gamma(\omega_{IJ}) \equiv \text{sign}[M_\gamma(\omega_{IJ})] = \pm 1. \tag{3.22}$$

Given this, it is pertinent to consider these phases closely, and, in particular, to consider (1) the possible distributions of phases over the $N(N-1)/2$ beat components modulating a given $I_\gamma(t)$ and (2) how these distributions change with band type (i.e., with $\gamma$). In what follows, we shall simply state results that have been derived elsewhere.

First, it is possible to show by using the orthonormality of the eigenvector matrix $C$ that each of the $N$ band types arising from an $N$-level system has a decay with a unique quantum beat phase distribution. That is, each decay type is different from the others by virtue of its quantum beat phases.

Second, $a$-type decays, by Eq. (3.18), are only modulated by positive cosine terms; all phases are $+1$. On the other hand, non-$a$-type decays must have some beat terms with $-1$ phases because of Eq. (3.21).

Third, although there are $N(N-1)/2$ beat frequencies in any given decay, which implies that for a particular $I_\gamma(t)$ there are $2^{N(N-1)/2}$ possible phase distributions [corresponding to all possible distributions of $+1$ and $-1$ in $N(N-1)/2$ "slots"], there are in fact only $2^{N-1}$ *allowed* distributions. This large reduction arises because there are only $(N-1)$ independent phases in any given decay. This, in turn, is a manifestation of the fact that

$$s_\gamma(\omega_{IK})s_\gamma(\omega_{JK}) = s_\gamma(\omega_{IJ}). \tag{3.23}$$

Fourth, Eq. (3.23) places restrictions on the phases of the quantum beat triplets defined by Eq. (3.15). In particular, the three phases $s_\gamma(\omega_{IJ})$, $s_\gamma(\omega_{JK})$, and $s_y(\omega_{IK})$ must be either all positive, or two of the phases must be negative and the remaining one positive.

Finally, one can show that decays with certain phase distributions cannot arise from the same $N$-level system as decays with certain other phase distributions, even though each of the distributions is one of the $2^{(N-1)}$ allowed ones. Such restrictions arise from the orthogonality of the columns (eigenvectors) of $C$.

At this point, it may seem to the reader that the detailed consideration of quantum beat phase distributions is a somewhat abstract exercise bearing little relation to IVR. We would justify our attention to the problem of phases by noting that the proper interpretation of experimental results from picosecond-jet experiments on IVR relies on the ability to determine how closely one's experimental conditions correspond to one's theoretical model of the experiment. A particularly convenient way to do this is by comparing phase characteristics from experiment with those from theory. In addition, phase characteristics are useful in helping one assign the various bands in a fluorescence spectrum to band types.

*f. Fluorescence Spectra of Band Types.* Any given zero-order vibrational level in an excited electronic-state manifold may have emission strength to more than just one $S_0$ level. Therefore, in the fluorescence spectrum arising from a coupled $N$-level system, one expects that there will be several bands present for each of the $N$ band types. Based on purely spectroscopic considerations, it is often possible to predict which bands in a spectrum are of the same type. It is also often possible to distinguish between $a$-type and non-$a$-type bands in a spectrum. Thus, one has spectroscopic guidelines as to where to observe particular types of decays. This is useful both as a guide in performing time-resolved experiments and as a check on time-resolved results.

A number of papers (e.g., Refs. 9–14, 24, and 25) have considered the form of fluorescence spectra arising from excited molecules in which IVR occurs. The standard approach is first to recognize that in a molecular electronic transition there are generally just a few vibrational modes that are optically active (i.e., are such that their quantum numbers can change in the electronic transition), and many others that are optically inactive (i.e., their quantum numbers cannot change). In the emission spectrum from any given $S_1$ zero-order state, only progressions of optically active modes can occur. These progressions are built on an "origin" band characteristic of the zero-order state and corresponding to the fluorescence transition from that state, in which transition no vibrational quantum numbers change, $\{\Delta v = 0\}$. This origin band occurs in the same spectral region as the $0^0_0$ band of the electronic transition, but not at exactly the same position, since vibrational frequencies are different in different electronic state manifolds. In general, the origin bands arising from different zero-order states also occur at different spectral positions from one another, since different vibrational modes have different frequency shifts upon electronic excitation.

The preceding discussion implies that, given knowledge of the optically active modes of a molecule, one has an accurate spectral means by which to ascertain which bands in the fluorescence spectrum arising from a coupled $N$-level system are of the same band type. The first step is to note the bands in the spectral region of the $0_0^0$ transition. These bands are likely to be $\{\Delta v = 0\}$ origin bands of different band types. The next step is to search for optically active progressions emanating from each origin band. Each group of bands so determined to correspond to a given origin band represents the set of bands of the same band type.

Besides being able to classify bands into band types using spectroscopic considerations, it is also possible by such considerations to assign the $a$-type and non-$a$-type bands in a spectrum. By definition, our $|a\rangle$ state is the only state with any appreciable absorption strength from the populated ground-state level $|g\rangle$. Now, in jet-cooled samples $|g\rangle$ is the vibrationless level of the ground electronic state, $|0\rangle$. The fact that $|S_1\gamma\rangle \leftarrow |S_0 0\rangle$ transitions only have intensity for $\gamma = a$ implies that $|S_1\gamma\rangle \rightarrow |S_0 0\rangle$ fluorescence transitions only have intensity for $\gamma = a$. Immediately one sees that the bluest fluorescence band, the fluorescence band at the excitation wavelength in a spectrum arising from the excitation of vibrationally cold molecules, must be $a$-type. Furthermore, progressions of optically active modes off of this blue-most band are $a$-type as well. On the other hand, bands that are shifted from the laser frequency by an interval that cannot be associated with optically active modes are likely to be non-$a$-type. The fact that $a$-type bands tend toward the blue end of a fluorescence spectrum leads to a rough division of an $N$-level fluorescence spectrum into two regions (Fig. 3). The so-called vibrationally unrelaxed region is that part of the spectrum that occurs to the blue of the $0_0^0$ band position. This region consists predominantly of $a$-type bands. To the red is the vibrationally relaxed region, which, when IVR is appreciable, is composed primarily of non-$a$-type bands.

*g. The Calculation of Vibrational Coupling Parameters Using Quantum Beat Results.* We have previously noted in conjunction with two-level IVR that quantum beat results can be used to determine the parameters of the coupling that gives rise to IVR. It turns out that this can be done in the $N$-level case as well. The method of calculation is based on the relation

$$\mathrm{CEC}^t = H, \tag{3.24}$$

where $C$ is given by Eq. (3.10), $C^t$ is the transpose of $C$, $E$ is the diagonal

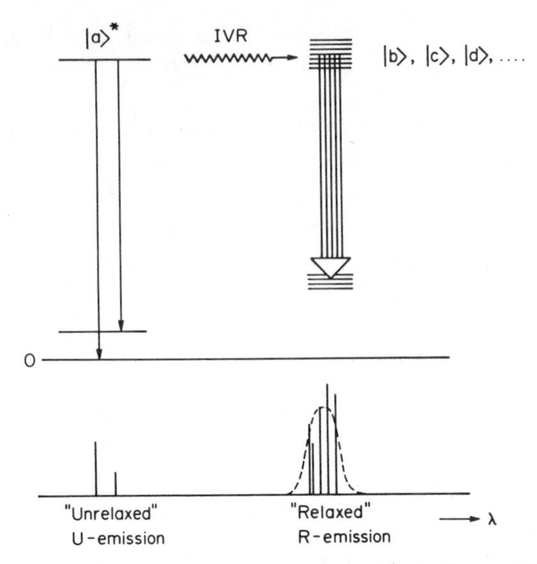


**Figure 3.** Level diagram and fluorescence spectrum representing the dispersed fluorescence characteristics to be expected from a set of coupled vibrational levels in $S_1$. Excitation prepares the zero-order $|a\rangle$ state (indicated by the asterisk), which then undergoes IVR. Emission gaining its strength from $|a\rangle$ is termed vibrationally unrelaxed and tends to occur in the blue region of the spectrum. Emission gaining its strength from $|b\rangle$, $|c\rangle$, ..., is termed vibrationally relaxed and occurs near and to the red of the $0^0_0$ transition energy of the molecule.

eigenvalue matrix, and $H$ is the Hamiltonian matrix of the system expressed in terms of the zero-order basis set of states. Since one wishes to find $H$, one must know something about $E$ and $C$.

To find $E$, one must know the energy eigenvalues of the eigenstates $|1\rangle$, $|2\rangle$, etc. In fact, one cannot determine the absolute eigenvalues of these eigenstates from quantum beat data. What one can determine, however, is that

$$E = E_0 I \pm \Delta E, \tag{3.25}$$

where either the plus or the minus sign obtains, $E_0$ is an unknown scalar constant, $I$ is the $N \times N$ identity matrix, and $\Delta E$ is an $N \times N$ matrix that can be found completely from quantum beat data.

To determine $\Delta E$, one assigns labels to the eigenstates such that energy-adjacent eigenstates have adjacent labels—that is, $|I\rangle$ is taken to lie in energy between $|I-1\rangle$ and $|I+1\rangle$. (Note that one cannot say whether $|I+1\rangle$ or $|I-1\rangle$ has the greater energy.) With such a labeling scheme, if one then arbitrarily assigns an energy $E_0$ to state $|1\rangle$, then any given state $|I\rangle$ has energy

$E_I = E_0 + \hbar\omega_{1I}$ or $E_I = E_0 - \hbar\omega_{1I}$, where either the former equation obtains for all $I$ or the latter one does. From this, it is clear that

$$\Delta E = \begin{bmatrix} 0 & 0 & 0 & \cdots \\ 0 & |\omega_{12}| & 0 & \cdots \\ 0 & 0 & |\omega_{13}| & \cdots \\ \vdots & \vdots & \vdots & \ddots \end{bmatrix}. \tag{3.26}$$

The task now becomes the proper assignment of observed beat frequencies to the $\omega_{1I}$. [We will assume that all $N(N-1)/2$ frequencies have been observed.] One starts with the fact that, by the state-labeling scheme used, the largest observed beat frequency must be $|\omega_{1N}|$. Next, one finds those $(N-2)$ triplets of beat frequencies defined by Eq. (3.15) that contain $|\omega_{1N}|$. The smallest frequency appearing in these triplets can be assigned to $\omega_{12}$. Having found $|\omega_{12}|$, one can assign all the other relevant frequencies by determining all the triplets to which $|\omega_{12}|$ belongs and comparing these triplets with those to which $|\omega_{1N}|$ belongs. Aside from $|\omega_{12}|$, $|\omega_{2N}|$, and $|\omega_{1N}|$, there are $(N-3)$ frequencies that can occur in *both* $|\omega_{12}|$-containing triplets *and* $|\omega_{1N}|$-containing triplets. These frequencies, in increasing order, are $|\omega_{13}|$, $|\omega_{14}|$, $\ldots$, $|\omega_{1,N-1}|$. Clearly, by Eq. (3.26), all of the frequencies needed to determine $\Delta E$ are obtained by this procedure.

The eigenvector matrix $C$ can be determined from relative values of quantum beat modulation depths, together with conditions associated with the fact that $C$ is orthonormal. The first step is to work with an $a$-type decay and to note that

$$\alpha_{Ia}^2 = \frac{M_a(\omega_{IK})}{M_a(\omega_{JK})}\alpha_{Ja}^2. \tag{3.27}$$

Using this equation, the $\alpha_{Ia}^2$ for all $I$ except one can be determined, the undetermined one being the one in terms of which all the others are expressed. The undetermined $\alpha_{Ia}^2$ can be found using the normalization condition for $|a\rangle$,

$$\sum_{I=1}^{N} \alpha_{Ia}^2 = 1. \tag{3.28}$$

Now, because eigenvector phases are arbitrary, one can choose all the $\alpha_{Ia}$ to be positive. Since one knows the $\alpha_{Ia}^2$, then by this choice of phase the $\alpha_{Ia}$ are known as well.

The second step in determining $C$ involves the decays of the non-$a$-type bands. The relevant $\alpha$'s are given by

$$\alpha_{I\gamma} = \left( \frac{M_\gamma(\omega_{IK})\alpha_{Ja}}{M_\gamma(\omega_{JK})\alpha_{Ia}} \right) \alpha_{J\gamma}, \tag{3.29}$$

where the values in parentheses are known from experiment and from Eqs. (3.27) and (3.28). Equation (3.29) plus the normalization condition

$$\sum_{I=1}^{N} \alpha_{I\gamma}^2 = 1 \tag{3.30}$$

give the $\alpha_{I\gamma}$ up to an overall sign. That is, the *relative* signs of the coefficients $\alpha_{I\gamma}$, $I = 1, N$ are fixed. Equations (3.29) and (3.30) can be applied to all non-$a$-type decays to determine each of the rows of $C$ up to an overall sign.

At this point it is worth noting that there are relations associated with the orthogonality of $C$ and with *absolute* quantum beat modulation depths that also can be used to find the elements of $C$. Clearly, from the above, one may not need these relations. However, the analysis we have presented presumes that a good deal of high-quality experimental data are available. This may not always be the case. When it is not, these "extra" relations may be particularly useful. They are useful, in any case, in providing consistency checks on the elements calculated for the matrix.

With $E$ and $C$ determined to the degree possible using beat parameters, $H$ can be found using Eq. (3.24). Given the ambiguities in $E$ and $C$, $H$ can be determined to be

$$H = E_0 I \pm \Delta H, \tag{3.31}$$

where (1) either the plus or the minus sign obtains, and (2) the magnitudes of the elements of $\Delta H$ are known, but (3) the signs of the off-diagonal elements of $\Delta H$ (the coupling matrix elements) are unknown.

*h. Dependence of Decay Behavior on N.* As we have stated above, a number of papers, basing their interpretations on measurements of time-integrated spectral properties, have considered the changing nature of IVR as a function of molecular vibrational energy. (For reviews, see Refs. 23 and 24.) In general, this work has indicated that as vibrational energy increases, IVR becomes faster and more extensive. This is reasonable since the average spacing of vibrational levels decreases with vibrational energy. This decrease in spacing, in turn, increases the opportunity for vibrational coupling between zero-order states.

In light of the time-integrated results, it is pertinent to consider how time-resolved measurements reflect changes in the number ($N$) of coupled vibrational levels involved in IVR. In particular, one is interested in general trends in the gross features of IVR-related decays. Guided by experimental results, together with the theory presented herein, we have found it useful to classify decays according to three ranges of $N$: (1) $N = 1$, (2) $N = 2-\sim 10$, and (3) $N \geq 10$. These ranges may be loosely identified with low-, intermediate-, and high-vibrational-energy regimes, respectively.

The case of $N = 1$ is trivial in a dynamics sense in that it corresponds to no IVR. A fluorescence spectrum belonging to this case consists entirely of vibrationally "unrelaxed" ($a$-type) bands. Each of these bands decays in the same manner. In most situations, these decays are unmodulated, single exponentials, although quantum beats and multiexponential decays arising from couplings other than those associated with IVR are possible.

The $N = 2-\sim 10$ case (the upper limit on $N$ is somewhat ill-defined) corresponds to what we call *restricted* IVR, the oscillatory flow of vibrational energy between a small number of vibrational levels. A fluorescence spectrum belonging to this case typically consists of an uncongested region of vibrationally unrelaxed ($a$-type) bands, and a somewhat more congested region consisting primarily of relaxed (non-$a$-type) bands. The decays of the various bands in the spectrum are quantum beat modulated with beat modulation depths and phases that are dependent on the type of band detected. The distinguishing feature of IVR in the restricted regime is that there are large-amplitude recurrences (on the time scale of the excited-state lifetime) in the distribution of energy over the coupled levels in the molecule. This is manifested in the fact that each of the fluorescence decays [the $I_\gamma(t)$] has a number of maxima, all of which approach the global maximum intensity of the decay. Thus, IVR is restricted in the sense that energy does not flow irreversibly from one vibration into others, but oscillates between a small number of coupled levels.

The case where $N$ is large corresponds to *dissipative* IVR, the irreversible flow of vibrational energy (irreversible on the time scale of the excited-state lifetime) from the initially prepared $|a\rangle$ state to the states with which $|a\rangle$ is coupled. Spectrally, this case is characterized by weak $a$-type fluorescence bands in the vibrationally unrelaxed region of the fluorescence spectrum and by a very congested relaxed spectral region, consisting primarily of non-$a$-type bands. Temporally, the individual $I_\gamma(t)$ in case of dissipative IVR are modulated by a large number of beat frequencies. These beat terms at times other than near $t = 0^+$ tend to destructively interfere and add to zero. At $t = 0^+$, however, all the cosine factors equal 1. Thus, the influence of the beat terms, while being small at most times, is substantial instantaneously after the excitation process and up until the time at which significant "dephasing" between the beat terms occurs. All this implies that a generic decay in the dissipative case will have an early time transient followed by an unmodulated, long time decay component. For *a*-type decays, in which all the beat terms have $+1$ phases, there is an initial, decaying transient, followed by a long component of comparatively smaller intensity. In non-*a*-type decays, because such decays have total modulation depths of $-1$, the initial transient is a rise of finite duration from zero intensity to a maximum, after which there is a long time decay. In both *a*-type and non-*a*-type decays, the possibility exists that small amplitude modulations may occur on the long decay components. However, because of the many beat components, dephasing is efficient enough to preclude anything close to a full recurrence in intensity on the time scale of the excited state. Relating decay behavior to IVR in the dissipative case, *a*-type decays reflect the irreversible flow of energy *out* of the initially prepared state $|a\rangle$. Non-*a*-type decays reflect the flow of vibrational energy *into* the states with which the $|a\rangle$ state is coupled.

The case of large $N$ corresponds to that regime of IVR in which kinetic rate equation models of the process are most likely to be valid. Lahmani et al.[31] have related decays associated with nonradiative processes in isolated molecules (in their case, intersystem crossing was the process) to a kinetic rate equation model. Applying these results to the decays associated with dissipative IVR yields the following expressions for *a*-type and non-*a*-type decays, respectively;

$$I_a(t) \sim \frac{1}{N}\{(N-1)e^{-(\Gamma+\Delta)t} + e^{-\Gamma t}\}, \tag{3.32}$$

$$\sum_{\gamma \neq a} I_\gamma(t) \sim \frac{N-1}{N}\{e^{-\Gamma t} - e^{-(\Gamma+\Delta)t}\}, \tag{3.33}$$

where $\Delta$ is a measure of the width of the distribution of beat frequencies and can be expressed in terms of vibrational coupling parameters. These equations, by virtue of their simplicity, serve several purposes. They show at a glance the general temporal behavior of IVR in the limit of large $N$. They serve as convenient functions with which one can fit experimental decays and thereby obtain parameters associated with IVR. They relate the rate of dissipative IVR to vibrational coupling parameters via $\Delta$. And, their derivation by means of a kinetic model reveals the close analogy between IVR and kinetic behavior. Still, one must not take Eqs. (3.32) and (3.33) too literally. As pointed out in Ref. 31, the equations exclude the possibility of quantum beats. Clearly, for this reason they are inapplicable to restricted IVR. Also, Eq. (3.32) gives the ratio of preexponential factors of fast versus slow fluorescence for an *a*-type decay as being equal to $N-1$, the number of states coupled to $|a\rangle$. However, one can see by Eq. (3.19) that $N-1$ is really just an upper limit to the ratio of preexponential factors. Anything less that perfectly uniform coupling (i.e., $|\alpha_{Ia}| = 1/N$ for all $I$) will give a ratio less than $N - 1$.

### C. Applications to Molecular Systems

In this section we review, in light of the theoretical results of the previous section, results of picosecond-beam studies applied directly to the study of IVR in excited electronic states. To date, a number of molecules have been subject to such study. Anthracene, being the first molecule in which phase-shifted quantum beats were discovered,[40] and being a molecule with high symmetry and well-characterized spectroscopy, is the most extensively studied. Quite detailed results,[42] including vibrational coupling parameters, are available for the molecule. Furthermore, trends found for anthracene[42] apply, at least in part, to other molecules as well. Hence, we regard anthracene as a prototypical molecule with respect to IVR and devote the first and largest subsection to it.

Picosecond-beam studies[44] of IVR in deuterated anthracenes have also been performed. The main impetus for performing such experiments was to assess the generality of the anthracene results and to determine the effects that changes in molecular symmetry and vibrational density of states have on IVR. We discuss results pertaining to these studies after the anthracene subsection.

Third, we review results[45] on *t*-stilbene. These results are of particular interest in that *t*-stilbene is significantly different from the anthracene species. It has a number of large-amplitude, low-frequency vibrational modes. It has low symmetry. And, at sufficiently high energies in the $S_1$ manifold, it undergoes photoisomerization on a subnanosecond time scale. Thus, the molecule is a species, the study of which allows one to test more rigorously the generality of the anthracene results.

Finally, we discuss time-resolved results on alkylaniline species.[46] One might expect IVR in such species to involve flow of energy from optically active ring modes to the bath of weakly coupled modes associated with the alkyl chain. Recognition of this led Powers et al.[10e] to their time-integrated studies of IVR as a function of alkyl chain length in the alkylanilines. Our picosecond results are of interest in that they provide points of comparison between time-resolved rate (and coherence) measurements versus inferences from spectral results.

Before addressing the experimental results, it is pertinent to summarize some details of our apparatus and procedures.[39,42] Experiments were performed on continuous, seeded, supersonic free jet expansions. Samples were typically heated in the preexpansion region (e.g., anthracene was heated to 180°C) so as to produce preexpansion vapor pressures in the range of 1 Torr for the molecule of interest. Heated samples were mixed in a glass nozzle with $P_0 = 20$–$50$ psig of carrier gas (almost always helium). The gaseous mixture was then allowed to expand through a $D \simeq 100$-$\mu$m-diameter pinhole into a vacuum chamber maintained at less than 1 mTorr. Picosecond-pulsed excitation of the free jet expansion typically occurred at a laser-to-nozzle distance of $x \geq 3$ mm (giving $x/D \geq 30$). Linearly polarized picosecond pulses in the ultraviolet were obtained by frequency doubling (with a 1-cm-long $LiIO_3$ crystal) the output of a synchronously pumped (argon ion laser as pump laser), cavity-dumped, dye laser (DCM or Rhodamine 590 as dye). The bandwidth of the excitation light under typical laser conditions (i.e., a three-plate birefringent filter and an ultrafine tuning etalon in the dye-laser cavity) was $\sim 2\ cm^{-1}$. Temporal pulse widths were on the order of 15 psec. Fluorescence was collected with right angle geometry and directed through a $\frac{1}{2}$ m monochromator (grating dispersion of 16 Å $mm^{-1}$, used in first order). Fluorescence photons were detected with a fast microchannel plate photomultiplier, the output of which was amplified before reaching the timing electronics. Decays were measured using time-correlated single-photon counting. The total temporal response of detection was typically 80 psec FWHM (full width at half maximum), as measured by scattering excitation pulses from the nozzle. (Response widths of $\sim$45 psec have been achieved with the apparatus by partially masking the monochromator grating.[47,48,50] Also, some data appearing in this review were taken before the use of the multichannel plate detector and correspond to a resolution of several hundreds of picoseconds.) Data analysis was performed with a PDP-11/23 computer. The analyses consisted of decay fits by nonlinear least squares routines and Fourier analysis[42] of beating decays.

### 1. *Anthracene*

As background material pertinent to the results of IVR studies of anthracene, Fig. 4 shows $S_1 \leftarrow S_0$ fluorescence excitation spectra of jet-cooled anthracene.[61] The two spectra correspond to nitrogen and helium jets, respectively. (The "extra" bands in the helium spectrum correspond to vibrational hot bands and sequence bands, helium being a less efficient cooler of anthracene vibrations than nitrogen is.) From the fairly sparse spectra, one surmises that, although the molecule has 66 normal modes, apparently only a few of these are optically active. We shall be interested in the IVR dynamics that occurs upon the excitation of several of the bands in the excitation spectrum. The assignments of these bands, along with assignments of $S_0$ vibrational levels, are discussed in Ref. 61.

A particularly convenient way by which to consider IVR in $S_1$ anthracene[42] is in terms of increasing $S_1$ vibrational energy. Three energy regimes—low, intermediate, and high—are pertinent. The presentation of IVR results in this subsection follows this organization. We shall treat in particular detail the level at $S_1 + 766\ cm^{-1}$ in the low-energy regime, the levels at $S_1 + 1380\ cm^{-1}$

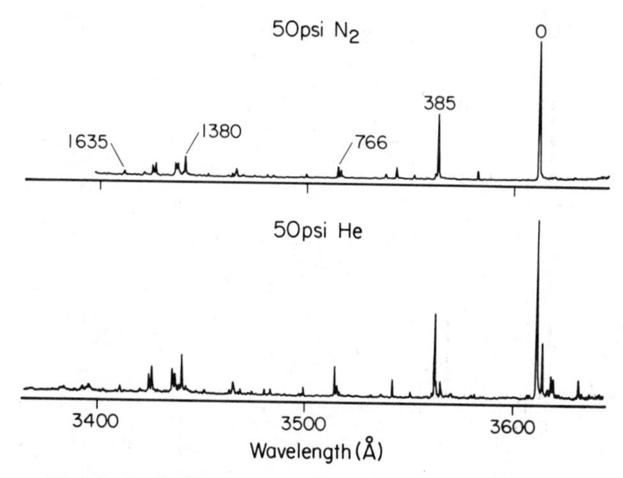


**Figure 4.** Normalized fluorescence excitation spectra of anthracene. Carrier gas parameters corresponding to the free jet expansion are given in the figure. Various bands are labeled in the $N_2$ spectrum with their energies in $cm^{-1}$ above the $S_1$ origin.

and $S_1 + 1420\ cm^{-1}$ in the intermediate-energy regime, and the level at $S_1 + 1792\ cm^{-1}$ in the high-energy regime.

*a. Low Vibrational Energy:* $E_{vib} = 0\text{–}1200\ cm^{-1}$. The fluorescence spectroscopy of jet-cooled anthracene excited to levels ranging from the $S_1 - 0^0_0$ level to levels near $S_1 + 1200\ cm^{-1}$ has been reported on in Ref. 61. Almost all of the spectra consist of resolvable bands that can be assigned as transitions from the optically prepared $S_1$ vibrational level to the manifold of $S_0$ vibrational levels. That is, these bands, based on spectroscopic considerations, are assignable as *a*-type bands. The exclusive appearance of *a*-type bands in any given fluorescence spectrum implies an absence of IVR.

Picosecond-beam studies[42] confirm the interpretation that IVR is absent in this low-energy regime. As an example, Fig. 5 shows the fluorescence spectrum that arises upon excitation of the $S_1 + 766\ cm^{-1}$ ($12^2$) level of anthracene and a decay of a band in that spectrum (the shift from the excitation wavelength, $\bar{\nu}_d$, for the band being $\bar{\nu}_d = 390\ cm^{-1}$). The spectrum is analyzed in Ref. 61. The decay is clearly unmodulated, is a single exponential with an 18 nsec lifetime, and is the same as the decays of the other bands in the spectrum. Such decay behavior is consistent with nonexistent IVR.

Decays of the spectrally resolved fluorescence from other $S_1$ levels in the low-energy regime of anthracene have also been measured. No beat-modulated, nonexponential decays have been observed, even with 80 psec temporal resolution. Both spectral and decay results indicate, therefore, that vibrational

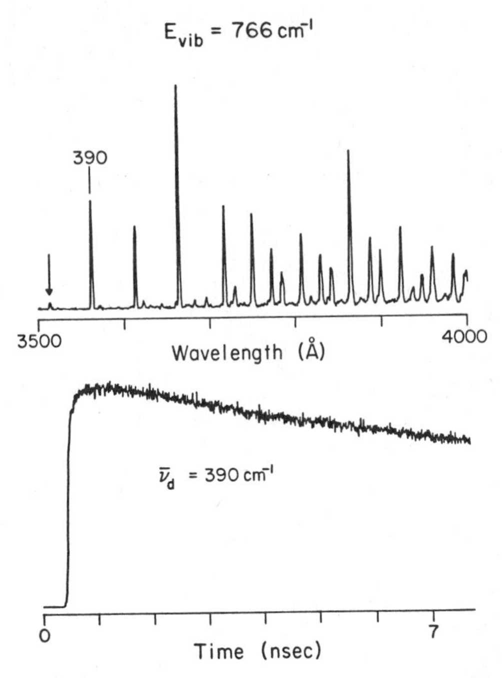


**Figure 5.** Dispersed fluorescence spectrum and fluorescence decay resulting from excitation of jet-cooled anthracene to $S_1 + 766$ cm$^{-1}$ ($12^2$). The fluorescence spectrum was obtained with 1.6 Å monochromator resolution ($\equiv R$). An arrow marks the excitation wavelength. The decay corresponds to detection of the $\bar{\nu}_d = 390$ cm$^{-1}$ band in the spectrum with $R = 3.2$ Å.

coupling leading to IVR does not occur to any appreciable extent in anthracene at $S_1$ vibrational energies less than $\sim 1200$ cm$^{-1}$.[61] (One must qualify this statement, however, by noting that the results do not exclude the possibility that strong Fermi resonances, resulting in splittings of several wavenumbers or more, are present in this energy regime. Indeed, some results[62] have provided evidence for such coupling in the higher energy portion of the region.)

*b. Intermediate Vibrational Energy:* $\sim$*1300–1514* cm$^{-1}$. Without exception, the excitation of jet-cooled anthracene at each of the five prominent vibronic bands between $S_1 + 1380$ and $S_1 + 1514$ cm$^{-1}$ in its excitation spectrum results in quantum beat-modulated fluorescence decays, the beat phases and modulation depths of which depend on the particular fluorescence band that is detected in the fluorescence spectrum.[42] (Here, it is pertinent to point out that recent work[44] has revealed beat-modulated decays upon excitation of anthracene to its $S_1 + 1290$ cm$^{-1}$ band as well.) We have analyzed in detail

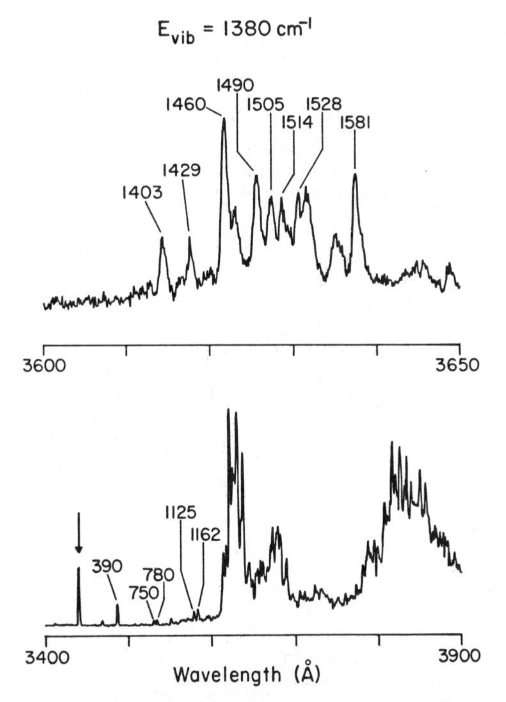


**Figure 6.** Dispersed fluorescence spectra resulting from excitation of jet-cooled anthracene to $S_1$ + 1380 cm$^{-1}$. The upper portion is a high resolution ($R$ = 0.5 Å) trace of the region in and about the wavelength of the $0_0^0$ transition of the molecule. The lower portion, taken with $R$ = 1.6 Å covers a wider range and includes the excitation wavelength (arrow). Various bands in the spectra are marked with their shifts in cm$^{-1}$ from the excitation energy.

the time-resolved results corresponding to two of these excitation energies: $S_1$ + 1380 cm$^{-1}$ ($6^1$-level) and $S_1$ + 1420 cm$^{-1}$ ($5^1$-level). These are the results to which we pay most attention.

*$6^1$-level:* The dispersed fluorescence spectrum arising from the $6^1$-level of anthracene, and a higher resolution blow-up of the same spectrum appear in Fig. 6.[63] The spectrum consists of a mixture of assignable ($a$-type) and unassignable (non-$a$-type) bands. One notes considerably more congestion in the spectra of Fig. 6 than in the $S_1$ + 766 cm$^{-1}$ spectrum of Fig. 5. Moreover, this congestion is principally near and to the red of the wavelength corresponding to the $S_1 - S_0$ $0_0^0$ transition of anthracene. As a whole, this spectral behavior indicates that the $6^1$-level participates in an IVR process.

The detailed nature of the IVR in which the $6^1$-level participates is revealed by time-resolved results. One finds that the decays of individual bands in the $6^1$ fluorescence spectrum are modulated by quantum beats, the phases and

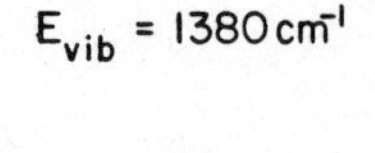


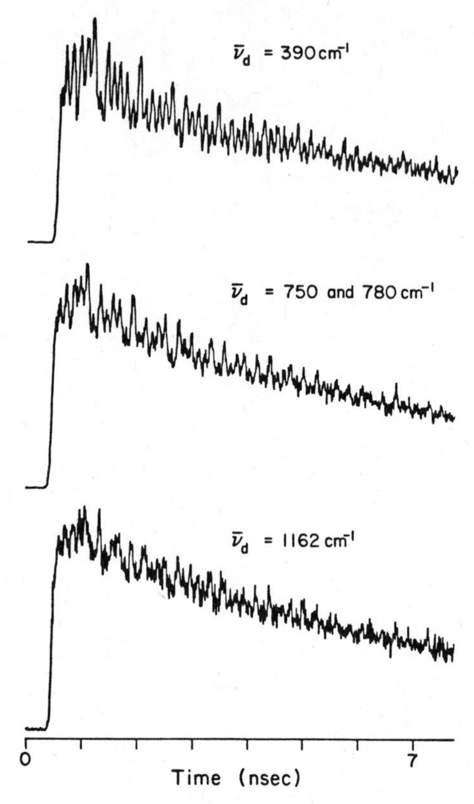


**Figure 7.** Fluorescence decays of "*a*-type" bands in the $E_{vib} = 1380\ cm^{-1}$ spectrum of anthracene. The shifts of the bands from the excitation energy are given in the figure. Since these bands are relatively isolated spectrally, low detection resolution was used to maximize the signal. From top to bottom $R = 24$, 16, and 5 Å.

modulation depths of which depend on the band detected. The particular characteristics of the decays indicate that the $6^1$ zero-order level undergoes IVR with two other levels. (It should be pointed out, though, that the analysis of the time-resolved results is complicated by the fact that there are two overlapping vibronic bands near $S_1 + 1380\ cm^{-1}$. Here, we consider just the stronger, bluer excitation band, which we take to be $6^1$. Details concerning the overlapping bands appear in Refs. 42 and 62.) Figure 7 shows decays associated with fluorescence bands that can be assigned, using only spectroscopic information, as *a*-type fluorescence. Fourier analysis of these decays

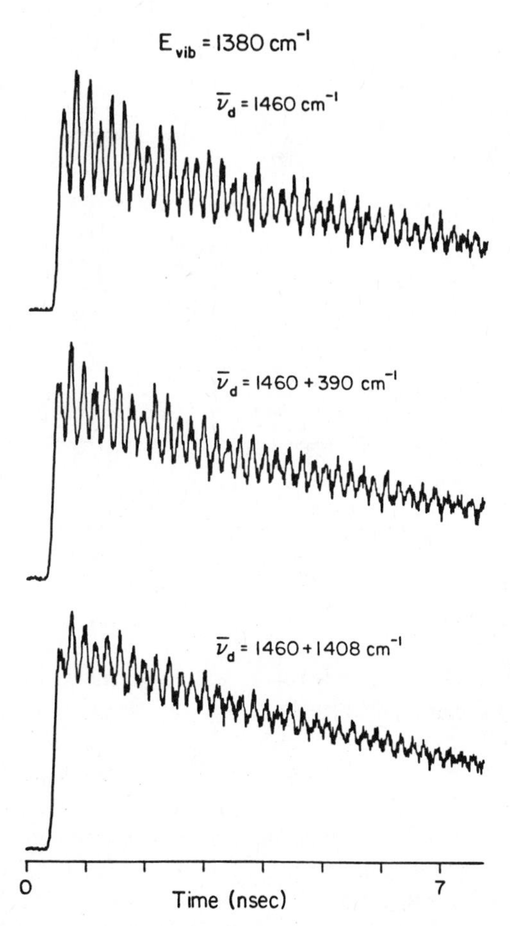


**Figure 8.** Fluorescence decays of "*b*-type" bands in the $E_{\text{vib}} = 1380\ \text{cm}^{-1}$ spectrum of anthracene. The shifts of the bands from the excitation energy are given in the figure. From top to bottom $R = 0.5$, 1.0, and 1.6 Å.

shows that they are modulated in the same way by the three beat components: $\omega/2\pi = 3.5$, 4.9, and 8.4 GHz. (It is important to note that $3.5 + 4.9 = 8.4$.) All three of these components have $+1$ phases. The decays are what one would expect for an *a*-type decay arising from a coupled three-level system.

Figure 8 shows decays of a second group of bands in the $6^1$ spectrum. The three decays, which one would assign to be of the same type based on one's knowledge of anthracene spectroscopy (390 and 1408 $\text{cm}^{-1}$ are intervals associated with strongly optically active $S_0$ modes of the molecule[61]), are indeed modulated similarly. A Fourier spectrum of the upper decay in Fig. 8

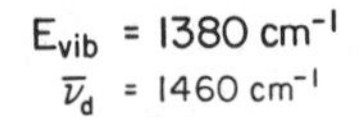


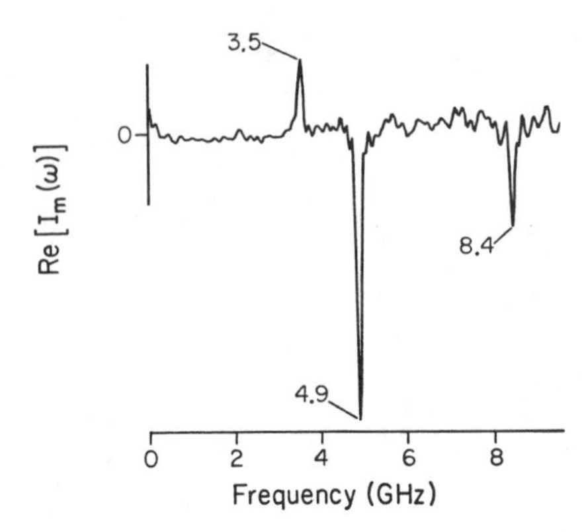


**Figure 9.** Fourier spectrum of the residual of the decay of Fig. 8 top. Fourier bands are labeled with their $\omega/2\pi$ frequency values in GHz. Note that two of these bands have $-1$ phases and one has a $+1$ phase.

is shown in Fig. 9. One notes (1) three beat components at 3.5, 4.9, and 8.4 GHz, (2) the phase behavior of the beat components, two $-1$ phases and one $+1$ phase, and (3) the fact that the sum of absolute modulation depths of the three beat components is $-0.75$ (note that this last point cannot be obtained from the figure). All of these characteristics are consistent with those of non-*a*-type fluorescence decays arising from a system undergoing three-level IVR. (The sum of modulation depths should be $-1.0$. There are experimental reasons why this is likely not to be strictly the case, however.)

Finally, Fig. 10 shows the decays of a third group of fluorescence bands in the $6^1$ spectrum. It is apparent from the figure that all four bands decay in a similar manner. Fourier analysis of the decays confirms that this is indeed so. Figure 11 shows the Fourier spectrum of the decay at the top of Fig. 10. One notes that the same three beat components that are present here are present in the other two decay-types. Moreover, one notes (1) two $-1$ beat phases and one $+1$ phase, (2) that the phase behavior in Fig. 11 *is different* from that in Fig. 9, and (3) that the sum of beat modulation depths is $-0.70$.

Taken together, the decay behavior of the three groups of fluorescence bands arising from the $6^1$-level is consistent with that expected from three anharmonically coupled vibrational levels. Accepting that this is indeed so, one can proceed further and, by using measured quantum beat parameters, calculate parameters associated with the vibrational coupling in this three-level system. Without presenting the details of the calculation, which appear

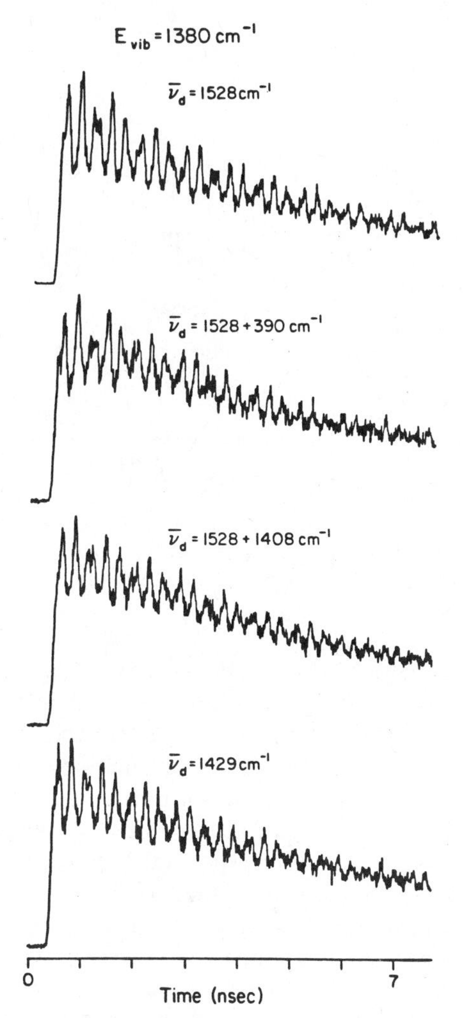


**Figure 10.** Fluorescence decays of "*c*-type" bands in the $E_{\text{vib}} = 1380\ \text{cm}^{-1}$ spectrum of anthracene. The shifts of the bands from the excitation energy are given in the figure. From top to bottom $R = 0.8$, 1.0, 1.0, and 1.3 Å.

in Ref. 42 and follow the general outline given in Section III B 2g, the result for the Hamiltonian matrix in the zero-order basis is

$$H^{1380} = E_0 I \pm \begin{bmatrix} 4.66 & -2.70 & 2.83 \\ -2.70 & 5.08 & -0.38 \\ 2.83 & -0.38 & 2.16 \end{bmatrix}, \qquad (3.34)$$

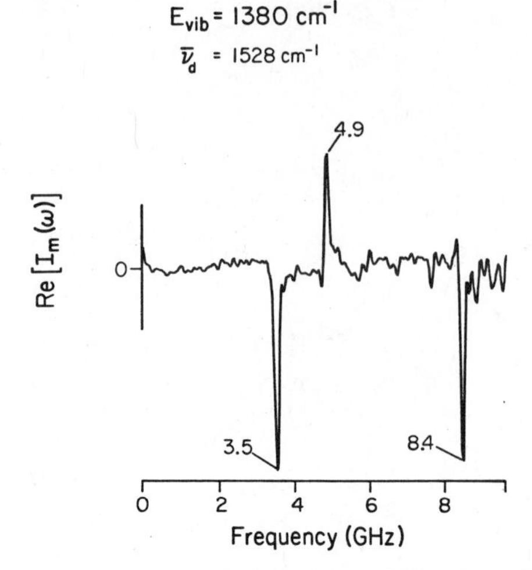


**Figure 11.** Fourier spectrum of the residual of the decay of Fig. 10 top. Bands are labeled with their frequency values in GHz. Note that two of the bands have $-1$ phases and one has a $+1$ phase. Note also, however, that the phase behavior is different from that in Fig. 9.

where the rows and columns are both labeled by *a*, *b*, and *c*, where the values are in GHz, and where we have assigned the decays in Figs. 7, 8, and 10 as *a*-type, *b*-type, and *c*-type decays, respectively. (Recall that the signs of the off-diagonal elements are not fixed by the calculation of *H*.)

$5^1$-*level:* The dispersed fluorescence spectrum resulting from the excitation of the $5^1$ ($S_1$ + 1420 cm$^{-1}$) level of anthracene is shown in Fig. 12 together with a high-resolution portion of the spectrum in the spectral region of the $S_1$ origin.[63] The general characteristics of the spectrum are quite similar to those of the $6^1$ spectrum. That is, there is a mix of assignable and unassignable bands, there is some resonance fluorescence, and there is resolvable spectral congestion near and to the red of the $S_1$ origin region.

Similar to the $6^1$-level case, the time-resolved results pertaining to $5^1$ excitation reveal fluorescence decays that are beat modulated and that depend on detection wavelength. Again, it is useful to consider groups of decays separately.

Figure 13 top shows the decay of the fluorescence band at $\bar{\nu}_d = 390$ cm$^{-1}$. Fourier analysis of the decay (Fig. 14 top) reveals three prominent beat components at $\omega/2\pi = 1.0$, 9.7, and 10.7 GHz. [Note that these three form a triplet of the form defined by Eq. (3.15).] All of these components have $+1$ phases. The $\bar{\nu}_d = 390$ cm$^{-1}$ decay is representative of a number of other decays, those of the $\bar{\nu}_d = 0$, 780, 1168, and 1480 cm$^{-1}$ bands. All of these bands are assignable in terms of intervals associated with optically active modes. Based

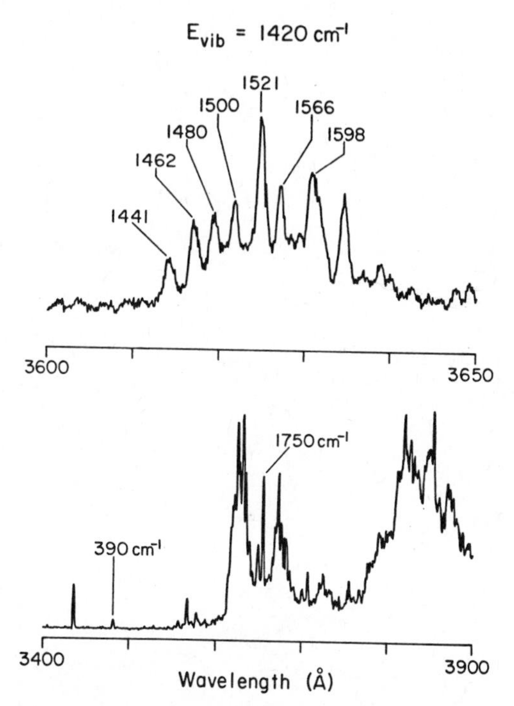


**Figure 12.** Dispersed fluorescence spectra resulting from excitation of jet-cooled anthracene to $S_1 + 1420\ \text{cm}^{-1}$. The upper portion was taken with $R = 0.6$ Å and the lower with $R = 1.6$ Å. Various bands in the spectra are marked with their shifts in $\text{cm}^{-1}$ from the excitation energy.

on the quantum beat characteristics of their decays, these bands would seem to be assignable as *a*-type bands arising from a coupled three-level system.

Figure 13 middle shows the decay of the $\bar{\nu}_d = 1750\ \text{cm}^{-1}$ band in the $5^1$ fluorescence spectrum. This decay clearly has a slow (1.0 GHz) component that is phase-shifted 180° from the 1.0 GHz component of the $\bar{\nu}_d = 390\ \text{cm}^{-1}$ decay. Fourier analysis of the $\bar{\nu}_d = 1750\ \text{cm}^{-1}$ decay (Fig. 14 middle) reveals that besides this beat component, the decay is also modulated by four other beat terms: the 9.7 and 10.7 GHz components found in the $\bar{\nu}_d = 390\ \text{cm}^{-1}$ decay and two more components at 3.5 and 4.5 GHz. The absolute modulation depths of these five components, in order of increasing frequency are $-0.74$, $-0.06$, 0.04, $-0.11$, and 0.08. (The sum of the modulation depths is $-0.79$.) Other bands that one would expect to decay in the same way as the $1750\ \text{cm}^{-1}$ band, that is, the bands at $\bar{\nu}_d = 1750 + 390\ \text{cm}^{-1}$ and $\bar{\nu}_d = 1750 + 1408\ \text{cm}^{-1}$, have been found (albeit with decreased temporal resolution) to decay in a manner consistent with this expectation. The fact that five beat components modulate the $1750\ \text{cm}^{-1}$ decay *and* that the 3.5 and 4.5 GHz components form

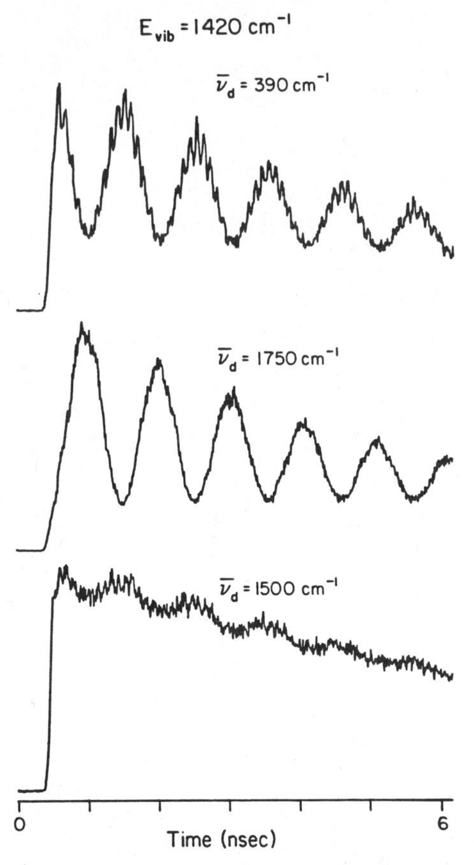


**Figure 13.** Representative decay types for fluorescence bands in the $E_{\text{vib}} = 1420\ \text{cm}^{-1}$ spectrum of anthracene. The wavenumber shifts of the bands from the excitation energy are given in the figure. From top to bottom $R = 16.0$, 1.6, and 1.6 Å.

a quantum beat triplet [Eq. (3.15)] with the 1.0-GHz component (which also appears in the $\bar{\nu}_d = 390\ \text{cm}^{-1}$ decay) indicates that the decay behavior of the $5^1$-level really is a manifestation of IVR between more than three levels. One would note, in particular, that the beat frequencies and the phase behavior exhibited in Fig. 14 middle are perfectly consistent with the decay behavior of a non-*a*-type band arising from a system undergoing four-level IVR (if one assumes that one of the six beat components expected from such a system is too weak to be observed).

A third band at $\bar{\nu}_d = 1500\ \text{cm}^{-1}$ exhibits different decay behavior than both the 390 and 1750 $\text{cm}^{-1}$-type bands. Figure 13 bottom shows the decay of this

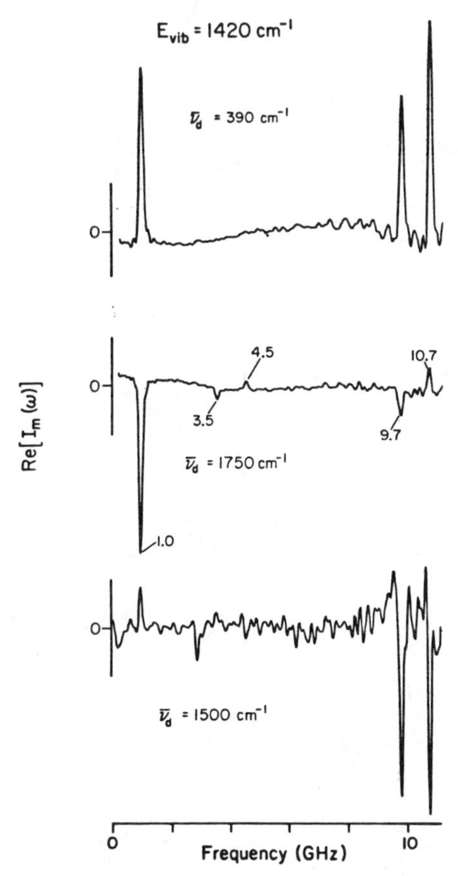


**Figure 14.** Fourier spectra of the residuals of the decays of Fig. 13. Bands in the middle spectrum are labeled with values in GHz. Although there appear to be more than three components in the lower spectrum, only the ones at 1.0, 9.7, and 10.7 GHz are reproducible.

band, and Fig. 14 bottom the results of Fourier analysis of the decay. Despite the noise in the Fourier spectrum, three beat components are clearly evident at $\omega/2\pi = 1.0, 9.7,$ and 10.7 GHz, with phases of $+1$, $-1$, and $-1$, respectively. Note that these characteristics are consistent with a different non-$a$-type decay (i.e., different than the type of the $\bar{v}_d = 1750$ cm$^{-1}$ decay) arising from the four-level system. (One must, of course, assume that three of the beat components are too weakly modulated to be observed.)

Finally, shown in Fig. 15 are decays for a number of other prominent bands that appear in the $5^1$ spectrum. It is evident that all of these decays are modulated to some extent by the 1.0 GHz beat component that is present in

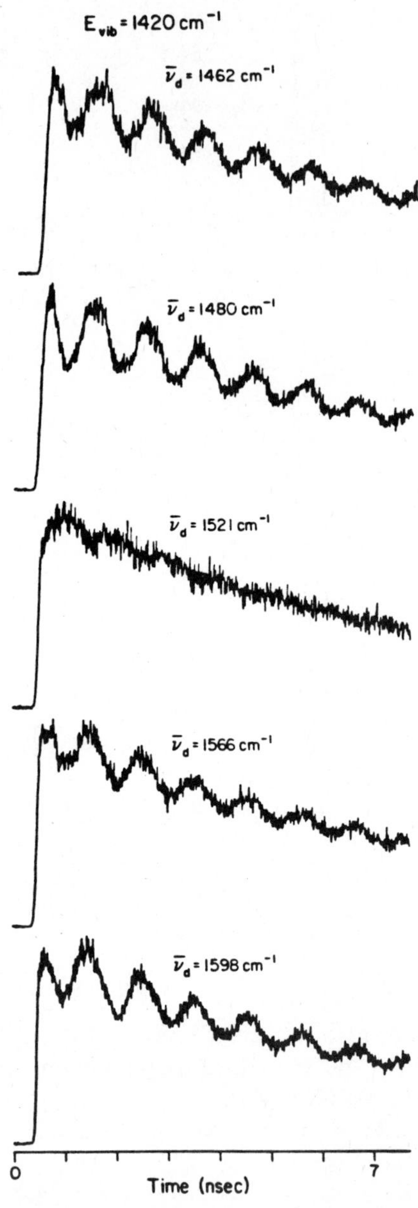


**Figure 15.** Fluorescence decays of bands in the $E_{vib} = 1420$ cm$^{-1}$ spectrum of anthracene that occur near the $0_0^0$ transition energy of the molecule (see the spectrum of Fig. 12 top). The wavenumber shifts of the bands from the excitation energy are given in the figure. From top to bottom $R = 1.6$, 1.6, 0.8, 1.6, and 1.6 Å

the decays of Fig. 13. Moreover, some are modulated with the 9.7 and 10.7 GHz components, as well.

Although detailed interpretation of the $5^1$-level decay results is somewhat ambiguous owing to experimental limitations of spectral and temporal resolution, the data are entirely consistent with one's expectations of the manifestations of restricted IVR. Moreover, they are, more likely than not, manifestations of IVR between four levels. Elsewhere,[42] assuming four-level IVR, making an assumption about the missing sixth beat frequency (of the two possible frequencies consistent with our data, we assumed that the lower frequency, $\omega/2\pi = 6.2$ GHz, was the one that actually applied), and making one assumption about one element of the $C$ matrix, we calculated from quantum beat data the Hamiltonian matrix for the coupling involving the $5^1$-level:

$$H^{1420} = E_0 I \pm \begin{bmatrix} 3.23 & -0.28 & -4.24 & -1.86 \\ -0.28 & 1.70 & 0.29 & 1.82 \\ -4.24 & 0.29 & 7.57 & 0.94 \\ -1.86 & 1.82 & 0.94 & 3.70 \end{bmatrix}, \tag{3.35}$$

where all the values are in GHz. While $H^{1420}$ may only be accurate in a semiquantitative sense, it certainly can serve as a useful indication of the magnitude of coupling matrix elements involved in IVR.

*Other excitation energies:* Other than the ones at $S_1 + 1380$ and $S_1 + 1420$ $\text{cm}^{-1}$, there are three prominent bands in the intermediate region of jet-cooled anthracene's excitation spectrum. Time- and frequency-resolved measurements subsequent to excitation of these bands have also been made. Without going into any detail concerning the results of these measurements, we do note that all three excitations give rise to quantum beat-modulated decays whose beat patterns (phases and modulation depths) depend on the fluorescence band detected.[42] Figure 16 shows an example of this behavior for excitation to $S_1 + 1514$ $\text{cm}^{-1}$. The two decays in the figure correspond to the detection of two different fluorescence bands in the $S_1 + 1514$ $\text{cm}^{-1}$ fluorescence spectrum.

c. *High Vibrational Energy.* Above $E_{\text{vib}} = 1514$ $\text{cm}^{-1}$ in $S_1$ anthracene, there are just a few excitation energies that give rise to fluorescence spectra that are amenable to the direct study of IVR through the measurement of the decays of individual fluorescence bands. (There is a good deal of congestion in these high-energy spectra.) The most amenable excitation, corresponding to $E_{\text{vib}} =$ 1792 $\text{cm}^{-1}$, is the one that has been studied most and is the one that we treat in this subsection.

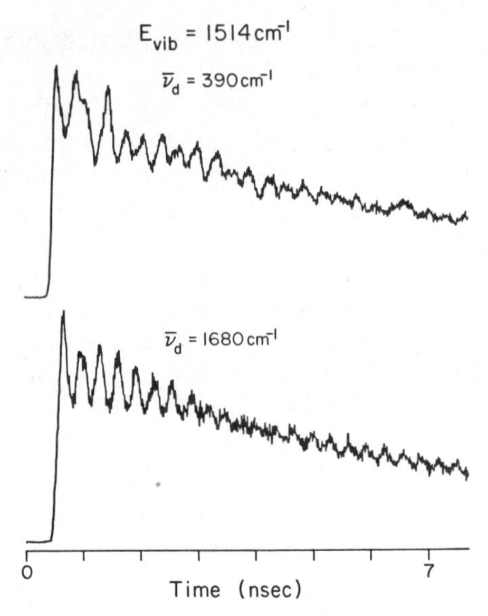


**Figure 16.** Fluorescence decays of two bands in the $E_{vib} = 1514$ cm$^{-1}$ spectrum of anthracene. The wavenumber shifts of the bands from the excitation energy are given in the figure. For the upper decay $R = 32$ Å, and for the lower one $R = 2.4$ Å.

Figure 17 shows the dispersed fluorescence characteristics of anthracene upon excitation to $S_1 + 1792$ cm$^{-1}$. One notes that the spectrum is quite congested relative to lower-energy spectra, but that, despite this, there are at least two discrete bands of weak intensity in the blue portion of the spectrum (Fig. 17 bottom) and some partially resolved structure in the spectral region of the $S_1$ $0_0^0$ band (Fig. 17 top). The two weak blue bands, being at $\bar{\nu}_d = 390$ cm$^{-1}$ and $\bar{\nu}_d = 780$ cm$^{-1}$, respectively, can be assigned as transitions from the zero-order state ($|a\rangle$) prepared by the laser pulse (i.e., as $a$-type bands). The bands near the $S_1 - 0_0^0$ energy probably arise from zero-order vibrational levels populated via IVR (non-$a$-type bands). Given this spectral indication of IVR, one wonders what the temporal characteristics of the bands are.

Figure 18 shows decays of the $\bar{\nu}_d = 390$ and 780 cm$^{-1}$ bands in the $E_{vib} = 1792$ cm$^{-1}$ fluorescence spectrum. (By several tests, these decays have been determined to be *isolated*-molecule decays, and to have a negligible contribution from scattered laser light.) Considering these decays in light of Section III B 2h [e.g., Eq. (3.32)] and noting the fact that Fourier analysis of the decays reveals that they are modulated by many beat frequencies, all of which have $+1$ phases, it is clear that the decays exhibit the behavior that one would expect in an $a$-type decay arising from a system undergoing dissipative IVR.

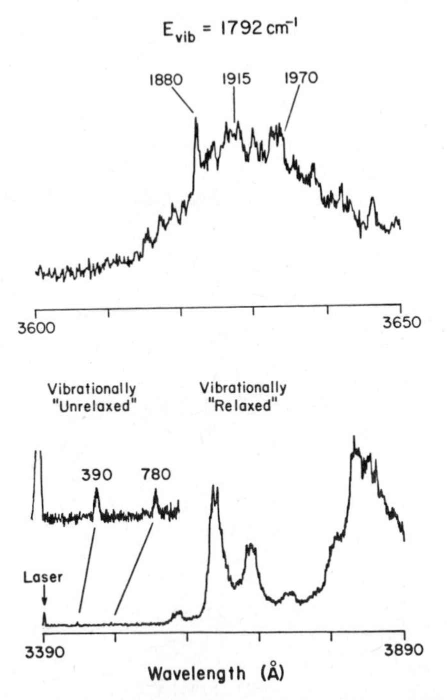


**Figure 17.** Dispersed fluorescence resulting from excitation of jet-cooled anthracene to $S_1$ + 1792 $cm^{-1}$. The upper portion was taken with $R = 0.5$ Å. For the lower portion $R = 1.6$ Å for the main spectrum and $R = 3.2$ Å for the inset. Various bands in the spectra are labeled with their wavenumber shifts from the excitation energy.

The short-time "spike" in the decay, which can be attributed to the dephasing of many quantum beat terms (all with $+1$ phases), represents the irreversible flow of vibrational energy out of the zero-order state prepared by the laser. The long-time component, although weakly modulated, represents an "equilibration" in the distribution of vibrational energy subsequent to the initial energy flow process.

Assuming the validity of Eq. (3.32), parameters associated with IVR can be obtained by fitting the observed decays of Fig. 18 to the functional form given in that equation. Figure 19 shows the result of fitting the lower decay in Fig. 18. (Convolution of the system temporal response was accounted for in the fit.) One notes that, except for the modulations on the long-time component, the fit matches the measured decay quite well. The fitted parameters indicate that IVR occurs on a fast time scale (22 psec) and is such that at least 18 levels are involved in the energy flow. Significantly, the long-time decay rate matches

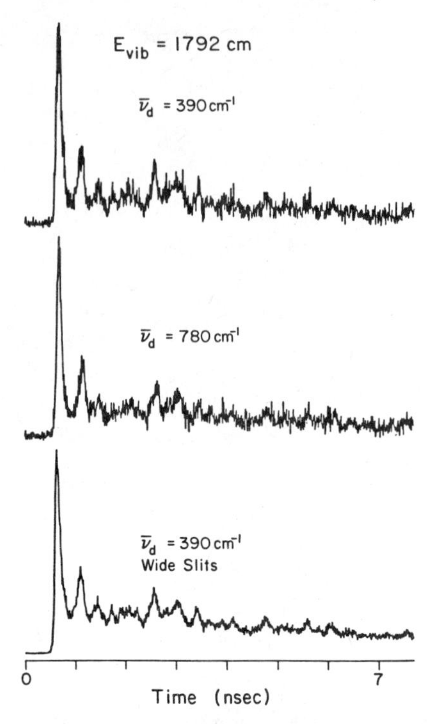


**Figure 18.** Fluorescence decays for the 390 and 780 $cm^{-1}$ bands in the spectrum of anthracene excited to $S_1$ + 1792 $cm^{-1}$. The top two decays were taken with $R = 3.2$ Å. For the bottom decay $R = 32$ Å. A slight increase in the relative intensity of the long time component is apparent in the lower decay compared with the upper two.

that of the decay of the *wavelength-unresolved* fluorescence of the $E_{vib} = 1792$ $cm^{-1}$ system.[63]

Further corroboration of the picture of IVR obtained from the *a*-type decays can be had from the decays of the bands in the congested region of the fluorescence spectrum. Examples of such decays are show in Fig. 20. One notes several points about these decays. First, upon fitting them each is found to have a long-time exponential decay constant matching those of the *a*-type decays. Second, the decays are modulated. Third, the decays all have finite rise times, and these rise times roughly match the ~20 psec fast decay time of the *a*-type decays. Finally, Fourier analysis of the decays reveals that they are modulated by many Fourier components, some of which have −1 phases. All of these characteristics are consistent with those expected for non-*a*-type decays that arise from the same set levels as the *a*-type decays of Fig. 18. Thus, the decays of Fig. 20 are pictures of the dissipative flow of energy into

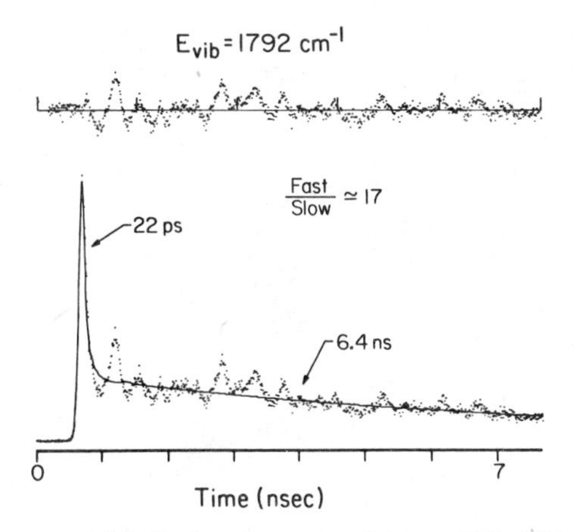


**Figure 19.** Double exponential fit (line) to the measured decay of Fig. 18 bottom. The weighted difference (residual) between the observed decay and the fitted curve appears at the top of the figure. Best-fit values of the two lifetimes and the ratio of fast to slow fluorescence are given in the figure. One notes that the best-fit fast-to-slow ratio is greater than what one would judge by eye from the decay because the finite temporal response of detection tends to reduce the apparent magnitude of the fast component.

zero-order states that are coupled with the optically prepared $|a\rangle$ state. (One notes that the presence of beats in the $a$-type and non-$a$-type decays implies that IVR is not complete subsequent to the initial dephasing process. Nevertheless, the modulations are weak enough that one can meaningfully speak of the energy flow as being "dissipative.")

*d. Conclusions Concerning IVR in $S_1$ Anthracene.* From the experimental results reviewed in this section a number of conclusions can be made concerning IVR in anthracene. A first conclusion, given the ready interpretation of the experimental results in light of the theory of Section III B, is that the physical situation in anthracene is a close match with the situation treated by the theory. This means that the two major assumptions of the theory—(1) that the zero-order vibrational levels are coupled by anharmonic coupling and (2) that *single* zero-order levels act as absorption and emission doorway states—would seem to apply to IVR in anthracene.

Second, from the results one has a picture of the changing nature of IVR as a function of vibrational energy in the molecule. In particular, the time-resolved results at low, intermediate, and high vibrational energy reveal the manifestations of nonexistent, restricted, and dissipative IVR, respectively. In the low-energy regime ($\leq 1200\ \text{cm}^{-1}$) vibrational coupling is not extensive, if present at all. Excitation to this regime prepares a vibrational eigenstate,

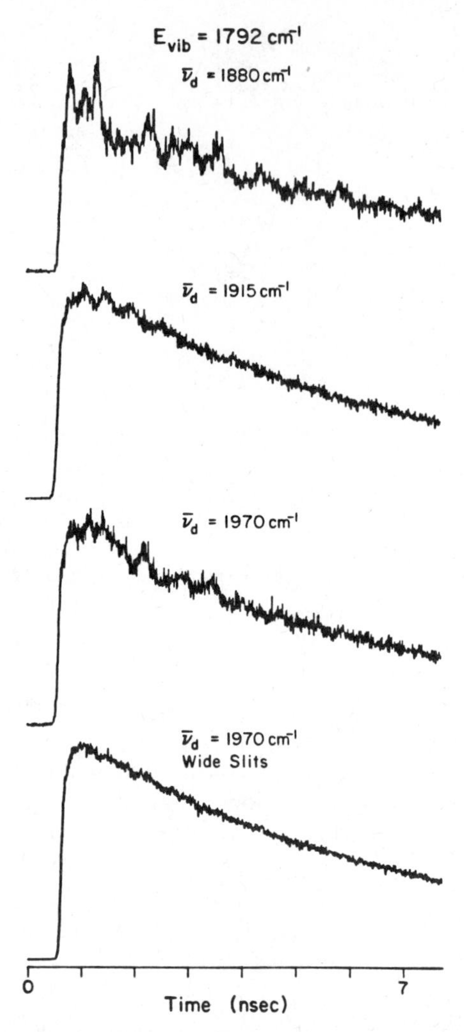


**Figure 20.** Fluorescence decays of bands in the vibrationally relaxed region of the $E_{vib} = 1792$ $cm^{-1}$ spectrum of anthracene (see Fig. 17 top). Wavenumber shifts of the detected bands from the excitation energy are given in the figure. From top to bottom, $R = 1.6$, 3.2, 1.6, and 16 Å.

which undergoes no evolution in its vibrational character. At somewhat higher vibrational energies in the $S_1$ manifold (1300–1514 $cm^{-1}$), vibrational coupling is appreciable; systems of coupled levels involving 2 to approximately 10 levels characterize the regime. Upon pulsed excitation to this regime, an optically active zero-order state is prepared. IVR then occurs, but is restricted in the sense that essentially full recurrences in the vibrational-energy distribution occur. Finally, at high energies (e.g., $E_{vib} = 1792\ cm^{-1}$) the extent of vibrational coupling is such that dissipative IVR occurs. That is, the optically prepared zero-order state (nearly) irreversibly loses its vibrational energy to other zero-order vibrational states with which it is coupled.

Third, one now has an idea of the relevant time scales for IVR in anthracene. In particular, in the restricted IVR regime, where the IVR time scale corresponds to the inverses of the frequencies of beat components, it is clear that significant vibrational-energy flow occurs on time scales at least as short as several hundreds of picoseconds (beat frequencies are on the order of 5 GHz). In the dissipative regime, the pertinent time scale for IVR is given by that of the initial, approximately exponential dephasing of the optically prepared state. The one data point we have obtained corresponding to dissipative IVR in anthracene gives $\tau_{IVR} \sim 20$ psec.

Fourth, it is of interest to correlate changes in the nature of IVR in anthracene with changes in the vibrational density of states ($\rho_{vib}$) in the $S_1$ manifold. We have calculated[42] $\rho_{vib}$ at various energies using a direct count procedure and the calculated frequencies of Refs. 64. The results are shown in Fig. 21, with energies and time scales pertinent to IVR in anthracene noted also. (In considering our calculated values for $\rho_{vib}$, one must note that we have based the calculations on $S_0$ vibrational frequencies. Given that $S_1$ frequencies are most likely less that $S_0$ ones on average, the curve in Fig. 21 should be construed as a lower limit to the "real" curve.) In the region of no IVR, $\rho_{vib}$ is less than ~10 per $cm^{-1}$. In the region of restricted IVR, $\rho_{vib}$ goes from ~25

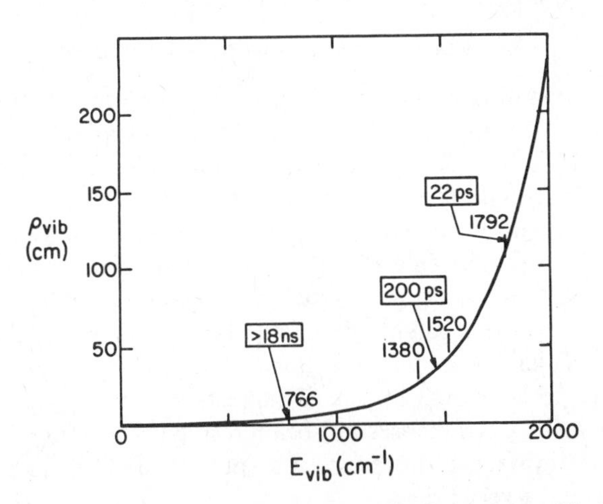


**Figure 21.** Vibrational density of states ($\rho_{vib}$) versus vibrational energy ($E_{vib}$) in anthracene calculated using a direct count method and the frequencies of Refs. 64. Given in the figure are characteristic times for IVR in the absent, restricted, and dissipative regimes.

to 40 per $cm^{-1}$. And, at $E_{vib} = 1800\ cm^{-1}$, $\rho_{vib} \simeq 120$ per $cm^{-1}$. Consideration of these values for $\rho_{vib}$, in conjunction with the extent of IVR at these various energies, allows one to make an important point: vibrational coupling is selective in the anthracene molecule. For instance, at $E_{vib} = 1380\ cm^{-1}$ there are three strongly coupled vibrational states within an 8.4 GHz energy interval. But if there were mixing between all states within this interval, then, since $\rho_{vib}(1380) \simeq 25\ cm^{-1}$, one would expect at least twice as many coupled levels. (Moreover, remember that our $\rho_{vib}$ values are lower limits.) A similar point may be made for the systems of coupled vibrational levels at $E_{vib} = 1420$ and 1792 $cm^{-1}$; the number of coupled levels is less than would be expected from considerations of $\rho_{vib}$. Of course, this selectivity in coupling is perfectly plausible given that anharmonic coupling is dominant in anthracene. For vibrational states to be coupled by anharmonic interactions, they must have the same vibrational symmetry. In $D_{2h}$ anthracene this means that any given vibrational level can couple anharmonically with only about one-eighth of the total number of other states in the molecule.[65]

Finally, from picosecond-beam results, one now has an idea of the magnitudes of the vibrational coupling matrix elements that significantly affect IVR. The restricted IVR Hamiltonian matrices [Eqs. (3.34) and (3.35)] show that coupling matrix elements from 0.3 to 4.2 GHz are important in IVR. In the dissipative $E_{vib} = 1792\ cm^{-1}$ case, the variance in coupling matrix elements, $\bar{V}^2$, can be calculated by using the expression[54] $1/\tau_{IVR} = 2\pi\bar{V}^2\rho$ and by assuming (because of symmetry selectivity) that $\rho = \rho_{vib}/8$ (or $\rho = 15$ per $cm^{-1}$). This gives $(\bar{V}^2)^{1/2} = 3.8$ GHz. Clearly, coupling matrix elements on the order of several GHz heavily influence IVR in anthracene.

### 2. *9-$d_1$-Anthracene and $d_{10}$-Anthracene*

As we have previously stated, the usefulness of obtaining picosecond-beam results relating to IVR in deuterated anthracenes is twofold. First, such results can provide an indication of how general vibrational coherence effects are. That is, if no such effects were to be observed in species so very similar to anthracene, then anthracene might justifiably be regarded as an oddity with respect to IVR. Second, while all the isotopic anthracenes might be expected to have very similar vibrational force fields, there are significant differences between the species that might affect IVR. The effects on IVR of such differences, in particular, differences in $\rho_{vib}$ versus $E_{vib}$ curves and possible differences in point group symmetry, are of interest.

As of this writing, our picosecond-beam results[44] on vibrational coherence in deuterated anthracenes have yet to be published. The treatment of these results here will be very brief.

*a. The Presence of Vibrational Coherence Effects.* One point that is clear from picosecond-jet studies of IVR in 9-$d_1$-anthracene and $d_{10}$-anthracene is

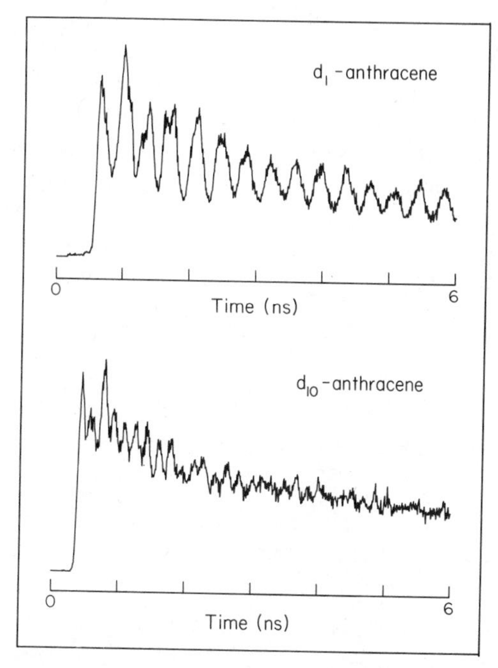


**Figure 22.** Time- and frequency-resolved fluorescence of jet-cooled $d_1$-anthracene (top) and $d_{10}$-anthracene (bottom). The $d_1$ decay corresponds to excitation to $S_1 + 1411\ \text{cm}^{-1}$ and detection of the 388 $\text{cm}^{-1}$ band in the fluorescence spectrum. The $d_{10}$ decay corresponds to excitation to $S_1 + 1452\ \text{cm}^{-1}$ and detection of the 372 $\text{cm}^{-1}$ band in the spectrum. Quantum beats are clearly present in both decays.

that the molecules exhibit vibrational coherence effects. As examples, Fig. 22 shows two decays, one corresponding to 9-$d_1$-anthracene excited to its $S_1$ + 1411 $\text{cm}^{-1}$ level with the $\bar{\nu}_d = 388\ \text{cm}^{-1}$ fluorescence band being detected, and the other corresponding to the $\bar{\nu}_d = 372\ \text{cm}^{-1}$ band in the fluorescence spectrum of the $S_1 + 1452\ \text{cm}^{-1}$ level of $d_{10}$-anthracene.

*b. IVR as a Function of* $E_{\text{vib}}$. An important second point about IVR in 9-$d_1$-anthracene, in particular, is that although the molecule exhibits behavior that is qualitatively similar to that exhibited by anthracene, IVR seems to "turn on" at lower energy in the deuterated molecule.[44] With regard to restricted IVR, the lowest excitation energy for which quantum beats have been observed in anthracene[44] is at $E_{\text{vib}} = 1290\ \text{cm}^{-1}$, there being a band at $E_{\text{vib}} = 1168\ \text{cm}^{-1}$ that does not give rise to beats. In contrast, excitation to $E_{\text{vib}} = 1173\ \text{cm}^{-1}$ in 9-$d_1$-anthracene does produce beat-modulated decays. A similar situation appears to apply to dissipative IVR. While the $E_{\text{vib}} = 1514$ $\text{cm}^{-1}$ level in anthracene exhibits decay behavior that places it in the restricted

IVR regime, the $E_{\text{vib}} = 1499\ \text{cm}^{-1}$ in the $d_1$ species gives rise to an $a$-type decay that appears to be a manifestation of IVR in a regime somewhere between restricted and dissipative. (That is, the $d_1$ decay has fairly well-modulated quantum beats, but also has a prominent early time component.)

The apparent difference in IVR behavior between anthracene and 9-$d_1$-anthracene might be reasonably attributed to the different point group symmetries of the molecules ($D_{2h}$ and $C_{2v}$, respectively). While at any given energy the overall density of vibrational levels in the two species differs by just a small multiplicative factor,[44] the number of levels per unit energy interval allowed by symmetry to couple anharmonically with each other is a factor of 2 greater in $C_{2v}$ molecules than in $D_{2h}$ species.[65] Therefore, the effective density of states available for vibrational coupling at a given $E_{\text{vib}}$ is roughly two times greater in the $d_1$ species relative to the $d_0$ species. This, of course, should shift the onset of IVR down in energy in the deuterated molecule.

### 3. *t-Stilbene*

The normalized fluorescence excitation spectrum of jet-cooled *t*-stilbene[66–68] (from Ref. 68) is shown in Fig. 23. One notes that the spectrum, even at lower

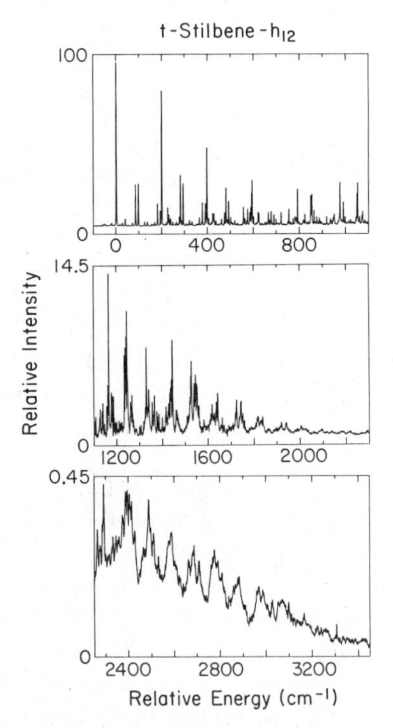


**Figure 23.** Fluorescence excitation spectra of *t*-stilbene-$h_{12}$ for the $S_1 \leftarrow S_0$ electronic transition. Spectra were recorded under expansion conditions of 50 psi He and 80°C sample temperature.

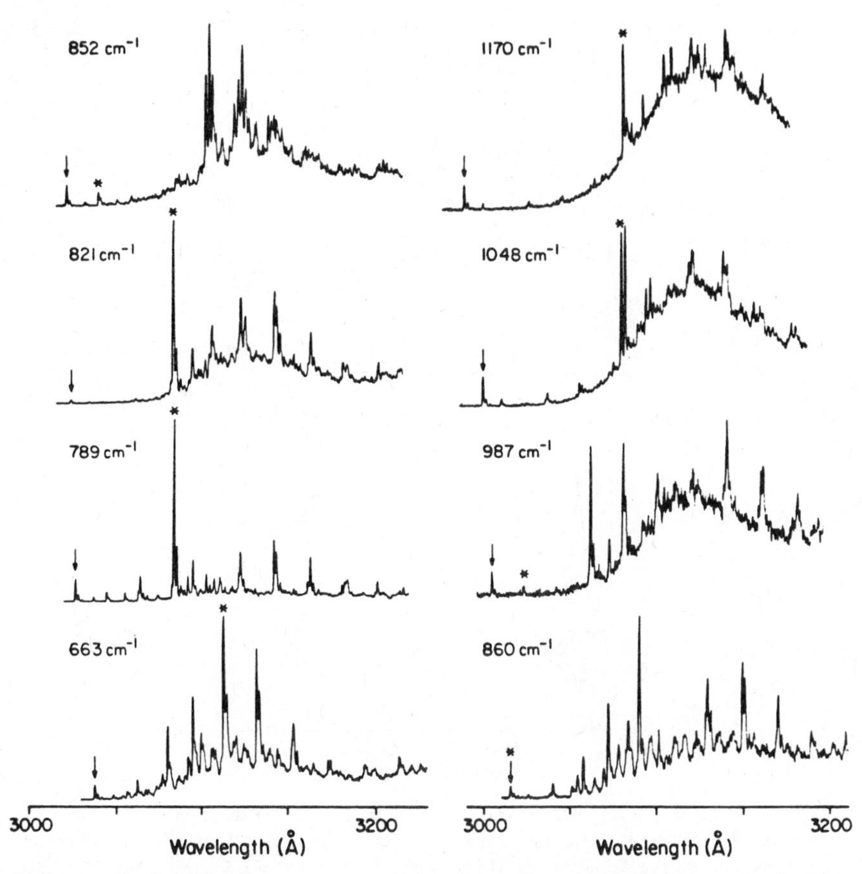


**Figure 24.** Dispersed fluorescence spectra of jet-cooled *t*-stilbene for excitation energies that have been observed to give rise to beat-modulated fluorescence decays. Excess energies in $S_1$ for the excitations are given in $cm^{-1}$ in the figure and the excitation wavelengths are marked with arrows in the spectra. The asterisks refer to detection wavelengths for the decays of Fig. 26. All of the spectra were obtained using similar experimental conditions: that is, $R = 0.5$ Å for most spectra, except $R = 0.65$ Å for the $E_{vib} = 821$ and 860 $cm^{-1}$ spectra and $R = 0.32$ Å for the $E_{vib} = 789$ $cm^{-1}$ spectrum.

energies, is considerably congested. This is in keeping with the low symmetry of the molecule and with the presence of a number of low-frequency vibrational modes.

Dispersed fluorescence spectra[68] (Figs. 24 and 25) obtained upon excitation of various bands in Fig. 23 reveal spectral trends similar to those of anthracene[42,61,63] and many other large molecules (e.g., Refs. 10–14, 23, and 24). Excitation of low-energy bands gives rise to spectra that are structured and that contain a significant fraction of vibrationally unrelaxed (*a*-type) fluores-

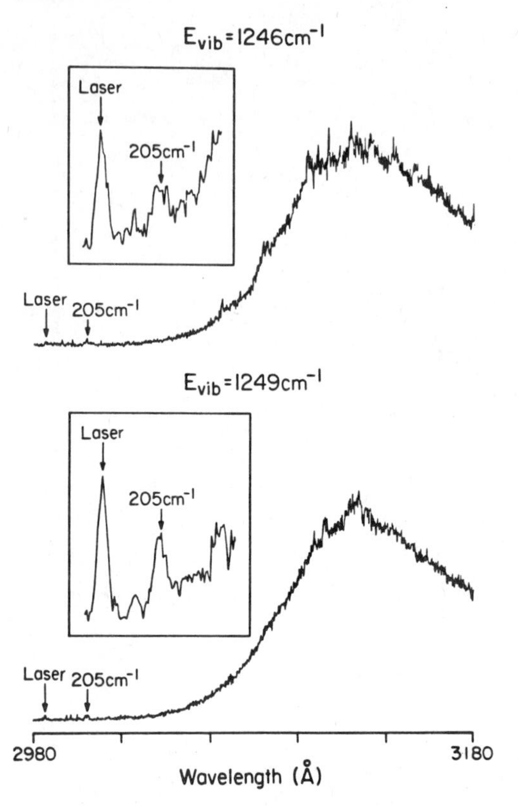


**Figure 25.** Dispersed fluorescence spectra of jet-cooled *t*-stilbene for excitation to $S_1$ + 1246 cm$^{-1}$ (top), and $S_1$ + 1249 cm$^{-1}$ (bottom). Insets are the blue portions of each spectrum. In all spectra the positions of the excitation wavelength and the 205 cm$^{-1}$ band are marked. Both the main spectra were obtained with $R = 0.64$ Å. For the insets $R = 3.2$ Å.

cence bands. As the excitation energy increases, spectra become more and more congested and begin to consist primarily of vibrationally relaxed (non-*a*-type) bands. These spectral trends provide an indication of the nature of IVR in the molecule. (For example, from Fig. 25 one would surmise that at $E_{vib} \simeq 1240$ cm$^{-1}$ in the $S_1$ manifold IVR is already very extensive.) Yet, we are most interested in the more detailed results available from time-resolved measurements of the decays of individual bands in the fluorescence spectra. To review these results,[45] we organize them as we organized the anthracene results, that is, according to the total vibrational energy to which they correspond.

*a. Low Vibrational Energy.* There many bands in the excitation spectrum of *t*-stilbene (Fig. 23) in the region from $S_1 + 0$ to 752 cm$^{-1}$. Measurements of time- and frequency-resolved fluorescence have been made for the most prominent of these bands in the region below $E_{vib} = 592\ cm^{-1}$ and for most of the bands with any intensity in the $E_{vib} = 592–752\ cm^{-1}$ region. This survey revealed only one excitation band giving rise to any beat-modulated fluorescence decays. (Naturally, this must be qualified with the fact that the observation of beats relies on sufficient temporal resolution.) This almost total absence of quantum beats implies the almost total absence of IVR in this low-vibrational-energy regime.

In contrast to other vibrational levels in its vicinity, excitation to the level at $E_{vib} = 663\ cm^{-1}$ gives rise to fluorescence bands, the decays of which are beat modulated. This is evident from Fig. 26 lower left, which shows the decay of the $\bar{\nu}_d = 800\ cm^{-1}$ band in the $E_{vib} = 663\ cm^{-1}$ spectrum. (The spectrum appears in Fig. 24 lower left.) Fourier analysis reveals that this decay is modulated by a 780-MHz beat component having a $+1$ phase. Significantly, at least one other band in the same spectrum ($\bar{\nu}_d = 585\ cm^{-1}$) is modulated at 780 MHz, but with the beat component having a $-1$ phase. Therefore, unlike levels of similar energy, it appears that the *t*-stilbene level undergoes restricted IVR.

*b. Intermediate Vibrational Energy.* In the region from $E_{vib} = 789–1170\ cm^{-1}$ a total of seven excitation energies have been shown to give rise to beat-modulated decays.[45] These energies correspond to $E_{vib} = 789$, 821, 852, 860, 987, 1048, and $1170\ cm^{-1}$. (The fluorescence spectra associated with these $S_1$ levels appear in Fig. 24.) They correspond to roughly one-half of the total number of prominent bands in the $E_{vib} = 780–1170\ cm^{-1}$ region of the excitation spectrum. Decays of fluorescence bands for other excitation bands in this region have been measured with 80 psec temporal resolution and have been found to exhibit no obvious beats.

Figure 26 shows representative decays for the seven beating excitations. It is important to point out that the degree of detection resolution was found to be critical to the observation of the beat modulations. This fact implies that decay behavior depends on detection wavelength, as one would expect for beats arising from IVR. Now, Fourier analysis of the decays of Fig. 26 reveals that all of the prominent beat frequencies in each of the decays occur with $+1$ phases. However, for three of the beating excitations ($E_{vib} = 789$, 860, and $987\ cm^{-1}$) the decays of some fluorescence bands have beat components with $-1$ phases. (An example of such phase-shift behavior is shown in Fig. 27, which shows the decays of the $\bar{\nu}_d = 610\ cm^{-1}$ and $\bar{\nu}_d = 700\ cm^{-1}$ bands that occur in the fluorescence spectrum arising from excitation to $E_{vib} = 789\ cm^{-1}$.) For the four beating excitation bands at $E_{vib} = 821$, 852, 1048, and $1170\ cm^{-1}$, the combination of weak fluorescence intensity and congested fluorescence spectra renders very difficult all but the most cursory experimental survey of decays versus detection wavelength. Thus, the fact that we have not observed phase-

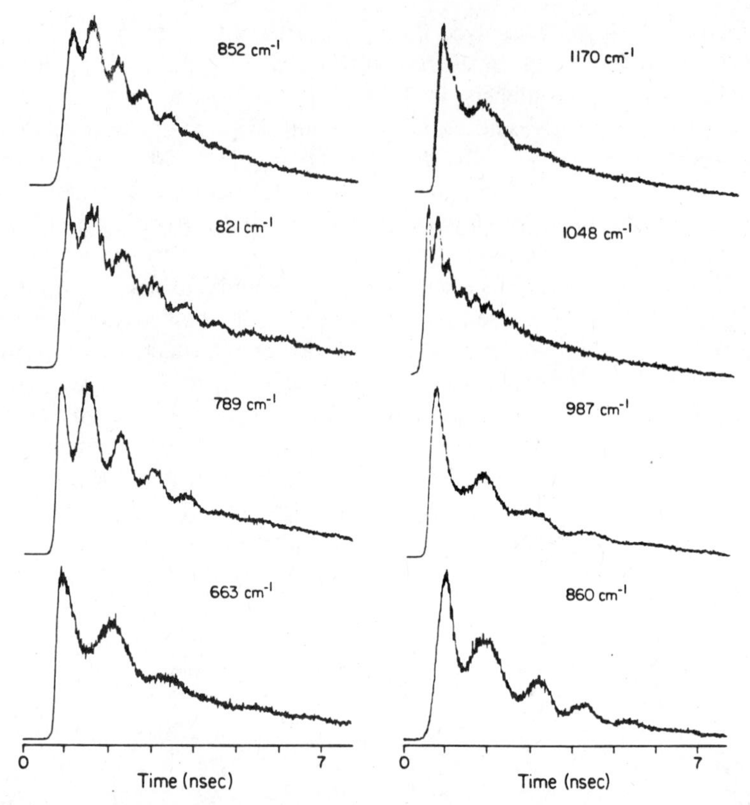


**Figure 26.** Quantum beat-modulated fluorescence decays observed for excitation of various bands (the excess $S_1$ vibrational energies are given in $\text{cm}^{-1}$ in the figure) of jet-cooled *t*-stilbene. The particular fluorescence band detected for each decay is given by an asterisk in the appropriate spectrum in Fig. 24. All decays were obtained with 80 psec temporal resolution except the ones corresponding to the $S_1 + 852$ and $860\ \text{cm}^{-1}$ excitations, which were measured with 300 psec resolution. $R$ for the decays was 1.6 Å, except the $S_1 + 821$ and $987\ \text{cm}^{-1}$ decays for which $R = 3.2$ and 16.0 Å, respectively.

shifted beats for these excitations does not mean that such beats are not present.

Given the beat-modulated decays and phase-shift behavior, it is apparent that the intermediate-energy regime we have defined for $S_1$ *t*-stilbene is one in which restricted IVR is prevalent. Nevertheless, it is also apparent that IVR in *t*-stilbene does not fit as "cleanly" into the theory of vibrational coherence as anthracene does. This point will be addressed in subsection *d*.

*c. High Vibrational Energy.* Excitation of *t*-stilbene to energies above $E_{\text{vib}} \sim 1230\ \text{cm}^{-1}$ in $S_1$ results in fluorescence spectra that bear the spectral charac-

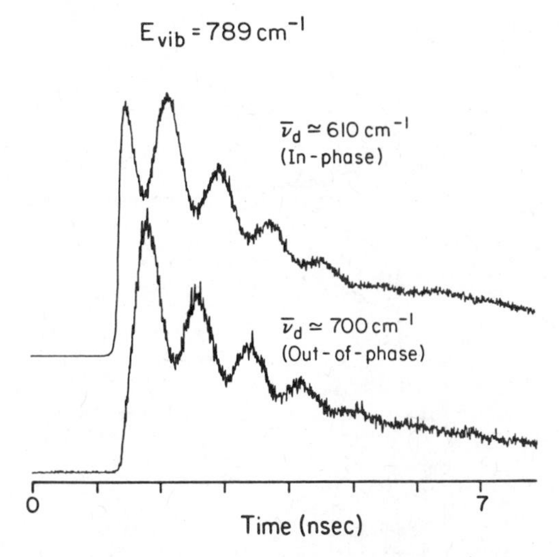


**Figure 27.** Fluorescence decays of two bands in the $E_{vib} = 789\ cm^{-1}$ spectrum of *t*-stilbene. The wavenumber shifts of the bands from the excitation energy are given in the figure. Note the clear phase shift of the prominent 1.3-GHz beat component in the lower relative to the upper decay. For the upper decay $R = 1.6$ Å, and for the lower one $R = 1.2$ Å.

teristics of levels that undergo dissipative IVR (examples are shown in Fig. 25). We have studied five excitation bands in this region, corresponding to $E_{vib} = 1237, 1241, 1246, 1249$, and $1332\ cm^{-1}$. In each of these five fluorescence spectra there is a weak band at $\bar{\nu}_d = 205\ cm^{-1}$ (see the insets in Fig. 25), a band that can be assigned (apart from any time-resolved results) with confidence as an *a*-type (vibrationally unrelaxed) band. Each of these spectra also has a broad, congested region near and to the red of the $S_1$ origin, a region very likely associated with vibrationally relaxed bands.

Figure 28 shows the decays that are observed when the $205\ cm^{-1}$ band in each of the five high-energy spectra is detected with 3 Å spectral resolution. Each decay, although being different from the others, clearly has the characteristics expected for an *a*-type decay arising from a system undergoing dissipative IVR. Double exponential fits of the decays (with convolution) give the lifetimes and fast/slow ($F/S$) intensity ratios shown in the figure. Of these values one notes first that the fast lifetime and the fast/slow ratios are of the same order of magnitude as analogous values derived from the dissipative decays of anthracene. A second point is that the long-component lifetime values are all consistent with the fluorescence lifetimes of the wavelength-unresolved fluorescence of the molecule at the pertinent excitation energies. Finally, one would note that one of the decays has a particularly long-lived early time component, the decay corresponding to $E_{vib} = 1249\ cm^{-1}$. This

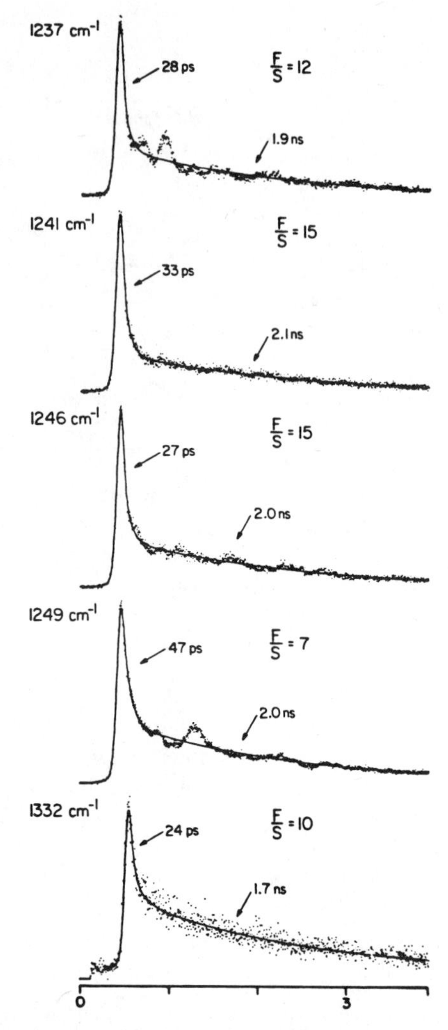


**Figure 28.** Fluorescence decays and double exponential fits corresponding to the 205 $cm^{-1}$ bands in the (from top to bottom) $E_{vib}$ = 1237, 1241, 1246, 1249, and 1332 $cm^{-1}$ spectra of jet-cooled *t*-stilbene. Given in the figure are the best-fit parameters for both the fast and slow lifetimes, and the ratio ($F/S$) of preexponential factors of fast versus slow fluorescence. All decays were obtained with $R = 3.2$ Å.

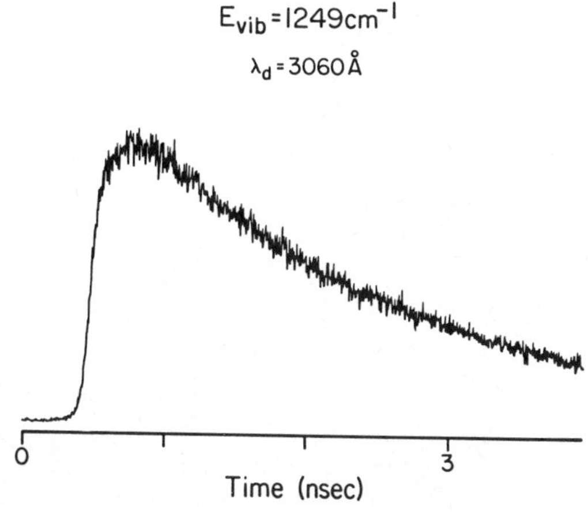


**Figure 29.** Measured fluorescence decay for detection in the blue part of the vibrationally relaxed region of the $E_{vib} = 1249\ cm^{-1}$ spectrum of *t*-stilbene. The detection wavelength is given in the figure. The decay has a rise time similar to the initial decay time (47 psec) of the 205 $cm^{-1}$ band in the same spectrum, and a long component having the same lifetime (2.0 nsec) as the long component of the 205 $cm^{-1}$ decay. The decay was measured with $R = 3.2$ Å.

point, which is treated more fully elsewhere, is a strong indication that there is some degree of vibrational mode selectivity in the coupling leading to dissipative IVR in the *t*-stilbene molecule.

While the double exponential form characterizes (approximately) the decays of the 205 $cm^{-1}$ bands in the high-energy spectra, decays in the vibrationally relaxed region of the spectra are different. These other decays, an example of which is shown in Fig. 29, are devoid of any early time spike, but have finite rise times and long-component lifetimes matching those of the long components of Fig. 28. This behavior is consistent with the decay of vibrationally relaxed fluorescence in the dissipative regime of IVR.

*d. Conclusions Concerning IVR in $S_1$ t-Stilbene.* From the results reviewed above, it is apparent that the general trend of IVR versus $E_{vib}$ in *t*-stilbene is similar to that in anthracene. At low energy ($\leq 750\ cm^{-1}$) the time-resolved data indicate an absence of IVR except for one vibrational level. At intermediate energies (789–1170 $cm^{-1}$) the presence of phase-shifted quantum beats points to the existence of restricted IVR. Finally, at energies greater that $\sim 1230\ cm^{-1}$, the decay behavior indicates that IVR is dissipative. In connection with this, however, one would note a point of difference between *t*-stilbene and anthracene. That is, the "onsets" of restricted and dissipative IVR in *t*-stilbene occur at significantly lower vibrational energies than in anthracene.

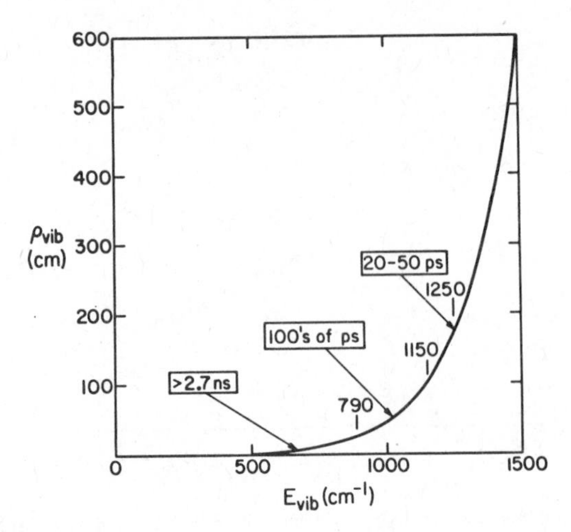


**Figure 30.** Vibrational density of states ($\rho_{vib}$) versus vibrational energy ($E_{vib}$) in $S_1$ *t*-stilbene, calculated as outlined in Ref. 45. In the figure the characteristic times for IVR in the absent, restricted, and dissipative regimes are given.

This difference is not surprising for two reasons. First, the greater number of vibrational modes, in particular low-frequency modes, in *t*-stilbene give rise to a $\rho_{vib}$ versus $E_{vib}$ curve[45] (Fig. 30) that is shifted to considerably lower energy than the anthracene curve (Fig. 21). Second, *t*-stilbene has lower point group symmetry than anthracene. As stated in conjunction with our discussion of 9-$d_1$-anthracene, this means that a greater fraction of the total number of states at a given energy can couple with one another. That is, even if $\rho_{vib}$ versus $E_{vib}$ were identical for *t*-stilbene and anthracene, the effective density of states available for coupling at a given $E_{vib}$ would still be higher for *t*-stilbene.

Besides the trend in the nature of IVR as a function of vibrational energy, the IVR time scales and vibrational coupling matrix elements manifested in *t*-stilbene decays are similar to those characterizing the process in anthracene. In the restricted regime recurrence times on the order of several hundreds of picoseconds seem to be the rule. In the dissipative regime, $\tau_{IVR}$ is on the order of tens of picoseconds. Regarding the vibrational coupling matrix elements manifested in the time-resolved results, the data for *t*-stilbene are not as amenable to quantitative analysis as those for anthracene. Nevertheless, in the restricted regime, the order of magnitude of the coupling matrix elements can be obtained from beat frequencies. These are similar to the anthracene values. In the dissipative regime, one can calculate the variance $\bar{V}^2$ in coupling matrix elements (as was done with anthracene) if one assumes $C_2$ or $C_i$

symmetry for the molecule and uses the $\rho_{vib}$ curve of Fig. 30 to obtain an effective $\rho = \rho_{vib}/2 = 80$ per $cm^{-1}$ for $E_{vib} = 1240\ cm^{-1}$. Using the $\tau_{IVR}$ values from Fig. 28 one finds $(\bar{V}^2)^{1/2} = 1.1–1.7$ GHz for the vibrational levels in the vicinity of $E_{vib} = 1240\ cm^{-1}$.

Finally, one must note that although there is much about the picosecond-jet results on *t*-stilbene that is similar to the anthracene results and that may be readily interpreted within the context of the theory reviewed in Section III B, there are two significant points of difference. One point is the fact that many prominent excitation bands in the energy regime corresponding to restricted IVR in *t*-stilbene are nonbeating excitations, while in the corresponding regime in anthracene, beating excitations are ubiquitous. The second point is that *t*-stilbene decays that do beat are less modulated than one might expect. Aside from "trivial" reasons for this behavior (i.e., reasons of insufficient temporal or spectral resolution[42]), there are a number of other possible causes. First, it may be that rotational levels play a large role in the vibrational coupling in *t*-stilbene. If Coriolis coupling occurs to a significant degree, a possibility neglected in Section III B, then many different beat frequencies will obtain in the decay of a band type, even if only two vibrational levels are coupled. It is likely that these many modulations would add in such a way as to render them undetectable. Or, even if Coriolis coupling is not important, differences in rotational constants between anharmonically coupled states can be important in washing out beats.[43] (This concept is addressed in Section III D.) Perhaps coupling between vibrational levels having significantly different rotational constants is prevalent in *t*-stilbene.

Second, owing to the reduced symmetry of *t*-stilbene, one expects a large fraction (relative to anthracene) of its vibrational modes to be optically active. Hence, the probability of encountering two or more coupled optically active levels is significant. The situation of coupled optically active levels (absorption doorway states) runs counter to the basic assumption of the theory of Section III B and has been shown[30b] to give rise to "undermodulated" beating decays.

Finally, in *t*-stilbene there may be large off-diagonal anharmonic coupling matrix elements that can couple widely separated zero-order states. This can lead to less than ideal modulation depths.

### 4. *Alkylanilines: "Ring and Tail" Systems*

Alkylanilines are particularly attractive to study by the picosecond-beam technique for a number of reasons. First, the stimulating (time-integrated) experiments of Smalley and co-workers[10e] have revealed spectral broadenings for some of the species at particular energies, which broadenings, when analyzed by means of a kinetic model similar to the one discussed in Section II A, led them and others to the conclusion that very rapid IVR occurs in the molecules at these energies. In order to unequivocally establish that this is

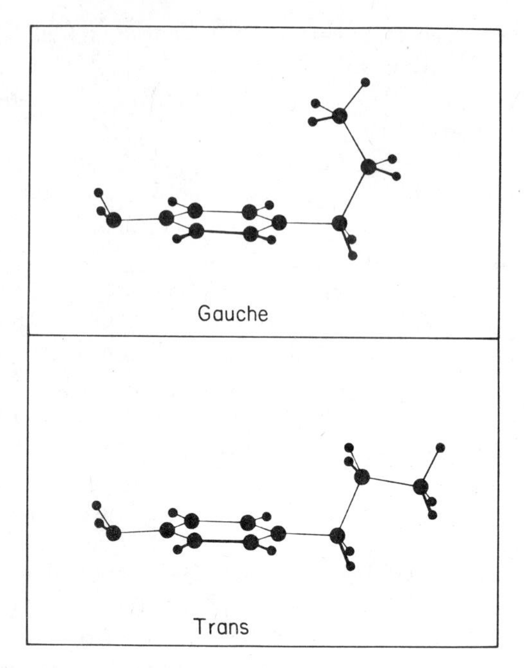


**Figure 31.** Gauche and trans conformations of *p*-propyl aniline. These illustrations are the result of structural optimization using a molecular mechanics computer program. Atoms were drawn (not to scale) in order to improve spatial perspective.

true, picosecond-resolved work is necessary. Second, the series offers an opportunity to examine the role of the chain length (and density of states) on IVR. Third, because some of the chain molecule species have different conformers (e.g., extended and coiled, Fig. 31), one can examine the dynamics for these different species.

*a. Time and Frequency-Resolved Spectra.* We have investigated[46] the spectral and temporal character of fluorescence arising upon excitation of (n) *para*-alkylanilines (from aniline to butylaniline) to single vibronic levels (SVL's) in their $^1B_2$ electronic states. Existing spectral data on aniline is reviewed by Chernoff and Rice[69] and serves as a basis for vibrational assignments by Powers et al.[10e] of the excitation spectra of the jet-cooled alkylanilines. The excitations of interest in the present work are to those SVL's assigned in Ref. 10*e* as $6a^1$, $I^2$, and $1^1$. (For propyl- and butylaniline these vibrational assignments include specification of molecular conformation.) Figure 32 shows the transients obtained for these excitations of propylaniline and the striking differences observed for its two conformers. For two of the excitations ($\nu_1$

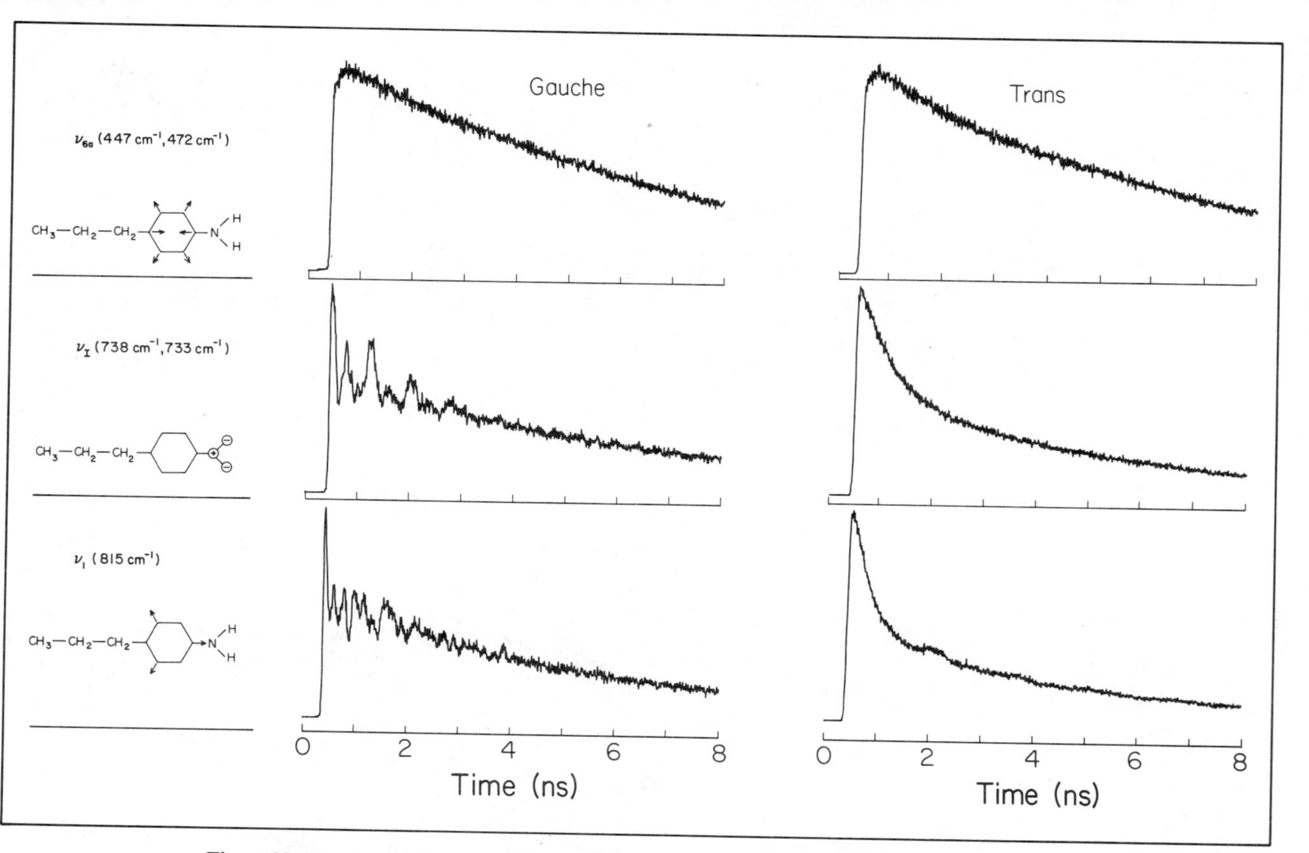


**Figure 32.** Temporal behavior versus excess energy in propylaniline. The vibrational modes excited are represented at left. $R = 1.2$ Å for the $\nu_I$ decay of the gauche conformer, $R = 0.8$ Å for $\nu_1$ trans, and 1.6 Å for all others. Fits of these data were performed using the methods described in Refs. 39 and 42.

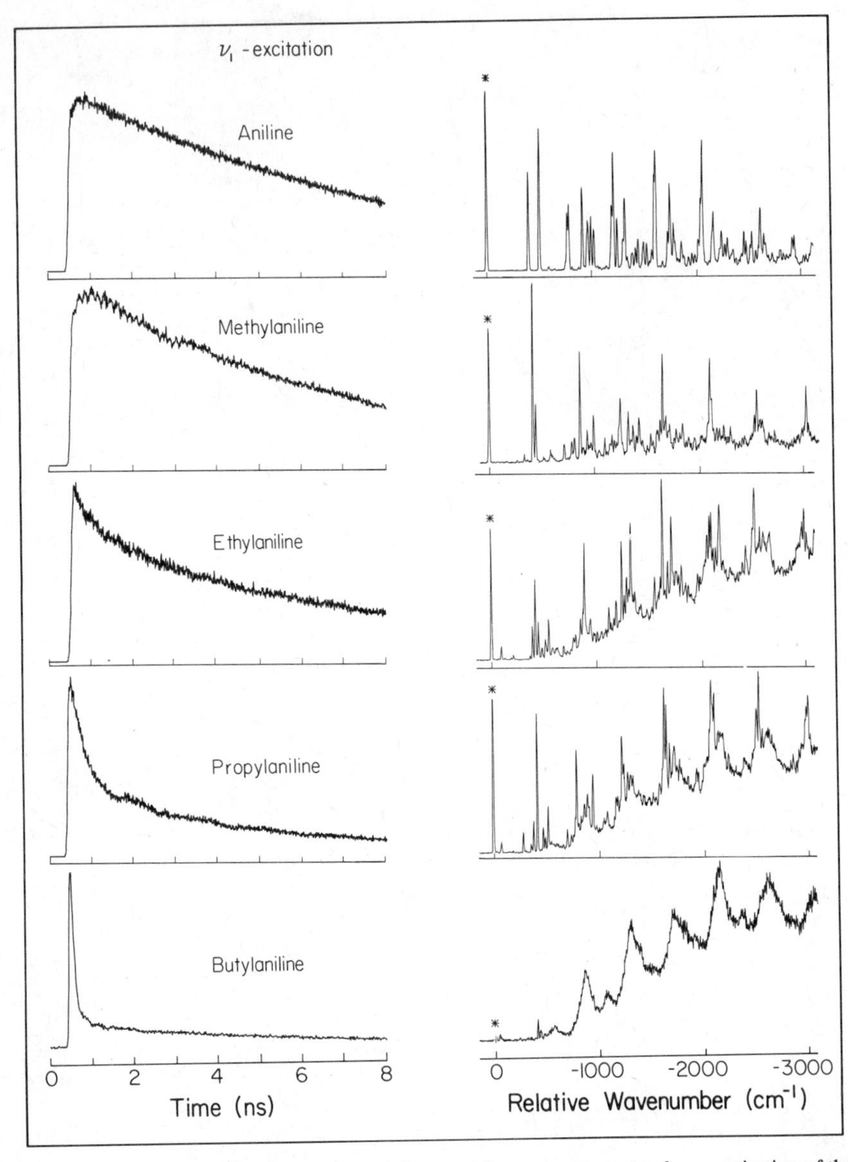


**Figure 33.** Time-resolved fluorescence and dispersed fluorescence spectra for $\nu_1$ excitation of the alkylanilines. The excitation frequencies of the spectra have been marked by an asterisk and aligned for comparison. The detection wavelengths were different from the excitation wavelengths for all decays. $R$ for the decays were 2.4 Å for aniline, 3.2 Å for butylaniline, and 1.6 Å for all others.

and $\nu_I$) the para position is not significantly involved in the initially excited vibrational motion. This fact is evidenced by the constancy of $\nu_1$ and $\nu_I$ frequencies throughout the series.

Figure 33 shows the dependence of the spectra (time- and frequency-resolved) on chain length ($\nu_1$ excitation). The dispersed fluorescence spectra obtained for the above excitations confirm the trend toward greater spectral broadening with increasing chain length and with increasing excess energy, the trends found in the earlier study of these molecules and in a similar study of alkylbenzenes.[10a–e]

In addition to the factors of chain length and energy, conformation was found to play a significant role in determining the degree of spectral broadening of the dispersed fluorescence. This striking effect, which was not reported in the earlier studies,[10e] is shown in Fig. 34 for the propylaniline $\nu_1$ excitation. The effect exists for the other excitations studied, too, including those of butylaniline. The sharp features evident in the trans spectrum are dramatically reduced in that of the gauche conformer. By contrast, analogous comparisons in the alkylbenzene series reveal a qualitatively opposite effect.[10a–d] That is,

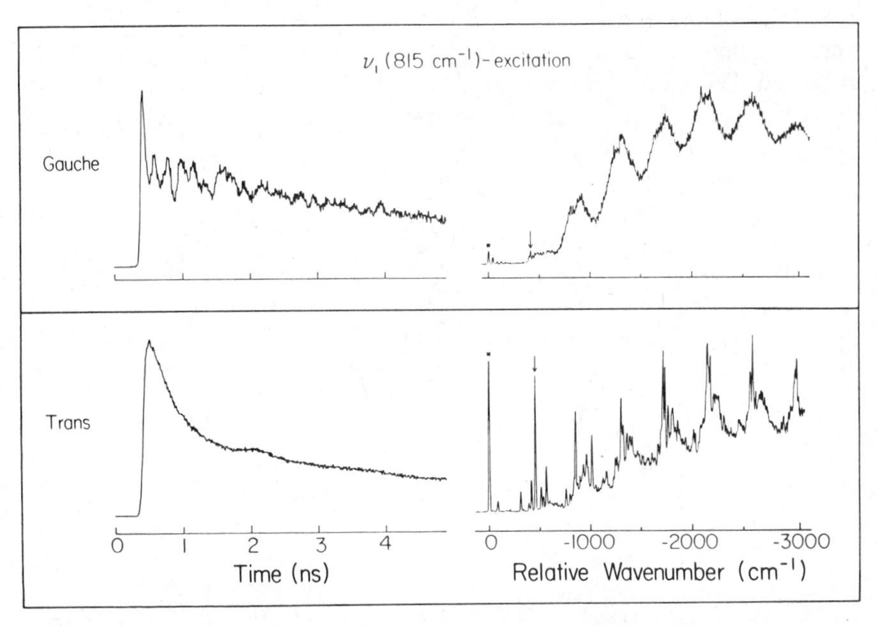


**Figure 34.** Time-resolved fluorescence and dispersed fluorescence spectra for $\nu_1$ excitation of the gauche and trans conformers of propylaniline. An asterisk marks the excitation frequency in each spectrum, while an arrow indicates the detection frequency associated with the decay at left. For the trans species, complementary build-ups at different detection wavelengths were observed. For the gauche species, the Fourier transform of the decay is very rich.

of the dispersed fluorescence spectra obtained upon $6b_0^1$ excitation of trans and gauche conformers of butyl- and pentylbenzene, *those of trans species show broader fluorescence.* (It should be noted, however, that these spectra are, on average, much sharper than the propyl- and butylaniline spectra of our study.)

For each of the dispersed fluorescence spectra previously described, we obtained the time-resolved spectra. A selection of our results is displayed in Figs. 32–34. The trans decays show again, as did the spectra, that the increase of either excess energy (Fig. 32) or chain length (Fig. 33) gives rise to similar effects. One sees a progression from single exponential to quasibiexponential behavior, with increasingly short initial dephasing times (down to ~50 psec in butylaniline). However, the gauche conformer of propylaniline again shows very different behavior from its trans sibling. The well-modulated quantum beat patterns seen in Figs. 32 and 34 show many frequencies ranging from a fraction of a GHz to the limit of the detection system resolution (~12 GHz). Both in-phase and out-of-phase beats[30,40] were observed, with the phase of a particular frequency component depending on detection wavelength.

To further explore the temporal behavior of trans-propyl aniline $\nu_1$ emission, a time-gating experiment was performed. Dispersed fluorescence spectra were obtained with acquisition of signal limited to the time range indicated in Fig. 35. The spectral range shown corresponds to ~775–~1850 $cm^{-1}$ on the energy scale of Fig. 34. The comparison of short- and long-time emission

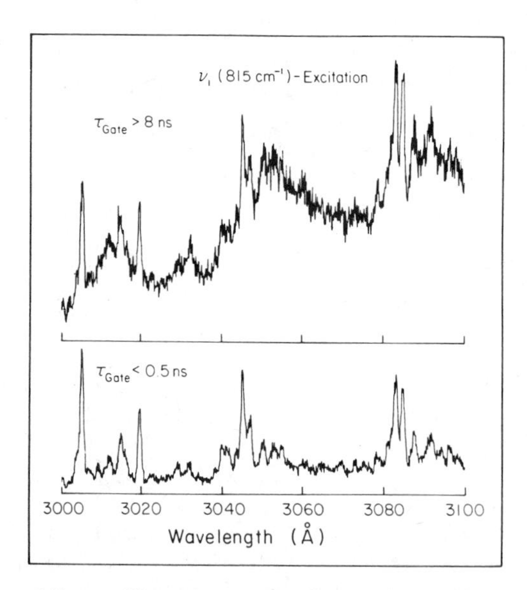


**Figure 35.** Time-gated dispersed fluorescence spectra for $\nu_1$ excitation of trans-propylaniline. The time ranges are as indicated. The top spectrum is the result of 12 scans, the bottom of 20 scans.

demonstrates clearly that the continuum fluorescence underlying the sharp spectral features arises on a time scale of nanoseconds after the laser excitation.

*b. p-Propylaniline and Mode Specificity.* The SVL fluorescence of the trans species excited to its $\nu_1$ mode (815 cm$^{-1}$) is quite sharp. For the corresponding $\nu_1$ excitation in the gauche species, the spectrum (Fig. 34) is, in comparison, very broad. The two molecules, having the same number of atoms and being excited to the same total energy, show different behavior. Based on this spectral behavior one would expect the trans species to show manifestations of restricted IVR (quantum beats) and the gauche species to exhibit IVR in the statistical limit (a single exponential decay with rate constant $k_{IVR} + k_f$). In contrast to these expectations, the results in Fig. 34 show that the time-resolved data corresponding to the gauche species exhibit quantum beats, while the trans decay shows a quasibiexponential decay, with the fast component having a time constant of typically 300 psec. Clearly, the correspondence between spectral broadening and time-resolved rates is not straightforward. Thus, the observed quantum beats for the gauche species indicate that energy does not dissipate from the *ring* to the *tail.* Moreover, using the kinetic model of Section II A and the integrated intensities from the spectra of Fig. 34 (gauche) one infers that $k_{IVR}^{-1} \sim 10$ psec, whereas the time-resolved data show recurrences occurring on a much different time scale.

The possibility of mode-specific dynamics was also investigated in the trans and gauche forms of propylaniline. The results in Fig. 32 indicate that there is no IVR (on our time scale) when the $\nu_{6a}$ mode is excited, whereas for both $\nu_1$ and $\nu_I$, "restricted" IVR is observed (gauche).

*c. Time-Gating and Chain-Length Effects.* As discussed elsewhere,[33] to distinguish between coherent and incoherent IVR dynamics, one must see the decay of the initial state $|a^*b\rangle$ and the complementary build-up of the $|ab^*\rangle$ state. With this in mind, we have gated a portion of the trans-propylaniline dispersed fluorescence for times less than 500 psec and greater than 8 nsec. As shown in Fig. 35, the sharp spectrum corresponds to the early time dynamics and the underlying broad spectrum corresponds to the long-time emission.

The chain length effect is very pronounced in the data shown in Fig. 33 for $\nu_1$ excitation. In aniline there is no IVR. In methylaniline, there are recurrences, but weak in modulation. In the ethyl species, there is the hint of the biexponential behavior that is fully present in the propyl decay. In butylaniline, the fast component dominates. Clearly, as the chain length increases, the dynamics of IVR changes.

*d. Model for IVR in Alkylanilines.* The key question regarding our work on alkylanilines, aside from the influence of spectral congestion on time-integrated rate measurements, is this: Does energy flow occur from the *pure* ring

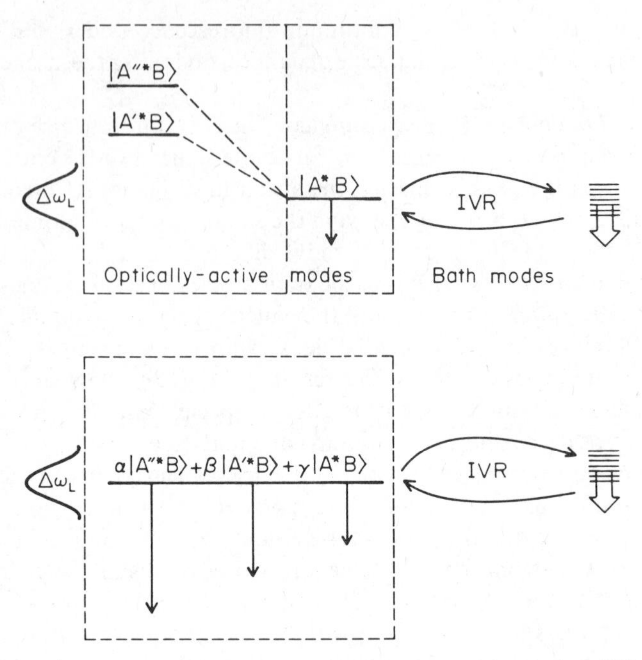


**Figure 36.** General level schematic pertaining to IVR and portraying the possibility of mixing between optically active states outside the laser bandwidth ($\Delta\omega_L$).

mode to the bath modes? The time-resolved data give us the dynamics of IVR from the states excited by the coherent bandwidth of the laser. We can assign this state by analogy to the parent aniline mode ($\nu_1$, $\nu_{6a}$, or $\nu_I$), but the degree of purity (i.e., mixing with other optically active modes) cannot be determined without further spectroscopic analysis.

The data on trans- and gauche-propylaniline are particularly revealing with respect to this. We consider the scheme shown in Fig. 36. The optically active $|a^*b\rangle$ state is coupled to bath modes by IVR. In addition, there are nearby $|a'^*b\rangle$-type states (optically active) that can couple to the $|a^*b\rangle$ state. The laser coherence width, $\Delta\omega_L$, excites the molecule in the energy region of the $|a^*b\rangle$ state. Even though the $|a'^*b\rangle$ states are not directly excited because of this energy restriction, they come into the picture because $|a^*b\rangle$ is contaminated via its long-range coupling with them. We note that this type of mixing leaves the absorption spectrum sharp and does not contribute to dynamical IVR. Only if $\Delta\omega_L$ spanned the energy range of the coupled states would they would be involved in the dynamics.

For propylaniline, it appears that the observed broadening in the gauche spectra can be accounted for if one assumes the prevalence of these off-resonance effects. Since the chain is "bent over" the ring, the molecule may have lower effective symmetry than the trans species. This can facilitate coupling between modes. Yet, the time-resolved data, which exhibit quantum beats, indicate that there is coupling between the mixed initial state and a limited number of bath modes. For trans-propylaniline, the off-resonance mixing should be less, leading to less spectral congestion. The density of bath modes, on the other hand, should be larger than in the gauche species because of the floppiness of the extended alkyl chain. The time-resolved spectra imply that IVR to the bath occurs with a time constant (ca. 300 psec) that reflects very weak coupling. This weak coupling could be due either to anharmonic interactions or Coriolis coupling.

Finally, a note about the linewidth/IVR relationship. The time scale for IVR in, for example, butylaniline (fastest decay measured in the series) translates to a linewidth of 0.2 $cm^{-1}$. The apparent absorption linewidth is much larger than this. Clearly, it would be a mistake to use the observed linewidth to obtain parameters associated with the dynamics.

*e. Conclusions.* The time- and frequency-resolved results on alkylanilines illustrate several points. First, spectral broadening in fluorescence does not give a direct view of IVR dynamics. Second, in cases where the spectra are completely congested, we observe behavior associated with restricted IVR–energy does not dissipate from ring to tail. Third, the value of $k_{IVR}$ obtained from spectral information by kinetic analysis is not consistent with the existence of these recurrences.

### D. Manifestations of Molecular Rotations

Throughout the preceding review of picosecond-beam work, we have consciously chosen to avoid detailed consideration of the influence of molecular rotations on our results. Thus, in Section III B we have considered vibrational states $|\gamma\rangle$ rather than ro-vibrational states $|\gamma; JKM\rangle$. Moreover, in conjunction with this, we have neglected the polarization properties of the exciting light and of the fluorescence, there being no point in accounting for such properties (in deriving fluorescence decay forms) when molecular rotations are neglected.

In large part, our seemingly drastic approximations related to rotations can be justified by the close agreement between theory and experiment. On the other hand, we have reported some results[39] that can only be interpreted by taking account of molecular rotations. Also, other groups (e.g., Ref. 19) have obtained results that point to significant rotational level effects on IVR. In view of such experimental results and in the interest of completeness, one wants to understand how rotations may be manifested in the results of picosecond-beam studies of IVR.

In our work, we have obtained theoretical and experimental results pertaining to two aspects of this problem. The first concerns rotational level effects on *anharmonic coupling* and on the quantum beat-modulated decays that arise from such coupled systems.[43] In some sense this work represents a "fine-tuning" of our theory of vibrational quantum beats[30b] so as to put that theory on firmer ground and to explain certain experimental results. The second aspect is more fundamental. Briefly, in exciting an isolated molecule with a picosecond pulse, one would expect to prepare a *rotational superposition state* composed of coherently prepared rotational eigenstates. Such a superposition state would be expected to give rise to quantum coherence effects in addition to and distinct from those arising from vibrational coupling. Our concern[47–50] has been to characterize such coherence effects and to investigate their influence in decays in which vibrational coherence effects are also manifest.

In this section, we review the results that we have obtained pertaining to the two aforementioned influences of rotational level structure in time-resolved dynamics experiments.

### 1. *Effects of Rotations on Anharmonic Coupling: Mismatches of Rotational Constants*

Anharmonic or Fermi resonance coupling between the rotational manifolds of $N$ zero-order vibrational levels ($|a\rangle, |b\rangle, |c\rangle$, etc.) has three notable characteristics with regard to the influence of rotational level structure on the coupling. First, symmetry restrictions and angular momentum selection rules limit the coupling to ro-vibrational levels having the same rotational quantum numbers.[58] Therefore, the coupling of the vibrational levels is described by an infinite set of distinct $N \times N$ matrices, one matrix in the set for each possible set of rotational quantum numbers. Second, coupling matrix elements do not depend on rotational level.[58] Therefore, the individual off-diagonal elements of the $N \times N$ Hamiltonian matrices do not vary with rotational quantum number. Third, as a consequence of the previous two characteristics, the only way in which anharmonic coupling can vary with rotational level is through variations in the differences between *diagonal* elements of the $N \times N$ Hamiltonian matrices. Such variations will occur if two or more of the zero-order vibrational levels have different rotational constants, that is, if there are "mismatches" of rotational constants between the levels.

The effect of rotational constant mismatches on vibrational quantum beats[43] is the subject of this subsection. We first review theoretical results that show that the qualitative effect of such mismatches is to increase the apparent damping rate of quantum beat envelopes relative to the decay rate of the unmodulated portion of a decay and that such beat damping rates increase with increasing rotational temperature. We then review results that show that such effects on beat damping are consistent with experiment.

*a. Theoretical Simulations.* Given $N$ anharmonically coupled vibrational levels *with* rotational manifolds, and assuming that, aside from the presence of the rotational levels, all else is the same as the situation treated in Section III B (i.e., there is a single absorption doorway vibrational state $|a\rangle$ and a single emission doorway state $|\gamma\rangle$), then the fluorescence of a $\gamma$-type band can be expressed as

$$I_\gamma(t) = \sum_{J=0}^{\infty} \sum_{K_a=-J}^{J} W(J, K_a, T) I_\gamma(J, K_a, t). \tag{3.36}$$

In this equation we have chosen to consider a near prolate symmetric top (like anthracene). The symbols $J$ and $K_a$ have their usual meanings as the rotational quantum numbers of such a molecule and refer in the equation to rotational levels in the manifold of the $|a\rangle$ zero-order vibrational level. As for the other factors in Eq. (3.36), $T$ is the rotational temperature of the sample, $W(J, K_a, T)$ is a weighting factor for each ro-vibrational level $|a, J, K_a\rangle$, and $I_\gamma(J, K_a, t)$ is the $\gamma$-type fluorescence decay which arises from the coupling of $|a, J, K_a\rangle$ with the same rotational levels of the other zero-order vibrational states. $I_\gamma(J, K_a, t)$ is just given by $I_\gamma(t)$ of Eq. (3.12). [Note that in deriving Eq. (3.36) we have neglected the possibility of any coherence effects arising from the coherent preparation of rotational levels within the same vibrational state. This possibility is the subject of Section III D 2.]

To calculate $I_\gamma(t)$ from Eq. (3.36) it is very convenient to make several approximations. The first has been mentioned already; the molecule is taken to be an approximate symmetric top. Thus, the rotational energy of rotational level $|JK_a\rangle$ in the manifold of the zero-order vibrational state $|\gamma\rangle$ is[58]

$$E_\gamma(J, K_a) = \tfrac{1}{2}(B_\gamma + C_\gamma)J(J+1) + [A_\gamma - \tfrac{1}{2}(B_\gamma + C_\gamma)]K_a^2, \tag{3.37}$$

where $A_\gamma$, $B_\gamma$, and $C_\gamma$ are the rotational constants of the molecule for the $|\gamma\rangle$ vibrational state. A second useful approximation is to assume that the thermal distribution of ground-state rotational levels that exists prior to the excitation of the sample is projected into the excited state upon excitation. One then has the following expression for the weighting factor in Eq. (3.36):

$$W(J, K_a, T) = g_N(J, K_a)(2J+1)\exp(-E(J, K_a)/k_B T), \tag{3.38}$$

where $g_N(J, K_a)$ is the nuclear spin statistical weight of level $|JK_a\rangle$ and $k_B$ is Boltzmann's constant. A third approximation involves limiting the sum over $J$ in Eq. (3.36). The calculated results we review here were obtained for $J \leq 30$. No qualitative changes in these results were found when the $J$ range was increased beyond $J_{max} = 60$.

Knowing both $W(J, K_a, T)$ and $I_\gamma(J, K_a, t)$ for all pertinent rotational levels,

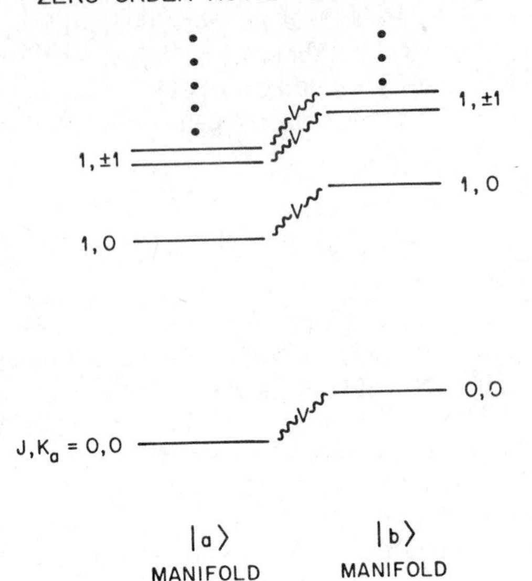


**Figure 37.** Schematic diagram of the anharmonic coupling between the rotational levels of two zero-order vibrational states—$|a\rangle$ (left) and $|b\rangle$ (right). Only those rotational levels having the same rotational quantum numbers ($J, K_a$) are coupled. Moreover, the coupling matrix element $V$ is constant for each pair of coupled ro-vibrational states. Coupling can vary up the rotational manifold only through differences in energy spacings between coupled states. (Such differences are not shown in the figure.) Note that there has been no attempt to draw the spacings between levels to scale.

$I_\gamma(t)$ can be calculated. Since we are interested in comparing simulated decays with observed ones, all the simulated decays presented were convoluted with a temporal response function characteristic of that of our experimental apparatus.[42] In addition, all simulations pertain to the anthracene molecule. Thus, all rotational constants used, $\frac{1}{2}(B + C)$ and $A - \frac{1}{2}(B + C)$, were near 0.415 and 1.74 GHz, respectively. While the rotational levels of anthracene fall into four groups having different $g_N(J, K_a)$ values, these differences are small enough that they were neglected in the simulations.

*Two-level vibrational coupling:* The first case we consider is the simple and illustrative case of two coupled vibrational levels, $|a\rangle$ and $|b\rangle$. The situation is represented schematically in Fig. 37. Let us consider it first in a somewhat qualitative light. If $|a\rangle$ and $|b\rangle$ have different rotational constants, then the zero-order energy differences between the pairs of coupled ro-vibrational levels up the two rotational manifolds (Fig. 37) will depend on $J$, $K_a$ and will be of the form

$$E_{ab}(J, K_a) = E_{ab}(0,0) + \Delta_{ab}J(J+1) + \Delta'_{ab}K_a^2, \tag{3.39}$$

where $E_{ab}(0,0)$ is the energy difference between ro-vibrational states $|a;0,0\rangle$ and $|b;0,0\rangle$, and where $\Delta_{ab} \equiv \frac{1}{2}[(B_a + C_a) - (B_b + C_b)]$ and $\Delta'_{ab} \equiv (A_a - A_b - \Delta_{ab})$ represent the mismatch in rotational constants between the two levels. If zero-order energy differences between pairs of coupled levels vary with rotational level, then the beat frequencies (and the $a$-type beat modulation depths) that arise from such coupled levels will also vary since

$$\frac{\omega(J, K_a)}{2\pi} = \{[E_{ab}(J, K_a)]^2 + 4V_{ab}^2\}^{1/2}, \tag{3.40}$$

where $\omega(J, K_a)$ is the rotational level-dependent beat frequency and $V_{ab}$ is the (rotational level-independent) anharmonic coupling matrix element between $|a\rangle$ and $|b\rangle$. Now, if the sample of interest is a sample of large molecules at finite temperature, one expects a large number of $J$, $K_a$ values to be relevant (i.e., to have significant population). This, in turn, means that a large number of different beat frequencies will modulate the thermally averaged $a$-type and $b$-type fluorescence decays arising from the sample. If the different beat frequencies are not too different, as will be the case if $\Delta_{ab}$ and $\Delta'_{ab}$ are small, then one might expect that the thermally averaged decays will appear as if they were modulated by a single quantum beat component, with the decay rate of the envelope of that beat component being greater than the decay rate of the unmodulated portion of the decay. In such a case, one also expects that the higher the temperature, the faster the apparent beat decay rate will be, since the width of the distribution of beat frequencies will increase. Figure 38 shows decays calculated[43] by using Eq. (3.36), which verify the qualitative arguments just presented. The decay simulations shown pertain to a two-level system with $E_{ab}(0,0)$ and $V_{ab}$ values of 2.24 and 1.0 GHz, respectively (to give an average beat frequency of $\omega/2\pi \simeq 3$ GHz). The other parameters used are included in the figure caption. One will clearly note from the figure that the beat envelope decay rate and its behavior as a function of temperature match one's intuitive expectations.

*Other simulations:* To aid direct comparison with experimental data, simulations have been performed[43] using the experimentally derived $4 \times 4$ and $3 \times 3$ Hamiltonian matrices [Eqs. (3.35) and (3.34)] that describe the coupling of $S_1$ anthracene levels at $E_{\text{vib}} = 1420$ and $1380\ \text{cm}^{-1}$, respectively. Figure 39 shows simulated decays as a function of temperature for the $b$-type bands of the $E_{\text{vib}} = 1420\ \text{cm}^{-1}$ system. The $b$-type band was treated because good time-resolved data for it are relatively easy to obtain and because its decay is dominated by a single beat component at 1.0 GHz ($-1$ phase). The rotational constants used in the calculation are given in the caption to Fig. 39. (Sets of

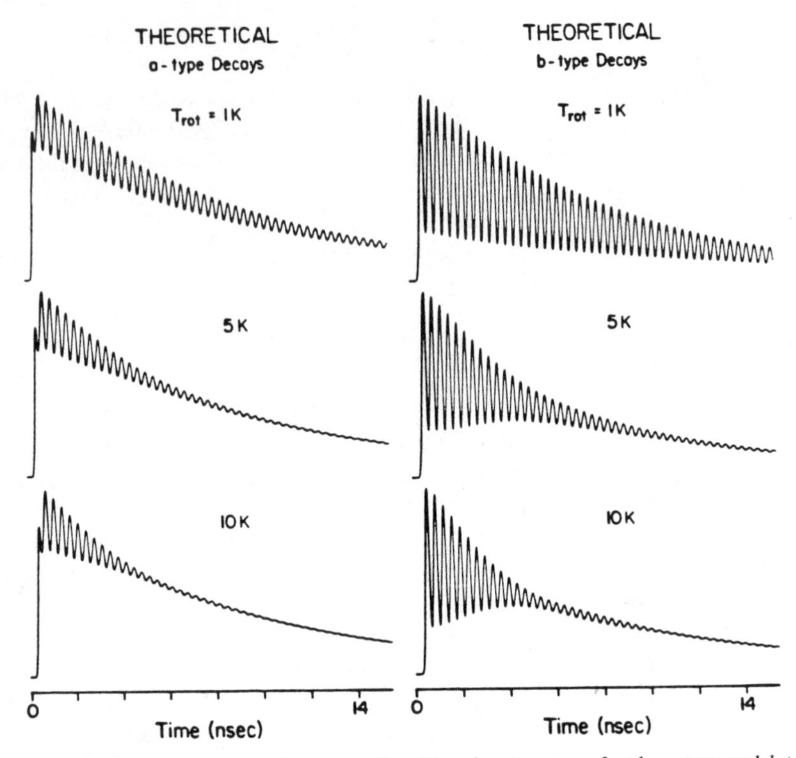


**Figure 38.** Simulated decays as a function of rotational temperature for the *a*-type and *b*-type decays of a coupled two-level system. The lowest rotational states of the two zero-order levels were taken to be spaced by 2.24 GHz (the $|b\rangle$ state being at higher energy) and the coupling matrix element was taken to be 1 GHz. The rotational constants used, $\frac{1}{2}(B + C)$ and $A - \frac{1}{2}(B + C)$, were 0.4119 and 1.7396 GHz for the $|a\rangle$ state; and 0.4116 and 1.7385 GHz for the $|b\rangle$ state. For other details see the text.

other rotational constants yielded similar results.) Clearly, there is a marked temperature effect on the decay, similar to the effect in the two-level simulation of Fig. 38. Although all of the decays are modulated similarly at early time, at later times the beats are increasingly washed out as the rotational temperature increases—the apparent decay rate of the beat envelope increases at higher rotational temperatures. Similar behavior characterizes calculated decays of the other band-types in the $E_{vib} = 1420$ cm$^{-1}$ spectrum.

Figure 40 shows calculated decays pertaining to *b*-type bands in the $E_{vib} = 1380$ cm$^{-1}$ fluorescence spectrum of anthracene. Such decays are modulated by beat components at $\omega/2\pi = 3.5$, 4.9, and 8.4 GHz, with phases of $+1$, $-1$, and $-1$, respectively (see Section III C 1b). Again, one can see an increase in beat envelope decay rate as the rotational temperature increases. As with the

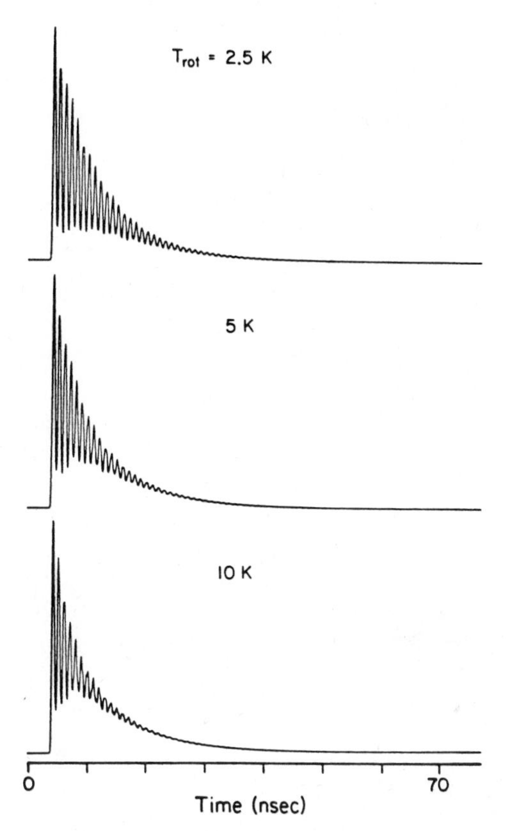


**Figure 39.** Simulated *b*-type decays as a function of rotational temperature for the coupled four-level system described by the Hamiltonian matrix of Eq. (3.35), which represents the coupling situation at $E_{vib} = 1420\ cm^{-1}$ in $S_1$ anthracene. The rotational constants $\frac{1}{2}(B+C)$ and $A - \frac{1}{2}(B+C)$ were for the $|a\rangle$, $|b\rangle$, $|c\rangle$, and $|d\rangle$ states, respectively, 0.4119 and 1.7396; 0.4124 and 1.7417; 0.4120 and 1.7401; and 0.4122 and 1.7407 GHz.

$E_{vib} = 1420\ cm^{-1}$ case, use of different sets of rotational constants yielded similar behavior.

*b. Experimental Results.* Experimental results were obtained by the picosecond-beam techniques used to obtain the results of Section III C. To assess the influence of rotational temperature on vibrational quantum beats, the rotational temperature of the free jet expansion was varied by changing carrier gases and carrier gas pressure. The carrier gases used were neon,

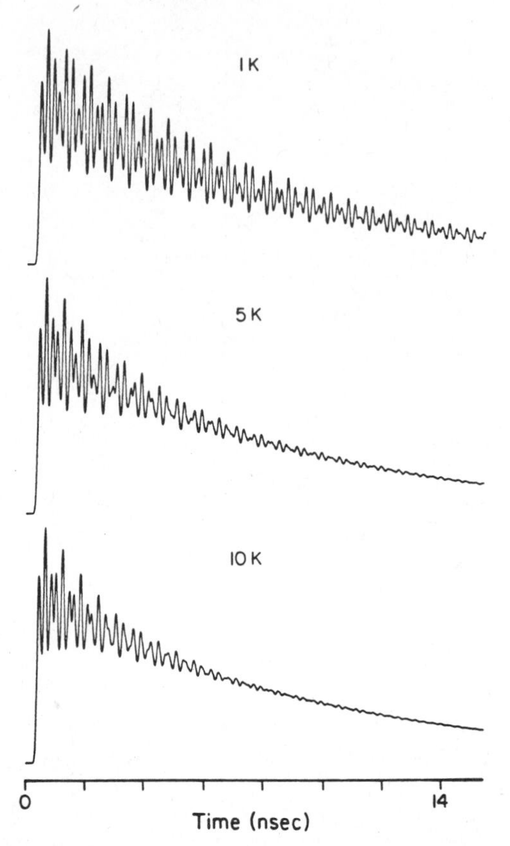


**Figure 40.** Simulated $b$-type decays as a function of rotational temperature for the coupled three-level system described by the Hamiltonian matrix of Eq. (3.34), which represents the coupling situation at $E_{vib} = 1380\ cm^{-1}$ in $S_1$ anthracene. The rotational constants $\frac{1}{2}(B + C)$ and $A - \frac{1}{2}(B + C)$ were for the $|a\rangle$, $|b\rangle$, and $|c\rangle$ states, respectively, 0.4127, 1.7428; 0.4119, 1.7395; and 0.4117, 1.7389 GHz.

helium, and nitrogen. By rotational band contour measurements[61,62] these gases have been shown to produce anthracene jets of different rotational temperature: neon producing the coldest jets, helium the next coldest, and nitrogen the warmest. For example, expansions using 50 psig Ne give rotational contours consistent with rotational temperatures of $\sim 1$ K, while 20 psig $N_2$ gives temperatures on the order of 10 K.

$5^1$-LEVEL OF ANTHRACENE. Figure 41 shows measured decays, as a function of carrier gas, of the $b$-type, $\bar{\nu}_d = 1750\ cm^{-1}$ band of anthracene excited to its

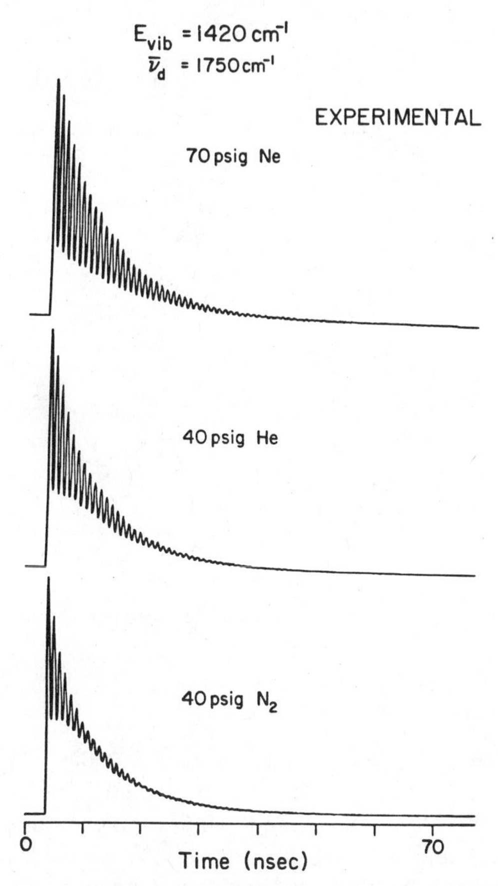


**Figure 41.** Measured fluorescence decays of the 1750 cm$^{-1}$ ($b$-type) band in the $E_{vib} = 1420$ cm$^{-1}$ spectrum of jet-cooled anthracene as a function of carrier gas parameters. Decays were measured under identical conditions except for carrier gas. For each decay $x = 6$ mm, the monochromator resolution $R = 3.2$ Å, and the laser bandwidth $BW \simeq 2$ cm$^{-1}$.

$E_{vib} = 1420$ cm$^{-1}$ ($5^1$) level. The carrier gas parameters used represent an increase in temperature as one looks at the figure from top to bottom.

The decays of Fig. 41 clearly show an increase in beat decay rate as rotational temperature increases. This is substantiated by more quantitative analysis of the decays. The presence of one dominant beat component (at 1 GHz) allows one to fit the decays to a function of the form

$$I(t) = A_1(e^{-\gamma_1 t} + A_2 e^{-\gamma_2 t} \cos \omega t) + A_3, \tag{3.41}$$

where $\omega$, $A_1$, $A_2$, $A_3$, $\gamma_1$, and $\gamma_2$ are fitting parameters. (Note that $A_2$ is the

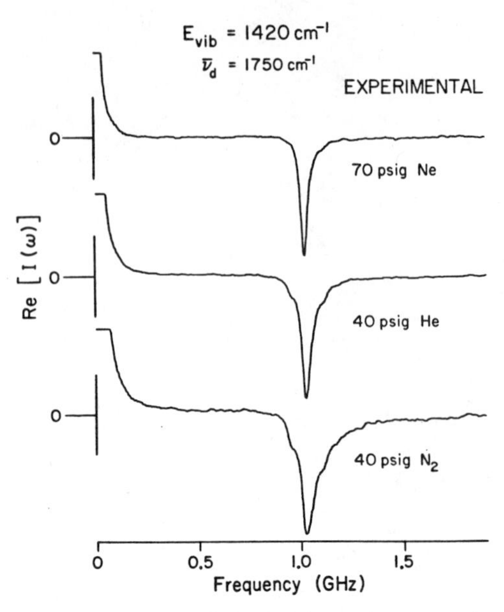


**Figure 42.** Fourier spectra of the decays of Fig. 41. The peaks are negative because the 1 GHz component in the decays has a −1 phase.

modulation depth of the beat component and $\gamma_2$ is the beat envelope decay rate.) Doing so, the modulation depths obtained for the three decays, from top to bottom, are −0.7, −0.67, and −0.61, and the quantum beat decay rates are 0.13, 0.18, and 0.31 GHz, respectively. While the values for the three modulation depths do not vary very widely, the values for $\gamma_2$ change from being very close to the unmodulated decay rate ($\gamma_1 = 0.11$ GHz) to a value that is a factor of ~3 greater than this. Notably, this trend in $\gamma_2$ is matched by behavior in the Fourier spectra of the decays (Fig. 42). In particular, the width of the dominant 1 GHz Fourier component clearly increases as the carrier gas changes from Ne to He to $N_2$ (i.e., as the rotational temperature increases).

The changes that occur in $\gamma_2$ as the carrier gas is changed also occur when carrier gas pressure is varied. In general, as the pressure increases (giving rise to greater rotational cooling), $\gamma_2$ decreases. This effect is not as pronounced as the effect produced when carrier gases are changed, but it is reproducible. In the case of *no* carrier gas, quantum beats are completely washed out (Fig. 43).

Besides the 1750 cm$^{-1}$ band, the carrier gas dependences of the decays of other bands in the $5^1$-level spectrum of anthracene have been determined. The effects are similar to those manifested in the decay of the 1750 cm$^{-1}$ band.

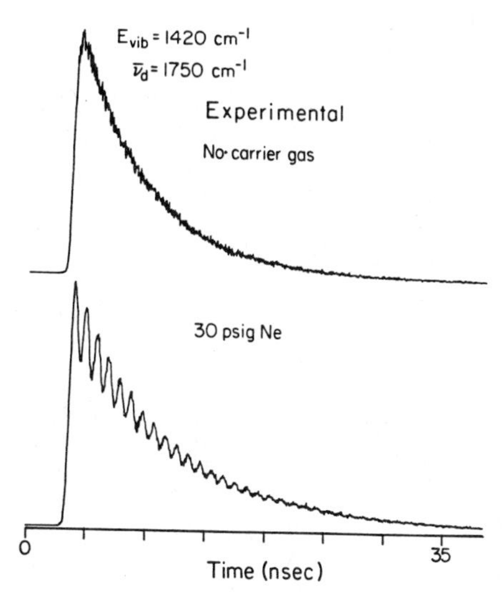


**Figure 43.** Measured fluorescence decays for detection of the 1750 cm$^{-1}$ band in the $E_{vib} = 1420$ cm$^{-1}$ spectrum of anthracene for no carrier gas (top) and for 30 psig neon (bottom), all other conditions being the same ($BW \simeq 3$ cm$^{-1}$, $R = 8.0$ Å, $x = 3$ mm). The relative lack of modulation in the neon decay compared to the decay of Fig. 41 top is primarily due to the poorer detection spectral resolution used in obtaining the former decay.

$6^1$-LEVEL OF ANTHRACENE. Decays measured for bands in the $E_{vib} = 1380$ cm$^{-1}$ ($6^1$) level fluorescence spectrum of anthracene reveal behavior entirely analogous with that observed for $E_{vib} = 1420$ cm$^{-1}$ excitation. Figure 44 shows decays of the $\bar{\nu}_d = 1460$ cm$^{-1}$ ($b$-type) band in the $6^1$ spectrum. Although the presence of three beat frequencies in the decays renders them difficult to fit, it is clear without fitting them that the beat modulations decay much more quickly in the higher-temperature nitrogen decay than in the neon decay. Similar behavior obtains for the decays of other bands in the $6^1$ spectrum.

c. *Conclusions.* Theoretical results show that physically reasonable rotational constant mismatches between anharmonically coupled vibrational levels have the major effect of increasing the apparent decay rates of the quantum beat modulations that arise in the fluorescence decays of such coupled systems. Moreover, these decay rates increase with the temperature of the sample. Experimental results on anthracene are entirely consistent with these theoretical results. This is apparent if one compares Figs. 39 (theory) and 41 (experiment), and Figs. 40 (theory) and 42 (experiment). Taken together, the results are significant in that (1) they reaffirm the theoretical results of Section III B at a higher level of approximation, (2) they emphasize the importance of cooling in determining the observability of vibrational

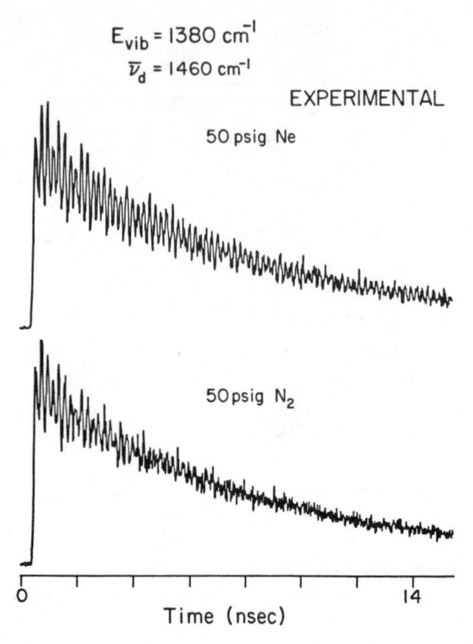


**Figure 44.** Measured fluorescence decays for detection of the 1460 $cm^{-1}$ (*b*-type) band in the $E_{vib} = 1380\ cm^{-1}$ spectrum of anthracene as a function of carrier gas parameters. For each decay $x = 3$ mm, $R = 0.8$ Å, and $BW \simeq 2\ cm^{-1}$.

quantum beats, and (3) they point to the predominance of anharmonic coupling in the IVR of anthracene.

### 2. *Rotational Coherence Effects*

As previously discussed, if two or more excited eigenstates can combine in absorption with a common ground-state level, then these eigenstates can be excited so as to form a coherent superposition state. The superposition state, in turn, can give rise to quantum beat-modulated fluorescence decays. All this, of course, lies at the heart of the theory of vibrational coherence effects. However, it also implies that the same experimental conditions under which vibrational coherence effects are observed should allow for the observation of *rotational* coherence effects. That is, since more than one rotational level in the manifold of an excited vibronic state can combine in absorption with a single ground-state ro-vibrational level, then in a picosecond-resolved fluorescence experiment rotational quantum beats should obtain.

As it turns out, in contradiction to the preceding considerations, rotational quantum beats have never been observed (to our knowledge). It also turns out that for large molecules there is a reasonable explanation for this fact.

The explanation has three facets. First, the values of the rotational quantum beat frequencies that can arise via excitation from a given ground-state rovibrational level depend on the rotational quantum numbers of the ground-state level. Second, in a gaseous sample of large molecules at finite temperature, there are many different ground-state rotational levels that have significant population. Finally, the first two points together imply that for a thermal sample many different rotational quantum beat frequencies will modulate measured fluorescence decays. In such a situation, a reasonable expectation is that the many modulations will destructively interfere and wash each other out. In this way, the negative experimental results can be explained.

Now, the argument just presented relies on the unproven assumption that rotational quantum beats arising from a thermal sample of isolated molecules will wash each other out. Recently, we examined this assumption by directly simulating the decays associated with thermally averaged rotational beats.[47] (Our initial motivation for this work was to try to explain the picosecond pump-probe results of Refs. 51 and 52, which results showed the existence of polarization-dependent early time transients in the decays of *t*-stilbene.) These theoretical simulations and subsequent picosecond-beam experiments[47–50] have revealed that the manifestations of rotational coherence in thermally averaged decays can, in fact, be observed. In this section, we briefly review these results and examine some of their implications with regard to time-resolved studies of IVR.

*a. Theory of Purely Rotational Coherence Effects.* We consider here the specific situation depicted in Fig. 45. We choose this case because it is the simplest one to treat and because it corresponds directly to experimental results. A more general theoretical treatment of rotational coherence effects may be found in Ref. 49. Figure 45 corresponds to the case of a prolate symmetric top molecule excited by a linearly polarized (polarization vector $\hat{e}_1$ taken to be parallel to the laboratory $Z$-axis), short pulse of light from the ground ro-vibronic state $|S_0 v_0; J_0 K_0 M_0\rangle$ to the rotational manifold of the excited vibronic state $|S_1 v_1\rangle$. We assume that the transition dipole moment for this process ($\boldsymbol{\mu}_1$) is parallel to the symmetric top axis so that the rotational selection rules are $\Delta J = 0, \pm 1, \Delta K = 0$, and $\Delta M = 0$.[58] The excitation process creates a coherent superposition state, $|\Psi(t)\rangle$, composed of those $|S_1 v_1\rangle$ rotational eigenstates whose excitation is consistent with the rotational selection rule[47]

$$|\Psi(t)\rangle = (c_1|J_0 - 1K_0M_0\rangle e^{-i2\pi(J_0-1)J_0Bt} + c_2|J_0K_0M_0\rangle e^{-i2\pi J_0(J_0+1)Bt} + c_3|J_0 + 1K_0M_0\rangle e^{-i2\pi(J_0+1)(J_0+2)Bt})e^{-i2\pi(\nu_{ev}+(A-B)K_0^2)t}e^{-\Gamma t/2}|S_1 v_1\rangle, \tag{3.42}$$

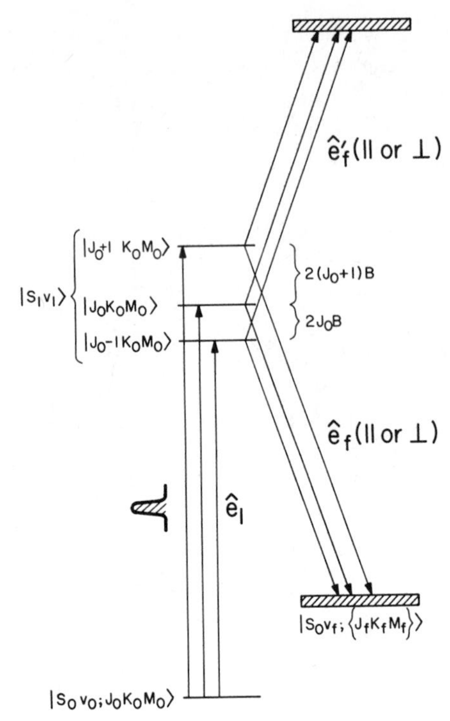


**Figure 45.** Schematic representation of the preparation and detection of rotational coherence in a molecule. The case depicted corresponds to the linearly polarized excitation (polarization vector $\hat{e}_1$) of a symmetric top molecule in ground-state ro-vibronic level $|S_0 v_0; J_0 K_0 M_0\rangle$ to those rotational levels of the excited vibronic state $|S_1 v_1\rangle$ allowed by the rotational selection rules germane to a parallel-type transition moment. The excitation process creates a superposition state of three rotational levels, the coherence properties of which can be probed by time resolving the polarized fluorescence (polarization $\hat{e}_f$) to the manifold of ground-state ro-vibronic levels $|S_0 v_f; J_f K_f M_f\rangle$, or by probing with a second, variably time-delayed laser pulse (polarization $\hat{e}'_f$).

where $A$ and $B$ have their usual meanings as rotational constants (in GHz), $\Gamma$ is the decay rate of the $|S_1 v_1\rangle$ vibronic state, $\nu_{ev}$ is the energy of $|S_1 v_1\rangle$, and $c_1$, $c_2$, and $c_3$ are constants that depend on the rotational quantum numbers $J_0$, $K_0$, and $M_0$.

Now consider the temporal behavior of polarization-analyzed (polarization vector $\hat{e}_f$) fluorescence from this superposition state to all possible rotational eigenstates of the ground vibronic level $|S_0 v_f\rangle$. This fluorescence decay is

$$I(t) \sim \sum_{J_f K_f M_f} |\langle \Psi(t) | \hat{e}_f \cdot \boldsymbol{\mu} | S_0 v_f; J_f K_f M_f \rangle|^2, \tag{3.43}$$

where $\boldsymbol{\mu}$ is the dipole moment operator. From Eq. (3.42) for $|\Psi(t)\rangle$, one can see that $I(t)$ will be modulated at the three beat frequencies $\omega_1/2\pi = 2BJ_0$, $\omega_2/2\pi = 2B(J_0 + 1)$, and $\omega_3/2\pi = (\omega_1 + \omega_2)/2\pi$. These *rotational quantum beats*, whose phases and modulation depths depend on the orientation of $\hat{e}_f$ with respect to $\hat{e}_1$, and on the orientation of the emission transition dipole $\boldsymbol{\mu}_f \equiv \langle S_1 v_1|\boldsymbol{\mu}|S_0 v_f\rangle$ with respect to $\boldsymbol{\mu}_1$, are the manifestations of purely rotational coherence in the molecule. If one now makes the assumption that one's sample of molecules is isotropic and that the emission transition moment of the $|S_1 v_1\rangle \rightarrow |S_0 v_f\rangle$ transition is parallel type, then it can be shown that the fluorescence decay is[47]

$$I(J_0, K_0, \hat{e}_f, t) = [\alpha(J_0, K_0, \hat{e}_f) + \beta(J_0, K_0, \hat{e}_f)\cos\omega_1 t + \gamma(J_0, K_0, \hat{e}_f)\cos\omega_2 t + \delta(J_0, K_0, \hat{e}_f)\cos\omega_3 t]e^{-\Gamma t}, \tag{3.44}$$

where $\alpha$, $\beta$, $\gamma$, and $\delta$ depend on $J_0$, $K_0$, and $\hat{e}_f$, and when $\beta$, $\gamma$, and $\delta$ have the very important properties that *if $\hat{e}_f \| \hat{e}_1$, they are all positive for all $J_0$, $K_0$, and if $\hat{e}_f \perp \hat{e}_1$, they are all negative for all $J_0$, $K_0$.*

In a real sample of large molecules, an observed decay will not be Eq. (3.44), but some thermal average of Eq. (3.44) over $J_0$ and $K_0$. That is, the observed decay will be

$$I(T, \hat{e}_f, t) = \eta \sum_{J_0 K_0} I(J_0, K_0, \hat{e}_f, t)e^{-[J_0(J_0+1)B + (A-B)K_0^2]/k_B T}, \tag{3.45}$$

where $\eta$ is a constant. It turns out, contrary to what one might think, that this thermal averaging does not necessarily preclude the observation of rotational coherence effects.[47] The reason for this has two parts. First, there are the aforementioned sign properties of the $\beta$, $\gamma$, and $\delta$ coefficients for particular orientations of $\hat{e}_f$. Second, $\omega_1/2\pi$, $\omega_2/2\pi$, and $\omega_3/2\pi$, regardless of the values of $J_0$ and $K_0$, are all integer multiples of $2B$ ($J_0$ is always an integer). These two aspects taken together imply that at times $n/2B$, $n = 0, 1, 2, \ldots,$ there will be a complete constructive interference between all beat terms, giving rise to recurrences in the fluorescence intensity at such times. For $\hat{e}_f \| \hat{e}_1$ these recurrences are positive, for $\hat{e}_f \perp \hat{e}_1$ they are negative. (We reiterate the fact that we have considered here a very specific physical situation. More general situations, involving asymmetric top molecules, and absorption and emission dipole moments that are not parallel type, etc., can also give rise to transients associated with thermally averaged rotational quantum beats.[49])

One can directly simulate the manifestations of thermally averaged rotational coherence in time-resolved fluorescence by using standard angular momentum algebra to find[47] the $\alpha$, $\beta$, $\gamma$, and $\delta$ coefficients of Eq. (3.44) and then by using Eq. (3.44) in Eq. (3.45), the values of $A$, $B$, and $T$ in the equations being chosen to correspond to the specific problem of interest, and the sum over $J_0$ in Eq. (3.45) being truncated at a physically reasonably value. We have performed such simulations for the $t$-stilbene molecule at temperatures corresponding to jet-cooled samples (5 K). The molecule is very nearly a prolate symmetric top with parallel-type $S_1 - S_0$ transition moment. Hence, it is appropriate to apply Eq. (3.44) to it. Figure 46 shows the results of the decay simulations. The top two traces were calculated with the assumptions of

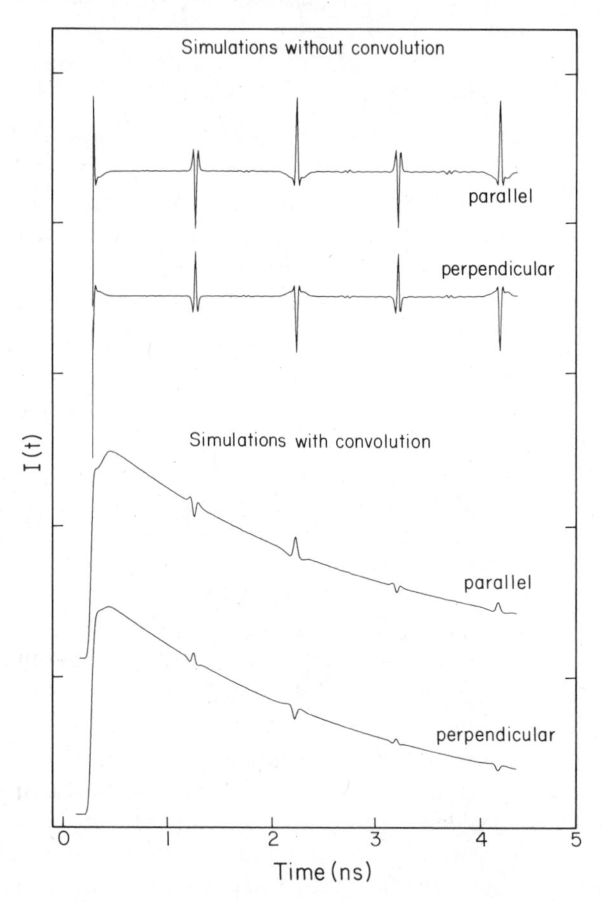


**Figure 46.** Simulated fluorescence decays showing the effects of purely rotational coherence. Decays were calculated using the results of Refs. 47 and 49 and parameters given in the text. The top two decays correspond to infinite temporal resolution and infinite fluorescence lifetime. "Parallel" and "perpendicular" refer to the orientation of $\hat{e}_f$ with respect to $\hat{e}_1$. The bottom two decays are just convolutions of the top two decays with a finite detection response function (80 psec FWHM), assuming, in addition, a 2.6 nsec fluorescence lifetime.

infinite temporal resolution and infinite fluorescence lifetime. In these decays one can clearly see the transients associated with thermally averaged rotational coherence. [Note the additional, negative-going transients in the $\hat{e}_f \| \hat{e}_1$ decay and the positive-going ones in the $\hat{e}_f \perp \hat{e}_1$ decay. These occur at times $m/4B$, $m = 1, 3, 5, \ldots$. They are manifestations of the constructive interference between a subset of all the cosine terms in Eq. (3.45),[47] unlike the $n/2B$ transients, which we have noted arise from total constructive interference.] The bottom two decays in Fig. 46 are just the upper two decays after a finite temporal response of 80 psec FWHM and a finite fluorescence lifetime of 2.6 nsec were accounted for. These lower decays correspond to what one would expect to see in an experiment using our apparatus. One can see that, although the transients are reduced in apparent intensity in the lower decays relative to those in the upper decays, they are still large enough that they should be observable.

*b. Rotational Coherence versus Vibrational Coherence.* Elsewhere[47] we have demonstrated and/or noted uses to which the rotational coherence phenomenon might be put. These uses include application to sub-Doppler high-resolution spectroscopy of large molecules,[48] studies of molecular collisions, and determinations of the symmetry properties of vibronic states. Yet, our original motivation to study rotational coherence (and our principal concern here as regards the phenomenon) was to understand the influence of such effects in picosecond studies of IVR. With regard to this aspect of rotational coherence, the results of our work thus far, can be summarized in two points.

The first point pertains to the potential application of time-resolved polarization spectroscopy to the study of intramolecular vibrational–rotational energy transfer (IVRET). (For example, energy transfer occurring as a result of Coriolis coupling.) Studies by Nathanson and McClelland[53] on the polarization properties of time-integrated fluorescence have indicated that such properties may serve as useful monitors of IVRET. Given this, it is natural to expect that *time-resolved* polarization measurements might yield even more information on such processes. In particular, one might expect to observe early-time, polarization-dependent transients as manifestations of rovibrational energy flow. Such transients were in fact observed in the earliest picosecond-resolved polarization studies of isolated large molecules.[51,52,70] However, from Fig. 46 top, it is apparent that purely rotational coherence effects, effects that reflect a time-dependent orientation of molecules, rather than any energy flow process, can be manifest as polarization-dependent, early-time transients.[47] (In conjunction with this, one would note that the more asymmetric toplike a molecule is, the smaller the recurrences associated with rotational coherence are. Yet, even as the recurrences are reduced in magnitude, the magnitude of the transient at early time changes little from the symmetric top value.) The upshot of this is that any interpretation of time-resolved polarization measurements must account for the possibility of purely rotational coherence effects. This signals the need for further theoretical work aimed at distinguishing between the manifestations of rotational coherence and those of IVRET in time-resolved studies.

The second point we would make concerning the influence of rotational coherence on time-resolved studies of IVR is that even if one studies a system of levels coupled exclusively by anharmonic coupling, one must worry about the polarization characteristics of one's experiment. In particular, assume that one performs the same experiment as that treated in Section III B. That is, assume that a time- and frequency-resolved fluorescence experiment is performed to probe vibrational energy flow in a system of $N$ anharmonically coupled $S_1$ vibrational levels. But, now, account for the rotational level structure of the vibrational levels by (1) allowing for a thermal distribution of ground-state levels in the sample of molecules prior to the excitation pulse and (2) allowing for the possibility of rotational quantum beats. Given such a situation, one can show that the observed fluorescence decay is of the form[49]

$$I_{\gamma,\mathrm{obs}}(T, \hat{e}_f, t) = I(T, \hat{e}_f, t) I_\gamma(t), \tag{3.46}$$

where $I(T, \hat{e}_f, t)$ is given by Eq. (3.45) (with $\Gamma = 0$) and is just the purely *rotational* coherence decay to be expected if only one $S_1$ vibrational level were being excited, and where $I_\gamma(t)$ is given by Eq. (3.12) and is just the *vibrational* coherence decay to be expected if one could neglect the rotational levels.

From Eq. (3.46), it is clear that rotational coherence effects can significantly influence measured decays. As an example, consider an $a$-type decay corresponding to dissipative IVR in $t$-stilbene. The vibrational factor $I_a(t)$ of the measured decay will be approximately a double exponential of the form of Eq. (3.32). The rotational factor $I(T, \hat{e}_f, t)$, assuming, say, $\hat{e}_f \| \hat{e}_1$, will be the topmost decay in Fig. 46, a decay with a prominent early-time transient decaying on a time scale of $\sim 10$ psec. Clearly, $I_{a,\mathrm{obs}}(T, \hat{e}_f, t)$ in this case will be significantly different than the IVR decay $I_a(t)$ in two respects. First, the fast-to-slow intensity ratio of the observed decay will be greater than that of $I_a(t)$. Second, the decay time of the observed early-time transient will be shifted from the $\tau_{\mathrm{IVR}}$ of $I_a(t)$ toward some value intermediate between $\tau_{\mathrm{IVR}}$ and the $\sim 10$ psec decay time of the purely rotational coherence. In the different situation corresponding to $\hat{e}_f \perp \hat{e}_1$, the pertinent $I(T, \hat{e}_f, t)$ is the decay second from the top in Fig. 46. In this case, the fast-to-slow ratio of $I_{a,\mathrm{obs}}(T, \hat{e}_f, t)$ will be less than that of $I_a(t)$, while the early-time transient of the observed decay will decay on a slower time scale than that of $I_a(t)$. (More precisely, the observed early time transient will be the product of a rise and a decay.)

Notably, experimental results[50] substantially confirm the theoretical points, as touched on below.

Fortunately, in the cases we have considered, it is easy to eliminate the contributions of rotational coherence to an observed decay. First, one notes that rotational coherence effects vanish when $\hat{e}_f$ is at an angle of 54.7° with respect to $\hat{e}_1$. Second, one notes that in Eq. (3.46), only $I(T, \hat{e}_f, t)$ depends on $\hat{e}_f$. Therefore, when $\hat{e}_f$ and $\hat{e}_1$ are at the magic angle

$$I_{\gamma,\mathrm{obs}}(T, \hat{e}_f, t) = KI_{\gamma}(t), \tag{3.47}$$

where $K$ is a constant.[49]

### c. *Experimental Results*

PURELY ROTATIONAL COHERENCE AND SUB-DOPPLER SPECTROSCOPY. Guided by the theoretical decay simulations of Fig. 46, the first unambiguous observation of thermally averaged rotational coherence effects was made for excitation and detection of the $S_1 - S_0 0_0^0$ band of jet-cooled *t*-stilbene.[47] Observed fluorescence decays are shown in Fig. 47; theory and experiment match very well. The recurrences associated with rotational coherence effects in fluorescence have been observed for a number of other species as well. Among these species are *t*-stilbene-$d_{12}$, *t*-stilbene–argon complexes,[48] and *t*-stilbene–helium complexes.[71] The recurrences allow the determination of the excited-state rotational constants to a high degree of accuracy. [For example, for *t*-stilbene we find $\frac{1}{2}(B + C)$ to be $0.00854 \pm 0.00004\ \mathrm{cm}^{-1}$.] The indications are that with currently available temporal resolution, rotational coherence effects should be observable in a multitude of species and should allow the accurate determination of such species' excited-state rotational constants.

ROTATIONAL VERSUS VIBRATIONAL COHERENCE. We have obtained experimental results[50] pertaining to the coexistence of vibrational and rotational coherence in spectrally resolved, polarization-analyzed fluorescence decays of *t*-stilbene in the intermediate- and high-energy regimes of the $S_1$ vibrational level structure (see Section III C 3). In Fig. 48 data are presented for the $E_{\mathrm{vib}} = 789\ \mathrm{cm}^{-1}$ excitation of the molecule. The vibrational beats, which are observed without the detection polarizer present and which correspond to IVR in the restricted regime, are clearly seen in the figure. In addition, the recurrences characteristic of purely rotational coherence in *t*-stilbene are also present in the decay. These recurrences match those observed for *t*-stilbene upon excitation of the $S_1$ origin band. Further analysis[50] lends support to our theoretical findings[49] (see the previous section) pertaining to situations in which rotational coher-

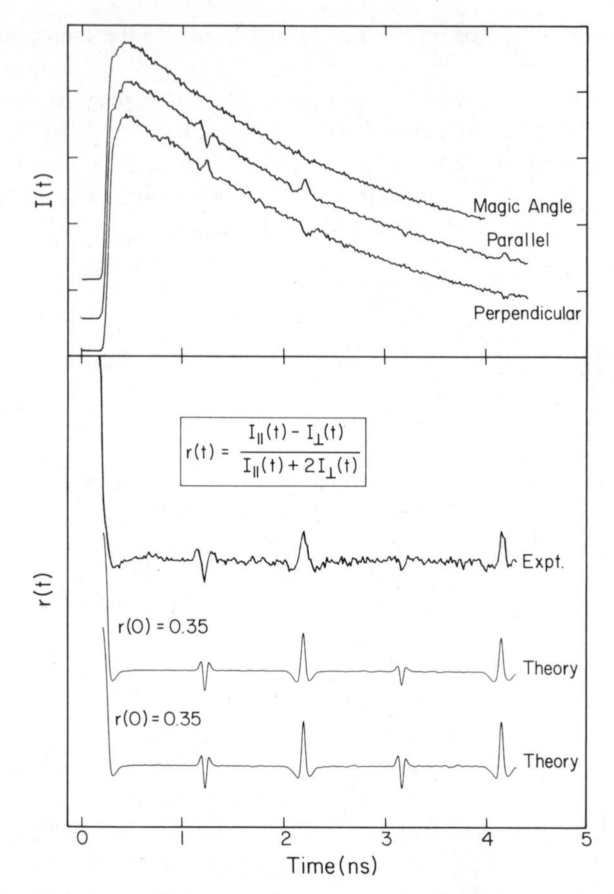


**Figure 47.** Top: Experimental fluorescence decays corresponding to the excitation and detection of the $S_1 - 0_0^0$ band of jet-cooled *t*-stilbene: expansion orifice 70 $\mu$m, 75 psig Ne backing pressure, nozzle $T \simeq 150°$C, laser-to-nozzle distance 3 mm. Bottom: Fluorescence anisotropies $r(t)$. The experimental trace was obtained directly from the parallel and perpendicular decays at the top of the figure using the expression for $r(t)$. The upper theoretical trace was obtained from decays calculated for an asymmetric top (rotational constants 2.678, 0.262, and 0.250 GHz) at 5 K with convolution of the experimental response function accounted for. The bottom trace was calculated from the bottom two decays (symmetric top) of Fig. 46.

ence effects and vibrational coherence effects arising from anharmonic coupling are present in the same decay.

In the dissipative IVR regime of *t*-stilbene (for example, upon excitation of the molecule to $E_{vib} = 1249$ cm$^{-1}$—see Fig. 49), one observes the quasi-biexponential decay characteristic of dissipative IVR, and, again, the recurrences associated with rotational coherence. Thus, in the dissipative regime

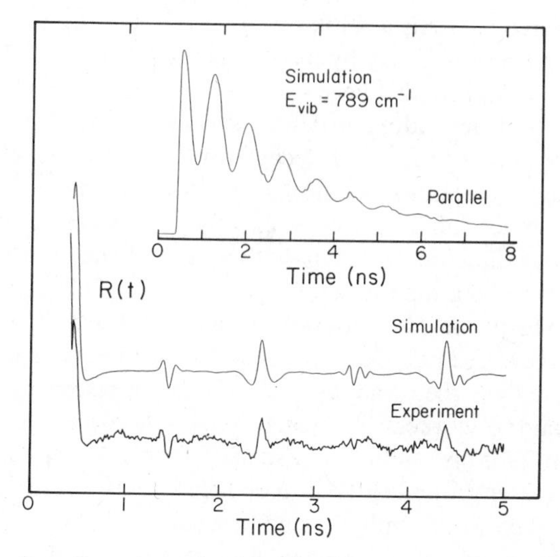


**Figure 48.** Comparison of experimental and simulated fluorescence polarization anisotropies for the $S_1 + 789\ \text{cm}^{-1}$ excitation of jet-cooled *t*-stilbene. The anisotropies include convolution effects associated with the finite excitation pulse width. The upper trace was calculated using the theoretical results of Ref. 49. The lower trace was obtained from experiment. The inset shows a simulated decay for this excitation band.

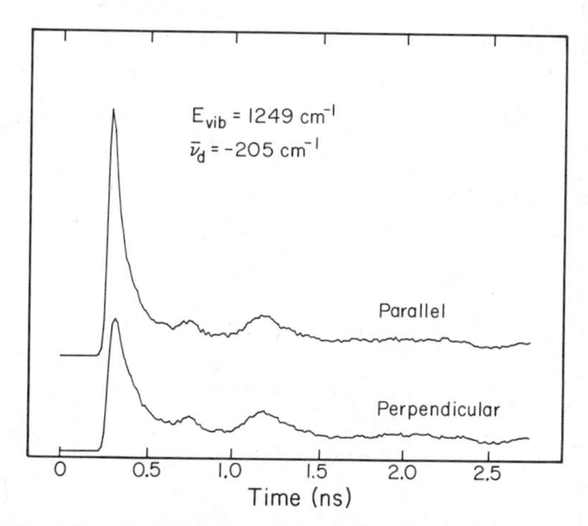


**Figure 49.** Polarization-analyzed fluorescence decays for $S_1 + 1249\ \text{cm}^{-1}$ excitation of jet-cooled *t*-stilbene. The fluorescence band detected was at a shift of $-205\ \text{cm}^{-1}$ from the excitation energy and was detected with $32\ \text{cm}^{-1}$ resolution.

rotational coherence persists. Notably, the "distortion" of the early time portion of the relaxation decay by the rotational coherence (as mentioned in the previous section) has also been demonstrated experimentally.[50] That is, we have shown that, depending on detection polarization, the fast lifetime and the fast-to-slow intensity ratio of decays like those of Fig. 49 will take on different values. This points up the fact that one must be concerned with rotational coherence influences in ostensibly dynamical decays. We would also note that our data are consistent with an early-time rotational coherence transient decaying on a time scale of 10 psec or less. This agrees with earlier (polarized) pump-probe photoionization experiments[51,52] on jet-cooled *t*-stilbene, which revealed polarization-dependent early-time transients (see Fig. 50) that decay in ~8 psec, and with later bulb fluorescence experiments by Hochstrasser and co-workers,[70a,c] which showed the existence of polarization-dependent transients on a similar time scale. (Other measurements reported[70b] by the Hochstrasser group led them to deduce a rotational relaxation time of 48 psec for *t*-stilbene in the bulb. In a private communication we have learned that this number corresponds to population decay and not rotational coher-

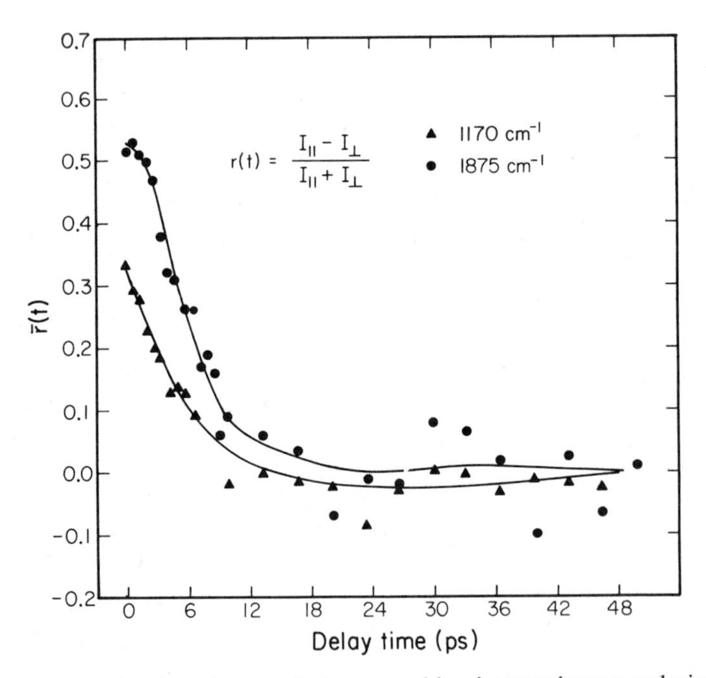


**Figure 50.** Polarization-dependent transients measured by picosecond pump-probe ionization techniques [$\bar{r}(t)$ is defined in the figure] as a function of the delay time and the excess vibrational energy of jet-cooled *t*-stilbene.

ence.) An important feature of the Hochstrasser group experiments is their careful measurement of the steady-state fluorescence anisotropy,[70c] a value pertinent in determining (as per the theory of Nathanson and McClelland[53]) whether *t*-stilbene is a "regular" or "irregular" rotator at any given excess vibrational energy.

Clearly, much activity may be expected in the specific area of rotational coherence and the more general one of time-resolved and polarization-analyzed fluorescence. One area of interest, not touched upon here, concerns the influence of rotational coherence in *electronic* relaxation processes. In this we regard it is pertinent to note the polarization-dependent decays reported first by Matsumoto et al.[72] in their studies on the nanosecond time scale of singlet–triplet coupling in pyrazine.

## IV. CONCLUDING REMARKS

In ending this Chapter, it is pertinent to reflect on some of the only recently known yet key features of IVR, coherence, and mode-selective dynamics in isolated large molecules. It is also useful to consider some problems that are still outstanding and some possible future directions in these areas.

Part of the general picture of IVR dynamics that is now emerging is that the dynamics is not governed by kinetics—IVR processes are coherent. Such processes depend on the state that is initially excited and on the excess vibrational energy in the molecule. Significantly, this picture is based on experimental observations made on a number of molecules, as well as being based on theory. Regarding the time scale of IVR in typical large molecules, time-resolved work has established it to be in the range of from tens of picoseconds to several nanoseconds, which translates to coupling matrix elements of hundredths to tenths of $cm^{-1}$ between "zero-order" vibrational modes. These coupling matrix elements, moreover, are not uniform. And, couplings between "bath" states are important in determining dynamics. Therefore, using an average coupling matrix element, and assuming that the only coupling that occurs is between the initially prepared state and the bath states, are both gross approximations with regard to IVR in real molecules. Notably, fluctuations in coupling matrix elements also have been found via high-resolution, time-integrated experiments pertaining to IVR in benzene[19] and to singlet–triplet coupling in pyrazine.[73]

The nature of the initially prepared state is of paramount importance in determining (1) the subsequent IVR dynamics of a species and (2) the way in which the dynamics is manifest in time-resolved and time-integrated fluorescence. The theoretical picture, reviewed in Section III B of the manifestations of IVR in time-resolved fluorescence relies on the assumption that single zero-order states act as "doorway" states in optical transitions from and to ground-state vibrational levels. Clearly, this need not always be the case. When it is not, behavior differing from the theoretical predictions can occur. Regarding time-integrated spectral studies of IVR, mixings between optically active states can give rise to spectral congestion in fluorescence that does not reflect dynamics. One must be very cautious in interpreting spectral results in terms of IVR. A more general model of vibrational couplings and IVR appears in Fig. 36, in which the preparation of the initial state by the laser and the possibility of multiple doorway states are indicated schematically.

Linewidth measurements (in absorption or excitation spectra) of IVR suffer from spectral congestion problems (inhomogeneous) and from different sources of homogeneous broadening. Such broadening mechanisms are particularly prevalent for large molecules at finite pressure and temperature. If, however, eigenstate spectra can be measured, and the homogeneous widths of the lines free from pure dephasing can be determined, then frequency domain results can be related to dynamics. High-resolution studies on pyrazine[73] and benzene[19] are examples of the potential of linewidth measurements in studies of molecular dynamics.

Recent results make the point increasingly more forcefully that the rotational level structure of a molecule can play an important role (1) in IVR processes and (2) in "skewing" the form of observables ostensibly associated with IVR. Here we have reviewed our work pertaining to the role of the rotational level structure in influencing anharmonic coupling between vibrational levels. Rotational constant mismatches between zero-order states can give rise to observable effects associated with IVR. Others have shown that in some cases rotations are involved in even more direct ways, that is, via Coriolis couplings.[19] A prime example of this in a reasonably large molecule has been revealed by the work of Riedle et al.[19a-d] on benzene. Finally, the coherent preparation of rotational levels by a laser pulse can give rise to polarization-dependent transients in time-resolved fluorescence, which transients have nothing to do with energy relaxation dynamics. Such transients, aside from representing a very useful, sub-Doppler spectroscopic tool, can interfere with the transients associated with IVR. One must account for this in the interpretation of time-resolved experiments. Thus, rotational coherence, as well as vibrational coherence, both observed by using picosecond-beam techniques, illustrates the crucial role played by the coherence of the laser excitation source in preparing the initial state.

Despite the progress of recent years in developing an understanding of IVR, there are a number of interesting problems that remain for future research. Perhaps these problems are best highlighted by casting them in the form of questions.

*First*, what determines the selectivity and magnitude of coupling between vibrational levels? In this regard, we would note the recent work of

Mukamel's group[74] in trying to understand the time- and frequency-resolved spectra of anthracene in terms of a vibrational force field for the molecule. Such studies have great potential for developing deeper insight into the coupling associated with IVR.

*Second*, what is the nature of IVR at higher vibrational energies in excited electronic states? While time-domain fluorescence techniques are not well suited to answer this question, pump-probe ionization experiments have the potential of doing so.

*Third*, what is the nature of IVR on ground state surfaces? Corresponding to the low-energy region of ground-state surfaces, where the vibrational modes are assigned and well defined, there is substantial body of information on the nature of IVR from the time-integrated (or low time-resolution) experiments by the groups of McDonald,[75] Yablonovitch–Bloembergen,[76] Mazur,[77] Kwok,[78] and others. It will be of great interest to learn of the results of picosecond time-resolved studies of IVR in ground electronic states. Several groups are initiating experiments in this direction and in the area of photodissociation of van der Waals molecules, which we did not touch on here.

*Fourth*, what is the nature of IVR near chemical activation energies? Pump-probe picosecond experiments at these energies on ground-state

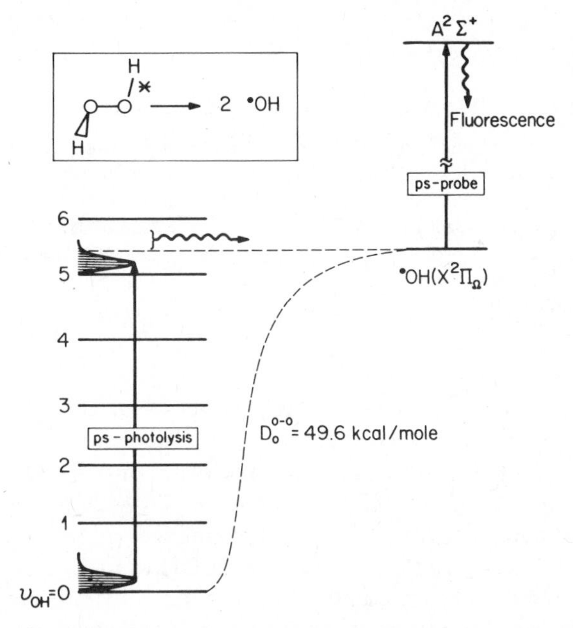


**Figure 51.** Schematic depicting the pump-probe-fluorescence detection scheme used to monitor the photodissociation of $H_2O_2$ subsequent to overtone excitation on the ground-state surface.

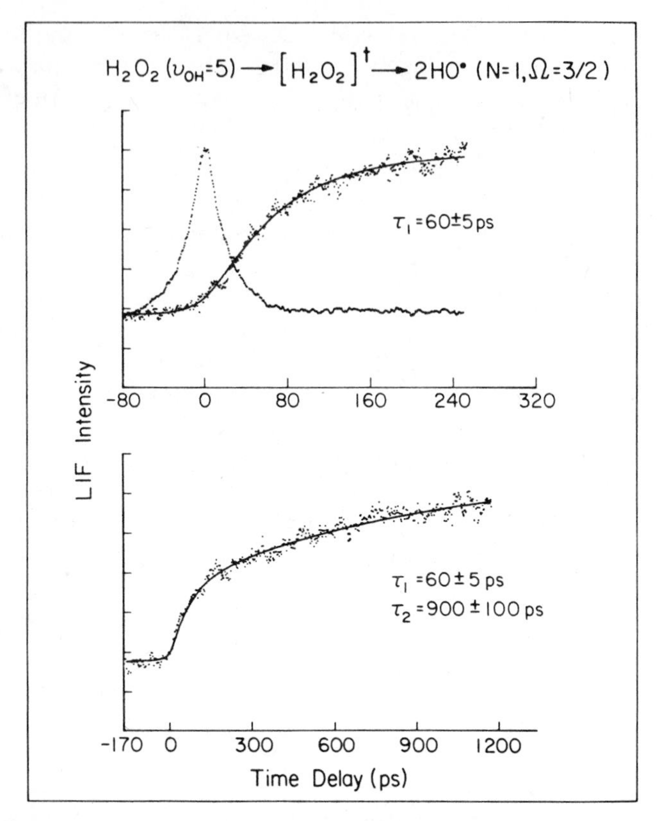


**Figure 52.** Time-resolved build-up of OH ($N = 1, \Omega = \frac{3}{2}$) fragments of $H_2O_2$ dissociation initiated by $v_{OH} = 5$ excitation. In the lower plot the points are the experimental data and the solid curve is a best-fit biexponential rise convoluted with the cross correlation function of the pulse. The upper plot is an expanded scale scan of the early-time portion of the build-up along with a fit of a rising exponential convoluted with the cross correlation function. Also shown on the same scale is the cross correlation function of the pump and probe pulses measured using difference frequency generation.

surfaces are difficult to perform (owing to very weak absorption strengths). Only recently have such experiments been successful.[79] For example, in one experiment, a pure OH overtone ($\Delta v = 5$) of $H_2O_2$ was excited, inducing the photodissociation reaction $H_2O_2 \rightarrow 2OH$ (see Figs. 51 and 52). The dynamics of O–O bond breaking was monitored by a pump-probe technique. In this case, IVR from OH to O–O is needed for this reaction to go. The transient of Fig. 52 reflects this process,[79] and more experiments along these lines are needed. Very interesting experiments in the frequency domain have also been performed on molecules at high ground-state vibrational energies by Field and Kinsey and their coworkers[80] using stimulated emission pumping. The technique has also been applied to *p*-difluorobenzene by Lawrance and Knight[81] and will continue to be an important source of information about ground state vibrational dynamics. Both time- and frequency-resolved experiments should eventually set the stage for a more detailed theoretical understanding of vibrational motion at high energies on ground-state surfaces. In particular, one wonders if the motion is chaotic or quasiperiodic, or if there are regions in which both types of motion occur. Many theoretical groups (for example, Refs. 26–28 and references therein) have contributed to this field of study and will continue to do so. Whether or not anthracene or *t*-stilbene (large molecules) behave quantally as, for example, the Henon–Heiles system, remains to be seen.

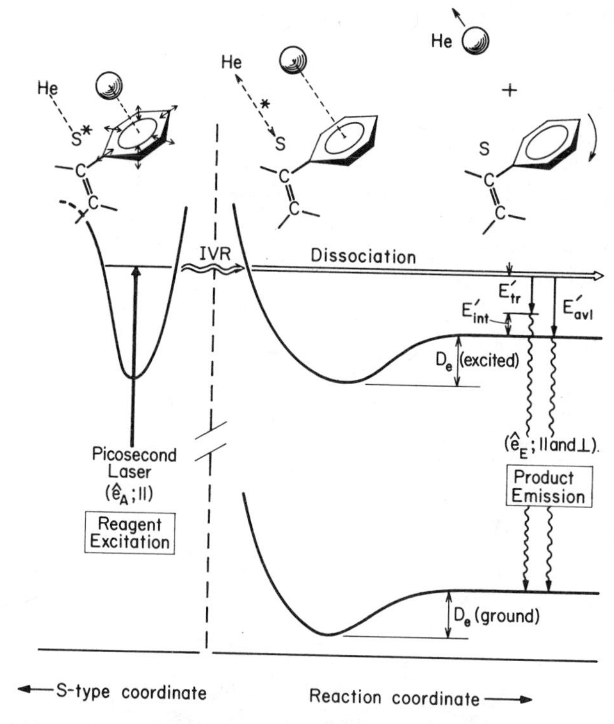


**Figure 53.** Schematic diagram depicting the time-resolved probing of the photodissociation and coherence retention of *t*-stilbene–He complexes. $\hat{e}_A$ and $\hat{e}_E$ are the polarization vectors for absorption and nascent product emission, respectively. $D_e$ is the binding energy of the complex. The optically excited stilbene-type mode of the complex is represented on the left side of the figure. Energy reaches the reaction coordinate (right half) via IVR. After dissociation (far right) the available energy is partitioned between translation ($E_{tr}$) and internal excitation of stilbene ($E'_{int}$).

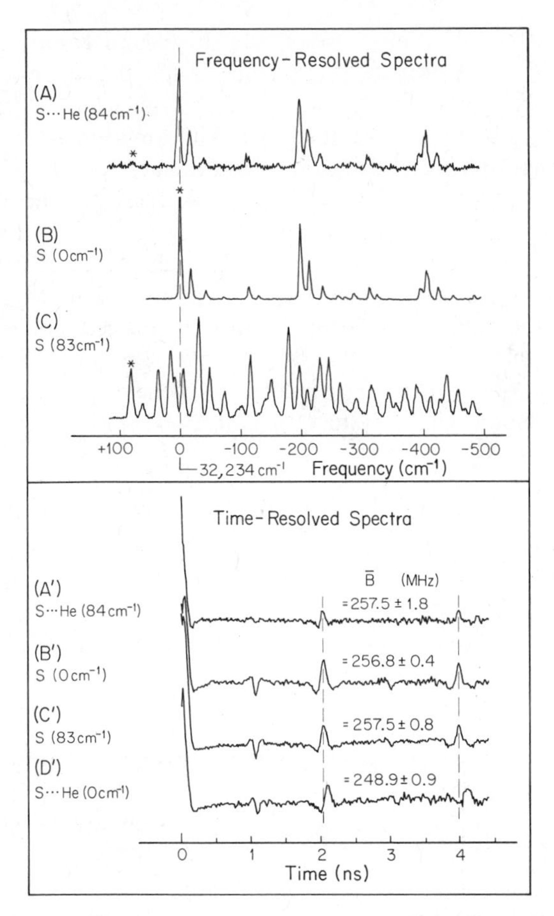


**Figure 54.** Time- and frequency-resolved spectra of stilbene and stilbene–He (1 : 1) complex. Excitation energies relative to the respective complex or bare molecule $0_0^0$ transition energies are given at left. Excitation frequencies are marked by an asterisk in the upper spectra. Spectral resolutions for (*A*), (*B*), and (*C*) are 4, 1, and 5 $cm^{-1}$, respectively. The difference in resolution accounts totally for the broadening of (*A*) relative to (*B*). The time-resolved data are presented in the form of polarization anisotropies. $\bar{B}$ is $\frac{1}{2}(B + C)$ of the complex or bare stilbene.

*Finally*, what is the role of IVR in photodissociation? Very recently, we monitored the photodissociation of *t*-stilbene–He(Ar) complexes in real time (see Fig. 53)[71] using polarized (coherent) excitation and analyzed product state distribution. The surprising finding was that the dissociation process, occurring subsequent to IVR, leaves bare *t*-stilbene in coherent rotational states (see Fig. 54 and compare the results with those in Fig. 47 for bare *t*-stilbene). The initial coherence induced in the complex by the photoexcitation process survives the photodissociation. Work will continue in this area to help elucidate the role of IVR in photodissociation dynamics.

In conclusion, picosecond time-resolved studies of IVR in beam-isolated molecules has been very fruitful over the past six years. Together with high-resolution spectroscopic studies, the approach should help us further in unraveling the details of the dynamics of vibrational motion in large molecules and in chemical processes at low and high energies, so that one may ultimately direct the fate of energy redistribution in laser chemistry experiments.

## Acknowledgments

This work is supported by grants from the National Science Foundation. Ahmed Zewail would also like to acknowledge helpful support from the Keck Foundation and the President's fund for some of the work reviewed here.

## References

1. See, for example, N. Bloembergen and A. H. Zewail, *J. Phys. Chem.* **88**, 5459 (1984).
2. See, for example, S. Buelow, M. Noble, G. Radhakrishnan, H. Reisler, C. Wittig, and G. Hancock, *J. Phys. Chem.* **90**, 1015 (1986).
3. See, for example, R. Rettschnick, in *Radiationless Transitions* (S. H. Lin, ed.). Academic, New York, 1980, p. 185.
4. (*a*). B. S. Rabinovitch, E. Tschuikow-Roux, and E. S. Schlag, *J. Am. Chem. Soc.* **81**, 1081 (1959); (*b*) D. W. Placzek, B. S. Rabinovitch, and E. H. Dorer, *J. Chem. Phys.* **44**, 279 (1966); (*c*) I. Oref, D. Schuetzle, and B. S. Rabinovitch, *J. Chem. Phys.* **54**, 575 (1971).
5. R. B. Bernstein and A. H. Zewail, *J. Phys. Chem.* **90**, 3467 (1986).
6. J. D. Rynbrandt and B. S. Rabinovitch, *J. Phys. Chem.* **75**, 2164 (1971).
7. See, for example, (*a*) S. Fischer and E. W. Schlag, *Chem. Phys. Lett.* **4**, 303 (1969); (*b*) S. Fischer, E. W. Schlag, and S. Schneider, *Chem. Phys. Lett.* **11**, 583 (1971); (*c*) E. W. Schlag, S. Schneider, and D. W. Chandler, *Chem. Phys. Lett.* **11**, 474 (1971); (*d*) J. C. Hsieh, C.-S. Huang, and E. C. Lim, *J. Chem. Phys.* **60**, 4345 (1974); (*e*) S. F. Fischer and E. C. Lim, *J. Chem. Phys.* **61**, 582 (1974); (*f*) T. A. Stephenson and S. A. Rice, *J. Chem. Phys.* **81**, 1073 (1984).
8. For a review see, for example, R. E. Smalley, L. Wharton, and D. H. Levy, *Accts. Chem. Res.* **10**, 139 (1977).
9. P. S. H. Fitch, L. Wharton, and D. Levy, *J. Chem. Phys.* **70**, 2018 (1979).
10. (*a*) J. B. Hopkins, D. E. Powers, and R. E. Smalley, *J. Chem. Phys.* **72**, 5039 (1980); (*b*) **72**, 5049 (1980); (*c*) **73**, 683 (1980); (*d*) **74**, 745 (1981); (*e*) D. E. Powers, J. B. Hopkins, and R. E. Smalley, *J. Chem. Phys.* **72**, 5721 (1980); (*f*) S. M. Beck, D. E. Powers, J. B. Hopkins, and R. E. Smalley, *J. Chem. Phys.* **73**, 2019 (1980); (*g*) S. M. Beck, J. B. Hopkins, D. E. Powers, and R. E. Smalley, *J. Chem. Phys.* **74**, 43 (1981).
11. A. Amirav, U. Even, and J. Jortner, *J. Chem. Phys.* **74**, 3745 (1981).
12. (*a*) C. Bouzou, C. Jouvet, J. B. Leblond, Ph. Millie, A. Tramer, and M. Sulkes, *Chem. Phys. Lett.* **97**, 61 (1983); (*b*) B. Fourmann, C. Jouvet, A. Tramer, J. M. LeBars, and Ph. Millie, *Chem. Phys.* **92**, 25 (1985).
13. M. Fujii, T. Ebata, N. Mikami, M. Ito, S. H. Kable, W. D. Lawrance, T. B. Parsons, and A. E. W. Knight, *J. Phys. Chem.* **88**, 2937 (1984).
14. (*a*) M. M. Doxtader, I. M. Gulis, S. A. Schwartz, and M. R. Topp, *Chem. Phys. Lett.* **112**, 483 (1984); (*b*) S. A. Schwartz and M. R. Topp, *Chem. Phys.* **86**, 245 (1984).

15. (*a*). J. A. Syage, W. R. Lambert, P. M. Felker, A. H. Zewail, and R. M. Hochstrasser, *Chem. Phys. Lett.* **88**, 266 (1982); (*b*) J. A. Syage, P. M. Felker, and A. H. Zewail, *J. Chem. Phys.* **81**, 4706 (1984); (*c*) P. M. Felker and A. H. Zewail, *J. Phys. Chem.* **89**, 5402 (1985).
16. (*a*) P. M. Felker, J. A. Syage, W. R. Lambert, and A. H. Zewail, *Chem. Phys. Lett.* **92**, 1 (1982); (*b*) J. A. Syage, P. M. Felker, and A. H. Zewail, *J. Chem. Phys.* **81**, 2233 (1984).
17. P. M. Felker and A. H. Zewail, *J. Chem. Phys.* **78**, 5266 (1983).
18. (*a*) A. Amirav, U. Even, and J. Jortner, *J. Chem. Phys.* **75**, 3770 (1981); (*b*) A. Amirav and J. Jortner, *Chem. Phys. Lett.* **94**, 545 (1983); (*c*) M. Sonnenschein, A. Amirav, and J. Jortner, *J. Phys. Chem.* **88**, 4214 (1984).
19. (*a*) E. Riedle, H. J. Neusser, and E. W. Schlag, *J. Chem. Phys.* **75**, 4231 (1981); (*b*) E. Riedle, H. J. Neusser, and E. W. Schlag, *J. Phys. Chem.* **86**, 4847 (1982); (*c*) E. Riedle and H. J. Neusser, *J. Chem. Phys.* **80**, 4686 (1984); (*d*) U. Schubert, E. Riedle, H. J. Neusser, and E. W. Schlag, *J. Chem. Phys.* **84**, 6182 (1986); (*e*) H. Saigusa, B. E. Forch, and E. C. Lim, *J. Chem. Phys.* **78**, 2795 (1983); (*f*) A. Amirav, *J. Chem. Phys.* **86**, 4706 (1987).
20. (*a*) A. Amirav and J. Jortner, *J. Chem. Phys.* **81**, 4200 (1984); (*b*) A. Amirav, M. Sonnenschein, and J. Jortner, *J. Phys. Chem.* **88**, 5593 (1984); (*c*) A. Amirav and J. Jortner, *J. Chem. Phys.* **82**, 4378 (1985).
21. W. R. Lambert, P. M. Felker, and A. H. Zewail, *J. Chem. Phys.* **75**, 5958 (1981).
22. For reviews see (*a*) A. H. Zewail, *Faraday Disc. Chem. Soc.* **75**, 315 (1983); (*b*) P. M. Felker, and A. H. Zewail, in *Applications of Picosecond Spectroscopy to Chemistry* (K. B. Eisenthal, ed.). Reidel, Dortrecht, 1984, p. 273; (*c*) see also the more recent references in this review by this group.
23. C. S. Parmenter, *Faraday Disc. Chem. Soc.* **75**, 7 (1983).
24. R. E. Smalley, *J. Phys. Chem.* **86**, 3504 (1982).
25. V. Bondybey, *Ann. Rev. Phys. Chem.* **35**, 591 (1984).
26. S. A. Rice, *Advances in Laser Chemistry* (A. H. Zewail, ed.). Springer, New York, 1978, p. 2.
27. D. W. Noid, M. L. Koszykowski, and R. A. Marcus, *Annu. Rev. Phys. Chem.* **32**, 267 (1981).
28. W. L. Hase, *J. Phys. Chem.* **90**, 365 (1986).
29. For reviews on radiationless transitions see, for example, (*a*) J. Jortner and S. Mukamel, in *The World of Quantum Chemistry* (R. Daudel and B. Pullman, eds.). Reidel, Dortrecht, 1976, p. 205; (*b*) K. F. Freed, *Top. Appl. Phys.* **15**, 24 (1976); (*c*) Ph. Avouris, W. M. Gelbart, and M. A. El-Sayed, *Chem. Rev.* **77**, 793 (1977); (*d*) A. Tramer and R. Voltz, in *Excited States* (E. C. Lim, ed.). Academic, New York, 1979, Vol. I, p. 281.
30. (*a*) P. M. Felker and A. H. Zewail, *Phys. Rev. Lett.* **53**, 501 (1984); (*b*) P. M. Felker and A. H. Zewail, *J. Chem. Phys.* **82**, 2975 (1985).
31. F. Lahmani, A. Tramer, and C. Tric, *J. Chem. Phys.* **60**, 4431 (1974).
32. (*a*) S. Mukamel, *J. Chem. Phys.* **82**, 2867 (1985); (*b*) S. Mukamel and R. E. Smalley, *J. Chem. Phys.* **73**, 4156 (1980).
33. A. H. Zewail, *Ber. Bunsenges. Phys. Chem.* **89**, 264 (1985).
34. (*a*). R. A. Coveleski, D. A. Dolson, and C. S. Parmenter, *J. Chem. Phys.* **72**, 5774 (1980); (*b*) R. A. Coveleski, D. A. Dolson, and C. S. Parmenter, *J. Phys. Chem.* **89**, 645 (1985); (*c*) **89**, 655 (1985); (*d*) K. W. Holtzclaw and C. S. Parmenter, *J. Chem. Phys.* **84**, 1099 (1986); (*e*) D. A. Dolson, K. W. Holtzclaw, D. B. Moss, and C. S. Parmenter, *J. Chem. Phys.* **84**, 1119 (1986); (*f*) D. L. Catlett, K. W. Holtzclaw, D. Krajnovich, D. B. Moss, C. S. Parmenter, W. D. Lawrance, and A. E. W. Knight, *J. Phys. Chem.* **89**, 1577 (1985).
35. D. A. Dolson, B. M. Stone, and C. S. Parmenter, *Chem. Phys. Lett.* **81** 360 (1981).
36. (*a*) R. Moore, F. E. Doany, E. J. Heilweil, and R. M. Hochstrasser, *Faraday Discuss. Chem. Soc.* **75**, 331 (1983); (*b*) R. Moore, F. E. Doany, E. J. Heilweil, and R. M. Hochstrasser, *J. Phys. Chem.* **88**, 876 (1984); (*c*) R. Moore and R. M. Hochstrasser, *Chem. Phys. Lett.* **105**, 359 (1984).

37. (*a*). J. Chaiken, T. Benson, M. Gurnick, and J. D. McDonald, *Chem. Phys. Lett.* **61**, 195 (1979); (*b*) J. Chaiken, M. Gurnick, and J. D. McDonald, *J. Chem. Phys.* **74**, 106 (1981); (*c*) J. Chaiken and J. D. McDonald, *J. Chem. Phys.* **77**, 669 (1982); (*d*) H. Henke, H. L. Selzle, T. R. Hays, S. H. Lin, and E. W. Schlag, *Chem. Phys. Lett.* **77**, 448 (1981).
38. For reviews of quantum beats and interference effects see (*a*) S. Haroche, in *High Resolution Laser Spectroscopy*. (K. Shimoda, ed.). Springer, New York, 1976, p. 254; (*b*) J. N. Dodd and G. W. Series, in *Progress in Atomic Spectroscopy*, (W. Hanle and H. Kleinpoppen, eds.). Plenum, New York, 1978, Part A; (*c*) R. N. Zare, *Accts. Chem. Res.* **4**, 361 (1971).
39. W. R. Lambert, P. M. Felker, and A. H. Zewail, *J. Chem. Phys.* **81**, 2217 (1984).
40. P. M. Felker and A. H. Zewail, *Chem. Phys. Lett.* **102**, 113 (1984).
41. P. M. Felker and A. H. Zewail, *Chem. Phys. Lett.* **108**, 303 (1984).
42. P. M. Felker and A. H. Zewail, *J. Chem. Phys.* **82**, 2975 (1985).
43. P. M. Felker and A. H. Zewail, *J. Chem. Phys.* **82**, 2994 (1985).
44. D. Semmes and A. H. Zewail (unpublished).
45. P. M. Felker, W. R. Lambert, and A. H. Zewail, *J. Chem. Phys.* **82**, 3003 (1985).
46. J. S. Baskin, M. Dantus, and A. H. Zewail, *Chem. Phys. Lett.* **130**, 473 (1986).
47. P. M. Felker, J. S. Baskin, and A. H. Zewail, *J. Phys. Chem.* **90**, 724 (1986).
48. J. S. Baskin, P. M. Felker, and A. H. Zewail, *J. Chem. Phys.* **84**, 4708 (1986).
49. P. M. Felker and A. H. Zewail, *J. Chem. Phys.* **86**, 2460 (1987).
50. J. S. Baskin, P. M. Felker, and A. H. Zewail, *J. Chem. Phys.* **86**, 2483 (1987).
51. (*a*) J. W. Perry, N. F. Scherer, and A. H. Zewail, *Chem. Phys. Lett.* **103**, 1 (1983); (*b*) N. F. Scherer, J. W. Perry, F. Doany, and A. H. Zewail, *J. Phys. Chem.* **89**, 894 (1985).
52. N. F. Scherer, J. F. Shepanski, and A. H. Zewail, *J. Chem. Phys.* **81**, 2181 (1984).
53. G. M. Nathanson and G. M. McClelland, *J. Chem. Phys.* **81**, 629 (1985).
54. C. Tric, *Chem. Phys.* **14**, 189 (1976).
55. W. M. Gelbart, D. F. Heller, and M. L. Elert, *Chem. Phys.* **7**, 116 (1975).
56. K. F. Freed, *Chem. Phys. Lett.* **42**, 600 (1976).
57. K. F. Freed and A. Nitzan, *J. Chem. Phys.* **73**, 4765 (1980).
58. G. Herzberg, *Infrared and Raman Spectra of Polyatomic Molecules*. Van Nostrand, Princeton, 1945.
59. (*a*) S. Okajima, H. Saigusa, and E. C. Lim, *J. Chem. Phys.* **76**, 2096 (1982); (*b*) B. J. van der Meer, H. T. Jonkman, G. M. ter Horst, and J. Kommandeur, *J. Chem. Phys.* **76**, 2099 (1982); (*c*) P. M. Felker, W. R. Lambert, and A. H. Zewail, *Chem. Phys. Lett.* **89**, 309 (1982).
60. (*a*) W. Sharfin, M. Ivanco, and S. C. Wallace, *J. Chem. Phys.* **76**, 2095 (1982); (*b*) M. Ivanco, J. Hager, W. Sharfin, and S. C. Wallace, *J. Chem. Phys.* **78**, 6531 (1983).
61 W. R. Lambert, P. M. Felker, J. A. Syage, and A. H. Zewail, *J. Chem. Phys.* **81**, 2195 (1984).
62. B. W. Keelan and A. H. Zewail, *J. Chem. Phys.* **82**, 3011 (1985).
63. W. R. Lambert, P. M. Felker, and A. H. Zewail, *J. Chem. Phys.* **81**, 2209 (1984).
64. (*a*) D. J. Evans and D. B. Scully, *Spectrochim. Acta* **20**, 891 (1964); (*b*) B. N. Cyvin and S. J. Cyvin, *J. Phys. Chem.* **73**, 1430 (1969).
65. See, for example, S. M. Lederman, J. H. Runnels, and R. A. Marcus, *J. Phys. Chem.* **87**, 4364 (1983).
66. A. Amirav and J. Jortner, *Chem. Phys. Lett.* **95**, 295 (1983).
67. T. S. Zwier, E. Carrasquillo M., and D. H. Levy, *J. Chem. Phys.* **78**, 5493 (1983).
68. J. A. Syage, P. M. Felker, and A. H. Zewail, *J. Chem. Phys.* **81**, 4686 (1984).
69. D. A. Chernoff and S. A. Rice, *J. Chem. Phys.* **70**, 2511 (1979).
70. (*a*) D. K. Negus, D. S. Green, and R. M. Hochstrasser, *Chem. Phys. Lett.* **117**, 409 (1985); (*b*) A. J. Bain, P. J. McCarthy, and R. M. Hochstrasser, *Chem. Phys. Lett.* **125**, 307 (1986); (*c*) A. B. Myers, P. L. Holt, M. A. Pereira, and R. M. Hochstrasser, *Chem. Phys. Lett.* **130**, 265 (1986).
71. J. S. Baskin, D. Semmes, and A. H. Zewail, *J. Chem. Phys.* **85**, 7488 (1986).

72. (*a*) Y. Matsumoto, L. H. Spangler, and D. W. Pratt, *Chem. Phys. Lett.* **95**, 343 (1983); (*b*) **98**, 333 (1983).
73. (*a*) B. J. van der Meer, H. T. Jonkman, and J. Kommandeur, *Laser Chem.* **2**, 77 (1983); (*b*) B. J. van der Meer, H. Th. Jonkman, J. Kommandeur, W. L. Meerts, and W. A. Majewski, *Chem. Phys. Lett.* **92**, 565 (1982).
74. S. Mukamel et al., (in this volume and *J. Chem. Phys.*, in press).
75. (*a*) G. M. Stewart and J. D. McDonald, *J. Chem. Phys.* **78**, 3907 (1983); (*b*) G. M. Stewart, M. D. Ensminger, T. J. Kulp, R. S. Ruoff, and J. D. McDonald, *J. Chem. Phys.* **79**, 3190 (1983); (*c*) G. M. Stewart, T. J. Kulp, R. S. Ruoff, and J. D. McDonald, *J. Chem. Phys.* **80**, 5353 (1984); T. J. Kulp, R. S. Ruoff, and J. D. McDonald, *J. Chem. Phys.* **80**, 5359 (1984).
76. See, N. Bloembergen and E. Yablonovitch, *Physics Today* **31**, 23 (1978), and references therein.
77. (*a*) E. Mazur, I. Burak, and N. Bloembergen, *Chem. Phys. Lett.* **105**, 258 (1984); (*b*) J. Wang, K.-H. Chen, and E. Mazur, *Phys. Rev. A* (to be published).
78. P. Mukherjee and H. S. Kwok, Proceedings of Ultrafast Phenomena, Snowmass, Colorado, p. 176, 1986.
79. N. F. Scherer, F. E. Doany, A. H. Zewail, and J. W. Perry, *J. Chem. Phys.* **84**, 1932 (1986); N. F. Scherer and A. H. Zewail, *J. Chem. Phys.* **87**, 97 (1987); see also L. Khundkar, J. Knce, and A. H. Zewail, *J. Chem. Phys.* **87**, 77 (1987); *J. Chem. Phys.* **87**, 115 (1987).
80. See, for example, (*a*) D. E. Reisner, P. H. Vaccaro, C. Kittrell, R. W. Field, J. L. Kinsey, and H. L. Dai, *J. Chem. Phys.* **77**, 573 (1982); (*b*) E. Abramson, R. W. Field, D. Imre, K. K. Innes, and J. L. Kinsey, *J. Chem. Phys.* **80**, 2298 (1984).
81. W. D. Lawrance and A. E. W. Knight, *J. Phys. Chem.* **87**, 389 (1982).

# CHAOS AND REACTION DYNAMICS

PAUL BRUMER

*Chemical Physics Theory Group, Department of Chemistry, University of Toronto, Toronto, Ontario, Canada M5S 1A1*

MOSHE SHAPIRO

*Department of Chemical Physics, Weizmann Institute of Science, Rehovot, Israel*

## CONTENTS

# I. INTRODUCTION

The expectation that classical mechanics provides a simple, deterministic, easily predictable view of the dynamics of few-body systems is now recognized as a gross oversimplification. Research over the past 20 years has shown that such systems are capable of displaying relaxation to equilibrium and extreme sensitivity to both initial conditions as well as system parameters. These features, quantified subsequently, are essential characteristics of what is now termed "chaotic behavior" in a conservative Hamiltonian system. The relationship between chaotic behavior in conservative systems and the reaction dynamics of isolated molecules is the subject of this chapter.

The reaction dynamics of isolated molecules is at the heart of chemistry, and enormous effort has been directed toward the goal of obtaining a theoretical understanding of the detailed nature of elementary molecular processes.[1] Work in this area has long recognized the need for simplifying theories that can be extended to the realm of large molecules, where most of traditional chemistry takes place. It is this need that justifies the major effort which has been directed toward developing and understanding the range of validity of simplified *statistical* theories of dynamics and intramolecular relaxation (IVR). It is primarily through these statistical theories that reaction dynamics links to chaotic behavior.

The essential nature of this relationship is clear; statistical theories are based on a number of simplifying assumptions consistent with chaotic behavior. Specifically,[2] any such theory must satisfy microscopic reversibility and the condition of "zero relevance." The latter condition requires that the final state be independent of all initial conditions other than conserved quantities, that is, from the viewpoint of classical mechanics, that the system display the relaxation characteristic of chaotic motion. We note, for reference, that this minimal set of requirements allows for the construction of a large number of theories,[3] the most prominant of which are the RRKM theory of unimolecular dissociation[4] and the phase space theory of bimolecular reactions.[5] Such theories have analogues, and in some cases their origins are in other areas such as nuclear physics.[6]

Recognition that the basis of statistical theories of dynamics lies in motion now termed chaotic allows for a serious effort toward providing a sound theoretical basis, rooted in dynamics, for the applicability of statistical theories of both reaction dynamics and IVR. Further stimulus comes from theoretical developments in nonlinear mechanics[7] as well as through experimental results[8] which indicate that phase space equilibration may not be achieved, even for long-lived systems. Such detailed studies of the dynamical origins of statistical behavior in isolated molecular processes, as well as studies in chaotic motion, have been ongoing for over a decade and, ideally, we should be able to provide a global picture of conditions under which chemical reactions, or IVR, displays statistical behavior. This is, unfortunately, not the case since such a picture requires the resolution of a number of fundamental unsolved problems. Specifically, a full theory would entail the following sequence of steps:

1. Since chemical dynamics is dictated by quantum mechanics, one begins with quantum scattering theories of reaction dynamics or molecular energy flow.
2. The exact dynamical theory is reduced to quantum statistical theory[4,5] by invoking a set of well-defined approximations. Since the origin of statistical concepts is in the "loss of information" regarding the initial system state through relaxation, one would expect to base this approximation route on concepts of chaos in quantum ergodic theory.[9] In doing so one would also identify quantum conditions for statistical relaxation dynamics.
3. Conditions on system properties, for example, potential surfaces, state densities, masses, and so forth, which are necessary for relaxation and which emerge from quantum ergodic theory, would be used to identify properties of molecular systems necessary and sufficient to ensure statistical reaction dynamics.
4. Since quantum mechanics is difficult to extend to large molecules, one can then, if desired, consider the class of statistical theories wherein the dynamics (but not the intra- and intermolecular forces) are treated classically. A sequence of approximations to the three steps above would be required:
   a. Quantum dynamics is replaced by classical mechanics, with the establishment of conditions under which this approximation is valid.
   b. Quantum statistical theories would be reduced, via a sequence of well-defined approximations, to derive classical statistical theories. This would be accompanied by a similar reduction of chaos concepts in quantum ergodic theory to related concepts in classical ergodic theory.
   c. The conditions which emerge from b, in conjunction with those for the validity of the classical approximation to quantum mechanics, would be used to identify conditions under which classical statistical theories are valid, that is, under which reaction dynamics and IVR display chaotic motion.

The number of difficulties associated with carrying out the preceding program is formidable. First and foremost, there is no satisfactory theory of chaotic motion in quantum ergodic theory, either for isolated bound or unbound, systems. Indeed, the qualitative concept of relaxation that we traditionally associate with statistical theories is only consistent with results for *ideal* systems in the *classical* ergodic theory of *bound* systems. Second, we possess surprisingly little insight into the issue of classical–quantum correspondence in the regime where classical mechanics displays ergodic motion. Thus, carrying out the program previously outlined is tantamount to resolving a number of fundamental issues in physics. As a consequence of these difficulties we have, at present, only a partial understanding of a few of the individual steps in the outlined program.

This chapter emphasizes a number of these results as well as the problems yet to be solved. We take as our basic processes both unimolecular decay, where a molecule $A$ has sufficient energy to dissociate, that is,

$$A \rightarrow B + C, \tag{1.1}$$

and bimolecular collisions:

$$F + G \rightarrow H + I. \tag{1.2}$$

Here letters denote arbitrary molecules. Furthermore, we shall assume that the potential energy surfaces for these processes are known. Bound system dynamics will be treated as well since it links, as shown below, to relaxation processes in either $A$ prior to decay, or in an intermediate formed in the $F + G$ collision.

The clearest results have been obtained for classical relaxation in bound systems where the full machinery of classical ergodic theory may be utilized. These concepts have been carried over empirically to molecular scattering and decay, where the phase space is not compact and hence the ergodic theory is not directly applicable. This classical approach is the subject of Section II. Less complete information is available on the classical–quantum correspondence, which underlies step 4. This is discussed in Section III where we introduce the Liouville approach to correspondence, which, we believe, provides a unified basis for future studies in this area. Finally, the quantum picture is beginning to emerge, and Section IV summarizes a number of recent approaches relevant for a quantum-mechanical understanding of relaxation phenomena and statistical behavior in bound systems and scattering.

The areas of research that influence considerations in this area are vast, ranging from issues of irreversibility to quantum chaos. The authors have been selective, rather than exhaustive, in their efforts to present a sketch of the current state-of-the-art.

## II. CLASSICAL MECHANICS

Two areas in classical mechanics have contributed greatly to developments in our understanding of molecular Hamiltonian dynamics. The first is nonlinear mechanics, which we comment upon only briefly to introduce the integrable system; a number of reviews of this topic are available.[7] The second is ergodic theory, some relevant features of which are introduced.

The simplest system in nonlinear mechanics is *integrable* (or regular). This is defined as an $N$ degree of freedom system that possesses $N$ independent single-valued integrals of the motion (or constants of the motion) $F_k(\mathbf{q}, \mathbf{p})$, $k = 1, \ldots, N$, where $\mathbf{q}$, $\mathbf{p}$ are canonical coordinates and momenta. This is, $\{F_k, H\} = 0$, where $H$ is the Hamiltonian and $\{\ ,\ \}$ is the Poisson bracket. In addition, these integrals are in involution, that is, $\{F_k, F_j\} = 0$. Often a canonical transformation is carried out so that the $F_k$ are the new (constant) momenta; choosing $F_k$ as the classical actions $I_k$ with conjugate angles $\theta_k$ is a familiar example. The motion generated by the resultant Hamiltonian $H(\mathbf{I})$ is *quasiperiodic*, that is, can be represented by a Fourier series in the $N$ frequencies $\omega_i = \partial H(\mathbf{I})/\partial I_i$. Since the motion is confined to the surface of a torus in phase space and is quasiperiodic, the dynamics of these systems are, in some sense, at an extreme opposite of that of chaotic behavior.

Classical mechanics provides the least ambiguous statement of the nature of chaotic motion, with chaos also defined through a heirarchy of ideal model systems. We note, at the outset, that isolated molecule dynamics relates to chaotic motion in *conservative* Hamiltonian systems. This is distinct from chaotic motion in dissipative systems where considerable simplifications result from the reduction in degrees of freedom during evolution[10] and where objects such as strange attractors and fractal dimensions play an important role.

Subsection A contains a summary of the formal definitions of chaotic behavior, derived from ergodic theory; detailed discussions of this topic may be found elsewhere.[11] We comment, in this section, on the gap that must be bridged in order to apply these concepts to chemical dynamics. Subsection B discusses some recent developments in computational signatures of chaos. In Subsection C we review a number of studies that have provided some of these links and that, in some instances, have resulted in new useful computational methods for treating the dynamics of reactions displaying chaotic dynamics. In addition, it includes a subsection on connection between formal ergodic theory and statistical behavior in unimolecular decay.

### A. Input from Classical Ergodic Theory: What Is Chaos?

Classical ergodic theory defines a number of formal *ideal* model *bound* systems that display a heirarchy of properties of increasing statistical nature. Specifically, consider a dynamical system defined on a compact phase space (e.g., a bound molecule) with coordinates and momenta $\mathbf{q}$, $\mathbf{p}$. Of interest is the dynamics of a state, defined as a distribution $f(\mathbf{p}, \mathbf{q}, t = 0)$ at time $t = 0$ that evolves under the influence of the Liouville operator (see Section III) $L_c$ to

$f(\mathbf{p}, \mathbf{q}, t)$ at time $t$. To be consistent with usage in physics we assume that the only isolated conserved integral of motion is the energy and denote the average over the energy hypersurface as $\langle f \rangle$. The first two members of the heirarchy, of importance for reaction dynamics, may be defined in terms of time-dependent dynamics as:

1. *Ergodic Systems:* An ergodic system displays the equality of space and time averages almost everywhere, that is,

$$\lim_{T \to \infty} \int_0^T f(\mathbf{q}(t), \mathbf{p}(t))\, dt = \langle f \rangle. \tag{2.1}$$

2. *Mixing Systems:* A mixing system displays the following properties:
   a. $\lim_{t \to \infty} f[\mathbf{q}(t), \mathbf{p}(t)] = \langle f \rangle$.
   b. The correlation between any two dynamical properties, that is,

$$\langle g[\mathbf{q}(t), \mathbf{p}(t)], f[\mathbf{q}(0), \mathbf{p}(0)] \rangle - \langle g[\mathbf{q}(t), \mathbf{p}(t)] \rangle \langle f[\mathbf{q}(0), \mathbf{p}(0)] \rangle \tag{2.2}$$

   goes to zero as $t \to \infty$.
   c. Subdivide the total phase space into regions of particular volume. Then the probability of going from region $i$ to region $j$ in the long-time limit depends only on the size of the phase space regions $i$ and $j$.

   A mixing system is ergodic.

Also of interest are higher-order statistical systems such as $K$ systems and $C$ systems.[11] The term chaotic, as used herein, is reserved for all systems that are at least mixing, that is, display long-time relaxation.

A time-independent definition of such systems is contained in the characteristic spectrum of $L_c$. For example, denoting the eigenvalue associated with the eigenvalue problem $L_c \rho = \lambda \rho$ as $\lambda$, we have[12]

1. *Ergodic:* The point spectrum[13] at $\lambda = 0$, that is, the stationary solution, is nondegenerate.
2. *Mixing:* In addition to ergodicity, $\lambda = 0$ is the only member of the point spectrum.
3. *K System:* In addition to mixing, the spectrum displays equal degeneracy of each eigenvalue.

There are a number of difficulties associated with transferring these concepts to molecular dynamics. First, these definitions relate to infinite-time properties. Molecular dynamics is, however, of concern only over a finite time, being limited by competitive processes such as the dissociation of an excited molecule, measurement, or spontaneous emission. Over such a finite-time scale a system that is formally mixing may appear quasiperiodic or vice versa. Second, proving that a particular system is ergodic, mixing, and so forth is extremely difficult and, at present, only a small number of physically relevant systems, such as the hard sphere gas[14] and stadium billiard,[15] have been classified according to this heirarchy. Finally, note that these concepts apply only to bound systems, rather than to the noncompact phase spaces associated with dissociation or scattering. For these reason, efforts have been directed toward identifying particular characteristics of chaotic systems that may serve as useful *computational* identifiers of chaotic behavior. Three specific computational tools are currently in use: local trajectory instability, erratic power spectra, and irregular Poincare surfaces of section. We focus on the former two, which are particularly relevant to the discussion below and which have recently been the subject of new developments. A discussion of the Poincaré surface of section can be found elsewhere.[7]

## B. Signatures of Chaos

*Local Trajectory Instability:* That is, the local (in phase space) exponential sensitivity of trajectories to changes in initial conditions. This property is the primary characteristic of a $C$ system. Specifically, consider, in an $N$ degree of freedom system, two trajectories with initial coordinates and momenta $\mathbf{q}_0$, $\mathbf{p}_0$ and $\mathbf{q}'_0$, $\mathbf{p}'_0$. For convenience we denote the column vector $[\mathbf{q}(t), \mathbf{p}(t)]$ associated with $\mathbf{p}_0$, $\mathbf{q}_0$ as $\mathbf{x}(t)$ and the trajectory $[\mathbf{q}'(t), \mathbf{p}'(t)]$ associated with $\mathbf{p}'_0$, $\mathbf{q}'_0$ as $\mathbf{x}'(t)$. The phase space separation $d(t)$ between trajectories is given by

$$d(t) = |\mathbf{d}(t)| = \left( \sum_i [x_i(t) - x'_i(t)]^2 \right)^{1/2}. \tag{2.3}$$

If the system displays local exponential instability, then

$$d(t) = d(0)\exp(ht), \tag{2.4}$$

where $h$ is discussed below. From the trajectory viewpoint it is this extreme sensitivity to initial conditions that is responsible for the loss of information during the course of the dynamics. That is, two initially close phase space points become rapidly dispersed during the course of the evolution.[16]

To quantify the rate of exponential growth (and contraction) in phase space consider the equation of motion for $d(t)$. Application to Hamilton's equations to the two trajectories $\mathbf{x}(t)$, $\mathbf{x}'(t)$ gives directly

$$\dot{\mathbf{d}} = \mathbf{S}(t)\mathbf{d}(t), \tag{2.5}$$

$$[\mathbf{S}(t)]_{ij} = \frac{\partial x_i(t)}{\partial x_j(0)}. \tag{2.6}$$

Here the overdot indicates a derivative with respect to time, and $\mathbf{S}(t)$ is the stability matrix, evaluated along $\mathbf{x}(t)$; the dependence of $\mathbf{S}(t)$ on $\mathbf{x}(0)$ has been suppressed. A quantitative measure of the growth rate of $\mathbf{d}(t)$ is given by the Lyapunov exponents[10,17] $\beta_i$, $i = 1, 2N$ defined as follows. Consider the matrix $\mathbf{B}$, with eigenvalues $b_i$, defined by

$$\mathbf{B} = \lim_{t\to\infty} [\mathbf{S}^+(t)\mathbf{S}(t)]^{1/(2t)}, \tag{2.7}$$

where the + superscript denotes the matrix transpose. Then

$$\beta_i = \log b_i. \tag{2.8}$$

The nature of Eq. (2.7) is such that only long-time exponential growth or contraction, observed in chaotic systems, yields nonzero $\beta_i$. Quasiperiodic systems, for example, display linear growth with time of the eigenvalues of $\mathbf{S}^+(t)\mathbf{S}(t)$, and hence zero $\beta_i$. For conservative Hamiltonian systems the $\beta_i$ come in pairs; both $\beta_i$ and $-|\beta_i|$ are eigenvalues. Thus, for each expansion direction there is a direction of contraction in phase space, the net result being conserved volume in accord with Liouville's theorem. Since the Lyapunov exponents are defined in the long-time limit, local (in time) exponential growth does not guarantee nonzero Lyapunov exponents. That is, the instantaneous directions of expansion and contraction in phase space are along the eigenvectors of $\mathbf{S}^+(t)\mathbf{S}(t)$, which change direction in time, so that local divergence may be followed by compensatory contraction, leading to a zero $\beta_i$. Nonetheless, there is reason to argue that exponential divergence over a time scale of interest suffices to label the system as "operationally chaotic."

A fundamental issue, as yet unresolved in chemical dynamics, is the question of the relationship among features of the potential energy hypersurface and chaotic behavior. It is clear from Eq. (2.5) that $\mathbf{S}(t)$ provides information on the *instantaneous* growth rate of $\mathbf{d}(t)$ and that this, in turn, is directly related to the eigenvalues of the $\mathbf{S}(t)$ *evaluated along the trajectory* $\mathbf{x}(t)$ *initiated at* $\mathbf{x}(0)$. An approximate variational equations method[18] eliminates this time dependence by linearizing Eq. (2.5) about $\mathbf{x}(t)$ and leads to the result that exponential divergence of adjacent trajectories occurs in the region of the potential surface where the eigenvalues of the force constant matrix $-(m_i m_j)^{-1/2}\, \partial^2 V/\partial q_i \partial q_j$ are real. Here the Hamiltonian is assumed to be written in cartesian coordinates, where the kinetic energy is a quadratic form in the momenta. This method links local exponential divergence to regions of negative curvature on the potential surface, in accord with some results on model $C$ systems. The approach has two advantages. First and foremost it allows one to examine the potential surface, in conjunction with the contour plot of the eigenvalues of $-(m_i m_j)^{-1/2}\, \partial^2 V/\partial q_i \partial q_j$ to ascertain, in a time independent way, the regions of coordinate space that induce exponential divergence. Doing so allows a prediction of the lowest energy (the so-called critical energy) at which one can begin to observe extensive exponential divergence.[19] Second, this approach predicts that exponential divergence will occur, along a given trajectory, when the trajectory enters a region of negative potential curvature. These features have been confirmed in computations on a number of systems.[20] Examples are shown in Figs. 1 and 2.

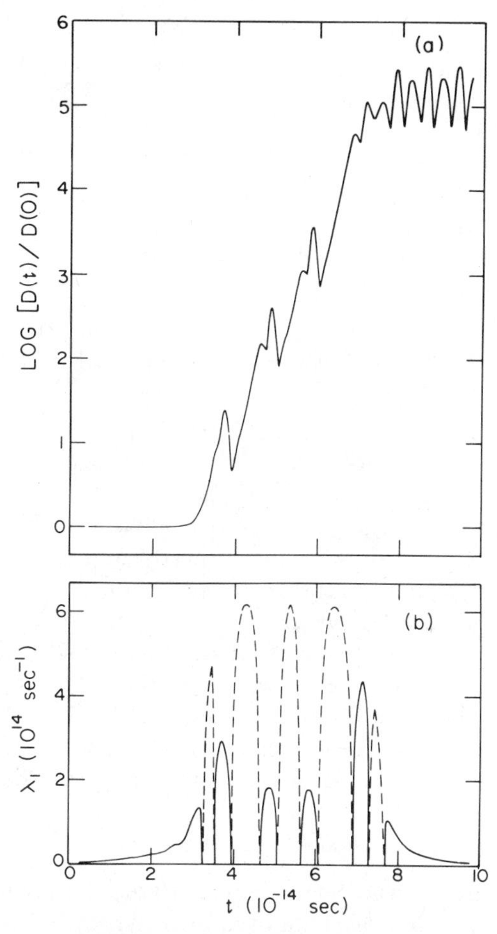


**Figure 1.** Typical long-lived trajectory at $E = 0.873$ in $H + H_2$ on the collinear Porter–Karplus potential surface. (*a*) $\log[d(t)/d(0)]$ versus $t$. (*b*) An eigenvalue of $\mathbf{S}(t)$ along the unperturbed trajectory yielding panel (*a*). Dashed curves denote a real $\mathbf{S}(t)$ eigenvalue, and solid curves denote an imaginary eigenvalue. Note the correspondence between exponential growth and times where the system encounters a real eigenvalue. (From Ref. 20.)

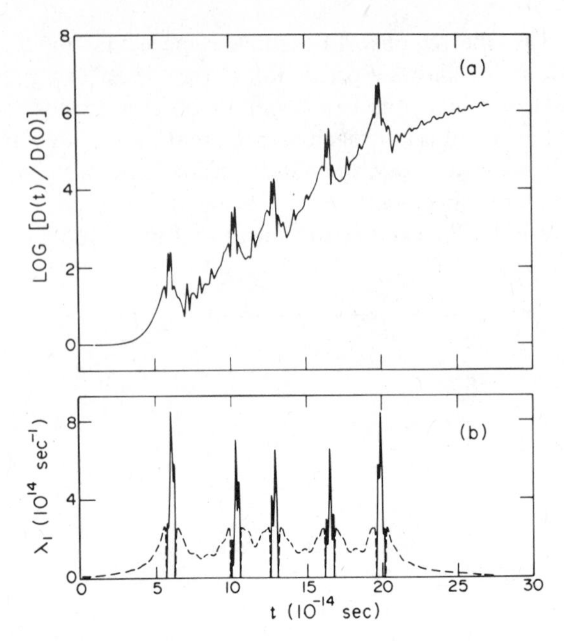


**Figure 2.** As in Fig. 1, but for the collinear coupled Morse system at translational energy $E_T = 0.1$ eV. (From Ref. 20.)

The variational equations approach, however, being local in time and resulting from a linearization of Eq. (2.5), has a number of disadvantages, which have been reviewed elsewhere.[7b] Specifically, the method fails to identify certain systems as chaotic although their exact dynamics reveals them as such. Furthermore, the method is insensitive to possible compensatory contraction, subsequent to expansion. Despite several modifications[21] designed to resolve these problems, no completely successful method for linking potential surface properties to exponential divergence is available. (This statement applies as well to the popular overlapping resonance method of Chirikov.[22] Since the extent of overlapping resonances depends on the arbitrary choice of a zeroth-order Hamiltonian $H_0$, overlapping resonances can simply result from an improper choice of $H_0$.)

A number of general comments on exponentiating trajectories are in order:

1. Exponential divergence in systems that are chaotic prevents accurate long-time trajectory calculations of their dynamics. That is, numerical errors[18a] propagate exponentially during the dynamics so that accuracy beyond 100 characteristic periods of motion is extraordinarily difficult to achieve; thus, accurate long-time dynamics is essentially "uncomputable" for chaotic classical systems. This serves as additional

motivation for the development of statistical theories and for alternatives to traditional trajectory techniques, such as those discussed later below.

2. Any arbitrary initial $d(0)$ will tend to grow with the maximum Lyapunov exponent. Thus, observing exponential growth of $d(t)$ only confirms the presence of one positive Lyapunov exponent amongst the possible $N$ positive $\beta_i$. Alternative numerical techniques[23] are available to compute all $N\beta$ values and have been applied[24,25] to model molecular dynamics. However, the suggestion that the observation of a zero value of $\beta_i$ implies an underlying constant of the motion[25] must be regarded as conjecture.[17]
3. There is a common misconception that the Lyapunov exponents, or the $K$-entropy defined below, relates directly to the rate of exponential relaxation of observables in chaotic systems. The actual situation is far more complex, as discussed in Section II C.
4. The $K$-entropy of a dynamical system[7b,10] is directly related to the Lyapunov exponents via

$$K(E) = \int d\mathbf{p}\, d\mathbf{q} \sum \beta_i \delta(E - H(\mathbf{p}, \mathbf{q})), \tag{2.9}$$

where the sum is over $\beta_i > 0$.

*"Grassy" Power Spectra:* The principal measure of the frequency content of a trajectory is the power spectrum defined as

$$I_T[\omega, \mathbf{x}(0)] = \left(\frac{1}{T}\right) \left| \int_0^T dt\, e^{-i\omega t} f[\mathbf{x}(t)] \right|^2, \tag{2.10}$$

where $f[\mathbf{x}(t)]$ is the dynamical variable $f$ evaluated along the trajectory initiated at $\mathbf{x}(0)$. The nature of $I_T[\omega, \mathbf{x}(0)]$ for quasiperiodic systems is well understood,[26] showing sharp peaks, which are fundamentals, overtones, and combinations of a set of $N$ frequencies. That is, the power spectrum reveals the discrete frequency spectrum associated with dynamics on the single torus to which the trajectory is confined (for a discussion of the complete frequency spectrum see Section III). The power spectrum for a chaotic system is, however, comprised of a typically large set of apparently unrelated peaks, that is, the spectrum appears "grassy." Dumont[27] has recently provided a careful analysis of the nature, and information content, of such chaotic spectra. Specifically, he finds, both formally and computationally, that a reasonable model of chaotic motion implies the following features of the power spectrum: (1) $I_T[\omega, \mathbf{x}(0)]$ and $I_T[\omega', \mathbf{x}(0)]$, $\omega \neq \omega'$, are asymptotically (i.e., large $T$)

independent for $|\omega - \omega'| > 1/T^{\varepsilon}$, $0 < \varepsilon < 1$. This implies that the spectrum shows increasingly grassy character as time progresses, with different values of $\omega$ being unrelated. (2) $I_T[\omega, \mathbf{x}(0)]$ and $I_{T'}[\omega, \mathbf{x}(0)]$, $T \neq T'$, are asymptotically independent if $T' > T^{\nu}$, $\nu > 2$, implying that the power spectrum does not converge with increasing time. (3) The probability of observing a given value of $I_T[\omega, \mathbf{x}(0)]$ with respect to a random sampling of $\mathbf{x}(0)$ is asymptotically Poissonian. The implication of these results is that the power spectrum for a chaotic system cannot provide detailed information on the dynamics, other than to indicate its chaotic character. For example, information such as relaxation rates cannot be extracted from the power spectrum of a single trajectory.

Qualitative justification for the highly random nature of the power spectrum in a chaotic system is readily given. That is, $I_T[\omega, \mathbf{x}(0)]$ is a function of a single trajectory. Since a single trajectory is known to depend sensitively on initial conditions, any such function is expected to show such extreme sensitivity as well. Nonetheless, it is necessary to reconcile this result with standard textbook statements[28] of the Wiener–Khinchin theorem,[29] which equate the power spectrum and spectral density $S(\omega, M)$:

$$S(\omega, M) = \int_0^{\infty} dt\, e^{i\omega t} \langle f | f(t) \rangle_M. \tag{2.11}$$

Here $\langle\ \rangle_M$ denotes an average over the manifold $M$ traced out by the trajectory initiated at $\mathbf{x}(0)$. Since the chaotic system is ensured of relaxing, $S(\omega, M)$ (and hence presumably $I_T$) should be *smooth*, reflecting the underlying continuous Liouville operator spectrum. Indeed, it is this relation of $I_T$ to $S(\omega, M)$ that leads one to anticipate some physical content, for example, relaxation times, in the power spectrum. However, the ragged $I_T$ discussed previously do not converge nor do they reveal a smooth underlying spectrum characteristic of $S(\omega, M)$. Dumont[27] has carefully examined this apparent contradiction and noted the following: The precise statement of the Wiener–Khinchin theorem[29] relating the power spectrum and spectral density is

$$\lim_{T \to \infty} \| I_T(\omega, \mathbf{x}(0)) - S(\omega, M) \|^2 = 0, \tag{2.12}$$

that is, that the power spectrum converges to the spectral density *in the mean.* In typical physical problems convergence in the mean is taken as pointwise convergence in $\omega$. In this case, however, where $I_T$ is highly nonconvergent, these two limits (convergence in the mean and pointwise convergence) are quite different. In essence[27] $I_T$ is comprised of a contribution from the spectral density plus ragged noise contributions that completely alter the character of $I_T$ relative to $S(\omega, M)$, the latter being the function which provides relaxation information. In addition, Dumont has shown[27] that although $S(\omega, M)$ is unrelated to $I_T$ for a single trajectory, an average over a set of trajectories eliminates the noise terms to reveal $S(\omega, M)$.

### C. Statistical Behavior and Chaos

Surface of section, exponentiating trajectories, and power spectra studies[30] on model molecular systems have shown that, typically, the dynamics at low vib-rotational energies resembles that of a quasiperiodic system, whereas at higher energies the systems show the characteristics associated with chaotic behavior. The pertinent question is whether these characteristics suffice to ensure that the dynamics behaves statistically. Research into this issue falls into three categories: (1) phenomenological studies of the relationship between chaotic properties and statistical theories of reactions; (2) numerical studies on real and model systems designed to expose the relationship between conditions of chaos and statistical behavior; and (3) studies on the formal link between ergodic theory and statistical theories of reaction dynamics. Results in each of these areas are discussed. The order of the topics is essentially historical.

#### *1. Phenomenological Criteria for Statistical Dynamics*

To what extent can characteristics of chaotic behavior be linked to the validity and utility of statistical theories of (isolated) unimolecular and bimolecular reaction dynamics? Exponentiating trajectories have served, in addressing this question, as the most useful of the three characteristics previously listed. In particular, we review results which indicate that the *extent* of exponential divergence during the course of the dynamics provides a means of identifying the subregions of phase space that evolve in accord with statistical theories. Furthermore, we note that this approach affords a means of replacing long-lived complicated trajectory dynamics by a method requiring only a minimal amount of computation.

In order to introduce notation and to emphasize the range of possible theories that can be termed statistical, we first summarize some essential aspects of statistical theories. An extension[31] of the bimolecular collision formulation of Wagner and Parks[3] provides a convenient starting point.

Consider an atom–diatom $A + BC(E_{BC}, j_{BC})$ collision at total energy $E$ with $AB(E_{AB}, j_{AB}) + C$ as one possible product. Here $E_{BC}, j_{BC}$ are the internal energy and rotational angular momentum of the $BC$ diatom and $E_{AB}, j_{AB}$ are the analogous $AB$ variables. The differential cross section for scattering into $dE_{AB}\, dj_{AB}$ (other cross sections have analogous formulations) is given by

$$
\begin{aligned}
\frac{d\sigma(E_{BC},j_{BC}\to E_{AB},j_{AB})}{dE_{AB}\,dj_{AB}} &= \frac{\pi}{2j_{BC}P_A^2}\int_0^\infty 2J\,dJ\int_{|J-j_{BC}|}^{J+j_{BC}} dl_A \\
&\times \int_0^{\tau_{BC}} (dt_{BC}/\tau_{BC})\int_0^{2\pi} (dq_{j_{BC}}/2\pi)\int_0^{2\pi} (dq_{l_A}/2\pi) \\
&\times \int_{|J-j_{AB}|}^{J+j_{AB}} dl_C \int_0^{\tau_{AB}} dt_{AB}\int_0^{2\pi} dq_{j_{AB}} \\
&\times \int_0^{2\pi} dq_{l_C}\,P^{EJ}(\Gamma_A;R_A|\Gamma_C,R_C). \qquad (2.13)
\end{aligned}
$$

Here $J$ is the total angular momentum and $l_A$ is the relative orbital angular momentum of $A$ with respect to $BC$. The variables $t_{BC}$ (of period $\tau_{BC}$), $q_{j_{BC}}$, $q_{l_A}$ are conjugate to $E_{BC}$, $j_{BC}$, $l_A$, and $\Gamma_A$ denotes the collection $(E_{BC}, j_{BC}, l_A, t_{BC}, q_{j_{BC}}, q_{l_A})$ appropriate to the $A + BC$ channel. The quantity $R_A$ is the $A$ to $BC$ distance and $P_A$ is its conjugate momentum. Equation (2.13) is a completely general expression, with $P^{EJ}(\Gamma_A; R_A|\Gamma_C, R_C)\,d\Gamma_C$ being the probability that a reactant trajectory, defined by $\Gamma_A$, $R_A$, will be found in the product region $d\Gamma_C$. This probability is zero unless $\Gamma_A$, $R_A$ and $\Gamma_C$, $R_C$ refer to the same trajectory. To define a statistical theory one first introduces the concept of a strong collision region (SCR) in phase space and partitions the set of trajectories associated with initial conditions $E_{BC}$, $j_{BC}$ into two sets, those which pass through the SCR and those which do not; the latter are termed direct. Then

$$
P^{EJ}(\Gamma_A;R_A|\Gamma_C,R_C) = P^{EJ}_{\text{dir}}(\Gamma_A;R_A|\Gamma_C,R_C) + P^{EJ}_{\text{stat}}(\Gamma_A;R_A|\Gamma_C,R_C). \qquad (2.14)
$$

Here $P^{EJ}_{\text{dir}}(\Gamma_A; R_A|\Gamma_C, R_C)$ is unity if the trajectory specified by $\Gamma_A$, $R_A$ does not pass through the (as yet unspecified) SCR and zero otherwise, while $P^{EJ}_{\text{stat}}(\Gamma_A; R_A|\Gamma_C, R_C)$ is unity if the trajectory passes through the SCR and zero otherwise. The differential cross section is then a sum of non-SCR and SCR contributions:

$$
\begin{aligned}
\frac{d\sigma(E_{BC},j_{BC}\to E_{AB},j_{AB})}{dE_{AB}\,dj_{AB}} &= \frac{d\sigma_{\text{dir}}(E_{BC},j_{BC}\to E_{AB},j_{AB})}{dE_{AB}\,dj_{AB}} \\
&\quad + \frac{d\sigma_{\text{stat}}(E_{BC},j_{BC}\to E_{AB},j_{AB})}{dE_{AB}\,dj_{AB}}. \qquad (2.15)
\end{aligned}
$$

The fundamental statistical approximation may be then introduced

$$
P^{EJ}_{\text{stat}}(\Gamma_A;R_A|\Gamma_C,R_C) = P^{EJ}_{\text{SCR}}(\Gamma_A;R_A)P^{EJ}_{\text{SCR}}(\Gamma_C;R_C)/C(E,J). \qquad (2.16)
$$

Here $C(E, J)$ is a normalization constant,[31] $P_{\text{SCR}}^{EJ}(\Gamma_A; R_A)$ is unity if the reactant trajectory specified by $\Gamma_A$; $R_A$ enters the SCR and zero otherwise, and $P_{\text{SCR}}^{EJ}(\Gamma_C; R_C)$ is unity if the product trajectory specified by $\Gamma_C$, $R_C$ emerges from the SCR; otherwise it is zero. Explicit note should be taken of the form of Eq. (2.16), which serves to partition the probability of reaction into a product of terms, one dependent on the reactants and one on the product. This separable form is the essence of all statistical approximations.

Utilizing Eqs. (2.15) and (2.16) gives, for Eq. (2.14)

$$\frac{d\sigma_{\text{stat}}}{dE_{AB}\,dj_{AB}} = \frac{\pi}{2j_{BC}P_A^2}\int_0^{\infty} 2J\,dJ \int_{|J-j_{BC}|}^{J+j_{BC}} dl_A\, T_{\text{SCR}}^{EJ}(E_{BC}, j_{BC}, l_A) \times \left( \int_{|J-j_{AB}|}^{J+j_{AB}} dl_C\, \bar{P}_{\text{SCR}}^{EJ}(E_{AB}, j_{AB}, l_C) / C(E, J) \right), \tag{2.17}$$

where

$$T_{\text{SCR}}^{EJ}(E_{BC}, j_{BC}, l_A) = \frac{1}{(2\pi)^2 \tau_{BC}} \int_0^{\tau_{BC}} dt_{BC} \int_0^{2\pi} dq_j \int_0^{2\pi} dq_{l_A} \times P_{\text{SCR}}^{EJ}(E_{BC}, j_{BC}, l_A, t_{BC}, q_{j_{BC}}, q_{l_A}; R_A),$$

$$\bar{P}_{\text{SCR}}^{EJ}(E_{AB}, j_{AB}, l_C) = \int_0^{\tau_{AB}} dt_{AB} \int_0^{2\pi} dq_j \int_0^{2\pi} dq_{l_C} \times P_{\text{SCR}}^{EJ}(E_{AB}, j_{AB}, l_C, t_{AB}, q_{j_{AB}}, q_{l_C}; R_C). \tag{2.18}$$

Here $T_{\text{SCR}}^{EJ}(E_{BC}, j_{BC}, l_A)$ is the probability that the reactants defined by $E$, $J$, $E_{BC}$, $j_{BC}$, and $l_A$ enter the SCR, whereas the bracketed term in Eq. (2.17) gives the distribution of products emerging from the SCR.

Several aspects of this formulation require emphasis. First, defining the SCR yields a *particular* statistical theory; hence, a wide variety of theories are possible. Second, one should anticipate that a typical scattering event will have both direct and statistical components. Third, an analysis of the degree of statistical behavior in an exact dynamical calculation requires three sequential steps: (1) the SCR must first be defined; (2) the component of the scattering that does not pass through the SCR is then eliminated from consideration; and (3) the product distribution associated with the component passing through the SCR is compared with the results predicted from Eq. (2.17); the latter often requires additional numerical computations.

Traditional statistical theories are based on strong collision regions that are well-defined regions in phase space, an approach which does not utilize insights afforded by studies of chaos. Consider, as an alternative, defining trajectories that enter the SCR as precisely those trajectories which display the characteristics previously described as chaotic. In doing so, two objectives are achieved. First, the SCR definition is linked directly to chaos properties. Second, one bypasses the need to specify a fixed SCR (which may in fact not apply to a particular system of interest[33]) for all systems, in favor of a more general statement that can be readily tested through the steps 1–3.

Of the surface of section, power spectra, and exponentiating trajectories, only the latter has thus far proven to be a useful computational measure for scattering systems with greater than a few degrees of freedom.[32] Initial studies[20] indicated that most scattering trajectories passing through a transition state are exponentially divergent from nearby neighbors. Thus, the fact that a region of phase space is characterized by exponentially growing $d(t)$ cannot suffice to allow the replacement of these trajectories by a statistical approximation. Furthermore, the rate of exponential divergence tends to be relatively uniform over all phase space. For these reasons, the following conjecture was extensively explored[20,31,33,34]: Those trajectories that display a specified *extent* of exponential divergence during the course of the dynamics are to be classified as trajectories which have passed through the SCR. Numerical evidence on a number of unimolecular and bimolecular models indicated that this approach is useful and that those trajectories which display $d(t)/d(0) > 10^3$ during the course of the dynamics do, indeed, evolve in accord with a statistical theory. Note, that since a typical $d(0) = 10^{-8}$ a.u., the requirement is for growth to $d(t) > 10^{-5}$ a.u., a minute fraction of the total available phase space. A number of examples are provided below.

Consider, as the first example, the collinear reaction[31]

$$F + H_2 \rightarrow H + HF.$$

Panels (*a*) and (*b*) of Fig. 3 show histograms of the translational energy distribution for reactive and nonreactive scattering of $F + H_2$ at the indicated energy. Histograms in panels (*c*) and (*d*) show the same distributions, but for the subset of trajectories characterized by $d(t)/d(0) > 10^3$ during the course of the collision. Both sets of histograms result from explicit trajectory calculations. These results can be compared to the solid lines in each panel, which display the results of two different statistical theories. Specifically, panel (*a*) and (*b*) contains solid lines obtained from Eq. (2.17) assuming an SCR defined by all asymptotic phase space. The results are similar to those of a simple prior statistical theory[35] and are in total disagreement with the computed histograms. However, if one assumes that the SCR is defined by those trajectories that display $> 10^3$ divergence, then the solid curves shown in panels (*c*) and (*d*) result. They are clearly in excellent agreement with the exact histogram calculations for the same set of trajectories. Since the "$> 10^3$"

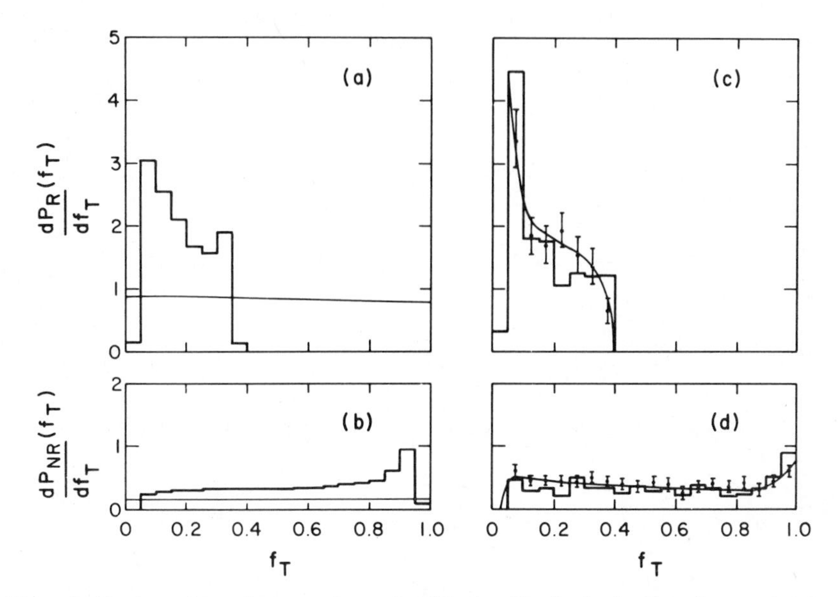


**Figure 3.** Final reactive and nonreactive probability densities for the fraction of energy in translation ($f_T$) for collinear F + $H_2$ ($v = 0$) at a total energy of $E = 0.375$ eV. (*a*) and (*b*): Reactive and nonreactive products using all trajectories (histograms). The solid curves are the corresponding predictions of a simple statistical theory. (*c*) and (*d*): Reactive and nonreactive products associated with trajectories with $d(t)/d(0) > 10^3$ (histograms). The solid curves, drawn through indicated calculated points, are the statistical prediction using the $10^3$ criteria to define the SCR . (From Ref. 31.)

component of the dynamics shows the same distributions when computed either exactly, or when computed with a statistical theory [Eq. (2.17)], these trajectories clearly may be classified as "statistical".

A similar result is shown (Fig. 4) for the set of nonreactive (and, less demanding, reactive) trajectories in the three-dimensional collision:

$$\mathrm{NaCl + K \leftrightarrow NaClK \rightarrow Na + KCl.}$$

In this instance, the solid curve is phase space theory, which turns out to be consistent with the product distributions associated with the $10^3$ criteria applied to Eq. (2.17). Each of the product distributions as well as the total reactive cross section is, in fact, given by the statistical theory based on an SCR characterized by trajectories that diverge by $>10^3$ from nearby neighbors in phase space.

The requirement that the final product distribution be independent of the initial conditions serves as an additional test of the statistical nature of the

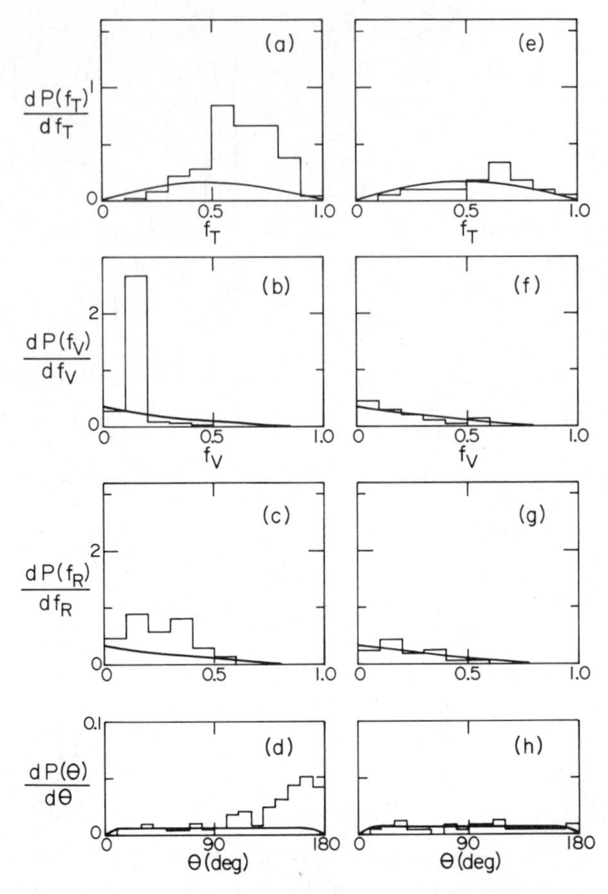


**Figure 4.** Probability densities for NaCl product translation ($f_T$), vibration ($f_v$), rotation ($f_r$), and scattering angle in NaCl + K at $E_t = 3.18$ kcal/mole shown as histograms for all trajectories [(*a*)–(*d*)] and trajectories with $d(t)/d(0) > 10^3$ [(*e*)–(*h*)]. Solid curves are phase space theory results. Note the different ordinate scales. (From Ref. 33.)

$>10^3$ component of the scattering. For example, in the case of applications to unimolecular decay in a number of model systems[34] (where results on product distributions similar to those discussed previously were obtained), the $>10^3$ component was shown to satisfy this zero relevance requirement. A sample result is shown in Fig. 5 for dissociation from a system of two collinearly coupled Morse oscillators with Hamiltonian

$$H = B(p_1^2 + p_2^2) - p_1 p_2/2 + [1 - \exp(-q_1)]^2 + [1 - \exp(-q_2)]^2. \tag{2.19}$$

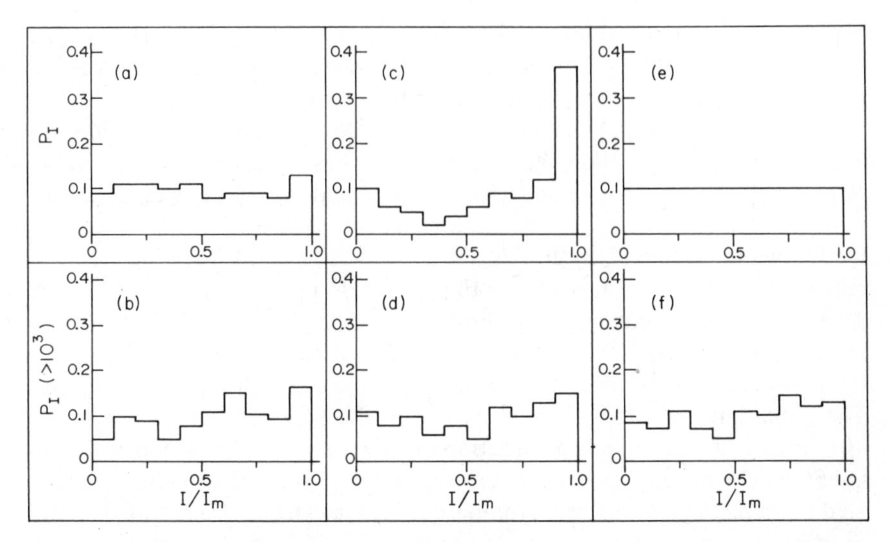


**Figure 5.** Product vibrational action distributions associated with the decay from (*a*) a distribution that is initially microcanonical within a critical surface in coordinate space and (*c*) an initial distribution consisting of a small square in coordinate space. (*b*) and (*d*) are the components of (*a*) and (*c*) that are characterized by $d(t)/d(0) > 10^3$. (*e*) is a distribution initially uniform in the asymptotic product channel with (*f*) being the subset of this distribution characterized by $d(t)/d(0) > 10^3$. For the case shown, $B = \frac{7}{16}$. (From Ref. 34.)

Panels (*a*) and (*c*) show product vibrational action distributions associated with two very different types of initial conditions. Although these two distributions yield considerably different product distributions, the component of each of these exact calculations that displays $d(t)/d(0) > 10^3$ results in the same product distribution (within computational accuracy), as shown in panels (*b*) and (*d*).

These studies indicate a direct connection between requirements of exponential divergence and adherence to statistical theories. Note, however, that the particular statistical theory obeyed by the dynamics need not be a simple analytic theory such as RRKM or phase space theory. It may appear, therefore, that the an essential simplicity of statistical theory, that is, the ability to bypass long-lived trajectory calculations in favor of an easily computed result, has been lost. This is indeed not the case, that is, it is easy to see that this approach affords a method for obtaining contributions from both the unstable (statistical) trajectory component as well the direct component with a minimum of computation. This technique, the "minimally dynamic"[33,34] approach, will now be sketched.

Consider Fig. 5*e*, which is a uniform asymptotic action distribution for the coupled Morse case. If one extracts the "$> 10^3$" component of this distribution identified by evolving short-lived trajectories *into* the region of the bound molecule, then Fig. 5*f* results. Note that this distribution is the same as those in Figs. 5*b* and 5*d*, which entail long-lived dynamics *outward* from the bound molecular region. This agreement, a consequence of time reversibility, has the important consequence that short-lived trajectories may be used to determine product distributions associated with the long-lived statistical dynamics. That is, for either bimolecular[33] or unimolecular[34] reactions, one proceeds in the following fashion. Trajectories are integrated forward from the reactant region to identify those trajectories that do, or do not, satisfy $d(t)/d(0) > 10^3$. This computation requires only short-lived trajectories. Those that do not exponentially diverge by this extent are termed direct trajectories and are integrated exactly to determine their contribution to the product distribution. The remaining trajectories, which pass into the SCR, are terminated when $d(t)/d(0) = 10^3$ and provide the fraction of the initial distribution which enters the SCR. The product distribution associated with these trajectories can also be determined by short-lived dynamics, if the trajectories are initiated in the product region, oriented so as to enter the SCR. The resultant distribution, combined with the previously determined probability of entering the SCR, provides the contribution of the statistical trajectories to the product distribution. In essence, then, the entire collision dynamics can be determined by using short-lived trajectories in conjunction with the statistical theory for a generalized SCR. The reader is referred to Refs. 33 and 34 for further computational details.

This minimally dynamic approach has been applied to both bimolecular and unimolecular reactions; a typical result for the latter case is shown in Fig. 6. In this case we consider the dissociation of CCH on two different potential surfaces due to Wolf and Hase.[36] These authors classified the first surface (their case IIC) as yielding RRKM dissociation, whereas their surface IIA yielded non-RRKM dynamics. The exact trajectory results for translational, vibrational, and rotational distributions for these two cases are shown as solid histograms in Fig. 6. The minimally dynamic construction, which requires only short-lived trajectory calculations, are shown as dashed histograms in the same figure and are seen to be in excellent agreement with the exact results.

These results provide a strong phenomenological link between the property of exponential divergence and statistical reaction dynamics. More work is required, however, to explain why the specific numerical value of $10^3$ divergence proves useful.

### 2. *Numerical Studies of Bound-State Relaxation*

The vast majority of the exponential divergence occurring during the course of the reactions previously discussed occurs during the time that the reactants

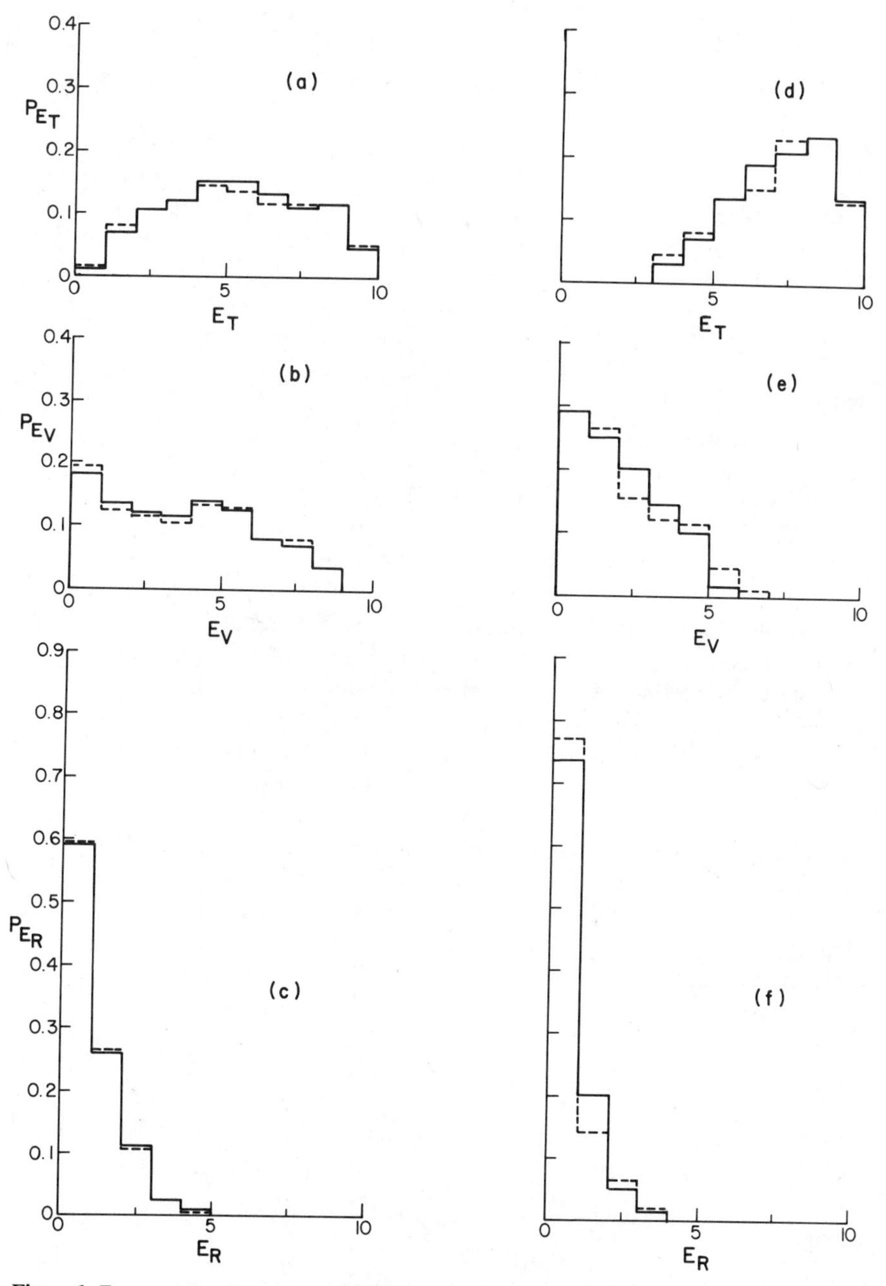


**Figure 6.** Exact product distributions (solid) versus minimally dynamic construction (dashed) for the Wolf–Hase case II C are shown in panels (*a*)–(*c*). Similar computations for case II A are shown in panels (*d*)–(*f*). (From Ref. 34.)

are in close proximity to one another. Thus, studies of bound-state chaos are expected to provide insight into statistical behavior and reaction dynamics as well as being directly relevant to the issue of bound-state intramolecular energy transfer.

Consideration of bound-state dynamics affords one advantage not shared by systems undergoing reaction or decay. Specifically, since formal ergodic conditions require a compact phase space, ideal chaotic systems exist for bound systems but not for bimolecular collisions or unimolecular decay. Studies of these ideal bound systems therefore provide a route for analyzing statistical behavior in circumstances where the system is fully characterized. Furthermore, these ideal system results can be compared with the behavior of model molecular systems to assess the degree to which realistic systems display chaotic relaxation.

Studies of this kind have been carried out in order to answer two types of question. First, the essence of chaotic behavior is relaxation [Eq. (2.2)] to a final state that is independent of initial conditions other than conserved integrals of motion. Since the rate of relaxation is of particular interest in chemical physics, efforts have been directed toward providing a *computationally feasible* method for estimating the relaxation time for systems behaving chaotically. For example, can the relaxation rate be related to features such as the $K$ entropy and Lyapunov exponents, each of which can at least be approximately estimated using short-time dynamics? Second, we also require a *feasible computational method* for identifying systems that do show mixing behavior. One may ask, for example, whether the observation of erratic surfaces of section, grassy power spectra, and exponentiating trajectories suffice to guarantee that the system relaxes to equilibrium.

Initial studies on the relationship between relaxation and chaos diagnostics focused on two degree of freedom systems[37] wherein computations are simplified and where the $K$ entropy, for systems which are chaotic over the entire energy hypersurface, equals the maximum Lyapunov exponent [see Eq. (2.9)]. We review the results for two flow cases, the Henon–Heiles potential,[38] which is of general interest in nonlinear mechanics and is given by

$$H = p_x^2/2\mu_x + p_y^2/2\mu_y + \tfrac{1}{2}(x^2 + y^2) + x^2 y - y^3/3, \qquad (2.20)$$

and the coupled Morse potential, which, in the original Thiele–Wilson form,[39] is

$$\begin{aligned} H = p_x^2/2\mu_x + p_y^2/2\mu_y &+ \{\exp[-\tfrac{1}{2}(y + x)] - 1\}^2 \\ &+ \{\exp[-\tfrac{1}{2}(y - x)] - 1\}^2. \end{aligned} \qquad (2.21)$$

The former is generally presumed highly chaotic at its dissociation energy, whereas the latter is known to have regions of chaos, mixed with regions of regular dynamics, even at energies above dissociation.[7b] In addition, we cite results[27c,37] for the Arnold Cat Map,[11] which is a $K$ system given by

$$\begin{aligned} x_{n+1} &= x_n + y_n, \\ y_{n+1} &= x_n + 2y_n. \end{aligned} \qquad [\text{mod } 1] \tag{2.22}$$

Studies on all three systems[37b] indicate that the qualitative nature of the time evolution of an initial distribution $\rho(x, y; t = 0)$ depends on the size of $\rho(x, y; t = 0)$ relative to the overall dimensions of the equilibrium state to which the system relaxes.[40] Distributions that are initially local in phase space show initial shearing with exponentially growing coverage of phase space. This growth is well characterized by the $K$ entropy and is most clearly manifest in plots of the coarse-grained information defined as follows. Consider a partition of phase space (e.g., a square grid in the two-dimensional case) with elements of the partition labeled by the index $i$ and where the equilibrium distribution is uniform on the partition. Then, the coarse-grained information associated with the evolution of an initial distribution defined on this partition is

$$I(t) = -\sum_i P_i(t) \log_2 P_i(t), \tag{2.23}$$

where $P_i(t)$ is the probability of occupying the $i$th partition element at time $t$. Figure 7$a$ shows a typical $I(t)$ for a localized initial distribution in the Cat Map and the Henon–Heiles along with ideal linear growth of $I(t)$ with rate $K$. The Cat Map shows excellent agreement with this growth until $I = 0$, at which point the system has reached equilibrium. Similarly, the Henon–Heiles shows qualitatively good agreement with this linear growth followed by a slower approach to equilibrium ($I = 0$). Detailed examination of the time evolution of the phase space density[37b] for the Henon–Heiles shows that the rapid growth of $I(t)$ corresponds to a region where the phase space density grows from a tight initial configuration to one covering all phase space, albeit nonuniformly. The slower $I(t)$ growth corresponds to the reorganization of this global density to that corresponding to an equilibrium microcanonical ensemble. Furthermore, the time scale $\tau$ associated with the initial region is in good agreement with the estimate of Sinai[41] and Zaslavsky[11b]: $\tau = -\log_2[\mu(U)]/K$, where $\mu(U)$ is the measure of the initial distribution. Similar behavior has been noted for a number of two-dimensional maps.[37a]

Distributions that are initially global in phase space do not display linear growth of $I(t)$ with $K$, as seen in Fig. 7$b$. Rather, the behavior is similar to that obtained at longer times. in Fig. 7$a$. In some sense the chosen measure

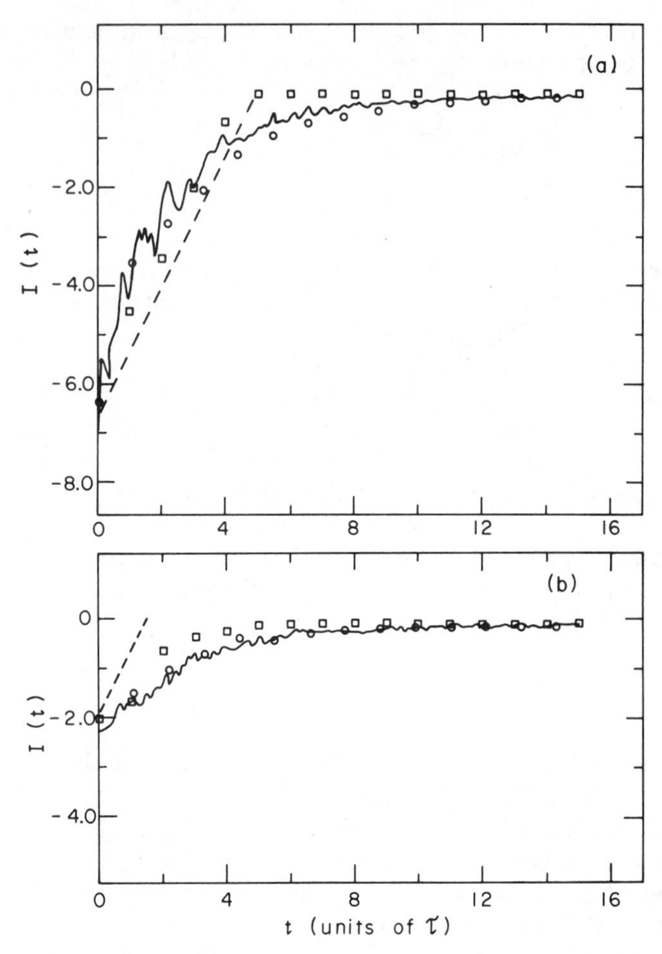


**Figure 7.** (*a*) Coarse-grained entropy $I(t)$ for Cat Map (boxes) and Henon–Heiles (solid curve) for an initially local distribution. The dashed line provides ideal $I(t)$ behavior with slope $K$. (*b*) As in panel (*a*) but for an initial global distribution consisting of a ring in phase space. (From Ref. 37*b*.)

of relaxation $I(t)$ is insufficiently sensitive to the difference between the equilibrium and the initial global nonuniform distribution to show details of the relaxation. For these cases, however, the $K$ entropy is apparant in the relaxation of the autocorrelation function of a property such as "energy in a mode." For example, with mode energies defined harmonically, for example,

$$E_{n2} = a_{2y}p_y^2 + b_{2y}y^2,$$
$$E_{n1} = a_{1x}p_x^2 + b_{1x}x^2 \tag{2.24}$$

(where the coefficients $a$, $b$ are given in Ref. 37$b$), the autocorrelation function $\langle E_{n2}(t)E_{n2}(0)/E_{n2}(0)^2\rangle$ in the Henon–Heiles system is typically that shown in Fig. 8$a$. Here the brackets denote a microcanonical average. Also shown in Fig. 8$a$ are the long-time statistical limit (arrow) and exponential falloff with rate $K$. The results are in good accord with this relaxation rate. Similar good agreement is obtained for the coupled Morse system[37b] as well as for the relaxation of average mode energies as a function of time. An example of the latter is shown in Fig. 8$b$, where the average value of $E_{n2}$ for an initial distribution corresponding to total energy $E = \frac{1}{6}$ and $E_{n2} = 0.1E$.

These results suggest a deceptively simple picture of relaxation in chaotic systems. Specifically, it appears that chaotic relaxation of simple properties, where the initial state is global, is exponential with rate $K$. Additional studies[42] have indicated, however, that simple exponential decay is not the rule. To confound the issue, Crawford and Cary[43] prove that autocorrelation functions in the Arnold Cat Map can display a *variety* of *exponential* decay rates, where the rate depends on the degree of differentiability of the initial distribution. If correct, this result presents a serious impediment to obtaining simple estimates of relaxation times, since arbitrarily small qualitative changes in the initial distribution could result in wildly different relaxation times. This being the case, Dumont and Brumer[27] undertook a detailed study of the relaxation rates in the Arnold Cat Map and in associated flow models which provide simple $K$ systems for analysis. The first result of this work was the recognition that the Cary–Crawford result only applies at long times, when the autocorrelation function may be arbitrarily close to the asymptotic statistical value. Thus, variations in decay rates due to the differentiability of the initial state can be irrelevant to a chemically meaningful relaxation time. The second result of this study was the realization that although a wide variety of functional forms are possible in relaxation (i.e., the exponential form need not occur over time scales relevant to physically meaningful relaxation), the relaxation time $\tau$, with appropriate interpretation, remains important. This interpretation of $\tau$ is the subject of a future publication.

These discussions apply to systems where the dynamics is guaranteed, either formally or computationally, to be operationally mixing, a requirement most easily satisfied for two degree of freedom systems. Efforts to extend an analysis of relaxation dynamics in bound systems with greater than two degrees of freedom have met with considerable difficulty. Specifically, we possess no reliable diagnostics to ensure that a system is indeed behaving in accord with ideal mixing conditions. The example of a model of OCS[44] is pertinent to this discussion. Specifically, an analysis of this model showed exponentiating trajectories, broadened power spectra, and a number of other diagnostics consistent with chaotic dynamics. However, an analysis of the relaxation of energy deposited initially in a single mode showed no significant relaxation over the $\tau$ time scale. A sample of this behavior, at high internal

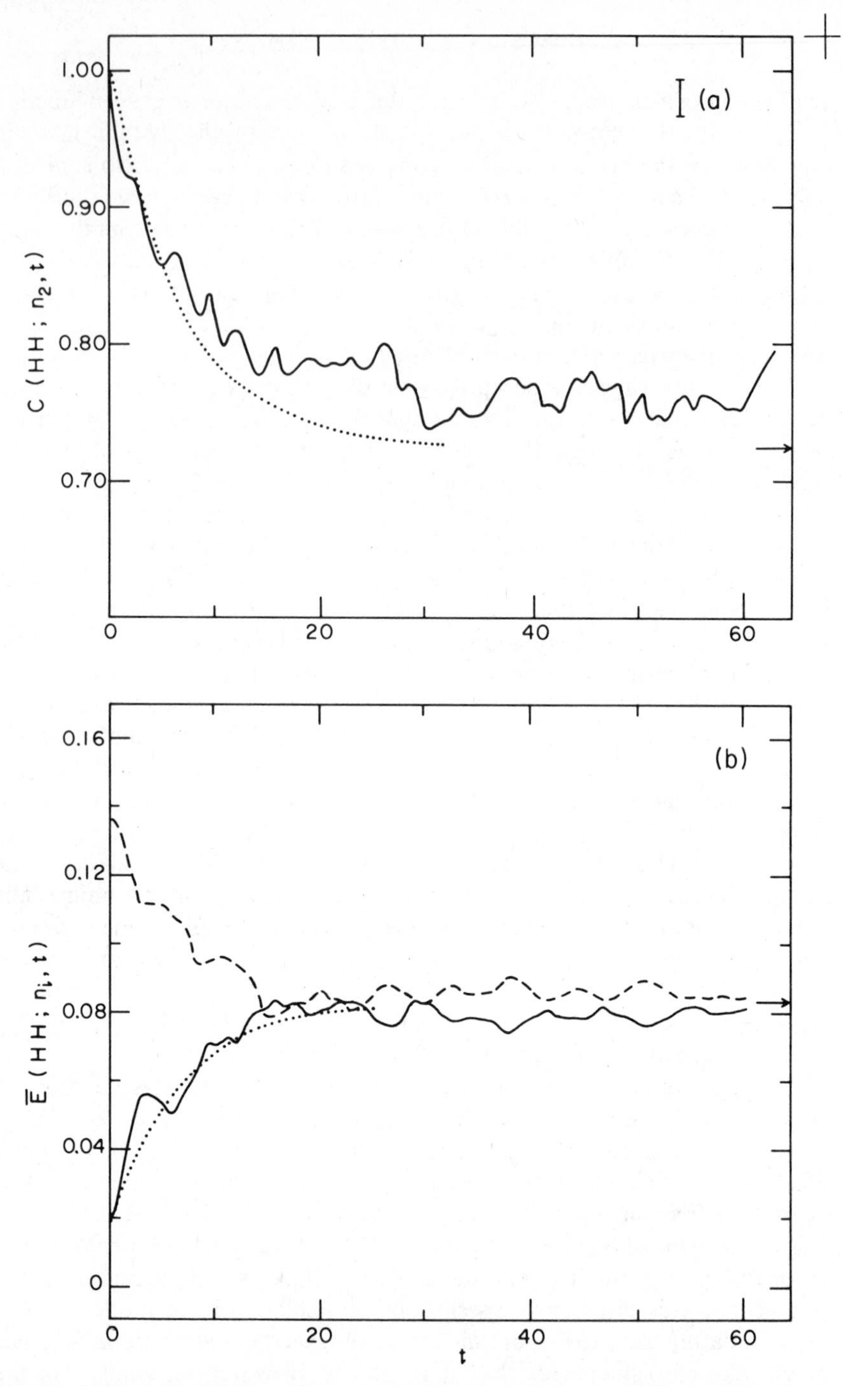


**Figure 8.** (*a*) Normalized autocorrelation function for $E_{n2}$ in the Henon–Heiles system at $E = \frac{1}{6}$ (solid line). Falloff with rate $K$ to the statistical value (arrow) is shown as a dotted curve. (*b*) Average value of $E_{n2}$ in the Henon–Heiles system at $E = \frac{1}{6}$ (solid curve) and of $E_{n1}$ (dashed curve). Exponential falloff with rate $K$ to statistical value (arrow) with rate $K$ is given by dotted curve. Initial conditions as described in test. (From Ref. 37*b*.)

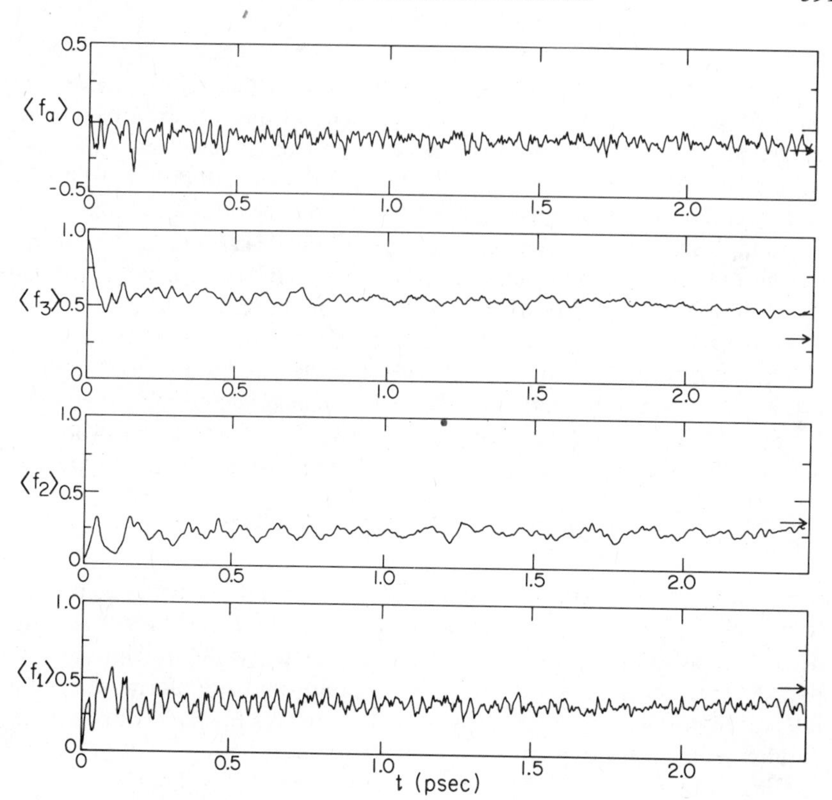


**Figure 9.** Time evolution of average fractional energy $\langle f_i \rangle$ in the $i$th normal mode and of the remaining anharmonic term $\langle f_a \rangle$ in model OCS. Here the total energy is 20,000 $cm^{-1}$ with almost all energy initially in the bend mode $\langle f_3 \rangle$. Microcanonical equilibrium average values are indicated by arrows. (From Ref. 44.)

OCS energy, is shown in Fig. 9. Further studies of this behavior in collinear OCS[24] suggest that the origin of this behavior is in the presence of local regions of phase space that serve as rate-determining steps in the relaxation. This behavior results in dramatically different values for the set of all Lyapunov exponents.

This result makes clear that considerably greater insight is necessary before a set of adequate diagnostics are available to define a multi-degree-of-freedom system as sufficiently chaotic to apply statistical models of bound relaxation.

### 3. *Formal Links to Chaos: Unimolecular Decay*

The computations previously described provide some insight into numerical conditions under which collisions, unimolecular decay, and bound-state dynamics satisfy requirements for statistical theories. They do not, however, provide a link between statistical theories and formal ergodic theory. For example, they do not suggest what formal conditions, for example, mixing, $K$-system properties, and so forth, are required to achieve statistical theory results. In this section we discuss the first steps toward that link; that is, we display sufficient conditions, within a formal theory of unimolecular decay, for unimolecular dissociation to show exponential decay with an RRKM rate. Furthermore, the method yields an extension of RRKM theory that properly accounts for direct dissociation, relaxation of the initial distribution, and yields the RRKM rate in the limit of small relaxation times.

There are two common assumptions associated with statistical theories of overall decay, specifically, that the decay is *exponential* and that, at energy $E$, it occurs with a statistical lifetime $\tau_s(E) = k_s^{-1}(E)$. That is, the probability density $P(t)$ for decay of the reactant $A$ at time $t$ is assumed to be given as

$$P(t) = P(0)e^{-k_s(E)t} = P(0)e^{-t/\tau_s(E)}, \tag{2.25}$$

where the essential statistical feature of $k_s(E)$ is that it is directly computable from phase space volumes at energy $E$. Typical arguments[45] for exponential decay are qualitative in that they simply assume, without relation to the dynamics, that the decay is somehow random, that is, uniform in any time interval. Below we summarize work[46] which relates this behavior to mixing requirements on the dynamics.

Consider unimolecular decay, $A \to$ products, treated within the framework of classical mechanics. The system phase space, confined to energy $E$, may be divided into three regions, denoted $R_-$, $R$, $R_+$, separated by transition states $S_-$, $S_+$ (see Fig. 10). Here $R$ denotes the bound molecular region associated with the unstable molecule $A$, $R_+$ is the region of separating products, and

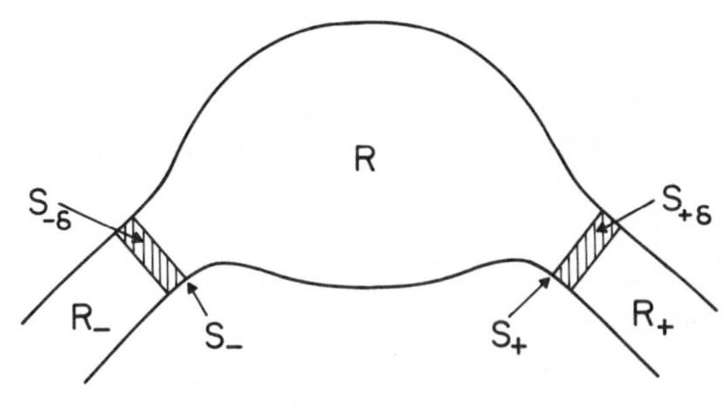


**Figure 10.** Schematic of unstable molecular region $R$, transition states $S_+$, $S_-$, and their $\delta$ broadenings. [Reprinted with permission from *J. Phys. Chem.* **90**, 3509. Copyright 1986 American Chemical Society.]

$R_-$ is the region of colliding reactants that give rise to all states of $A$. These definitions are such that $R_-$ and $R_+$ are related by time reversal and $R$ is time reversal symmetric. In terms of these sets, traditional unimolecular decay rate constant formulations essentially consider the evolution of an initially uniform distribution on $R$, into $R_+$.

We first reformulate unimolecular decay in terms of symbolic dynamics so as to permit utilization of modern concepts in ergodic theory. In doing so we, at least initially, replace the continuous time dynamics by a discrete time mapping. Specifically, we consider dynamics at multiples of a fixed time increment $\delta$, defining $T^n\mathbf{x}$ as the propagation of a phase space point $\mathbf{x}$ for $n$ time increments [i.e., $\mathbf{x}(t = n\delta) = T^n\mathbf{x}$]. In what follows, time parameters associated with the discrete dynamics are measured in units of $\delta$. These include $t$, $t'$, $\tau$, and $\tau_<$, which are also used in connection with the flow. In the later context the conversion to $\delta$-independent units is implicit. Note that within the assumed discrete dynamics, $S_-$ and $S_+$ are broadened from surfaces to volumes $S_{-\delta}$ and $S_{+\delta}$ comprising all points that enter or have left $R$ during a time interval $\delta$.

It is first necessary to impose conditions on the three phase space regions in Fig. 10. Specifically, we require that the transition states $S_\pm$ be defined so that recrossing from $R_+$ to $R$ or $R$ to $R_-$ does not occur. In addition, it proves necessary to impose additional conditions on $R$ to ensure statistical decay to $R_+$. Details are provided elsewhere,[46] but, qualitatively, they are designed, in the first instance, to exclude direct trajectories from consideration.

It suffices to characterize the dynamics, for purposes of examining decay, in terms of the location of the trajectories in $R_+$, $R_-$, or $R$ as a function of time. Specifically, each phase space point is classified, through its trajectory, as a member of a set

$$(A_0, A_1, \ldots, A_m) = \{\mathbf{x} | \mathbf{x} \in A_0, T\mathbf{x} \in A_1, \ldots, T^m\mathbf{x} \in A_m\}, \tag{2.26}$$

where $A_j$ is one of either $R_+$, $R_-$, or $R$. For example, $(R_-, R, R, R_+)$ denotes the set of points that are in $R_-$ at time zero and that will be in $R$ at times $\delta$ and $2\delta$ and in $R_+$ at time $3\delta$. For convenience we denote the set of points that are initiated in $R$ and remain there for the following $(t - 1)$ time steps before entering $R_+$ as $r^t$ and the set of points that are initiated in $R_-$ and remain in $R$ for $t - 1$ time steps before passing into $R_+$ as $s^t_{-\delta}$; that is,

$$\begin{aligned} r^t &= (R, \ldots, R, R_+), \quad R \text{ appearing } t \text{ times}; \\ s^t_{-\delta} &= (R_-, R, \ldots, R, R_+), \quad R \text{ appearing } (t - 1) \text{ times}. \end{aligned} \tag{2.27}$$

The spaces $S_{-\delta}$ and $R$ are thus partitioned according to the time required to reach $R_+$. That is,

$$R = \bigcup_{t=1}^{\infty} r^t; \qquad S_{-\delta} = \bigcup_{t=1}^{\infty} s_{-\delta}^t.$$

The initial relation obtained through this approach is that between the gap distribution $P_g(t)$ and the lifetime distribution $P(t)$. Here $P_g(t)$ is the distribution of times associated with bound molecular motion between the transition states $S_+$ and $S_-$.[47] Specifically, with the delta-broadened versions of these distributions defined as

$$P_{g,\delta}(t) = \mu(s_{-\delta}^t)/\mu(S_{-\delta}),$$
$$P_\delta(t) = \mu(r^t)/\mu(R),$$

we have

$$P_\delta(t+1) - P_\delta(t) = -k_{s,\delta}(R)P_{g,\delta}(t)$$

in the discrete time case and, in the continuous time case,

$$\frac{dP(t)}{dt} = -k_s(R)P_g(t), \tag{2.28}$$

where

$$k_s(R) = \tau_s^{-1}(R) = \lim_{\delta \to 0} k_{s,\delta}(R)\delta^{-1} \tag{2.29}$$

and

$$k_{s,\delta}(R) = \mu(S_{-\delta})/\mu(R) = \mu(S_{+\delta})/\mu(R).$$

It is convenient to introduce an additional measure on $S_+$ and $S_-$. Specifically, let $B$ be a Borel set[48] in $S_+$ (or $S_-$) so that $B_\delta = \bigcup_{t=0}^{1} T^{-t}B$ is a Borel set in $R$ (or $R_-$), that is, the $\delta$ broadening of $B$. We define the measure $\mu_g(B)$ as

$$\mu_g(B) = \lim_{\delta \to 0} \mu(B_\delta)/\delta. \tag{2.30}$$

Using Eq. (2.30), Eq. (2.29) assumes the form

$$k_s(R) = \tau_s^{-1}(R) = \mu_g(S_+)/\mu(R). \tag{2.31}$$

In terms of standard chemical literature notation Eq. (2.31) assumes the form[27c]

$$k_s(R) = \frac{\int_{S_+} \hat{\mathbf{n}} \cdot \mathbf{v}\, dx' / \|\nabla H\|}{\int_R dx/\|\nabla H\|}, \tag{2.32}$$

where $dx$, $dx'$ are volume elements on $R$ and $S_+$, respectively, $\hat{\mathbf{n}}$ is the unit normal to $S_+$, $\mathbf{v}$ is the *phase space velocity* $d\mathbf{x}/dt$, and $H$ is the Hamiltonian. Thus, Eq. (2.31) indicates that $k_s$ = flux/volume. Although this is the traditional form, it includes two generalizations: that is, it applies to any choice of $R$, and associated $S_-$, $S_+$, and the transition state can be a general phase space dividing surface.

Further characterization of $P(t)$ and $P_g(t)$ requires details of the dynamics, here formulated in terms of ergodic-theory-based assumptions. However, since ergodic theory[11] considers dynamics on a bounded manifold, it is not directly applicable to the unbounded phase space associated with molecular decay. To resolve this problem we first introduce a related auxiliary bounded system upon which conditions of chaos are imposed, and then determine their effect on the molecular decay; details of this construction are provided elsewhere.[46] What we show is that adopting this condition leads to a new model for decay, the delayed lifetime gap model (DLGM) for $P(t)$ and $P_g(t)$. The simple statistical theory assumption that $P_g(t)$ and $P(t)$ are exponential with rate $k_s(E)$ is shown to arise only as a limiting case.

The mixing condition assumed is

$$\mu(r^t \cap T^{t'} S_{-\delta}) = \mu(r^t)\mu(S_{-\delta}), \quad t' = \tau = \tau_<, \tag{2.33}$$

where $\tau_<$ is the smallest gap time. A number of qualitative remarks regarding this central condition are in order.

1. This statement includes the requirement that the relaxation time equal or exceed the smallest gap time. This condition formalizes the two-component nature (i.e., direct and indirect) of molecular dissociation discussed above. Specifically, only the indirect component may be treated within a statistical model. Equation (2.33) provides a model for the dynamics of this component, which consists of a modified region $R$ that contains no trajectories with gap times less than $\tau$. It is important to note that the definition of $R$ is naturally reflected in $k_s(R)$.

2. Equation (2.33) in no way suggests that relaxation applies at one time only. Rather, it should be interpreted as a specific mixing requirement by time $\tau_<$, or, equivalently, $\tau$.

3. Equation (2.33) makes precise the notion of relaxation within the mole-

cule $R$ prior to dissociation. Note that Eq. (2.33) is essentially a condition on the relaxation of a distribution initiated on the incoming transition state, as opposed to a condition on points initiated within $R$. This is consistent with the picture advocated here wherein all dynamics is viewed in terms of the partitions of the phase space or, equivalently, as trajectories occupying these regions as geometrically complete entities. From the viewpoint of the ergodic theorist, however, Eq. (2.33) is somewhat unsatisfactory as a mixing condition, since $r^t$ hides a considerable amount of the dynamics. We note, therefore, that Eq. (2.33) can be recast as a mixing condition involving $S_{-\delta}$ and its propagation alone.[27c]

4. Equation (2.33) provides a formal mixing condition, whose consequences, for $P(t)$ and $P_g(t)$, are discussed subsequently. Two alternate viewpoints may be adopted. In the first, Eq. (2.33), in conjunction with the already defined $R$, $R_+$, $R_-$ provides a formal model for statistical unimolecular decay. In this case the degree to which a real molecular system displays the $P(t)$ and $P_g(t)$ obtained (later below) through this assumption provides a measure of the statisticality of the system. This route fails to recognize that typical molecular systems possess both regions of direct as well as indirect trajectories, where only the latter are expected to be treatable within a statistical model. Recognition of this fact leads to an alternate *theoretical* viewpoint wherein one tailors $R$, $R_+$, $R_-$, to agree *qualitatively* with the conditions, the remainder of the dissociating molecular system requiring direct trajectory methods. In this case the extent of *quantitative* agreement between computed trajectory results initiated in this modified $R$, and this model signifies the extent of statistical system behavior.

Application[46] of Eq. (2.33) yields a new model [Eq. (2.27)] for the $P(t)$ and $P_g(t)$ distributions *associated with decay of the statistical* component, that is, trajectories such that $\tau_< = \tau$. Specifically, noting in Eq. (2.33) that

$$r^t \cap T^\tau S_{-\delta} = s_{-\delta}^{t+\tau} \tag{2.34}$$

gives

$$\mu(s_{-\delta}^{t+\tau})/\mu(S_{-\delta}) = \mu(r^t). \tag{2.35}$$

Since $\mu(R)$ has been normalized to unity, this implies that in the $\delta \to 0$ limit

$$P_g(t) = P(t - \tau), \qquad t \geq \tau. \tag{2.36}$$

Since $\tau = \tau_<$, $P_g(t) = 0$ for $t < \tau$. Setting $P(t) = 0$ for $t < 0$ allows us to rewrite Eq. (2.36) as

$$P_g(t) = P(t - \tau), \qquad t \geq 0. \tag{2.37}$$

Inserting Eq. (2.37) into Eq. (2.28) gives the delay differential equation:

$$\frac{dP(t)}{dt} = -k_s P(t - \tau), \qquad t \geq \tau, \tag{2.38}$$

$$\frac{dP(t)}{dt} = 0, \qquad t < \tau, \tag{2.39}$$

where here, and in the following, we suppress the $R$ dependence of $k_s(R)$. These equations, in conjunction with initial conditions[46] $P(t) = k_s$ for $0 \leq t \leq \tau P(t)$, provide the delayed lifetime gap model for $P(t)$ and $P_g(t)$, which has an analytic solution. Specifically, Equations (2.38) and (2.39) have a meaningful solution only if $\tau < (k_s e)^{-1}$. Two analytic expressions may be obtained for $P(t)$, with the more computationally useful form being given by

$$P(t'' + n\tau) = k_s \sum_{j=0}^{n} [-k_s\{t'' + (n - j)\tau\}]^j/j!, \tag{2.40}$$

where $n$ is a nonnegative integer and $0 \leq t'' \leq \tau$. The second form is considerably more complicated but provides direct insight into the rate constant for exponential decay of $P(t)$. Specifically, $P(t)$ is found to behave, in the long-time limit, as

$$P(t) \sim \frac{k_s e^{-\kappa t}}{(1 - \kappa\tau)}, \qquad t \gg \tau, \tag{2.41}$$

and $P_g(t)$, via Eq. (2.38), as

$$P_g(t) \sim \frac{\kappa e^{-\kappa t}}{(1 - \kappa\tau)}, \qquad t \gg \tau. \tag{2.42}$$

Here $\kappa$ is the smallest solution to the nonlinear equation

$$\kappa = k_s e^{\kappa}\tau. \tag{2.43}$$

A useful approximation to $\kappa$, given by $\kappa = k_s + \tau k_s^2$, obtains when $k_s\tau$ is small. Note that although Eqs. (2.41) and (2.42) are asymptotic forms, numerical results[46] indicate that they are good approximations at shorter times as well.

Equation (2.41) implies that the system dissociates with rate $K$, rather than $k_s$ as expected from simplified statistical theories. Furthermore, examination of Eq. (2.43) indicates that the range of possible $K$ values is large. That is,

$$k_s < K < ek_s \leq 1/\tau, \tag{2.44}$$

where

$$\begin{aligned} K &\to k_s, \quad \text{as } k_s\tau \to 0; \\ K &\to 1/\tau, \quad \text{as } k_s\tau \to 1/e. \end{aligned} \tag{2.45}$$

Several features of this result are of considerable importance. First, the simple statistical rate $k_s(E)$ obtains when $R$ may be taken as the full energy hypersurface and when $\tau$ is much shorter than the mean gap time $\tau_s(R)$. Second, even though the system is assumed to display mixing properties with respect to reaction, the rate constant for decay can be substantially faster than $k_s$. Observation of faster than statistical decay is often attributed to the participation of "fewer degrees of freedom" than the full molecular complement. Equation (2.41) suggests, however, an important alternative as well as the possibility of extracting $\tau$ from the experimental rate of exponential decay. Finally note that only when $\tau$ is very small does the random gap assumption of similar exponential falloff behavior for both $P(t)$ and $P_g(t)$, over all $t$, obtain.

These results constitute the first major steps in formalizing statistical theories of reaction dynamics and relating statistical molecular behavior to ergodic theory. Specifically, they demonstrate that by invoking a mixing condition on a well-chosen $R$ we obtain an analytically soluble model for $P(t)$ which is asymptotically well approximated by exponential decay with rate $K$. The rate of decay is directly affected by the relaxation time $\tau$ and equals $k_s(R)$ in the limit $\tau \to 0$. A similar approach can be used[46] to provide an ergodic theory basis for product distributions.

## III. CLASSICAL–QUANTUM CORRESPONDENCE

### A. Why Should Classical Mechanics Work?

It is clear, from the discussion thus far, that typical molecular Hamiltonians display features characteristic of chaotic classical motion. The logical order followed previously, that is, the introduction of well-defined concepts of chaotic behavior in classical ideal systems, followed by an examination of "realistic" molecular models, does not follow through to quantum mechanics. The primary difficulty is that quantum mechanics always predicts, for bound-state dynamics, quasiperiodic motion. Several aspects of quantum chaos are discussed in Section IV. We note at this point, however, that this quantum–

classical discrepancy has generated renewed interest in quantum–classical correspondence in the chaotic regime. Indeed, our lack of understanding of quantum–classical correspondence in this regime motivates a complete reexamination in the general question of the validity of classical mechanics for bound-state molecular dynamics.

To emphasize the difficulty consider, for example, the dynamics of a two-level quantum system. Here an initial pure state is given by

$$|\psi(t=0)\rangle = c_1|\phi_1\rangle + c_2|\phi_2\rangle, \tag{3.1}$$

where $|\phi_i\rangle$ is an eigenstate of the Hamiltonian with eigenvalue $E_i$. The system evolves in time to give

$$|\psi(t)\rangle = c_1|\phi_1\rangle \exp(-iE_1 t/\hbar) + c_2|\phi_2\rangle \exp(-iE_2 t/\hbar). \tag{3.2}$$

The average value of any property $B$ is given by

$$\begin{aligned}\langle B(t)\rangle = |c_1|^2\langle\phi_1|B|\phi_1\rangle + |c_2|^2\langle\phi_2|B|\phi_2\rangle + c_1c_2^*\langle\phi_2|B|\phi_1\rangle \\ \times \exp[-i(E_1 - E_2)t/\hbar] + c_2c_1^*\langle\phi_1|B|\phi_2\rangle \\ \times \exp[-i(E_2 - E_1)t/\hbar]\rangle.\end{aligned} \tag{3.3}$$

That is, time evolution occurs via phase interference between energy eigenstates. One hardly expects that traditional classical mechanics, viewed in terms of trajectories, will adequately describe this two-level dynamics. Indeed, one would immediately claim that the system is purely quantum mechanical and assign it presumably quantum-mechanical attributes such as:

1. Time dependence is due to phase interference between eigenstates of the Hamiltonian.
2. The population $|c_i|^2$ of each quantum energy eigenstate is constant in time.
3. The structure is such that time is explicitly separated from the remainder of the problem.
4. Regions of, for example, coordinate space that can be occupied at $t > 0$, are implicitly determined by the spatial character of the states included in the initial $t = 0$ superposition.
5. Spatial localization of the initial state is accomplished by superposing states, that is, via spatial interference.

Consider now the case of bound molecular dynamics wherein classical mechanics is often tacitly assumed valid. If the rate of spontaneous emission is smaller than that which induces overlapping levels (i.e., the actual spectrum of the system is discrete), then, qualitatively, spontaneous emission may be ignored. The initial molecular state, assumed pure, is given by the extension of Eq. (3.1) to $N$ levels, that is,

$$|\psi(t=0)\rangle = \sum_{j=1}^{N} c_j|\phi_j\rangle,$$

$$|\psi(t)\rangle = \sum_{j=1}^{N} c_j|\phi_j\rangle \exp(-iE_j t/\hbar). \tag{3.4}$$

Despite the increase in number of levels, the list of quantum properties assigned to the two-level problem also applies to this case. How then can we propose to approximate the dynamics of this system using classical mechanics?

We will argue that "straw men" of the kind thus constructed arise from the use of a qualitative picture based on *wavefunctions* in quantum mechanics and *trajectories* in classical mechanics. That is, the use of two different types of descriptions in classical and quantum dynamics leads to conceptual problems such as those previously described. We propose, in place of this view, the use of a uniform description based on the Liouville operator and phase space distributions. (Reference to Section II indicates that such a reliance on distributions in classical mechanics, rather than trajectories, is also consistent with the approach adopted in ergodic theory.) Furthermore, we argue that the essential objects for examination are the eigenfunctions (eigendistributions) of the Liouville operator. Although this approach has thus far only provided limited insight into the problem of correspondence in the chaotic regime, we are able to resolve the apparent problems raised above as well as provide a framework within which, we believe, classical–quantum correspondence, under all circumstances, should be viewed. The importance of this issue to the use of classical mechanics in chemical dynamics, as well as the view of classical relaxation afforded by this approach, justifies our apparent excursion away from the topic of this review.

### B. The Liouville Approach: Motivation

Longstanding interest in classical–quantum correspondence results from the fact that many processes involving the dynamics of nuclei occur in a regime suitable to classical and semiclassical approximations. Many recent studies have focused on the embedding of classical dynamics in appropriate quantum-like formulas to produce semiclassical approximations.[49] Although useful for semiclassical developments, these approaches have several disadvantages for studies of the classical–quantum correspondence. The most obvious of these is the absence of the analogue to the probabilty amplitude $\psi$ in classical mechanics. Less obvious are the conceptual deficiencies of the trajectory viewpoint in classical mechanics. For example, arguments based on the dynamics of individual trajectories might lead one to conclude that quasiperiodic systems show no relaxation, whereas chaotic systems relax. This is indeed not the case, that is, relaxation is a feature of both systems, as will be discussed.

In this section we advocate a far more advantageous route to studying conceptual features of the classical–quantum correspondence, and indeed for each mechanics independently, in which phase space distributions are used in both classical and quantum mechanics, that is, classical Liouville dynamics[50] in the former and the Wigner–Weyl representation in the latter. This approach provides, as will be demonstrated, powerful conceptual insights into the relationship between classical and quantum mechanics. The essential point of this section is easily stated: using similar mathematics in both quantum and classical mechanics results in a similar qualitative picture of the dynamics.

The starting point for a unified view of classical and quantum mechanics is the Hilbert space of distributions with abstract element $\rho$. In both mechanics, time evolution is governed by the Liouville equation

$$L\rho(t) = \frac{i\partial\rho(t)}{\partial t} \tag{3.5}$$

and averages of properties are defined as

$$\langle F(t) \rangle = \mathrm{Tr}(F\rho(t)). \tag{3.6}$$

In classical mechanics the abstract Hermitian operator $L$, defined as operating on the Hilbert space of distributions, is $L_c = -i\{\ ,H\}$, whereas it is $L_q = h^{-1}[\ ,H]$ in quantum mechanics. Here $\{\ ,\ \}$ denotes a Poisson bracket and $[\ ,\ ]$ denotes a commutator.

The nature of Eq. (3.5) admits, for time independent $H$, solutions of the form

$$\rho_\lambda(t) = \rho_\lambda e^{i\lambda t}, \tag{3.7}$$

where $\rho_\lambda$ and $\lambda$ are eigendistributions and eigenvalues of $L$, that is,

$$L\rho_\lambda = \lambda\rho_\lambda. \tag{3.8}$$

As discussed subsequently, introduction of the standard $p, q$ representation in classical mechanics, and of the Wigner–Weyl representation in quantum mechanics, defines densities $\rho(\mathbf{p}, \mathbf{q}) = (\mathbf{p}, \mathbf{q}|\rho)$ that both lie in the *same* Hilbert space. Thus, the essential difference between quantum and classical mechanics relates to the spectrum of $L$ and those external conditions, imposed by the particular mechanics, that define acceptable physical distributions $\rho(\mathbf{p}, \mathbf{q})$.

### 1. *Classical Phase Space Distributions*

Contact with functions on $2N$-dimensional classical phase space is obtained by introducing the $\mathbf{p}$, $\mathbf{q}$ representation of the abstract Hilbert space vector of $\rho$ and operator $L$. Specifically, $(\mathbf{p}, \mathbf{q}|\rho) = \rho(\mathbf{p}, \mathbf{q})$ and

$$L = -i\{\ , H\} = -i\left(\sum_j \frac{\partial H}{\partial p_j}\frac{\partial}{\partial q_j} - \frac{\partial H}{\partial q_j}\frac{\partial}{\partial p_j}\right), \tag{3.9}$$

where $L$ is an Hermitian operator on the Hilbert space of $L^2$ functions with inner product

$$\langle \rho_1(\mathbf{p}, \mathbf{q}), \rho_2(\mathbf{p}, \mathbf{q})\rangle = \int d\mathbf{p}\, d\mathbf{q}\, \rho_1^*(\mathbf{p}, \mathbf{q})\rho_2(\mathbf{p}, \mathbf{q}). \tag{3.10}$$

Alternative representations other than $p$, $q$ can be conveniently obtained via unitary transformations.

The Hermitian operator $L$ acting on this Hilbert space of square integrable functions possesses a complete set of eigendistributions $\rho_{\boldsymbol{\alpha}}(\mathbf{p}, \mathbf{q})$ with associated eigenvalue $\lambda_{\boldsymbol{\alpha}}$. For an $N$ degree of freedom quasiperiodic system with Hamiltonian $H(\mathbf{I})$, action-angle variables $(\mathbf{I}, \boldsymbol{\theta})$ and associated frequencies $\omega(\mathbf{I}) = \partial H/\partial \mathbf{I}$, the eigenvalue problem $L\rho_\lambda = \lambda\rho_\lambda$ becomes

$$-i\left(\sum_j \frac{\partial H}{\partial I_j}\frac{\partial}{\partial \theta_j}\right)\rho_{\boldsymbol{\alpha}} = \lambda_{\boldsymbol{\alpha}}\rho_{\boldsymbol{\alpha}},$$

with complex solutions

$$\rho_{\boldsymbol{\alpha}} = \rho_{\mathbf{I}',\mathbf{n}}(\mathbf{p}, \mathbf{q}) = (2\pi)^{-N}\delta(\mathbf{I}(\mathbf{p}, \mathbf{q}) - \mathbf{I}')\exp(i\mathbf{n}\cdot\boldsymbol{\theta}(\mathbf{p}, \mathbf{q})), \tag{3.11}$$

$$\lambda_{\boldsymbol{\alpha}} = \lambda_{\mathbf{I}',\mathbf{n}} = \mathbf{n}\cdot\boldsymbol{\omega}(\mathbf{I}'), \tag{3.12}$$

where $\mathbf{n} = (n_1, \ldots, n_N)$ are integers. Here the eigenfunction label $\boldsymbol{\alpha} = \mathbf{I}'$, $\mathbf{n}$, and the spectrum $\mathbf{n}\cdot\boldsymbol{\omega}(\mathbf{I}')$ is continuous due to the continuous nature of $\mathbf{I}'$. Correspondingly, the eigenfunctions $\rho_{\boldsymbol{\alpha}}$ are "improper," that is, not square integrable functions.

Note that eigenfunctions with $\lambda_{\mathbf{I}',\mathbf{n}} = 0$ are stationary eigenfunctions and those with $\lambda_{\mathbf{I}',\mathbf{n}} \neq 0$ are nonstationary. Furthermore, the stationary $\rho_{\mathbf{I}',\mathbf{n}}$ with $\mathbf{n} = 0$ are uniform distributions on the $\mathbf{I}(\mathbf{p}, \mathbf{q}) = \mathbf{I}'$ torus, whereas the non-

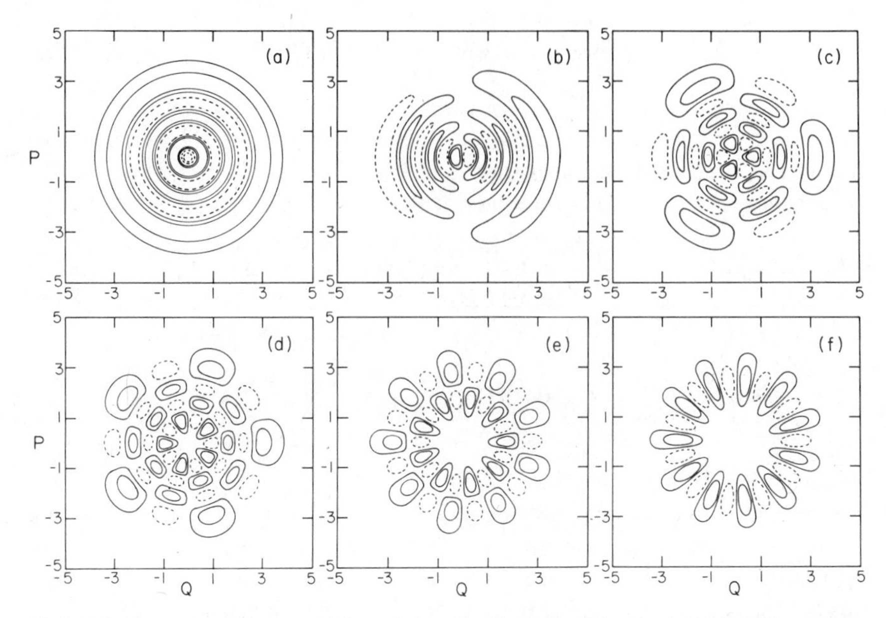


**Figure 11.** $[\rho_{n,j} + \rho_{n,-j}]$ superpositions of eigendistributions of the classical Liouville operator for the harmonic oscillator. Dashed contour is $-0.05$. Solid contours are at 0.005 and 0.05. (*a*) $n = 10, j = 0$; (*b*) $n = 11, j = 1$; (*c*) $n = 11, j = 3$; (*d*) $n = 11, j = 5$; (*e*) $n = 11, j = 9$; (*f*) $n = 11, j = 11$. [Reprinted with permission from *J. Phys. Chem.* **88**, 4829. Copyright 1984 American Chemical Society.]

stationary are nonuniform distributions on tori. An example, for the harmonic oscillator, is shown in Fig. 11.

In accord with Eq. (3.7) the solutions to Eq. (3.5) can be written as linear combinations of the time-dependent basis functions

$$\rho_{\mathbf{I}',\mathbf{n}}(\mathbf{p},\mathbf{q};t) = (2\pi)^{-N}\delta(\mathbf{I}(\mathbf{p},\mathbf{q}) - \mathbf{I}')e^{i\mathbf{n}\cdot\boldsymbol{\theta}(p,q)}e^{-i\lambda_{\mathbf{I}'\mathbf{n}}t}. \tag{3.13}$$

Eigendistributions for ergodic systems may also be readily obtained, although they prove less useful than those obtained for the regular case. Specifically, stationary eigenfunctions that are $L^2$ on the energy hypersurface for this case are given by

$$\rho_{E,0}(\mathbf{p},\mathbf{q}) = \delta(E - H(\mathbf{p},\mathbf{q}))/N, \tag{3.14}$$

where $N$ is a normalization integral. Nonstationary eigenfunctions have the form

$$\rho_\lambda(\mathbf{p},\mathbf{q}) \;\alpha\; \exp(-i\lambda\tau(\mathbf{p},\mathbf{q})), \tag{3.15}$$

where $\tau(\mathbf{p}, \mathbf{q})$ is the variable conjugate to the Hamiltonian and is a highly singular function of $(\mathbf{p}, \mathbf{q})$.

Considering our interest in relaxation phenomena, it is important to note that classical relaxation occurs in both regular as well as chaotic classical systems, as long as dynamics is not deliberately restricted to a torus in the former case. This is a direct consequence of contributions from the continuous spectrum, a feature of both regular and chaotic systems. To see this consider the correlation functions $P_{ab}(t) = \mathrm{tr}[\rho^a(t)\rho^{b*}]$, where $\rho^b$ is an arbitrary density, $\rho^b = \sum_{\boldsymbol{\beta}} b_{\boldsymbol{\beta}}\rho_{\boldsymbol{\beta}}$, or in terms of the time evolution of a property $B$, that is, $B(t) = \mathrm{tr}[B\rho^a(t)]$. With $\rho^a(t) = \sum_{\boldsymbol{\alpha}} a_{\boldsymbol{\alpha}}\rho_{\boldsymbol{\alpha}} \exp(-i\lambda_{\boldsymbol{\alpha}} t)$, we have

$$P_{ab}(t) = \sum_{\boldsymbol{\alpha}} a_{\boldsymbol{\alpha}} b_{\boldsymbol{\alpha}}^* e^{-i\lambda_{\boldsymbol{\alpha}} t} = \sum_{\boldsymbol{\alpha},\, \lambda_{\boldsymbol{\alpha}}=0} a_{\boldsymbol{\alpha}} b_{\boldsymbol{\alpha}}^* + \sum_{\boldsymbol{\alpha},\, \lambda_{\boldsymbol{\alpha}} \neq 0} a_{\boldsymbol{\alpha}} b_{\boldsymbol{\alpha}}^* e^{-i\lambda_{\boldsymbol{\alpha}} t}, \tag{3.16}$$

with a similar form for $B(t)$. Two qualitatively different forms of the long-time limit of $P_{ab}(t)$ are possible. In one case, restricted to regular dynamics on isolated tori, the $a_{\boldsymbol{\alpha}} b_{\boldsymbol{\alpha}}^*$ terms are such that only discrete $\lambda_{\boldsymbol{\alpha}}$ contribute. Then $P_{ab}(t)$ oscillates indefinitely. In the second case $a_{\boldsymbol{\alpha}} b_{\boldsymbol{\alpha}}^*$ terms contain only continuous $\lambda_{\boldsymbol{\alpha}} \neq 0$ contributions, with a long-time limit given by $\sum_{\boldsymbol{\alpha},\, \lambda_{\boldsymbol{\alpha}}=0} a_{\boldsymbol{\alpha}} b_{\boldsymbol{\alpha}}^*$, that is, a sum over stationary states. This is the typical result, for regular or chaotic systems, in classical mechanics. Note that, in either case, short-time observations can reveal apparent relaxation behavior, dephasing in the former and true relaxation in the latter. Furthermore, note that the nature of the time evolution depends on both the spectrum of $L$ and the nature of $\rho^a$, $\rho^b$ as reflected in the coefficients $a_{\boldsymbol{\alpha}} b_{\boldsymbol{\alpha}}^*$.

A basic distinction between regular and statistical systems lies in the types of terms that contribute to the first sum in Eq. (3.16). For chaotic systems the only stationary states in the discrete spectrum are $\delta(E - H)$; hence, the sum of $\lambda_{\boldsymbol{\alpha}} = 0$ terms corresponds to a weighted sum of uniform distributions over each energy shell represented in the initial distribution. In *principle* this is not substantially different from the long-time limit in regular cases involving continuous $\lambda_{\boldsymbol{\alpha}}$ where the long-time limit is also a weighted sum over stationary Liouville eigenstates. The difference, however, is of extreme practical significance since integrals of motion other than simple symmetry-related properties are rarely shown.

It is important to note that, in general, the $\rho_{\boldsymbol{\alpha}}(\mathbf{p}, \mathbf{q})$ are complex functions. Additional conditions on acceptable distributions must therefore be imposed, as discussed subsequently, to recover traditional classical mechanics. What is clear at this point is that several of the features normally attributed to quantum mechanics, *specifically all of the presumed quantum properties numbered (1)–(5) in the previously given list* are, in fact, characteristics of classical mechanics in the Liouville formulation. Further identification of classical and quantum mechanics results from recasting quantum mechanics in terms of phase space densities.

### 2. *Quantum "Phase Space" Distribution*

The Wigner–Weyl formulation of quantum mechanics provides an appropriate quantum analogue to classical phase space dynamics and consists of a *particular representation* of quantum mechanics.

Quantum mechanics is generally formulated on the Hilbert space of vectors $|\psi\rangle$. The functions $\psi(\mathbf{q}) = \langle \mathbf{q}|\psi\rangle$ and $\psi(\mathbf{p}) = \langle \mathbf{p}|\psi\rangle$ are the expansion coefficients of $|\psi\rangle$ in an appropriate complete orthonormal basis, that is,

$$|\psi\rangle = \int d\mathbf{q}\, |\mathbf{q}\rangle\langle \mathbf{q}|\psi\rangle = \int d\mathbf{q}\, |\mathbf{q}\rangle\psi(\mathbf{q}),$$

$$|\psi\rangle = \int d\mathbf{p}\, |\mathbf{p}\rangle\langle \mathbf{p}|\psi\rangle = \int d\mathbf{p}\, |\mathbf{p}\rangle\psi(\mathbf{p}). \tag{3.17}$$

Quantum mechanics can also be formulated in terms of an alternate Hilbert space whose elements are operators, with the density operator $\rho$ and the typical measurables $F$ among them. A variety of complete sets of basis operators in the space may be constructed, for example, as the tensor product of a basis set of the Hilbert space vectors $|\psi\rangle$ and its dual, yielding elements of the form $|\psi\rangle\langle\psi'|$.

The Wigner–Weyl representation of quantum mechanics is that obtained by the expansion of operators in an orthogonal set of operators $\Delta(\mathbf{p}, \mathbf{q})$; the resultant $\rho^w(\mathbf{p}, \mathbf{q})$ lie within a Hilbert space of $L^2$ functions. Specifically,[51]

$$\Delta(\mathbf{p}, \mathbf{q}) = \int d\mathbf{v} \exp(-i\mathbf{p}\cdot\mathbf{v}/\hbar)|\mathbf{q} - \tfrac{1}{2}\mathbf{v}\rangle\langle \mathbf{q} + \tfrac{1}{2}\mathbf{v}|, \tag{3.18}$$

with

$$\mathrm{Tr}[\Delta(\mathbf{p}, \mathbf{q})\Delta(\mathbf{p}', \mathbf{q}')] = h^N\delta(\mathbf{q}' - \mathbf{q})\delta(\mathbf{p}' - \mathbf{p}). \tag{3.19}$$

Any operator may be expanded in this basis, for example, for the scaled density operator $(\rho/h^N)$ or a measurable $F$, we have

$$\rho/h^N = h^{-N}\int d\mathbf{p}\, d\mathbf{q}\, \mathrm{Tr}(\Delta(\mathbf{p}, \mathbf{q})\rho/h^N)\Delta(\mathbf{p}, \mathbf{q}),$$

$$F = h^{-N}\int d\mathbf{p}\, d\mathbf{q}\, \mathrm{Tr}(\Delta(\mathbf{p}, \mathbf{q})F)\Delta(\mathbf{p}, \mathbf{q}). \tag{3.20}$$

The resultant expansion coefficients

$$\rho^w(\mathbf{p},\mathbf{q}) = \mathrm{Tr}[\Delta(\mathbf{p},\mathbf{q})\rho/h^N] = (\pi\hbar)^N \int d\mathbf{v}\, e^{-2i\mathbf{p}\cdot\mathbf{v}/\hbar}\langle \mathbf{q}-\mathbf{v}|\rho|\mathbf{q}+\mathbf{v}\rangle, \quad (3.21)$$

$$F_w(\mathbf{p},\mathbf{q}) = \mathrm{Tr}[\Delta(\mathbf{p},\mathbf{q})F] = 2^N \int d\mathbf{v}\, e^{-2i\mathbf{p}\cdot\mathbf{v}/\hbar}\langle \mathbf{q}-\mathbf{v}|F|\mathbf{q}+\mathbf{v}\rangle, \quad (3.22)$$

are the Wigner function associated with $\rho$ and the Wigner–Weyl representation of $F$, respectively. Note we use a superscript $w$ to denote the extra $h^{-N}$ needed to yield Wigner's original function associated with the pure state density $|\psi\rangle\langle\psi|$.

The algebra of the Wigner–Weyl representation has been extensively reviewed[51] and we only call attention to the form that Eqs. (3.5) and (3.6) take in this representation. The Liouville equation becomes

$$\frac{\partial\rho^w(\mathbf{p},\mathbf{q};t)}{\partial t} = L_w\rho^w(\mathbf{p},\mathbf{q};t),$$

where

$$L_w(\mathbf{p},\mathbf{q}) = (2/\hbar)\sin\left[(\hbar/2)\left(\frac{\partial^{(H)}}{\partial\mathbf{q}}\frac{\partial^{(\rho)}}{\partial\mathbf{p}} - \frac{\partial^{(H)}}{\partial\mathbf{p}}\frac{\partial^{(\rho)}}{\partial\mathbf{q}}\right)\right]H_w(\mathbf{p},\mathbf{q}) \quad (3.23)$$

and where $\partial^{(H)}/\partial x$ operates on $H_w(\mathbf{p},\mathbf{q})$, $\partial^{(\rho)}/\partial x$ on $\rho^w(\mathbf{p},\mathbf{q};t)$. $H_w(\mathbf{p},\mathbf{q})$ is the Wigner–Weyl transform $\mathrm{Tr}(H\Delta(\mathbf{p},\mathbf{q}))$ of the Hamiltonian. Average values become

$$\langle F\rangle = \mathrm{Tr}(F_w\rho^w) = \int d\mathbf{p}\,d\mathbf{q}\,F_w(\mathbf{p},\mathbf{q})\rho^w(\mathbf{p},\mathbf{q}). \quad (3.24)$$

Eigendistributions of $L_w(\mathbf{p},\mathbf{q})$, which span the Hilbert space, are readily constructed from eigenstates $|\psi_i\rangle$ of $H$ ($H|\psi_i\rangle = E_i|\psi\rangle$) and from the operators $\rho_{ij} = |\psi_i\rangle\langle\psi_j|$ that satisfy the time-independent quantum Liouville equation

$$L\rho_{ij} = (E_j - E_i)/\hbar\,\rho_{ij}. \quad (3.25)$$

The Wigner–Weyl transform of $\rho_{ij}/h^N$, that is,

$$\rho_{ij}^w = \mathrm{Tr}(|\psi_i\rangle\langle\psi_j|\Delta(\mathbf{p},\mathbf{q})/h^N), \quad (3.26)$$

then satisfies

$$L_w \rho_{\mathbf{ij}}^w(\mathbf{p}, \mathbf{q}) = [(E_{\mathbf{j}} - E_{\mathbf{i}})/\hbar]\rho_{\mathbf{ij}}^w(\mathbf{p}, \mathbf{q}). \tag{3.27}$$

Thus $\rho_{\mathbf{ij}}^w(\mathbf{p}, \mathbf{q})$ is an eigendistribution of $L_w$ with eigenvalue

$$\lambda_{\mathbf{ij}} = (E_{\mathbf{j}} - E_{\mathbf{i}})/\hbar \tag{3.28}$$

and form a set of basis functions for the $L^2$ Hilbert space of density operators in the Wigner–Weyl representation. They satisfy orthogonality and completeness as

$$(\rho_{\mathbf{ij}}^w(\mathbf{p}, \mathbf{q}), \rho_{\mathbf{lm}}^w(\mathbf{p}, \mathbf{q})) = h^N \int d\mathbf{p}\, d\mathbf{q}\, \rho_{\mathbf{ij}}^{w*}(\mathbf{p}, \mathbf{q})\rho_{\mathbf{lm}}^w(\mathbf{p}, \mathbf{q}) = \delta_{\mathbf{il}}\delta_{\mathbf{jm}}, \tag{3.29}$$

$$h^N \sum_{\mathbf{i},\mathbf{j}} \rho_{\mathbf{ij}}^{w*}(\mathbf{p}', \mathbf{q}')\rho_{\mathbf{ij}}^w(\mathbf{p}, \mathbf{q}) = \delta(\mathbf{p} - \mathbf{q}')\delta(\mathbf{q} - \mathbf{q}'). \tag{3.30}$$

Eigenstates of the time-dependent Liouville Equation [Eq. (3.5)] are given via Eq. (3.7) and $\rho_{\mathbf{ij}}$, $\mathbf{i} \neq \mathbf{j}$, correspond to nonstationary distributions.

### 3. *Classical versus Quantum Mechanics: Time-Independent Structure*

Most assuredly, the individual eigenfunctions $\rho_{\mathbf{ij}}$ of $L_Q$ and $\rho_\alpha$ of $L_c$ differ quantitatively. However, the formal structure set up thus far provides little insight into their qualitative differences since both classical and quantum mechanics have been formulated in terms of formally identical[52] Hilbert spaces of $L^2$ functions that are spanned by both the eigendistributions $\rho_\alpha$ of $L_c$ and $\rho_{\mathbf{i},\mathbf{j}}$ eigendistributions of $L_Q$. Completeness of both bases ensures that *any* $L^2$ function $g(\mathbf{p}, \mathbf{q})$ may be written as linear combinations of *either* set of basis vectors. Indeed, it is the auxiliary conditions which arise from the *interpretation* of distributions in classical and quantum mechanics that are *fundamental* to the qualitative distinction between classical and quantum mechanics.

In classical mechanics a distribution, denoted $\rho_c(\mathbf{p}, \mathbf{q})$, is interpreted as a probability density which requires that a classically acceptable $\rho_c(\mathbf{p}, \mathbf{q})$ satisfy

$$\mathrm{Tr}[\rho_c(\mathbf{p}, \mathbf{q})] = \int d\mathbf{p}\, d\mathbf{q}\, \rho_c(\mathbf{p}, \mathbf{q}) = 1, \tag{3.31a}$$

$$\rho_c(\mathbf{p}, \mathbf{q}) \geq 0 \quad \text{for all } (\mathbf{p}, \mathbf{q}), \tag{3.31b}$$

$$\bar{F} = \mathrm{Tr}[F(\mathbf{p}, \mathbf{q})\rho_c(\mathbf{p}, \mathbf{q})] < \infty. \tag{3.31c}$$

One additional condition, on the norm $\mathrm{Tr}[\rho^2(\mathbf{p},\mathbf{q})]$, must be considered. Adherence to square integrability would require that this trace be finite. However, the $\delta$ function nature of eigenfunctions of $L_c$ indicate that singular distributions appear naturally in classical mechanics and that the original Hilbert space on $L^2$ functions should be enlarged to include such improper functions. Hence, we assume

$$0 \leq \mathrm{Tr}[\rho^2(\mathbf{p},\mathbf{q})] \leq \infty. \tag{3.31d}$$

Note that these auxiliary conditions, particularly (31b), are the reason why the nontraditional (e.g., complex) nature of the underlying classical eigendistributions is suppressed in classical mechanics. Such features are, however, vital for an understanding of the classical–quantum correspondence from the distributions viewpoint.

Quantum distributions, denoted $\rho_Q^w(\mathbf{p},\mathbf{q})$, do not have a probabilistic interpretation. Rather, the conditions closest to those in Eq. (3.31) are

$$\mathrm{Tr}[\rho_Q^w(\mathbf{p},\mathbf{q})] = 1, \tag{3.32a}$$

$$\int d\mathbf{p}\, \rho_Q^w(\mathbf{p},\mathbf{q}) \geq 0 \quad \text{for all } \mathbf{q},$$

$$\int d\mathbf{q}\, \rho_Q^w(\mathbf{p},\mathbf{q}) \geq 0 \quad \text{for all } \mathbf{p}, \tag{3.32b}$$

$$\bar{F} = \mathrm{Tr}[F_w(\mathbf{p},\mathbf{q})\rho_Q^w(\mathbf{p},\mathbf{q})] < \infty, \tag{3.32c}$$

and

$$0 \leq \mathrm{Tr}[\rho_Q^{w2}(\mathbf{p},\mathbf{q})] \leq h^{-N}. \tag{3.32d}$$

Equation (3.32a) implies normalization, and Eq. (3.32b) contains the essential probabilistic interpretation of the projections onto the momenta or coordinates. The last condition, a natural consequence of the definitions of the density matrix and the Wigner–Weyl transform, explicitly eliminates the singular distributions allowed in Eq. (3.31d). That is, although the completeness of the quantum $\rho_{\mathbf{ij}}^w(\mathbf{p},\mathbf{q})$ basis permits the construction of $\delta$ function distributions, they make, unlike classical mechanics, no natural appearance in quantum mechanics wherein eigenfunctions of $L_Q$ are square integrable and such singular distributions are explicitly excluded in Eq. (3.32d).

The possible distributions $\rho(\mathbf{p},\mathbf{q})$ that satisfy *both* quantum and classical conditions is rather restricted. Specifically, distributions with any negative character are "quantum only," and those nonnegative $\rho(\mathbf{p},\mathbf{q})$ with $\mathrm{Tr}[\rho^2(\mathbf{p},\mathbf{q})] > (h)^{-N}$

are "classical only." Only nonnegative distributions satisfying Eq. (3.28d) are acceptable to both mechanics. Of particular interest is the proof[53] that the only nonnegative *pure* quantum states [i.e., those $\rho^w$ with $\mathrm{Tr}(\rho^{w2}) = (2\pi h)^{-N}$] are comprised of gaussian wavefunctions.[54] Thus, both quantum and classical mechanics accept *pure* states of this kind or nonnegative *mixed* states satisfying Eq. (3.32d). Nonetheless, it is possible to construct $\rho_Q^w$, expand it in the classical $\rho_{\boldsymbol{\alpha}}(\mathbf{p}, \mathbf{q})$, and propagate the system classically. This constitutes an extension of classical mechanics, discussed subsequently, in which the auxiliary conditions, but not the dynamics, are taken as those due to quantum mechanics.

### 4. *Classical versus Quantum Mechanics: Time Evolution*

An essential qualitative difference between the quantum and classical boundstate dynamics is in the nature of the spectrum of the Liouville operators. In the former case, the spectrum is discrete, resulting in quasiperiodic motion. In the latter case it is continuous, leading to relaxation [see Eq. (3.16)]. This discrepancy is the basis of the "quantum chaos" problem (see Section IV). We only note, at this point, that a similar problem arises in both integrable and chaotic systems insofar as the former displays classical relaxation. Thus, insight into the role of spectral continuity in classical mechanics versus the discrete spectrum in quantum mechanics can be gained from studies of integrable systems. Work in this area is ongoing.[55] At present it is clear from the literature[56] that there is an interplay between the time scale of observation and the discrete versus continuous spectrum. That is, the classical spectrum will appear, over time scales shorter than the *frequency* level spacing, to be observationally continuous. Times longer than this will display the characteristic discrete spectrum of the quantum system. Some time-dependent classical–quantum comparisons are discussed in Section III B (6).

### 5. *Classical–Quantum Correspondence*[57]

Our emphasis has been on the fundamental role of the eigendistributions of $L_c$ and $L_Q$, and it is within this eigenfunction basis that classical–quantum correspondence assumes its most simple form. At present the correspondence is clearly understood for stationary and nonstationary eigendistributions in regular systems and for stationary eigendistributions in the chaotic case. Further work is necessary to clarify the picture for nonstationary chaotic eigendistributions.

Specifically, the following general results have been proven[57,58] for regular systems:

$$\lim_{\hbar \to 0} \rho_{\mathbf{nn}}^w(\mathbf{p}, \mathbf{q}) = \delta(\mathbf{I}(\mathbf{q}, \mathbf{p}) - \mathbf{I_n}) = \rho_{\mathbf{I_n}, 0}(\mathbf{p}, \mathbf{q}), \tag{3.33}$$

$$\lim_{\hbar \to 0} \rho^{w}_{\mathbf{nm}}(\mathbf{p}, \mathbf{q}) = \rho_{\mathbf{I}_{\mathbf{nm}}, (\mathbf{m}-\mathbf{n})}$$

$$\lambda_{\mathbf{nm}} \to \lambda_{\mathbf{I}_{\mathbf{nm}}, (\mathbf{n}-\mathbf{m})} \tag{3.34}$$

where $\mathbf{I}_{\mathbf{nm}} = \mathbf{I}_{(\mathbf{m}+\mathbf{n})/2} = [\frac{1}{2}(\mathbf{m} + \mathbf{n}) + \boldsymbol{\beta}]\hbar$. Here $\mathbf{I}_{\mathbf{j}} = (\mathbf{j} + \boldsymbol{\beta})\hbar$ is the semiclassically quantized action with Maslov index $\boldsymbol{\beta}$.

Equations (3.33) and (3.34) make a succinct statement about quantum–classical correspondence. Specifically, in the classical limit, stationary quantum eigendistributions correspond to stationary classical eigendistributions at the related quantized action. Secondly, nonstationary quantum eigendistributions $\rho_{\mathbf{n},\mathbf{m}}(\mathbf{I}_{\mathbf{j}}, \boldsymbol{\theta})$ correspond to nonstationary classical eigendistributions. These consist of a nonuniform distribution on a torus at intermediate action $\mathbf{I}_{(\mathbf{m}+\mathbf{n})/2}$ with the nonuniformity determined by the difference $(\mathbf{n} - \mathbf{m})$ times the angles on this torus.[59] Thus, from this viewpoint there is a direct link between eigendistributions in quantum and classical mechanics in the $\hbar \to 0$ limit.

If the correspondence between stationary eigenstates of the quantum and classical eigendistributions is made for the ergodic case as well, we obtain a trivial qualitative route to the results obtained previously[60] by extensive mathematical considerations. Specifically, one would qualitatively expect the classical limit of $\rho_{\mathrm{ii}}$, with energy average $E_{\mathrm{i}}$, to reduce, as $\hbar \to 0$, to the classical Liouville eigenfunction at $E_{\mathrm{i}}$. From Eq. (3.14) one would expect

$$\lim_{\hbar \to 0} \rho_{\mathrm{ii}}(\mathbf{p}, \mathbf{q}) = \delta(E_{\mathrm{i}} - H)/N, \tag{3.35}$$

in accord with formal results[60] that utilize a smoothing over quantum oscillations. A similar correspondence may be conjectured in the nonstationary case, but further studies are necessary to accommodate the highly singular nature of the classical eigendistributions.

This correspondence route, with the recognition that a principal difference between the time dependence of classical and quantum bound-state dynamics lies in the nature of the spectrum suggests a classical *analogue*[57] to quantum mechanics that bridges the gap between them.

Consider once again a pure quantum superposition state consisting of $M$ levels:

$$|\psi\rangle = \sum_{\mathbf{n}}^{M} d_{\mathbf{n}} |\mathbf{n}\rangle \tag{3.36}$$

with associated Wigner–Weyl density:

$$\rho(\mathbf{p}, \mathbf{q}) = \sum_{\mathbf{n},\mathbf{m}}^{N} c_{\mathbf{nm}} \rho^{w}_{\mathbf{nm}}(\mathbf{p}, \mathbf{q}); \qquad c_{\mathbf{nm}} = d_{\mathbf{n}} d^{*}_{\mathbf{m}}. \tag{3.37}$$

Such states arise in any conservative problem involving time evolution, in which case $\rho(t) = |\psi(t)\rangle\langle\psi(t)|$ is the time-dependent density matrix and

$$F(t) = \mathrm{Tr}[F\rho(t)] \tag{3.38}$$

gives the time evolution of any observable $F$. We can now define a classical analogue $\rho_{ca}(\mathbf{p}, \mathbf{q})$ of the quantum density [Eq. (3.37)] associated with such a superposition state by replacing each $\rho_{\mathbf{nm}}(\mathbf{p}, \mathbf{q})$ by its classical limit $\rho_{\mathbf{I}_{\mathbf{nm}}, (\mathbf{n}-\mathbf{m})}$, $\mathbf{I}_{\mathbf{nm}} = \frac{1}{2}(\mathbf{I}_{\mathbf{n}} + \mathbf{I}_{\mathbf{m}})$ and by retaining the same expansion coefficients. That is,

$$\rho_{ca}(\mathbf{p}, \mathbf{q}) = \sum_{\mathbf{nm}} c_{\mathbf{nm}} \rho_{\mathbf{I}_{\mathbf{nm}}, (\mathbf{n}-\mathbf{m})}(\mathbf{p}, \mathbf{q}). \tag{3.39}$$

Thus, the classical analogue of Eq. (3.37) is a set of $M$ tori with uniform stationary density, weighted by $c_{\mathbf{nn}}$ and a set of $M(M-1)/2$ tori with nonuniform density weighted by $c_{\mathbf{nm}}$. Subsequent time evolution is solely classical, that is, using classical Liouville eigenvalues. The definition of the classical analogue invokes the WKB quantization conditions to identify the stationary and nonstationary classical eigendistributions that are associated with the stationary and nonstationary contributions to the quantum density. This approach has been used[61] to examine the nature of the time evolution of simple two-level states as well as to provide a classical analogue to phenomena such as quantum beats and objects such as the wavefunction autocorrelation function discussed in Section IV.

### 6. *Sample Computation: Stadium Dynamics*

The preceding discussion emphasizes the formal similarities between classical and quantum mechanics from the phase space viewpoint. Here we demonstrate, with an example relevant to classical versus quantum chaos, that this carries over to qualitative similarities in the dynamics. The example is the stadium billiard,[15] a classical $K$ system, consisting of a mass $m$ particle confined to a potential free region by infinite potential walls, the boundary being comprised of two parallel sides of length $2a$ closed by two semicircles of radius $r$. We choose the typical parameters $a = r = \frac{1}{2}$ and $\hbar = m = 1$. The phase space dynamics of various initial distributions in the stadium were determined[62] and compared by examining projections of the time-dependent distributions in coordinate and momentum space. In this case no attempt was made to choose parameters that enhance classical–quantum agreement (e.g., large quantum numbers were not required). A typical coordinate and momentum space comparison, for an initial gaussian phase space distribution, is shown in Figs. 12 and 13. In both cases the time development of the distributions is quite similar, with the classical converging to a smooth long-time

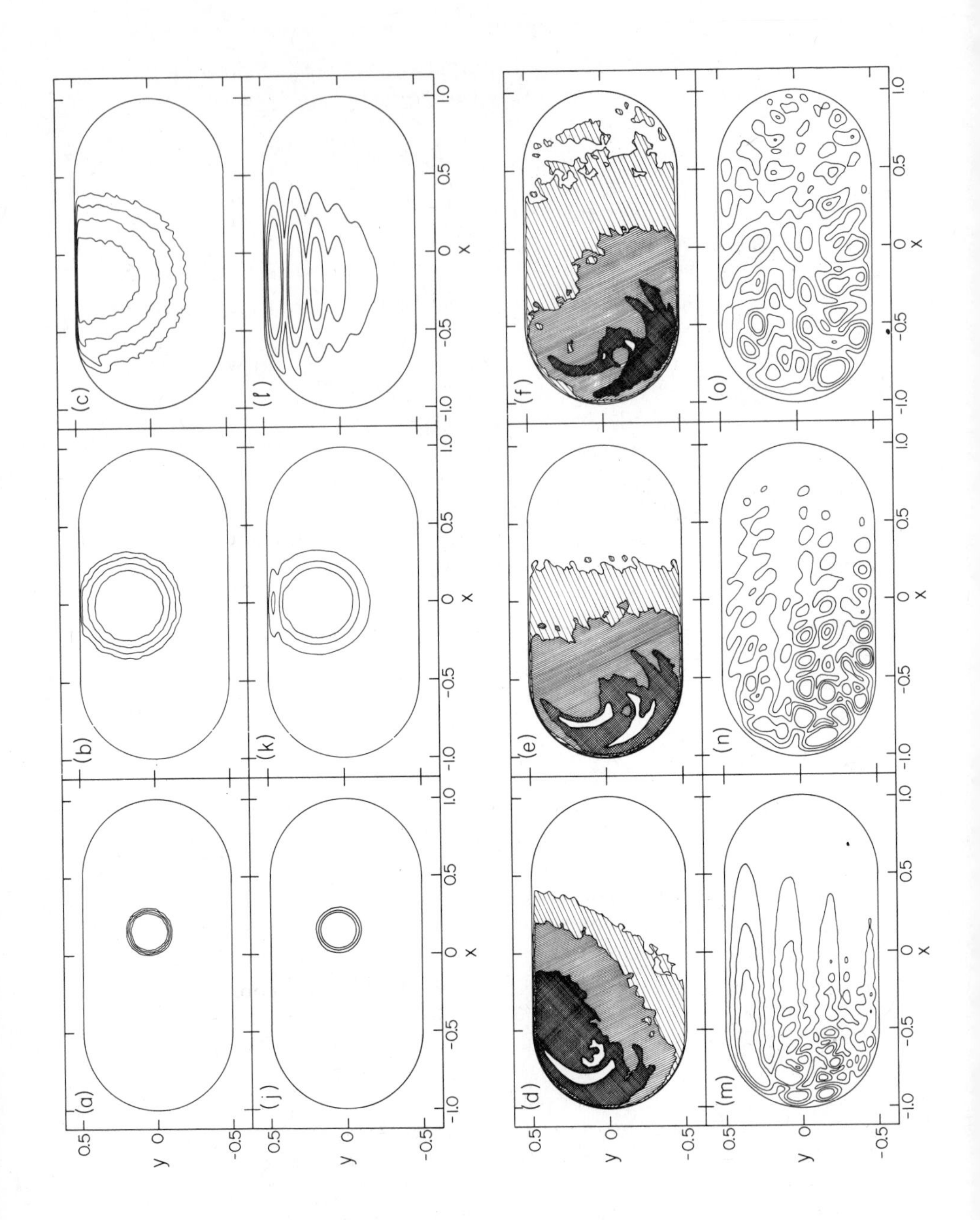

(a)
(b)
(c)
(j)
(k)
(ℓ)
(d)
(e)
(f)
(m)
(n)
(o)
x
y
-1.0
-0.5
0
0.5
1.0

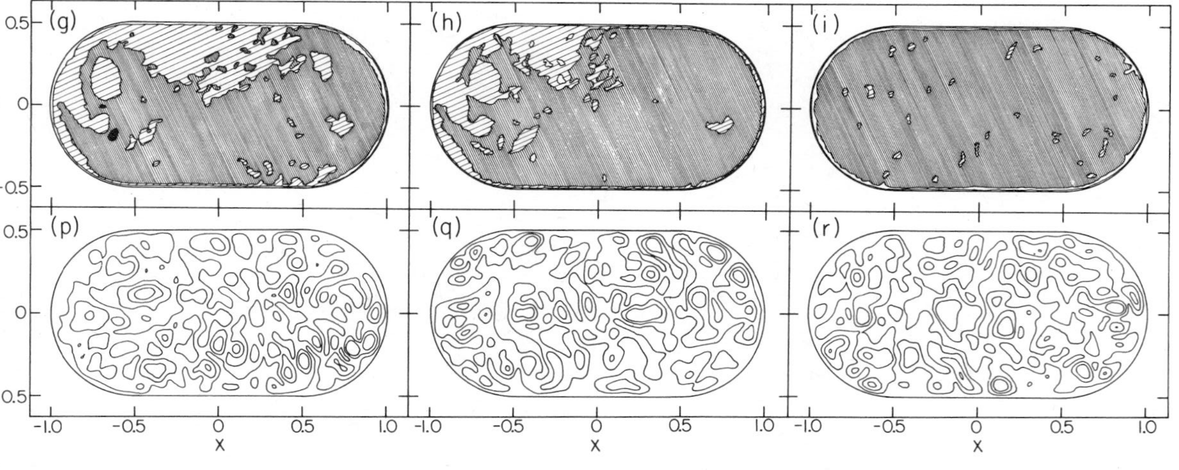


**Figure 12.** Contour plot of the coordinate space projection of a typical distribution evolving in the stadium. Classical results are shown in panels (*a*)–(*i*) and quantum results are shown in (*j*)–(*r*). Times shown are $t = 0.0, 0.01, 0.02, 0.04, 0.06, 0.08, 0.12, 0.16, 0.8$. Quantum contours are 0.25, 1.0, and 2.0; classical contours are the same with the addition of a contour at 0.5. In the later classical panels contours are replaced, for clarity, by increasing levels of shading indicating probabilities of 0.25–0.5, 0.5–1.0, 1.0–2.0. Panels (*d*) and (*e*) contain a white region embedded in the darkest regions; these are of maximum probability $>2.0$. (From Ref. 62*a*.)

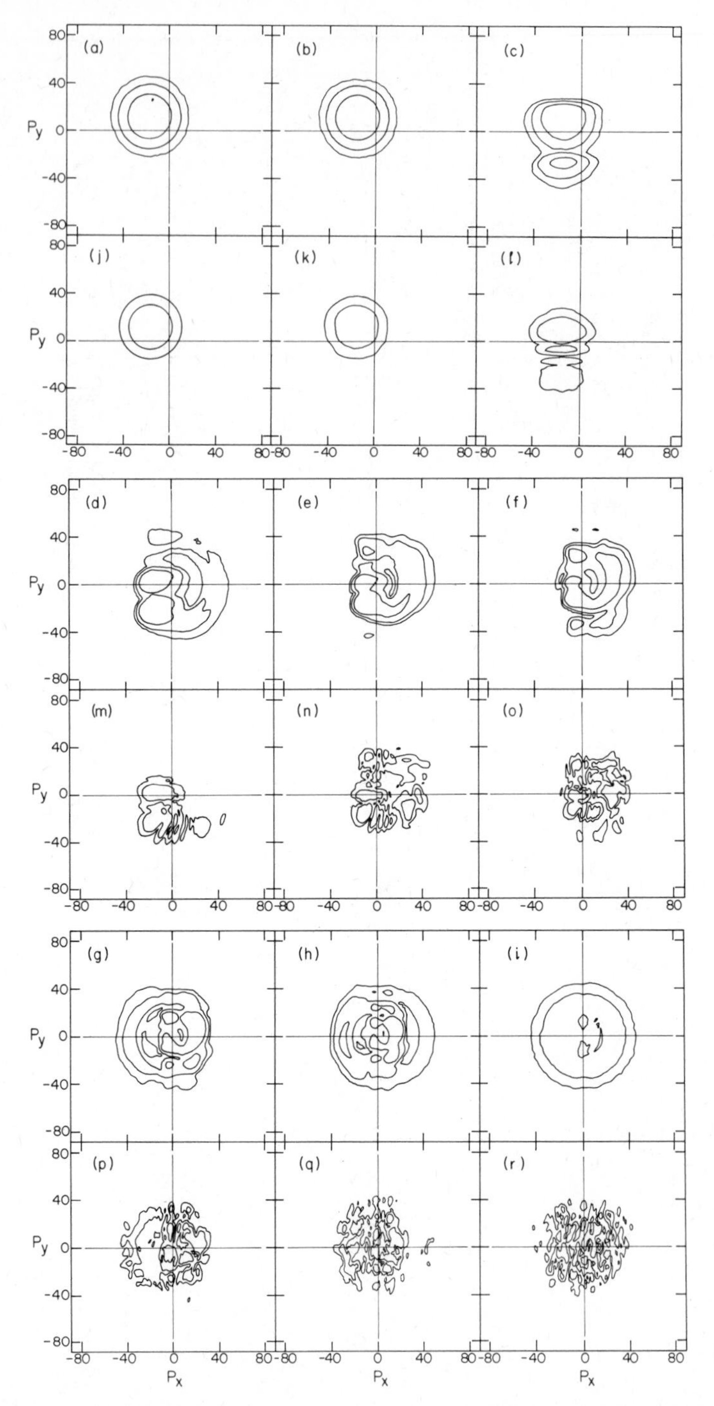


**Figure 13.** Contour plot of the momentum space projection of a distribution (as in Fig. 12) during its time evolution. Classical results are shown in panels (*a*)–(*i*) and quantum results in (*j*)–(*r*) with times shown as in Fig. 12. Quantum contours are at $8 \times 10^{-5}$ and $3.2 \times 10^{-4}$. Classical contours are at the same values with an additional contour at $2 \times 10^{-5}$. (From Ref. 62*a*.)

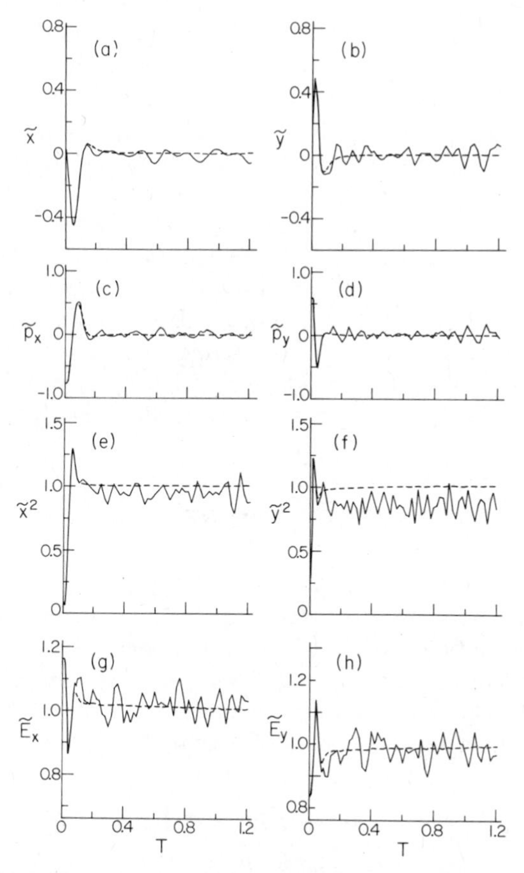


**Figure 14.** Time dependence of eight properties during the time evolution of the distribution shown in Figs. 12 and 13. Solid curves denote quantum results, dashed (visible only where they deviate from the quantum) are the classical results. Properties are shown scaled, hence the tildes, as follows: second moments $x^2$, $y^2$, $E_x$, $E_y$ are divided by their long-time statistical values. The quantities $x$ and $y$ are divided by the stadium size in the $x$ or $y$ directions and $p_x$, $p_y$ are divided by the initial value $\{(p_x^0)^2 + (p_y^0)^2\}^{1/2}$. (From Ref. 62*a*.)

distribution, while the quantum displays long-time oscillatory components required by complementarity. Careful examination of the classical versus quantum results clearly shows that with small coarse graining they are in remarkable agreement. This is further evidenced by a comparison of typical time-dependent averages shown in Fig. 14, which are computed either classically or quantally. Once again the only major difference is the small variation of the average values at long times in the quantum case, whereas the classical converges to a constant.

In summary, then, the Liouville approach demonstrates that the central *conceptual* difference between quantum and classical mechanics stems from the imposition of auxiliary conditions which extract allowed phase space distributions from the common Hilbert space that these two mechanics share. The correspondence limit identifies specific stationary and nonstationary eigendistributions of the classical Liouville operator as being associated with stationary and nonstationary quantum Liouville eigendistributions. Furthermore, this approach suggests a discretized classical mechanics that is based solely on the classical eigendistributions arising in the correspondence limit. We have also shown,[57] via this approach, that semiclassical mechanics in action-angle variables is operationally identical to the preceding discretized classical mechanics. Thus, this version of semiclassical mechanics corresponds to extracting a discrete set of classical eigenvalues *and* eigendistributions, from the classical continuum.

## IV. QUANTUM CHAOS AND REACTION DYNAMICS

Consider now bound-state dynamics from the viewpoint of quantum mechanics. The essential problem associated with chaotic dynamics has previously been alluded to, that is, since the quantum Liouville spectrum is discrete, the dynamics cannot display true chaotic relaxation. A number of proposals have been made as to what constitutes the quantum analogue of classical chaos.[63] We now focus on an approach we have recently been developing and discuss its link to statistical behavior in reaction dynamics.

### A. Chaos and Spatial Correlation Functions

#### 1. *Stationary Eigenfunctions*

As discussed in Section IV B, statisticality results from the combined action of individual energy eigenstates, some of which lead to statistical behavior and others of which do not. The spatial correlation function approach[64] focuses on properties of *individual* (stationary as well as nonstationary) quantum states and allows a unique labeling of quantum states as chaotic or regular, in a manner that links directly to both classical chaos as well as the statisticality of observables, such as reaction rates.

In this approach a wavefunction, $\psi(\mathbf{q})$, is termed *chaotic* if its autocorrelation function, defined as,

$$F(\boldsymbol{\delta}) = \int d\mathbf{q}\, \psi^*(\mathbf{q})\psi(\mathbf{q} + \boldsymbol{\delta}), \tag{4.1}$$

is aperiodic and decaying. The $\mathbf{q}$ integration in Eq. (4.1) can be performed over a "space filling" path or on a grid spanning part of, or the entire, configuration space.

Numerical results[64] on the stadium-billiard problem[15,65] strongly support this assignment of the label "chaotic": Low-lying states, with "regular" nodal patterns,[66] are found[64] to give rise to nondecaying, quasiperiodic correlation functions. In contrast, high-lying, "erratic" looking states are characterized by rapidly decaying correlation functions that fluctuate about zero for all values of $\delta \gg \lambda$, where $\lambda$ is an average deBroglie wavelength. Examples of these two cases[64] are given in Figs. 15*a* and 15*b*.

The correlation function method quantifies previous suggestions[63] that quantum eigenstates, in the classically chaotic regime, display erratic nodal patterns which are devoid of identifiable global structure. Physically, the decay of the correlation function in the chaotic regime implies that a slight degree of coarse graining results in a complete loss of correlation between the probabilities of observing the system at two points "sufficiently" removed from one another. This feature is in accord with a qualitative understanding of the nature of classically chaotic systems.

A direct quantitative connection between this labeling and classical concepts can also be established.[67] This results from the observation that the correlation function of Eq. (4.1) can be viewed as the expectation value of the shift operator,

$$\begin{aligned} F_{qu}(\boldsymbol{\delta}) &= \int d\mathbf{q}\, \psi^*(\mathbf{q}) \exp(i\boldsymbol{\delta}\cdot\mathbf{p}_{qu}/\hbar)\psi(\mathbf{q}) \\ &= \int d\mathbf{p}\, d\mathbf{q}\, \rho^w(\mathbf{p},\mathbf{q}) \exp(i\boldsymbol{\delta}\cdot\mathbf{p}/\hbar), \end{aligned} \tag{4.2}$$

where $\rho^w(\mathbf{p},\mathbf{q})$ is the Wigner function associated with $\psi$. The classical spatial correlation of interest is obtained by replacing $\rho^w(\mathbf{p},\mathbf{q})$ by an appropriate classical density $\rho(\mathbf{p},\mathbf{q})$. The resultant expression is the classical expectation value of the shift operator, that is,

$$F_{cl}(\boldsymbol{\delta}) = \int d\mathbf{p}\, d\mathbf{q}\, \rho(\mathbf{p},\mathbf{q}) \exp(i\boldsymbol{\delta}\cdot\mathbf{p}/\hbar), \tag{4.3}$$

where $\hbar$ is now regarded as a scaling parameter ensuring the same units for $\delta$ as in the quantum expression.

Consider then $F_{cl}(\boldsymbol{\delta})$ for the case of a classically ergodic (the "weakest" form of chaos) distribution at fixed energy. The latter is given by Eq. (3.14), where the Hamiltonian is of the form

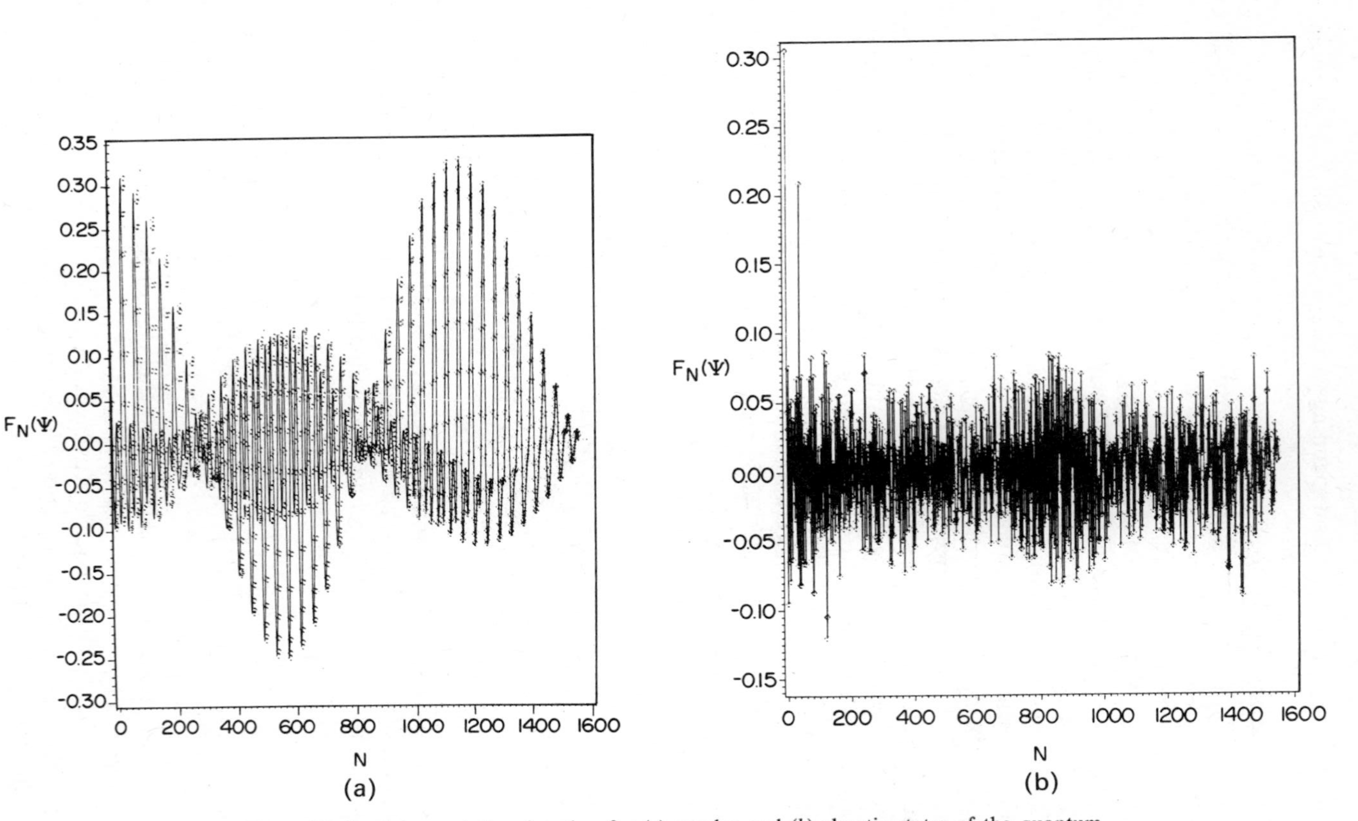


**Figure 15.** Spatial correlation function for (*a*) regular and (*b*) chaotic states of the quantum billiard problem. (From Ref. 64.)

$$H = \mathbf{p}^2/2m + V(\mathbf{q}),$$

where $m$ is the mass. Substituting this density into Eq. (4.3) gives the classically *ergodic* correlation function,

$$F_{cl}^{\text{erg}}(\boldsymbol{\delta}) = \frac{1}{N}\int d\mathbf{p}\,d\mathbf{q}\,\delta(E - H(\mathbf{p},\mathbf{q}))\exp(i\boldsymbol{\delta}\cdot\mathbf{p}/\hbar)$$

$$= \frac{1}{N}\int d\mathbf{q}\,d\hat{\mathbf{p}}\exp\{i\boldsymbol{\delta}\cdot\hat{\mathbf{p}}[2m(E - V(\mathbf{q}))]^{1/2}/\hbar\}, \tag{4.4}$$

where $\hat{\mathbf{p}}$ is a unit vector directed along $\mathbf{p}$ and $d\hat{\mathbf{p}}$ includes the appropriate volume element for the angular integration.

For two-dimensional problems, polar coordinates ($\mathbf{p} = |\mathbf{p}|, \theta$), with $\boldsymbol{\delta}$ directed along the $x$ axis, allow one to perform the angular integral[68] directly, to obtain,

$$F_{cl}^{\text{erg}}(\boldsymbol{\delta}) = \frac{1}{A}\int d\mathbf{q}\,J_0\{\delta\,[2m(E - V(\mathbf{q}))]^{1/2}/\hbar\}, \tag{4.5}$$

where $J_0(x)$ is the zero cylindrical Bessel function, and $A$ is the area in configuration space accessible at energy $E$.

It is instructive to examine properties of spatial correlation functions for specific cases. The stadium-billiard problem, discussed in previous sections in the context of classical–quantum correspondence, is a useful example. There is, in this case, no potential within the boundary, that is, the boundary alone induces ergodicity. Therefore, the $q$ integration is readily done, yielding,

$$F_{cl}^{\text{stadium}}(\delta) = J_0\{\delta\,[2mE]^{1/2}/\hbar\}. \tag{4.6}$$

Figure 16 provides[67] a comparison between the classical [Eq. (4.6)], and quantum [Eq. (4.1)] spatial correlation functions for four chaotic stadium eigenfunctions. The classical and quantum correlation functions are seen to agree very well for all distances other than those comparable to the size of the stadium's linear dimension. Thus, the rapid decay of the quantum spatial autocorrelation is a measure relating directly to classical ergodic behavior.

The classical–quantum correspondence also results in useful scaling laws for chaotic states. For the stadium problem the classical correlation functions scale as $(2mE)^{-1/2}$, a feature which is solely a consequence of classical ergodicity. As shown in Fig. 17, quantum states labeled "chaotic" (according to the aperiodic nature of their correlation function) do obey this scaling relation. Specifically, Fig. 17 displays the correlation lengths ($\Delta_{1/2}$), defined as

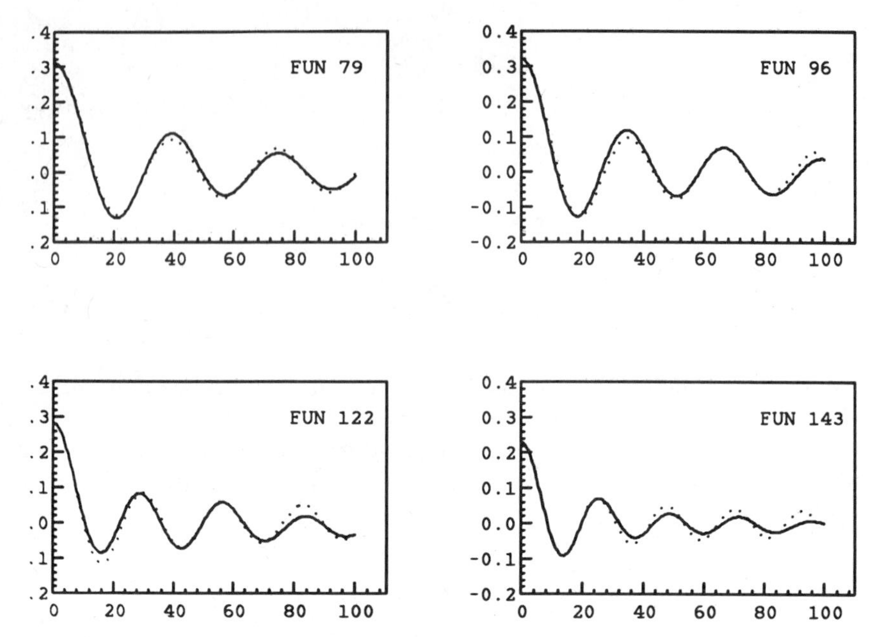


**Figure 16.** Quantum (dotted) and classical (dashed) $F(\delta)$ spatial correlation functions, at four different energies (of the 79th, 96th, 122nd, and 143rd quantum levels). $\delta$ is expressed in increments of $5 \times 10^{-3}$ a.u. (for comparison note that the stadium long axis = 2.6530 a.u.). (From Ref. 67.)

the smallest distance for which $F(\Delta_{1/2}) = \frac{1}{2}F(0)$, for 160 states of the stadium billiard. The correlation lengths of states 20 through 160 are seen to follow closely the classical scaling behavior. The major exceptions are regular states (marked as * and +) whose correlation functions are quasiperiodic.

The appearance of regular states embedded in a "sea" of chaotic states occurs for all quantum systems and may be attributed to near adiabaticity.[66,69–73] This concept of adiabaticity, which is crucial for the understanding of many observations on realistic systems, is discussed subsequently. At this point we just note that in the stadium case the regular states are the quantum analogues of the (zero measure) classical periodic orbits. We will present, later in this chapter an heuristic argument for the existence of regular states for *all* quantum systems that have no continuous spectrum.

### 2. *Time-Dependent Results*

The spatial correlation approach is readily extended to the nonstationary case by defining time-dependent (classical and quantum) spatial correlation function and correlation length; that is, one applies Eqs. (4.2) and (4.3) to time-dependent densities. An example, in which the time dependence of

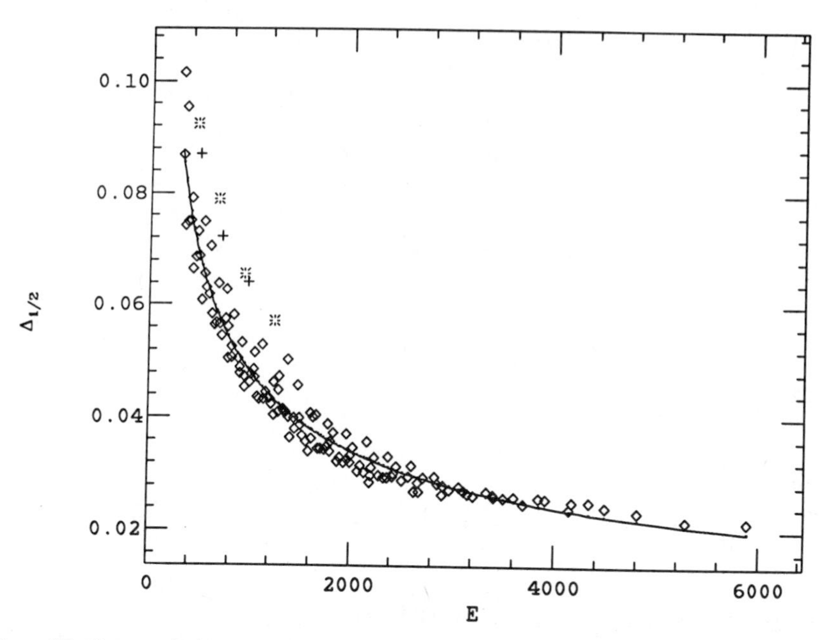


**Figure 17.** The correlation length as a function of energy for the classical ergodic distributions (dashed) and ergodic quantum states (diamonds). States marked as (* and +) are regular by the adiabatic criterion.[66] (From Ref. 67.)

the correlation lengths of quantum wavepackets moving in a stadium was computed,[67] is given in Fig. 18. At low average energies, $E = 40$ (top curve), wide fluctuations in the correlation length are observed. As the average energy increases, $E = 80$ (middle curve), the wavepacket encompasses more and more chaotic energy eignestates. As a result, the correlation length decreases, and its temporal fluctuations decay. At high average energy, $E = 160$ (bottom curve) the wavepacket is comprised of a majority ($\sim 85\%$) of chaotic eigenstates. The correlation length then *becomes a near constant of the motion.* This result can be understood by examining the origin of the time variation of the correlation functions. One may readily show that correlation function time dependence can only arise from the nonvanishing of the cross correlation functions $F_{n,m}(\boldsymbol{\delta})$, between *different* energy eigenstates. This cross correlation function $F_{n,m}(\boldsymbol{\delta})$, a generalization of the autocorrelation function of Eq. (4.1), is given by

$$F_{n,m}(\boldsymbol{\delta}) = \int d\mathbf{q}\, \psi_n^*(\mathbf{q})\psi_m(\mathbf{q} + \boldsymbol{\delta}). \tag{4.7}$$

Thus the near constancy of the correlation length at high energies is only

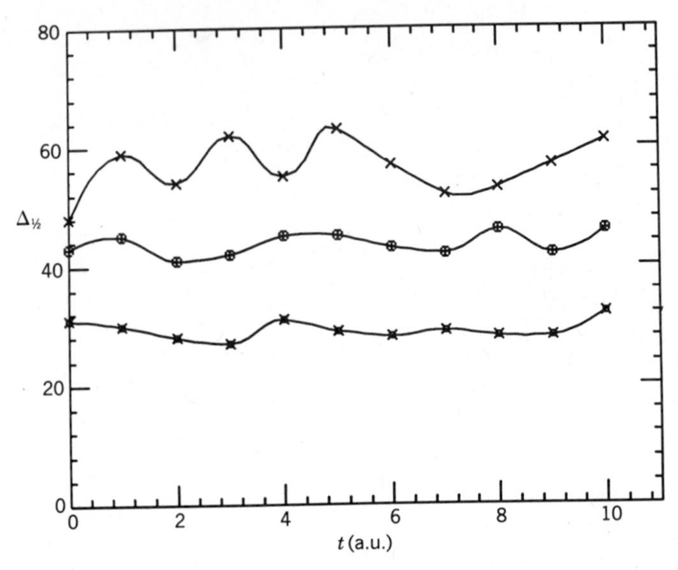


**Figure 18.** Temporal dependence of quantum correlation lengths for wavepackets of different average energies: top curve, $E = 40$ a.u.; middle curve, $E = 80$ a.u.; bottom curve, $E = 160$ a.u. (From Ref. 67.)

consistent with the near vanishing of the $n \neq m$ cross correlations between chaotic states, *for all values of* $\delta$. This is in fact the case,[67] as shown in Fig. 19 where cross correlation functions for various pairs of chaotic eigenstates are displayed.

This finding, which supplements the autocorrelation results, may well be a manifestation of *mixing* dynamics. Specifically, Kay[74] has suggested that the semiclassical analogue of classical mixing is the vanishing of off-diagonal matrix elements of "acceptable" operators. The shift operator is "acceptable," that is, is well defined in the $\hbar \to 0$ limit, provided we consider an ever decreasing range of $\delta$ values, namely, $\delta = O(\hbar)$.

An important aspect of this observation is that similar constant correlation lengths are also obtained classically.[67] In the classical case, however, the correlation lengths attain a constant value only after a short relaxation time. The very existence of classical relaxation is the major difference with respect to the quantum behavior. It results, as discussed in previous sections, from the continuous spectrum for this classical system: the classical cross correlation functions are not identically zero. Rather we conjecture that they rapidly decay to zero as $|\lambda - \lambda'|$ increases, where $\lambda$ and $\lambda'$ are eigenvalues of the classical Liouville operator. In contrast, the quantum cross correlation functions are essentially zero whenever (the discrete) $n - m \neq 0$. One further

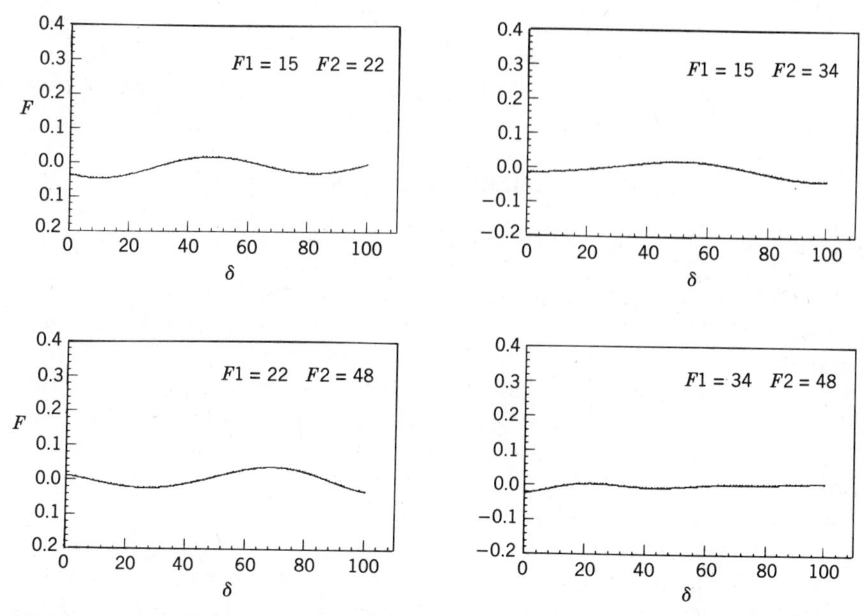


**Figure 19.** Cross correlation functions for the (15, 22), (15, 34), (22, 48), and (34, 48) pairs of quantum states. (From Ref. 67.)

classical–quantum difference can be noted. That is, the long-time limit of the classical correlation lengths are somewhat lower than the quantum ones. This behavior arises even at $E = 160$, since there is some ($\sim 15\%$) "contamination" of the quantum wavepacket by regular eigenstates.

It is easy to present an heuristic argument which shows that all quantum systems must possess regular states. This follows because the cross correlation functions for "mixing" states, which are the off-diagonal matrix elements of the shift operator, vanish. Thus, if *all* states were mixing, the shift operator would be diagonal in the energy representation, that is, the Hamiltonian would be translationally invariant. Since the potential part is not translationally invariant, we must conclude that there must exist states for which the cross correlation functions *do not* vanish. These states could belong to the continuous part of the energy spectrum, but in the absence of a continuum, as in the various particle-in-a-box problems, there must be discrete states whose cross correlation functions do not vanish. We term such states "nonmixing." The adiabatic states, mentioned previously, to be discussed more fully, are one example of a "nonmixing" state. Similarly, the states that were labeled regular (i.e., they have quasiperiodic autocorrelation function) show nonvanishing cross correlation functions.[67]

In this section we have discussed the concept of chaotic versus regular states, from the spatial correlation function viewpoint. Below, we make use of the tools developed in this section to examine the role played by these qualitatively different states in reaction dynamics.

## B. Statistical Reactions and Chaos[75]

### 1. *Resonance-Dominated Behavior*

In order to make connections between chaotic properties and statistical behavior in reaction dynamics, one must first define chaotic properties of *open* systems, since all chemical reactions involve unbounded motion in at least one coordinate. A way of linking chaotic behavior in bound systems to that in open systems was discussed previously for classical unimolecular decay. However, in the quantum case, we do not attempt a similar link but rather establish the circumstances under which chaos in closed systems implies statistical behavior in open systems.

Consider an isolated system prepared at $t = 0$ in a well-defined state $|s\rangle$. The probability of observing a particular product channel, that is, $|E, \mathbf{n}_0\rangle$, in the infinite time limit is obtained by expanding the state $|s\rangle$ in the $|E, \mathbf{n}^-\rangle$ scattering states,

$$|s\rangle = |\psi(0)\rangle = \int dE \sum_{\mathbf{n}} |E, \mathbf{n}^-\rangle \langle E, \mathbf{n}^- | s\rangle. \tag{4.8}$$

We use the $|E, \mathbf{n}^-\rangle$ (rather than the $|E, \mathbf{n}^+\rangle$) states, because they approach, as $t \to \infty$, a single well-defined asymptotic state $|E, \mathbf{n}_0\rangle$. Hence, $|\langle E, \mathbf{n}^- | s\rangle|^2$ is the probability of observing the state of interest, $|E, \mathbf{n}_0\rangle$, at the end of the reactive event; $\mathbf{n}$ denotes all quantum numbers associated with the reaction products (also including recoil directions), except the total energy $E$.

Statisticality of the product state distribution implies its independence of $\mathbf{n}$ (i.e., all final channels are equally probable). Statisticality of *total* decay rates, as defined below, means that their energy dependence is determined solely by the number of $n$ channels accessible at a given energy $E$.

The criterion outlined previously clearly expresses statisticality as a property of continuum states, and as such has little to do with bound-state dynamics. However, if the process is dominated by a resonance, or a number of resonances, then contact can be made with chaotic behavior in bound systems.[75] Denoting $Q$ as a projection operator onto a bound manifold and $P = I - Q$ as its orthogonal projector (which includes the unbound manifold), it is always possible to express $|E, \mathbf{n}^-\rangle$ in terms of the scattering solutions in the $P$ space, as [with $G = (E - i\varepsilon - PHP)^{-1}$]:

$$|E,\mathbf{n}^-\rangle = P|E,\mathbf{n}_1^-\rangle + (I + GPH)Q[E - QHQ - QHP \times (E - i\varepsilon - PHP)^{-1}PHQ]^{-1}QHP|E,\mathbf{n}_1^-\rangle, \tag{4.9}$$

where $|E,\mathbf{n}_1^-\rangle$ are solutions to the $P$-space scattering problem,

$$(E - i\varepsilon - PHP)|E,\mathbf{n}_1^-\rangle = 0 \tag{4.10}$$

in the absence of coupling between the $P$ and $Q$ subspaces.

Let us first consider the case where $|E,\mathbf{n}^-\rangle$ is dominated by a single resonance. Equation (4.9) now reduces to

$$|E,\mathbf{n}^-\rangle = |E,\mathbf{n}_1^-\rangle + (I + GPH)Q|\psi_i\rangle\langle\psi_i|QHP|E,\mathbf{n}_1^-\rangle/(E - E_i - i\Gamma_i/2). \tag{4.11}$$

Here $|\psi_i\rangle$ is a bound solution in the decoupled $Q$ space,

$$(E_i^0 - QHQ)Q|\psi_i\rangle = 0. \tag{4.12}$$

The resonance position $E_i$ is related, in a well-known way,[76] to the bound-state energy $E_i^0$.

The reaction (decay) rate is $\Gamma_i/\hbar$. The total resonance width $\Gamma_i$ is given as

$$\Gamma_i = 2\pi \sum_{\mathbf{n}} |\gamma_{i\mathbf{n}}|^2, \tag{4.13a}$$

where, $\gamma_{i\mathbf{n}}$—the partial width amplitudes—are given as

$$\gamma_{i\mathbf{n}} = \langle\psi_i|QHP|E,\mathbf{n}_1^-\rangle. \tag{4.13b}$$

Using Eq. (4.12), the probability amplitude for observing a product state $|E,\mathbf{n}_0\rangle$ becomes

$$\langle s|E,\mathbf{n}^-\rangle = \langle s|E,\mathbf{n}_1^-\rangle + \gamma_{si}\gamma_{i\mathbf{n}}/(E - E_i - i\Gamma_i/2), \tag{4.14}$$

where

$$\gamma_{si} = \langle s|(I + GPH)Q|\psi_i\rangle \tag{4.15}$$

is the amplitude for forming the $i$th resonance state.

At energies sufficiently close to the (assumed) narrow resonance, the second term of Eq. (4.14) dominates and the product distribution depends solely on the variation of $\gamma_{i\mathbf{n}}$ on $\mathbf{n}$.

Consistent with the assumption that at the energies considered the full scattering states $|E, \mathbf{n}^-\rangle$ are dominated by a single (isolated) resonance is the assumption that $P|E, \mathbf{n}_1^-\rangle$ represent "simple" dynamics. Otherwise, $P$ itself would contain some resonant behavior and the isolated resonance assumption is inapplicable. Under these circumstances, the variation of $\gamma_{i\mathbf{n}}$ with $\mathbf{n}$ depends solely on the nature of $|\psi_i\rangle$, and specifically on whether it is ergodic or not: If $|\psi_i\rangle$ has a long correlation length or a (quasi) periodic spatial correlation function, we expect the $\gamma_{i\mathbf{n}}$ integral to be influenced by the whole range of integration, consisting of the region where the intersection between $\psi_i$ and $V|E, \mathbf{n}_1^-\rangle$ is nonzero. On the other hand, if $\psi_i$ is ergodic, namely, it has a rapidly decaying correlation function, the $\gamma_{i\mathbf{n}}$ integral will be determined by some very localized regions in configuration space, where a chance stationary phase, combined of the local phase of $\psi_i$ and $V|E, \mathbf{n}_1^-\rangle$, exists. The decay of the correlation function guarantees that if such stationary phase regions exist, they will be very localized. Moreover, these regions will vary randomly for different $|E, \mathbf{n}_1^-\rangle$ states, that is, for different final channels $\mathbf{n}$. Under these circumstances, it is reasonable to expect a statistical, randomlike, final-state distribution.

Consider next the more general case of overlapping resonances. The $Q$ space is now multidimensional and the full effective Hamiltonian matrix of Eq. (4.9) needs to be considered. However, if all the states comprising the $Q$ space are "mixing," that is, having vanishing mutual cross correlation functions, than the off-diagonal matrix element of the effective Hamiltonian will also vanish, and Eq. (4.14) is replaced by

$$\langle s|E, \mathbf{n}^-\rangle = \langle s|E, \mathbf{n}_1^-\rangle + \sum_i \gamma_{si}\gamma_{i\mathbf{n}}/(E - E_i - \Gamma_i). \tag{4.16}$$

The final-state distribution is no longer independent of the preparation process, given by $\gamma_{si}$, since we need to sum over $i$—the contributing resonance states. However, if the $\psi_i$ states are mixing, then both $\gamma_{si}$ and $\gamma_{i\mathbf{n}}$ vary randomly about some average. This will give rise to statisticality in the final-state $n$ distribution. By the same token, $\Gamma_i$—the total width (and $\hbar$ times the decay rate)—will simply be proportional to the number of available final states.

This result is only true if the states comprising the $Q$ space are chaotic. If some of the states are regular, we can no longer ignore the off-diagonal coupling terms of the effective Hamiltonian. Moreover, some of the width amplitudes will not be statistically distributed. This may cause some final channels to be dynamically preferred. Moreover, large variation of the total rates with energy may occur. Thus, the effect of regular states may be profound.

Contrary to classical mechanics, quantum mechanics is never purely chaotic; contamination by regular states exists for even the most chaotic systems.[64,66]

This contamination is expected to gradually decrease, if the system is classically chaotic, as the energy is raised. However, at finite energies, the existence of regular states can deeply affect reaction dynamics. It was pointed out that unless the Hamiltonian is translationally invariant some regular states must exist. This argument does not tell us how numerous the regular states are nor to what extent they persist at high energies. The reason for persistence of regular states is best explained in terms of quantum adiabaticity, to be developed subsequently.

### 2. *Quantum Adiabaticity*

The adiabatic approximation, which is to be applied to bound-state dynamics, involves the separation of fast and slow motions. However, unlike the original application of the Born–Oppenheimer approximation,[77] in which fast electronic motion is separated from the slower nuclear one, there is no clear separation of time scales for nuclear motions. Nevertheless, there is now ample evidence to show that there are always states, to be termed "adiabatic," which are accurately described by an adiabatic approximation.[66,69–71,78]

Adiabatic states are separable and as such are regular according to Section IV A. Regular states, in the strict meaning of Section IV A, are the analogy, in quantum mechanics, of quasiperiodic trajectories in classical mechanics and, as in classical chaotic systems, are embedded in a host of chaotic states. Indeed, it is possible to show that in the classical limit, adiabatic states become quasiperiodic orbits. Unfortunately, it is not always possible to obtain *all* the quasiperiodic orbits of the system from adiabatic states belonging to a single type of adiabatic separation. Rather, all conceivable separations may have to be investigated, and even then there is no guarantee that *all* quasiperiodic orbits will result. Nevertheless, this appears a reasonable assumption.

In this section we outline the reason for the appearance and persistence of adiabatic states. In the next section we discuss their importance in accelerating and blocking-off certain dynamical processes.

Consider for simplicity a two-dimensional Hamiltonian, $H(R, r)$, written as

$$H(R, r) = K(R) + K(r) + V(r, R), \tag{4.17}$$

where $K(R)$, $K(r)$ are kinetic energy operators for the corresponding coordinates, and $V(R, r)$ is a potential.

We wish to solve the eigenvalue problem

$$[E_i - H(R, r)]\psi_i = 0. \tag{4.18}$$

In the spirit of the Born–Oppenheimer approximation, we introduce an adiabatic basis, defined as,

$$[\varepsilon_n(R) - K(r) - V(R,r)]\phi_n(R|r) = 0. \qquad (4.19)$$

In this way $r$ is singled out as the fast "electronic" coordinate. In the absence of prior knowledge we can, equally well, designate $R$ as the fast coordinate. This gives rise to an alternative adiabatic breakdown of the Schrödinger equation, and exemplifies the nonuniqueness associated with the adiabatic separation of bound-state nuclear dynamics.

We can use the adiabatic basis to solve for $\psi_i$ exactly, via the following expansion:

$$\psi_i(R,r) = \sum_n \phi_n(R|r)G_n(R), \qquad (4.20)$$

where $G_n(R)$ are wavefunction components satisfying a set of coupled channels equations:

$$[E_i - \varepsilon_n(R) - K(R)]G_n(R) = \sum_j [B_{nj}G_j(R) + A_{nj}(R)\,dG_j(R)/dR]. \qquad (4.21)$$

$A_{nj}(R)$ and $B_{nj}(R)$ are nonadiabatic couplings, defined as,

$$A_{nj}(R) = -\frac{\hbar^2}{m}\int dr\,\phi_n(R|r)\frac{d\phi_j(R|r)}{dR}, \qquad (4.22a)$$

$$B_{nj}(R) = -\frac{\hbar^2}{2m}\int dr\,\phi_n(R|r)\frac{d^2\phi_j(R|r)}{dR^2}. \qquad (4.22b)$$

The adiabatic approximation is simply the reduction of the expansion of Eq. (4.20) to a *single* term,

$$\psi_i = \phi_n(R|r)F_{n,m}(R), \qquad (4.23)$$

where $F_{n,m}(R)$ are solutions of a one-dimensional differential equation,

$$\left[E_{n,m} - \varepsilon_n(R) + \left(\frac{\hbar^2}{2m}\right)\frac{d^2}{dR^2} - A_{nn}(R)\frac{d}{dR} - B_{nn}(R)\right]F_{n,m}(R) = 0, \qquad (4.24)$$

obtained by neglecting the off-diagonal elements of Eq. (4.21).

Besides the fundamental importance of adiabatic states, they are of great utility because they form an excellent basis set[66,69,71] for the exact solution of Eq. (4.21). Moreover, for many problems of current interest, the first step in the factorization, that is, the calculation of the adiabatic basis set [Eq. (4.19)], can be done analytically.

We shall discuss adiabatic states by giving two specific examples, the

Henon–Heiles and stadium Hamiltonians, which serve to explain the appearance and persistence of adiabatic states.

The *Henon–Heiles* potential[38]

$$V(R, r) = \tfrac{1}{2}(r^2 + R^2) + r^2 R - R^3/3 \tag{4.25}$$

allows for an immediate solution of Eq. (4.19), which becomes that of a parametric harmonic oscillator. The adiabatic potentials $\varepsilon_n$ are therefore given as[69]

$$\varepsilon_n(R) = (n + \tfrac{1}{2})\hbar[(1 + 2R)/m]^{1/2} + \tfrac{1}{2}R^2 - R^3/3. \tag{4.26}$$

The nonadiabatic coupling terms can also be evaluated analytically to yield

$$A_{nj}(R) = \hbar^2/[8m(1 + 2R)]\{(n + 1)^{1/2}(n + 2)^{1/2}\delta_{n+2,j} + \delta_{n,j} - n^{1/2}(n - 1)^{1/2}\}, \tag{4.27}$$

with a similar but slightly more complicated expression holding for $B_{nj}(R)$.

The final stage in the adiabatic reduction is the solution of Eq. (4.24). Given the adiabatic potential of Eq. (4.26) this cannot be done analytically, but the resulting ordinary differential equation may be solved numerically using the finite difference method. As an example, we show in Fig. 20 a comparison between the even-parity adiabatic eigenvalues and the exact ones, obtained by solving the full coupled channels expansion, using the artificial channel method.[69]

Clearly, the adiabatic approximation is on the whole very good (rms deviation with respect to the exact eigenvalues of 1.8%). Of greater significance for the present discussion is that *some* eigenvalues, no matter how high the energy, are very accurately given by the adiabatic approximation. Such states we term "adiabatic states."

As an example, consider the $(n, 0)$ and $(0, n)$ "extreme motion" sequences. As shown in Ref. 69, state No. 15 with the adiabatic assignment of (5, 0) has an adiabatic eigenvalue of 5.8168 as compared to the exact value of 5.8170, and state No. 66 with adiabatic assignment of (11, 0) has adiabatic eigenvalue of 11.12 versus 11.15 for the exact energy, and so forth. The "extreme motion states" are adiabatic,[70] because they involve a (large-amplitude) fast motion in one coordinate and a (small-amplitude) slow motion in the other coordinate, which are precisely the conditions for adiabatic separability.

The degree of accuracy of the adiabatic approximation can be inferred from Eq. (4.27). The nonadiabatic terms are proportional to the adiabatic channel index $n$ and to $\hbar^2/m$. For fixed $E$, the electronic quantum index is proportional to $Em^{1/2}/\hbar$, hence the nonadiabatic terms are proportional to $E\hbar/m^{1/2}$.

We can define $\alpha$, a perturbation parameter, as the magnitude of the nonadi-

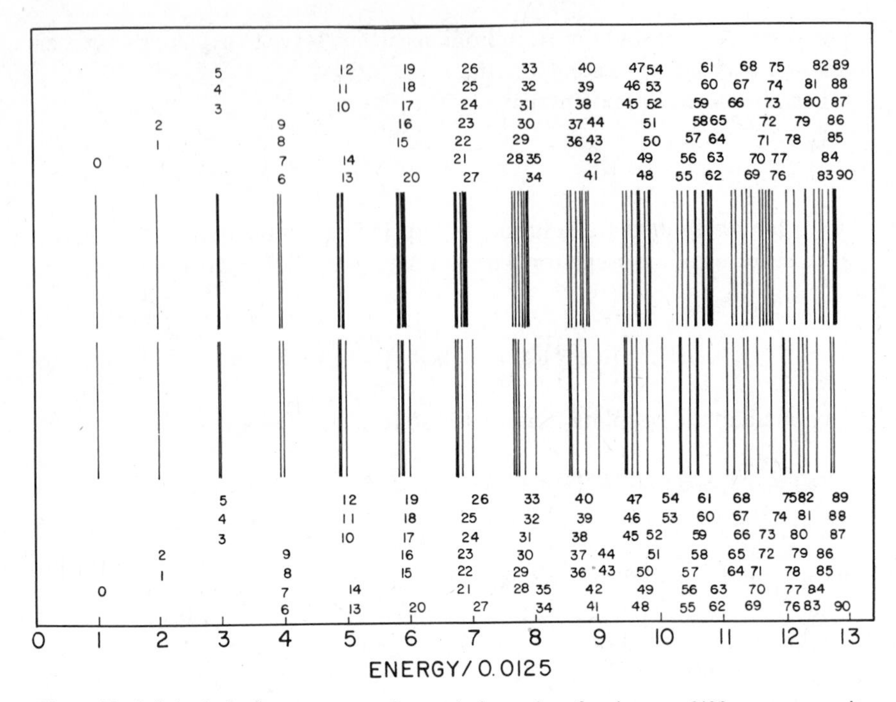


**Figure 20.** Adiabatic (top) versus exact (bottom) eigenvalues for the $m = 6400$ a.u. even parity states of the Henon–Heiles potential. The numbers on top and bottom are eigenvalue indices and their horizontal position matches that of the eigenvalue, so as to discern between eigenvalues that are almost degenerate. (From Ref. 69.)

abatic terms divided by the spacing between neighboring channel potentials, of the same parity:

$$\alpha_{n,j} = |A_{nj}(R)/(\varepsilon_n - \varepsilon_j)|. \tag{4.28}$$

Since the spacing is inversely proportional to $m^{1/2}/\hbar$, the perturbation parameter is simply proportional to $E$ and is independent of $m$ or $\hbar$. As a result, at a fixed energy, we expect the relative abundance of adiabatic states to be maintained in the classical limit. As we raise the energy, $\alpha$ increases linearly with $E$. We therefore expect the relative abundance of the adiabatic states to decrease.

This argument does not apply to individual states, with $n$ fixed. The perturbation parameter associated with the fixed $n$ subspace decreases, in the classical limit, as $\hbar/m^{1/2}$. This means that even for high $E$, in the classical limit, we can always find adiabatic states, albeit in smaller relative abundance. The importance of this fact to dynamics will be discussed.

The second example to be discussed in this section is that of the stadium billiard problem. The reader is reminded that this is a particle in a region made up of two parallel linear segments of length $2a$ connected by two semicircles of radius $R$. Denoting $x$, the direction parallel to the linear segments and $y$, the direction perpendicular to the linear segments, the adiabatic basis functions can be defined[66] as the solutions of a particle in a one-dimensional box whose size

$$l(y) = 2[a + (R^2 - y^2)^{1/2}]$$

is the stadium width at height $y$. Hence, the adiabatic basis is given by

$$\phi_n(y|x) = [2/1(y)]^{1/2} \sin[n\pi x/l(y) + n\pi/2], \quad -l(y) < 2x < l(y). \quad (4.29)$$

The eigenvalues (adiabatic channel potentials) are given as

$$\varepsilon_n(y) = [\hbar\pi n/l(y)]^2/2m. \quad (4.30)$$

The nonadiabatic coupling terms are given as[66]

$$A_{nj}(y) = -(\hbar^2/m)(1 - \delta_{nj})[1 + (-1)^{n+j}](4nj/y)/[\beta l(n^2 - j^2)] \quad (4.31)$$

with $\beta = 2a - l(y)$. A slightly more complicated expression holds for $B_{nj}(y)$.[66]

The existence of adiabatic states in the stadium has been shown by a number of investigators.[66,70–73] As in the Henon–Heiles case, it can be attributed to the smallness of the interaction between special type of states. However, in general, the average perturbation parameter $\alpha$ is not small. Because both the channel potential difference and the nonadiabatic coupling between adjacent channels are proportional to $n$, the average perturbation parameter is $n$ independent. Thus, contrary to the Henon–Heiles case there is a constant average perturbation between adiabatic levels. This fact may be connected to the difference in the classical dynamics of the two systems insofar as the Henon–Heiles system undergoes a transition from a regular motion to chaotic motion, whereas the stadium system is always chaotic.

Despite the persistence of a constant average perturbation between adiabatic levels, perturbations between individual levels may be weak. There are, in fact, an infinite number of adiabatic states, because whenever we open a new channel the only levels close enough to the ground state, $F_{n,0}(y)$, of this new channel are high-lying states of lower channels. As $n$ increases, matrix elements such as $\langle F_{n-1,m}| A_{n-1,n}|dF_{n,0}/dy\rangle$, become negligibly small, owing to the mismatch in deBroglie wavelengths between the two wavefunctions. This again results in a case of "extreme motion" adiabatic states.

It is easy to see that the $\phi_n(y|x)\, F_{n,0}(y)$ states become the periodic orbits along the $x$ axis of the stadium in the classical limit; the $F_{n,0}(y)$ function "sinks" to the bottom of the well as $\hbar^2/m \to 0$. Simultaneously, $n$ must be made to increase (proportional to $\hbar/m^{1/2}$) to maintain a constant energy. As a result, the motion in the $x$ coordinate becomes increasingly classical. As we increase $n$, more channels open up, each containing more states. The measure of the $F_{n,0}(y)$ states thus becomes vanishingly small.

### 3. *Statistical and Nonstatistical Decay Rates*

There is a general view that, for most molecules of intermediate and large sizes, decay rates may be described by a statistical theory, such as the RRKM model. However, modern experiments on decay rates of "single vibronic levels" (SVL) often show serious deviations from statistical behavior, even at energies where fast "intramolecular vibrational relaxation" (IVR) is presumed to exist. This suggests that apparent statisticality in previous experiments may have resulted from poor resolution. However, further work is necessary, since few SVL experiments have been performed thus far.

As was pointed out, statisticality implies that the partial rates, $|\gamma_{i\mathbf{n}}|^2/\hbar$, are indifferent to the final state **n**. Under these circumstances the total rate should only depend on the number of final **n** channels accessible at a given total energy $E_i$, relative to the size of the $Q$ space. In SVL experiments, the main deviations from statisticality are observed in the variation of the total decay rates $\Gamma_i$ [Eq. (4.13a)] as a function of the level (resonance) energy $E_i$. These fluctuations are sometimes regarded as evidence to "mode specificity," which in a sense implies the existence of regular states among the chaotic states. Indeed, for many well-characterized systems, decay rates of even neighboring states are found to deviate by orders of magnitudes. Such deviations exist not only for intermediate systems, such as the HF[79] and the NO[80] dimers, the formaldehyde photodissociation,[81,82] but also for substantially larger systems, such as the ethylene and ammonia dimers,[83] and the $p$-difluorobenzene–Ar complex.[84]

For intermediate systems, such as the HF dimer, "mode specificity" can be attributed to the differences in bonding of the two hydrogens.[85] Likewise, in the NO dimer the two neighboring states with drastically different lifetimes correspond to two types of stretch vibrations. In the larger systems, however, such direct identification of the regular states is not always possible. Moreover, it is clear[83,86] that the slowly decaying states are superimposed on other states with much faster decay, a behavior that can be anticipated theoretically.

The decay processes considered theoretically in the literature can be loosely classified as being controlled by *tunneling* or *energy transfer*. Decay processes of the tunneling type have been considered in the context of formaldehyde photodissociation,[87] and the dissociation of a Henon–Heles system[89] and the $H_3^+$ molecule.[88] In many of these systems definite mode specificity was observed[89] and was shown to occur as a result of near adiabaticity.[70] Decay via tunneling is however only weakly dependent on bound-state dynamics, as the phenomenon is almost entirely controlled by the tunneling process. In the language of Section IV B 3, the partial widths $\gamma_{in}$ are mainly affected by properties of the $P$ space continuum wavefunctions $|E, \mathbf{n}_1^-\rangle$ and are only marginally sensitive to intramolecular bound-state dynamics. Thus, the chaotic nature of $\psi_i$ may be masked by the fact that the tunneling rate increases (exponentially) as the energy is raised. In order to study chaotic behavior in such cases it is first necessary to unfold the dominant contribution of tunneling.

The case of decay by *energy transfer* akin to "Feshbach-type resonance(s)" in collision theory, presents a probe of greater sensitivity to bound-state dynamics. This is clearly exemplified by the "doorway" channel model.[69] In this model, the bound manifold is coupled to a dissociative manifold via a *single* ("doorway") channel. This forces the system, if excited to some arbitrary state, to "diffuse" to the subspace spanned by the doorway channel in order to dissociate.

As an example of doorway channel calculations consider the dissociation of the Henon–Heiles system. The potential used in this model[69] is a modification of the Henon–Heiles potential. Specifically, in order to eliminate the competing effects of tunneling (which, as previously noted, tend to mask intramolecular effects), Eq. (4.25) was modified as follows;

$$V(R, r) = \tfrac{1}{2}(r^2 + 1) + r^2 - \tfrac{1}{3} \qquad \text{for } R > 1 \tag{4.32a}$$

and

$$V(R, r) = \tfrac{1}{2}R^2 - \tfrac{1}{3}R^3 \qquad \text{for } R < -\tfrac{1}{2}. \tag{4.32b}$$

At all other values of $R$, $V(R, r)$ is given by Eq. (4.25).

The system described by the modified potential of Eqs. (4.25) and (4.32) is incapable of dissociating; hence, it is coupled to a dissociative channel here described by an $r$-independent potential:

$$V_{c,c}(R) = 0.005 \exp(-0.1R). \tag{4.33}$$

Of all the channels spanning the full two-dimensional problem, only one, the so-called "doorway" ($d$) channel, is coupled to the dissociative channel. The coupling potential, $V_{c,d}(R)$, is chosen equal to $V_{c,c}(R)$. [Note that the term channel is taken to mean a state spanning only *part* of configuration space, for example, the adiabatic $\phi_n(R|r)$ basis set.]

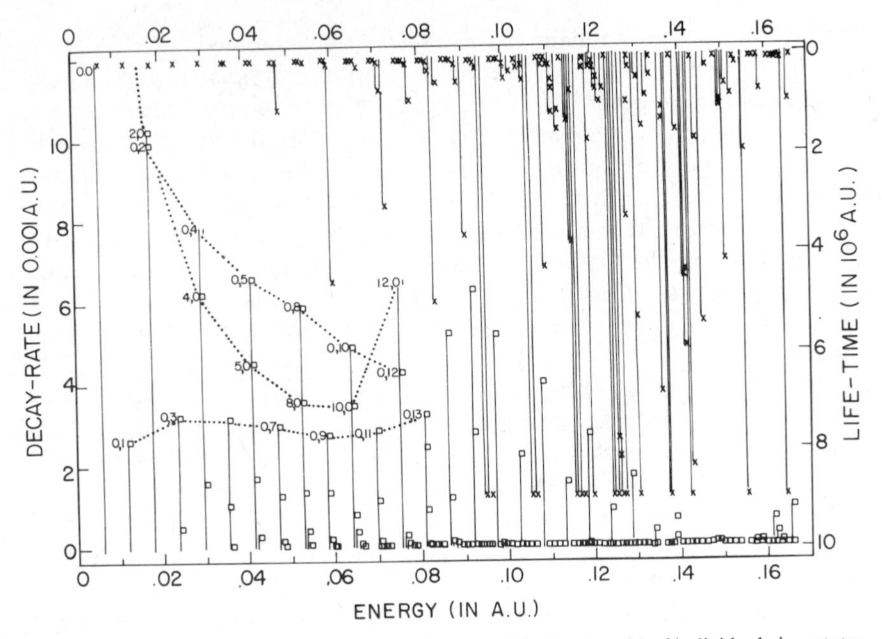


**Figure 21.** Decay rates (left-hand scale) and lifetimes (right-hand scale) of individual eigenstates of the modified Henon–Heiles system. Adiabatic assignment of the lower eigenstates are given as a pair of $n$, $v$ quantum numbers. (From Ref. 69.)

In this case the doorway component of the total wavefunction determines the dissociation rates. Based on our previous discussion, we anticipate that the decay rates will fluctuate considerably for the regular states, since the relative magnitude of the doorway channel component is expected to vary drastically from one state to another. In contrast, chaotic states should have a more homogeneous representation of the doorway component. As a result, chaotic states yield a diminished sensitivity of the decay rate to the exact eigenstate studied. Furthermore, individual $\Gamma_i$ are expected to be insensitive to the identity of the doorway channel.

The results of this computation, shown in Figs. 21 and 22, display a rich structure of regular and chaotic states. In the low-energy regime, the pattern of decay rates shows extreme state specificity, clearly indicating that all states are regular. As the energy is increased, a "forest" of slowly decaying states appears. These states coexist with the ever decreasing number of rapidly decaying regular states. However, not all the slowly decaying states are chaotic. In fact, as shown in Fig. 22, the results, even in the high-energy regime, deviate from statisticality. These deviations persist even if additional coarse graining is introduced. As shown in Fig. 22, where the individual decay rates

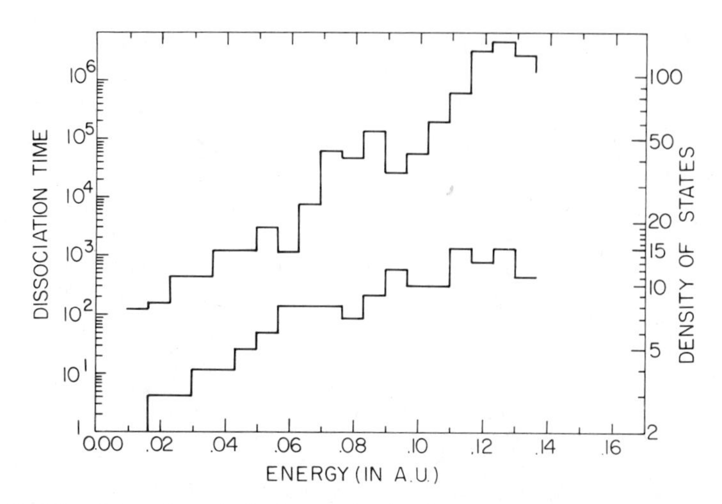


**Figure 22.** Dissociation time of mixed states (left-hand scale and upper curve) and density of states (right-hand scale and lower curve) as a function of energy. (From Ref. 69.)

were averaged over a bunch of states, the resulting "mixed" states[90] decay rates do not display statisticality in that the decay rates decrease faster than expected on the basis of the statistical theory. This effect is due to the existence of a large number of stable states whose decay rates are lower, by as much as four orders of magnitude, than those of the surrounding states.

Some of the stable states are clearly regular. These include some "extreme motion" adiabatic states, discussed previously, that have no direct coupling to the doorway channel. In order to decay, such states must undergo multiple quantum transitions (often as many as 10), until they "diffuse" to the doorway channel. Figures 21 and 22 show that many of the adiabatic states are very resistant to diffusion and serve as bottlenecks for dissociation.

Appearances of bottlenecks for energy transfer,[91] ionization,[92] and multiphoton dissociation[93] have recently been discussed in great detail. The existence of localization[91] in quantum systems, as opposed to free diffusion in classically chaotic systems, is closely related to the appearance of adiabatic states discussed here.

## V. SUMMARY

This chapter has covered a number of topics relating to the interplay between classical chaos and molecular reaction dynamics, a necessarily quantum phenomenon. It is clear that a host of problems remain to be resolved and that results obtained thus far have laid the groundwork for substantial progress in this area. Much of this future work will continue to be guided by the recognition that a large number of interrelated fundamental problems, such as classical–quantum correspondence and classical versus quantum chaos, are embodied in an understanding of elementary molecular phenomena.

## Acknowledgment

P. Brumer thanks the donors of the Petroleum Research Fund, administered by the American Chemical Society as well as NSERC Canada for support of this research. He is also grateful to Mr. and Mrs. M. Cohen-Nehemiah, Canadian Center for the Alexander Technique, for insights on body mechanics leading to control of debilitating back pain.

## References

1. See, for example, R. D. Levine and R. B. Bernstein, *Molecular Reaction Dynamics*. Oxford, New York, 1974. E. E. Nikitin, *Theory of Elementary Atomic and Molecular Processes in Gases*. Oxford University Press, Oxford, 1974.
2. R. D. Levine and R. B. Bernstein, *Adv. At. Mol. Phys.* **11**, 216 (1975).
3. For a classical formulation that admits a host of alternate statistical theories of bimolecular reactions see A. F. Wagner and E. K. Parks, *J. Chem. Phys.* **65**, 4343 (1976).
4. R. A. Marcus and O. K. Rice, *J. Phys. Coll. Chem.* **55**, 894 (1951).
5. J. C. Light, *Disc. Faraday Society* **44**, 14 (1968).
6. See, for example, the "Bohr assumption" in J. M. Blatt and V. F. Weisskopf, *Theoretical Nuclear Physics*. Wiley, New York, 1952.
7. See, for example, (*a*) M. V. Berry, in *Topics in Nonlinear Mechanics* (S. Jorna, ed.). AIP, New York, 1978; (*b*) P. Brumer, *Adv. Chem. Phys.* **47**, 201 (1981); (*c*) S. A. Rice, *Adv. Chem. Phys.* **47**, 117 (1981); (*d*) M. Tabor, *Adv. Chem. Phys.* **46**, 73 (1981).
8. See, for example, Far. Disc. Chem. Soc. **75** (1983).
9. For example, R. Jancel, *Foundations of Classical and Quantum Statistical Mechanics*. Pergamon, New York, 1969.
10. J. D. Eckmann and D. Ruelle, *Rev. Mod. Phys.* **57**, 617 (1985).
11. (*a*) V. I. Arnold and A. Avez, *Ergodic Problems of Classical Mechanics*. Benjamin, New York, 1968; J. P. Cornfeld, S. V. Fomin, and Ya. G. Sinai, *Ergodic Theory*. Springer-Verlag, New York, 1980. (*b*) For a review of less formal aspects of chaos in physics see G. M. Zaslavsky, *Chaos in Dynamical Systems*. Harwood, New York, 1985.
12. P. R. Halmos, *Lectures on Ergodic Theory*. Mathematical Society of Japan, Tokyo, 1953.
13. Our usage of point and continuous spectrum is in accord with formal definitions. For a lucid discussion see B. Friedman, *Principles and Techniques of Applied Mathematics*. Wiley, New York, 1956. Qualitatively, point eigenvalues are associated with $L^2$ eigenfunctions, whereas continuous eigenvalues are associated with improper eigenfunctions.
14. Ya. G. Sinai, *Sov. Math. Dokl.* **4**, 1818 (1963).
15. L. A. Buminovitch, *Funct. Anal. Appl.* **8**, 254 (1974).
16. For thoughtful consideration of the relationship between exponential instability and "loss of information" see N. S. Krylov, *Works on the Foundations of Statistical Physics*. Princeton University Press, New Jersey, 1979.
17. For a review of Lyapunov exponents for Hamiltonian systems see H.-D. Meyer, *J. Chem. Phys.* **84**, 3147 (1986). The formal theory is described in Eckmann and Ruelle in Ref. 10.
18. (*a*) P. Brumer, *J. Computational Phys.* **14**, 391 (1973); (*b*) M. Toda, *Phys. Lett. A* **48**, 335 (1974); (*c*) P. Brumer and J. W. Duff, *J. Chem. Phys.* **65**, 3566 (1976).
19. It is generally accepted that typical molecular Hamiltonians will display exponential diver-

gence at all energies. The critical energy therefore refers to an energy at which exponential divergence becomes manifest over a large region of phase space. This transition, from small to large regions of exponential divergence behavior, generally occurs abruptly as a function of energy [See I. Hamilton and P. Brumer, *Phys. Rev. A* **23**, 1941 (1981).]

20. See, for example, J. W. Duff and P. Brumer, *J. Chem. Phys.* **67**, 4898 (1977).
21. C. Cerjan and W. P. Reinhardt, *J. Chem. Phys.* **71**, 1819 (1979); R. Kosloff and S. A. Rice *J. Chem. Phys.* **74**, 1947 (1981).
22. G. M. Zaslavskii and B. V. Chirikov, *Sov. Phys. Usp.* **14**, 549 (1972); B. V. Chirikov, *Phys. Reports* **52**, 263 (1979). The initial application to chaotic behavior in molecules is due to D. W. Oxtoby and S. A. Rice, *J. Chem. Phys.* **65**, 1676 (1976).
23. G. Bennettin, L. Galgani, A. Giorgilli, and J. M. Strelcyn, *Meccanica* **15**, 9, (1980); **15**, 20 (1980).
24. M. J. Davis, *Chem. Phys. Lett.* **110**, 491 (1984).
25. M. V. Kuzmin, I. V. Nemov, A. A. Stuchebrukhov, V. N. Bagarashvili, and V. S. Letokhov, *Chem. Phys. Lett.* **124**, 522 (1986).
26. D. W. Noid, M. L. Koszykowski, and R. A. Marcus, *J. Chem. Phys.* **67**, 404 (1977).
27. (*a*) R. S. Dumont and P. Brumer (unpublished); (*b*) R. S. Dumont, Proceedings of the Canadian Mathematical Society-1986 (in press), (*c*) R. S. Dumont, Ph.D. Dissertation, University of Toronto, 1987.
28. See, for example, D. A. McQuarrie, *Statistical Mechanics*. Harper and Row, New York, 1976; L. E. Riechl, *A Modern Course in Statistical Physics*. University of Texas, Austin, 1980.
29. N. Wiener, *Acta Math.* **55**, 117 (1930); A. Khinchin, *Math. Annalen* **109**, 604 (1934).
30. A large number of model systems have been treated. See Refs. 7*b*, 7*c*, and 49*b* for references.
31. J. W. Duff and P. Brumer, *J. Chem. Phys.* **71**, 3895 (1979).
32. Note that power spectra during the course of unimolecular decay have been computed [J. D. McDonald and R. A. Marcus, *J. Chem. Phys.* **65**, 2180 (1976)]. Attributes of these spectra, from the viewpoint of statistical theories were not, however, examined.
33. J. W. Duff and P. Brumer, *J. Chem. Phys.* **71**, 2693 (1979).
34. I. Hamilton and P. Brumer, *J. Chem. Phys.* **82**, 1937 (1985).
35. J. S. Hutchinson and R. E. Wyatt, *J. Chem. Phys.* **70**, 3509 (1979). Reference 33 corrects the conclusions of this paper.
36. R. J. Wolf and W. L. Hase, *J. Chem. Phys.* **72**, 316 (1980); **73**, 3010 (1980); **73**, 3779 (1980); **75**, 3809 (1981).
37. (*a*) I. Hamilton and P. Brumer, *Phys. Rev. A* **25**, 3457 (1982); (*b*) I. Hamilton and P. Brumer, *J. Chem. Phys.* **78**, 2682 (1983); (*c*) see also Ref. 30.
38. M. Henon and C. Heiles, *Astron. J.* **69**, 73 (1984). Of interest is that D. Rod has recently announced a proof that the Henon–Heiles system is formally nonintegrable at all energies.
39. E. Thiele and D. J. Wilson, *J. Chem. Phys.* **35**, 1256 (1961).
40. More generally, one expects the dynamics to depend on some quantitative measure of the "distance" of the initial state from the final equilibrium state.
41. Ya. G. Sinai, *Acta Physica Austraica Suppl.* **X**, 575 (1973).
42. I. Hamilton, Ph.D. Dissertation, University of Toronto (1982); R. Dumont, Ph.D. Dissertation, University of Toronto (1986).
43. J. R. Crawford and J. D. Cary, *Physica D* **6**, 223 (1983).
44. D. Carter and P. Brumer, *J. Chem. Phys.* **77**, 4208 (1982) [Erratum: *J. Chem. Phys.* **78**, 2104 (1983)].
45. See, for example, W. Forst, *Theory of Unimolecular Reactions*. Academic Press, New York, 1973; N. B. Slater, *Theory of Unimolecular Reactions*. Cornell University Press, Ithaca, 1959; E. Thiele, *J. Chem. Phys.* **36**, 1466 (1962); W. L. Hase, in *Dynamics of Molecular Collisions, Part B*. (W. H. Miller, ed.). Plenum Press, New York, 1976.

46. R. S. Dumont and P. Brumer, *J. Phys. Chem.* **90**, 3509 (1986).
47. Another typical assumption is that $P_g(t)$ decays exponentially, the so-called random gap assumption. This assumption is not well justified in standard treatments and is not necessarily observed. See, for example, the calculation on decay from a punctured stadium [R. S. Dumont and P. Brumer (unpublished)].
48. Borel sets are a class of sets obtained by taking countable unions and intersections of open or closed sets. All Borel sets are measurable.
49. (*a*) W. H. Miller, *Adv. Chem. Phys.* **25**, 69 (1974); (*b*) D. W. Noid, M. L. Koszykowski, and R. A. Marcus, *Annu. Rev. Phys. Chem.* **32**, 267 (1981).
50. C. Jaffe and P. Brumer, *J. Phys. Chem.* **88**, 4829 (1984). This work contains a large number of references to previous work on classical Liouville mechanics.
51. For an excellent introduction see S. R. DeGroot and L. G. Suttorp, *Foundations of Electrodynamics*. North-Holland, Amsterdam, 1972. See also Y. M. Shirokov, *Sov. J. Part. Nucl.* **10**, 1 (1979).
52. R. T. Prosser, *J. Math. Phys.* **24**, 548 (1983).
53. F. Soto and P. Claveire, *J. Math. Phys.* **24**, 97 (1983).
54. One particular approach to classical–quantum correspondence deals with the behavior of coherent states as $\hbar$ approaches zero. It is interesting to note that this approach then deals with states which are, for all $\hbar$, physically acceptable in both quantum and classical mechanics.
55. S. Kanfer and P. Brumer (unpublished).
56. See, for example, M. J. Davis and E. J. Heller, *J. Chem. Phys.* **80**, 5036 (1984).
57. C. Jaffe and P. Brumer, *J. Chem. Phys.* **82**, 2330 (1985).
58. M. V. Berry, *Phil. Trans. Roy. Soc. London* **287**, 237 (1977).
59. This latter result constitutes the distribution dynamics statement of Heisenberg's original route to quantum mechanics.
60. A. Voros, *Ann. Inst. H. Poincare* **XXIV**, 31 (1976); M. V. Berry, in *Chaotic Behavior of Deterministic Systems* (E. Iooss, R. H. G. Hellemann, and R. Stora, eds.). North-Holland, Amsterdam, 1983.
61. C. Jaffe, S. Kanfer, and P. Brumer, *Phys. Rev. Lett.* **54**, 8 (1985).
62. (*a*) K. M. Christoffel and P. Brumer, *Phys. Rev. A* **33**, 1309 (1986); (*b*) K. M. Christoffel and P. Brumer, *Phys. Rev. A* **31**, 3466 (1985).
63. See, for example, K. S. J. Nordholm and S. A. Rice, *J. Chem. Phys.* **61**, 203, (1974); **61**, 768 (1974); I. C. Percival, *Adv. Chem. Phys.* **36**, 1 (1977); M. V. Berry, *Philos. Trans. Soc. London, Ser. A* **287**, 237 (1977); R. M. Stratt, N. C. Handy, and W. H. Miller, *J. Chem. Phys.* **71**, 3311 (1979); D. W. Noid, M. L. Koszykowski, and R. A. Marcus, *Ann. Rev. Phys. Chem.* **32**, 267 (1981); G. Hose and H. S. Taylor *J. Chem. Phys.* **76**, 5356 (1982); H. J. Korsch, *Physics Letters* **97A**, 77 (1983); K. G. Kay, *J. Chem. Phys.* **79**, 3026 (1983).
64. M. Shapiro and G. Goelman, *Phys. Rev. Lett.* **53**, 1714 (1984).
65. S. W. McDonald and A. N. Kaufman, *Phys. Rev. Lett.* **42**, 1189 (1979).
66. M. Shapiro, R. D. Taylor, and P. Brumer, *Chem. Phys. Lett.* **106**, 325 (1984).
67. M. Shapiro, J. Ronkin, and P. Brumer (unpublished); J. Ronkin, M. Sc. Thesis, The Weizmann Institute, 1986.
68. M. V. Berry, *J. Phys. A* **10**, 2083 (1977).
69. M. Shapiro and M. S. Child, *J. Chem. Phys.* **76**, 6176 (1982).
70. G. Hose, H. S. Taylor, and Y. Yan Bai, *J. Chem. Phys.* **80**, 4363 (1984).
71. G. Ezra (unpublished).
72. R. D. Taylor and P. Brumer, *Faraday Disc. Chem. Soc.* **75**, 117 (1983).
73. E. J. Heller, *Phys. Rev. Lett.* **53**, 1515 (1984); *Annu. Rev. Phys. Chem.* **35**, 563 (1984).
74. K. G. Kay, *J. Chem. Phys.* **79**, 3026 (1983); B. Ramachandran and K. G. Kay, *J. Chem. Phys.* **83**, 6316 (1985).

75. P. Brumer and M. Shapiro, *J. Chem. Phys.* **80**, 4567 (1984); P. Brumer and M. Shapiro, *Adv. Chem. Phys.* **60**, 371 (1985).
76. R. D. Levine, *Quantum Mechanics of Molecular Rate Processes.* Oxford University Press, London, 1969.
77. M. Born and R. Oppenheimer, *Ann. Phys. Leipzig* **84**, 461 (1927).
78. E. Segev and M. Shapiro, *J. Chem. Phys.* **78**, 4969 (1983).
79. A. S. Pine and W. J. Lafferty, *J. Chem. Phys.* **78**, 2154 (1983).
80. M. P. Casassa, J. C. Stephenson, and D. S. King, *Faraday Disc. Chem. Soc.* **82** (paper 18, 1986).
81. H. Dai, R. W. Field, and J. L. Kinsey, *J. Chem. Phys.* **82**, 1606 (1985).
82. C. B. Moore and J. C. Weisshaar, *Ann. Rev. Phys. Chem.* **34**, 525 (1983); D. R. Guyer, W. F. Polik, and C. B. Moore, *J. Chem. Phys.* **84**, 6519 (1986); H. Bitto, D. R. Guyer, W. F. Polik, and C. B. Moore, *Faraday Disc. Chem. Soc.* **82** (paper 8, 1986).
83. M. Snels, R. Fantoni, M. Zen, S. Stolte, and J. Reuss, *Chem. Phys. Lett.* **124**, 1 (1986); M. Snels, Ph.D. Thesis, Katholic University, Nijmegen (1986).
84. K. W. Butz, D. L. Catlett Jr., G. E. Ewing, D. Krajnovich, and C. S. Parmenter, *J. Phys. Chem.* **90**, 3533 (1986).
85. N. Halberstadt, P. Brechignac, J. A. Beswick, and M. Shapiro, *J. Chem. Phys.* **84**, 170 (1986).
86. C. M. Western, M. P. Casassa, and K. C. Janda, *J. Chem. Phys.* **80**, 4781 (1984).
87. W. H. Miller, *J. Am. Chem. Soc.* **101**, 6810 (1979); S. K. Gray, W. H. Miller, Y. Yamaguchi, and H. F. Schaefer, *J. Am. Chem. Soc.* **103**, 1900 (1981); B. A. Waite, S. K. Gray, and W. H. Miller, *J. Chem. Phys.* **78**, 259 (1983).
88. M. S. Child, *J. Phys. Chem.* **90**, 3595 (1986).
89. B. A. Waite and W. H. Miller, *J. Chem. Phys.* **73**, 3713 (1980); *J. Chem. Phys.* **74**, 3910 (1981).
90. K. G. Kay, *J. Chem. Phys.* **72**, 5955 (1980).
91. S. Fishman, D. R. Grempel, and R. E. Prange, *Phys. Rev. Lett.* **49**, 509 (1982).
92. R. Blumel and U. Smilansky, *Phys. Rev. Lett.* **52**, 137 (1984).
93. R. C. Brown and R. E. Wyatt, *J. Phys. Chem.* **90**, 3590 (1986).

# COHERENT PULSE SEQUENCE CONTROL OF PRODUCT FORMATION IN CHEMICAL REACTIONS

DAVID J. TANNOR[†] AND STUART A. RICE

*The Department of Chemistry and The James Franck Institute, The University of Chicago, Chicago, Illinois 60637*

## CONTENTS



[†] Present address: Department of Chemistry, Illinois Institute of Technology, Chicago, IL 60616.

# I. INTRODUCTION

Since the invention of the tunable laser chemists have dreamed of using its characteristic high intensity and spectral purity to control the selectivity of chemical reactions. When such selectivity can be achieved via use of spectral resolution, for example, photodissociation of different isotopic variants of the same chemical species, the desired control of product formation has been demonstrated. On the other hand, when such selectivity is sought via concentration of energy in particular bonds, rapid intramolecular energy transfer has prevented, for the examples studied to date, the desired control of product formation.

The designs of the previously mentioned selectivity schemes ignore the possibility of control of the evolution of excitation energy via exploitation of the coherence properties of the coupled matter–electromagnetic field system. Several schemes that do exploit the coherence of the time evolution of a wavepacket excitation have recently been proposed. This chapter is concerned with one of these schemes, namely, the use of coherent pulse sequences to control product formation in chemical reactions. We shall see that this scheme follows naturally from an understanding of the characteristics of time-delayed coherent anti-Stokes Raman spectroscopy (CARS) and of photon echo spectroscopy.

Heller and co-workers have shown the utility of formulating a variety of photochemical and spectroscopic processes in the time domain and have interpreted the time-domain formulas in terms of wavepacket dynamics.[1–4] In photodissociation the wavepacket originates in the interaction region—the Franck–Condon region—on a repulsive potential energy surface and evolves to the asymptotic region(s) of the potential energy surface to form products. In nondissociative photoexcitation processes, the wavepacket originates in the Franck–Condon region and executes a Lissajous-type motion, generalized in that anharmonicities may lead to a complicated trajectory. Conventional spectroscopic quantities may be obtained by appropriate projection of the time-evolving wavepacket. In particular, single-frequency or collision-energy cross sections require an energy projection (i.e., a Fourier transform) of the wavepacket amplitude. Heller et al. have exploited the energy projection methodology to obtain total and partial photodissociation cross sections and absorption and resonance Raman spectra.

Recently, in this laboratory, we have applied time-dependent quantum mechanics–wavepacket dynamics to several bona fide time-domain spectroscopies. Specifically, we have formulated time-dependent theories of coherent-pulse-sequence-induced control of photochemical reaction, picosecond CARS spectroscopy, and photon echoes. These processes all involve multiple pulse sequences in which the pulses are short or comparable in time scale to the molecular dynamics of interest. Time-dependent quantum mechanics provides the natural formulation of these processes, whether the field is weak, and a perturbative treatment is appropriate, or the field is strong and a nonperturbative treatment is necessary. In both cases we treat the time dependence of the field and the molecular evolution as embodied in the wavepacket dynamics within a single coherent framework. The interaction of time-dependent fields with polyatomics signifies a new and fruitful application of wavepacket technology to chemical problems.

Time-domain spectroscopies entail a major shift in emphasis from traditional spectroscopies, since the experimenter can control, in principle, the duration, shape, and sequence of pulses. One may say that traditional, CW spectroscopy, is passive—the experimenter attempts to study static properties of a particular molecule. Coherent pulse experiments are active in that, given a set of molecular properties (which may in fact be known from various spectroscopies), one tries to arrange for a desired chemical product, or to design a pulse sequence that will probe new molecular properties. The time-dependent quantum mechanics–wavepacket dynamics approach developed here is a natural framework for formulating and interpreting new multiple pulse experiments. Femtosecond experiments yield to a particularly simple interpretation within our approach.

## II. WAVEPACKET DYNAMICS

In this section we review the time-dependent wavepacket representation of photodissociation, photoabsorption, and resonance Raman scattering introduced by Heller and co-workers. Our approach to controlling the selectivity of a chemical reaction builds on concepts taken from all three of these areas, so an introductory review is appropriate before proceeding further. The reader is referred to Heller's beautiful review article[1] and to the original literature[2–4] for further details.

### A. Photodissociation

In a photodissociation reaction it is usual for the initial state of the molecule to be the ground vibrational state of the ground electronic state. The incident radiation is resonant with an excitation to an electronic state that is dissociative (repulsive potential energy surface) or predissociative (the optically allowed transition is to a bound-state potential energy surface that intersects a repulsive surface). In the Franck–Condon picture, the electrons respond "instantaneously" to the incident light, while the relatively massive nuclei respond only slowly. Hence, on absorption of a photon the nuclear wavefunction retains its shape but is projected up to the dissociative electronic state. In the traditional approach to the calculation of the photodissociation cross section, it is then necessary to evaluate Franck–Condon factors between the initial vibrational state and the continuum states of the excited-state surface, namely,

$$\sigma(\omega) = k\omega\rho(E)|\langle\psi_E|\mu|\psi_i\rangle|^2. \tag{2.1}$$

The states $\langle\psi_E|$ are the continuum eigenstates of the excited-state potential energy surface (labeled "$b$"),

$$H_b|\psi_E\rangle = E|\psi_E\rangle, \tag{2.2}$$

and $k$ is a constant, $E = \hbar\tilde{\omega} = \hbar(\omega + \omega_i)$, where $\hbar\omega$ is the photon energy and $\hbar\omega_i$ the energy of $\psi_i$.

An alternative representation of the photodissociation cross section can be obtained from the following considerations. Although the initial-state vibrational wavefunction is an eigenfunction of the ground-electronic-state Hamiltonian, it is a nonstationary state of the excited-state Hamiltonian. Therefore, immediately following the electronic transition the nuclei are displaced from the equilibrium geometry corresponding to the excited state potential energy surface, and they begin to move downhill, initially along the coordinate of steepest descent. Pictures of the time-evolving wavefunction on two different excited-state potential energy surfaces are shown in Fig. 1. By Ehrenfest's theorem, the center of the wavepacket obeys Hamilton's equations of motion, generally for several vibrational periods. The wavepacket spreads (or possibly contracts) in the course of its evolution, but for smooth potential energy surfaces and reasonably direct paths to the asymptotic geometry, the wavepacket will remain localized out into the product region of the excited-state potential energy surface. The initial conditions on the wavepacket are fully determined by the ground- and excited-state potential energy surfaces and the Franck–Condon principle. The ensuing wavepacket evolution, governed by the time-dependent Schrödinger equation, determines the total photodissociation cross section as well as the vibrational, rotational, and electronic-state distribution of the photofragments.

The total cross section for photodissociation is given by[2]

$$\sigma(\omega) = k\omega \int_{-\infty}^{\infty} \langle\phi|\phi(t)\rangle e^{i\tilde{\omega}t}\, dt, \tag{2.3}$$

where

$$|\phi\rangle = \mu|\psi_i\rangle \tag{2.4}$$

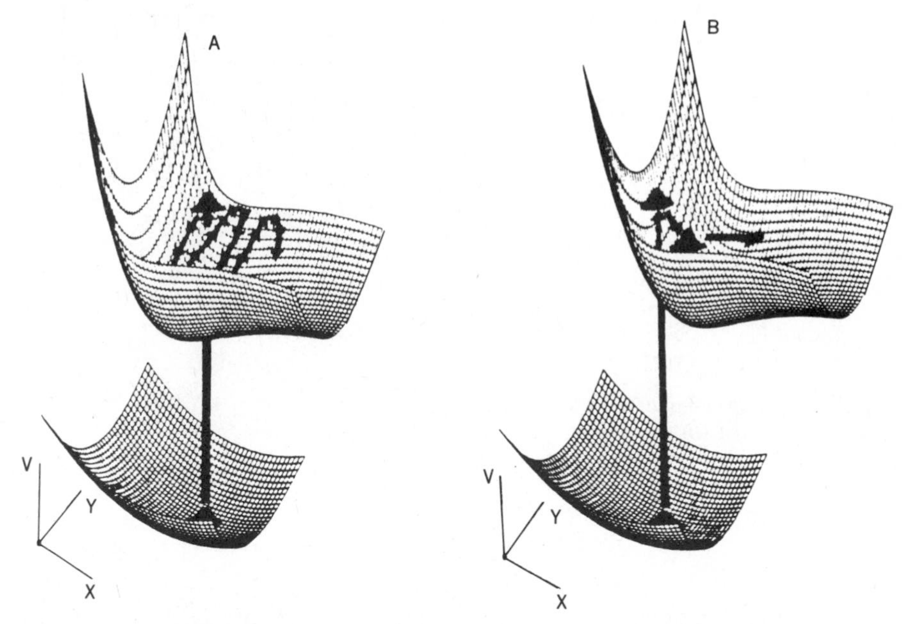


**Figure 1.** Photoabsorption between two Born–Oppenheimer potential energy surfaces. (*A*) The Franck–Condon wavepacket, $\phi(t)$, arising out of $\phi = \mu\chi$ [$\chi$ is shown on the lower surface, $\phi$, and $\phi(t)$ on the upper], grazes $\phi(0)$ several times on the way to dissociation. The result is an absorption band with some limited vibrational structure. (*B*) Direct dissociation leading to a broad, featureless absorption band. (Reproduced, with permission, from Ref. 1.)

and

$$|\phi(t)\rangle = e^{-iH_b t/\hbar}|\phi\rangle. \tag{2.5}$$

Thus, the total photodissociation cross section is given by the Fourier transform of the autocorrelation of the time-evolving wavepacket. Figures 1*a* and 1*b* suggest two possible cases. In Fig. 1*a*, as the wavepacket leaves the Franck–Condon region its autocorrelation function decays; however, the wavepacket revisits the Franck–Condon region before its ultimate exit to form products (partial recurrence), which generates the later structure in the autocorrelation function in Fig. 2*a*. The Fourier transform of the autocorrelation function of Fig. 2*a* is shown in Fig. 2*b*. The initial decay time determines the overall width of the photodissociation spectrum, while the recurrence in the time domain leads to the vibrational structure in the cross section. Note that the recurrence time determines the spacing of vibrational features in the spectrum. Figure 3*a* shows the autocorrelation function for the wavepacket evolution in Fig. 1*b*. This wavepacket makes a direct exit to

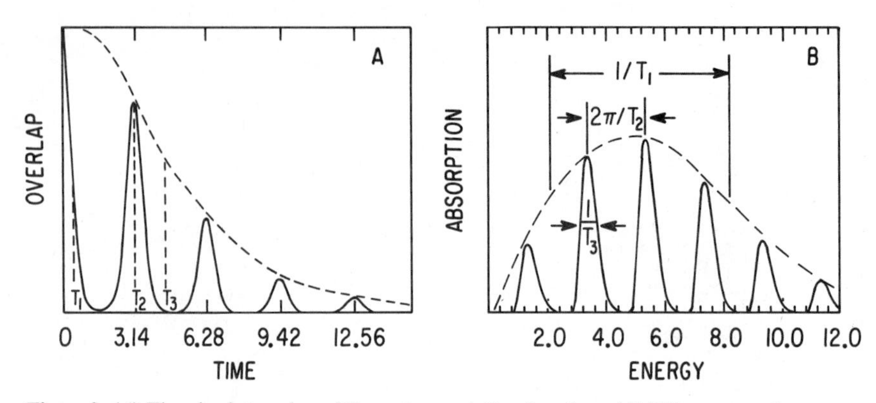


**Figure 2.** (*A*) The absolute value of the autocorrelation function, $\langle \phi | \phi(t) \rangle$, versus $t$ for a case similar to Fig. 1*A*. (*B*) The absorption spectrum corresponding to Fig. 1*A*, given by the Fourier transform of $\langle \phi | \phi(t) \rangle$. (Reproduced, with permission, from Ref. 1.)

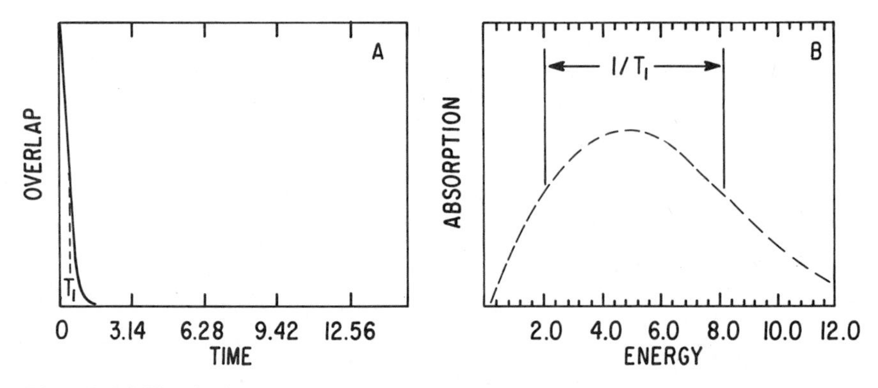


**Figure 3.** (*A*) The absolute value of the autocorrelation function, $\langle \phi | \phi(t) \rangle$, versus $t$ for Fig. 1*B*. (*B*) The absorption spectrum corresponding to Fig. 1*B*. (Modified, with permission, from Ref. 1.)

form products, and the autocorrelation function decays monotonically. The Fourier transform of this autocorrelation is shown in Fig. 3*b*; it is structureless.

In both cases, 1*a* and 1*b*, the total photodissociation cross section is completely determined by the short-time dynamics in the Franck–Condon region. In contrast, the partial cross sections, which determine the vibrational, rotational, and electronic-state distributions of the products, involves longer time dynamics. To obtain all of the relevant information about the reaction, the wavepacket evolution must be followed out into the product region of the potential energy surface and projected onto the various different vibrational and rotational states of the fragments. The partial cross section for scattering to a particular vibrational or ro-vibrational final state of the photofragments is given by[2]

$$\sigma_n(\omega) = k\omega_I \rho(E - E_n) \lim_{t\to\infty} |\langle \psi_{k,n}^{(-)} | \phi(t) \rangle|^2, \tag{2.6}$$

where $E_n$ is the internal energy and $\rho(E - E_n)$ is the density of translational states for the fragments.

As long as the photodissociation reaction is fairly direct, the time-dependent formulation is fruitful and provides insight into both the process itself and the relationship of the final-state distributions to the absorption spectrum features. Moreover, solution of the time-dependent Schrödinger equation is feasible for these short-time evolutions, and total and partial cross sections may be calculated numerically.[5] Finally, in those cases where the wavepacket remains well localized during the entire photodissociation process, a semiclassical gaussian wavepacket propagation will yield accurate results for the various physical quantities of interest.[6]

### B. Photoabsorption

The formalism for photoabsorption (and photoemission) is virtually identical to that for the total photodissociation cross section except that one is generally concerned with long-time evolution on a bound excited electronic-state potential energy surface.[2] Several processes can limit the lifetime in the excited state, notably radiative decay (if no other processes intercede) and nonradiative decay (intersystem crossing or internal conversion). In the absence of any decay processes, one could, in principle, follow the autocorrelation function of the wavepacket for an infinite length of time; the Fourier transform of the autocorrelation function would then yield delta functions at the positions of the molecular eigenstates—an infinite resolution absorption spectrum. In practice, for large molecules with many ($\geq 10$) modes undergoing displacement in the excited electronic state, there will be an enormous density of vibrational states at, say, 30,000 $cm^{-1}$ of vibrational energy. One cannot possibly be interested in the identification and assignment of transitions to each of these states. In a limited resolution experiment, one will observe "features' in the absorption spectrum, consisting of transitions to very many individual molecular eigenstates. The interpretation of these coarser features, and their numerical modeling, is often conveniently performed in the time domain with the spectrum recovered via Fourier transform.[3]

### C. Resonance Raman Scattering

The discussion we have given of the wavepacket representation of photodissociation and photoabsorption is within the framework of first-order

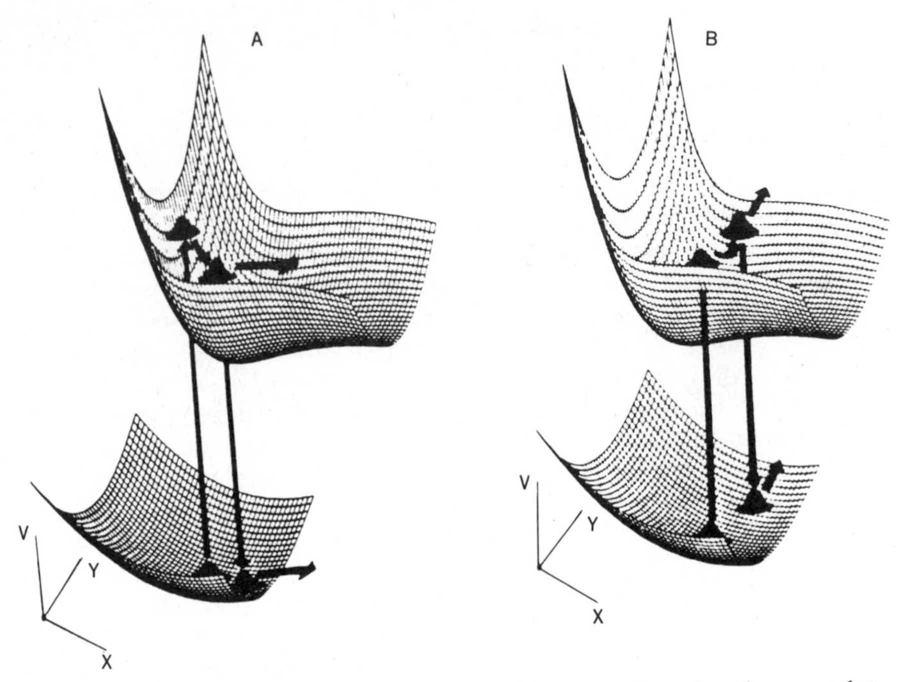


**Figure 4.** (*A*) Raman scattering from an upper potential energy surface, where the wavepacket acquires motion in the $x$ direction initially; $x$-mode overtones will be enhanced in the Raman as resonance is approached. (*B*) $y$ modes enhanced due to $y$ motion initially. (Reproduced, with permission, from Ref. 1.)

perturbation theory, and refers to one-photon processes. In the same spirit, one may treat two-photon processes and thereby arrive at a time-dependent wavepacket description of resonance fluorescence, resonance Raman scattering, two-photon absorption, and stimulated emission processes (Fig. 4). We begin with the time-dependent perturbation theory expression for the second-order amplitude, $\psi_a^{(2)}$, on the ground-electronic-state pontential energy surface,[4] $a$:

$$\psi_a^{(2)}(T) = \frac{-1}{\hbar^2} \int_{-\infty}^{T} dt_2 \int_{-\infty}^{t_2} dt_1\, e^{-iH_a(T-t_2)/\hbar} \mu E(t_2) e^{-iH_b(t_2-t_1)/\hbar} e^{-\Gamma(t_2-t_1)} \times \mu E(t_1) e^{-iH_a t_1/\hbar} |\psi_i\rangle + \text{nonresonant term.} \tag{2.7}$$

This expression is valid for arbitrary waveforms of the incident and scattered (or stimulating) field, $E(t_1)$ and $E(t_2)$, respectively; it will be examined again in its general form in Section 5C.1. For now we specialize to the case of a single-frequency incident and (spontaneously) scattered photon, that is, $E(t_1) = e^{-i\omega_I t_1}$ and $E(t_2) = e^{i\omega_s t_2}$. The Raman amplitude, $\alpha_{fi}$, for scattering from initial state $\psi_i$, to final state $\psi_f$ is given by the $T \to \infty$ limit of the projection of $\psi_a^{(2)}(T)$ onto $\psi_f$, a vibrational eigenstate of the ground-electronic-state surface:

$$\alpha_{fi}(\omega) = \lim_{T\to\infty} \langle \psi_f | \psi_a^{(2)}(T) \rangle. \tag{2.8}$$

Substituting Eq. (2.7) into (2.8) and introducing the new variable, $t = t_2 - t_1$, yields

$$\alpha_{fi}(\omega) = \int_0^\infty \langle \phi_f | \phi_i(t) \rangle e^{i\tilde{\omega}t} e^{-\Gamma t}\, dt + \text{nonresonant term}, \tag{2.9}$$

$|\phi_i(t)\rangle$ is defined as previously,

$$|\phi_f\rangle = \mu |\psi_f\rangle. \tag{2.10}$$

The nonresonant term may be obtained from the resonant term by the replacement $\omega_I \to -\omega_S$, and, henceforth, will be neglected. Equation (2.9) states that the scattering amplitude is the half-Fourier transform of the overlap of the time-evolving wavepacket with the final state of interest (multiplied by the transition moment). Equation (2.9) bears a close resemblance to Eq. (2.3) for the absorption cross section, but there are three differences to note: (1) the cross-correlation function of the moving wavepacket with the final vibrational state of interest is required, rather than the autocorrelation function; (2) an integral over the range $[0, \infty]$, not $[-\infty, \infty]$, is required for the Raman amplitude; (3) The cross-section $I^{i\to f}(\omega)$ is proportional to the absolute value squared of $\alpha$:

$$I^{i\to f}(\omega) = k\omega\omega_S^3 |\alpha_{fi}(\omega)|^2. \tag{2.11}$$

Although, formally, the integral in Eq. (2.9) is over the range $[0, \infty]$, the domain of integration may be shortened via three mechanisms.[4] First, the effective lifetime of the wavepacket on the excited-state potential energy surface is limited by radiative decay rate and/or the collisional deactivation rate of the excited electronic state; these effects can be represented by a phenomenological lifetime, $\Gamma^{-1}$. Second $\langle \phi_f | \phi_i(t) \rangle$ has an intrinsic decay that may result from (1) photodissociation on the excited-electronic-state potential energy surface; (2) intersystem crossing, or internal conversion to a weakly allowed state; or (3) significant displacement in many modes so that, with small

Franck–Condon factors, a recurrence to a particular $\phi_f$ becomes unlikely. Third, if the incident photon is off-resonance with the excited state, the effective lifetime, $\tau$, is given by

$$\tau = \frac{1}{\Delta\omega}, \tag{2.12}$$

where $\Delta\omega$ is the frequency mismatch between the incident light frequency, $\omega_I$, and excited-state absorption maximum (Fig. 5). In ordinary Raman scattering, $\tau \cong 0$; transitions are possible (and only weakly so, relative to the resonance case) via the coordinate dependence of the transition moment (or, equivalently, the polarizability). If we imagine tuning $\omega_I$ continuously from off-resonance to on-resonance, where long-time dynamics may be significant, we pass through an intermediate regime where short-time dynamics govern the Raman intensities. This regime corresponds to preresonance Raman scattering, for which $\omega_I$ is just below the 0–0 transition in a discrete absorption spectrum or in the wings of a continuous absorption spectrum. When

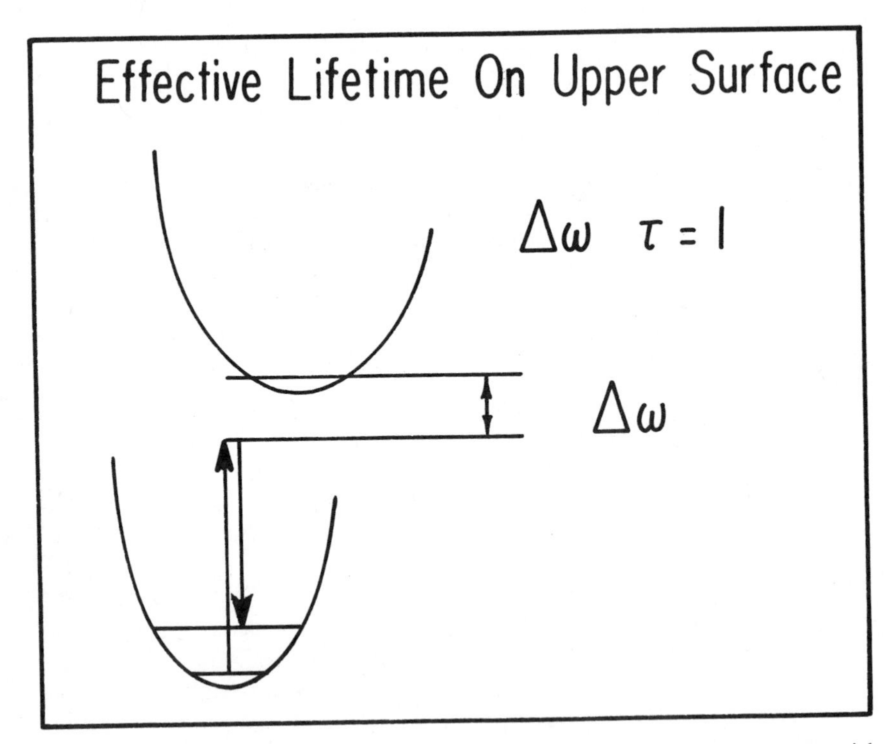


**Figure 5.** Uncertainty principle diagram showing detuning from the excited state potential energy surface by an amount $\Delta\omega$.

any of the above three mechanisms limits the effective lifetime in the excited electronic state to several vibrational periods or fewer, the time-dependent formulation is fruitful. The interpretation of intensities in experimental spectra often is straightforward in terms of the short-time wavepacket dynamics on the excited-electronic-state potential energy surface. As in the case of photodissociation, either semiclassical gaussian wavepacket methods[6] or numerical integration of the time-dependent Schrödinger equation[5] may be used to calculate, for example, cross sections. The preference for one approach over the other depends on the degree of anharmonicity of the excited-state potential energy surface, the size of the initial wavepacket, the time scale required for the propagation, and the number of degrees of freedom involved.

The first lifetime-limiting mechanism discussed is well illustrated in the experiments of Paramenter and co-workers on the temporal evolution of fluorescence from *p*-difluorobenzene.[7] As the pressure of a quencher gas is increased, thereby increasing $\Gamma$, the pattern of the dispersed fluorescence shifts to weight emission from states that are reached in the early-time dynamics. (Note that early, in this context, is on the order of 100 vibrational periods.)

The second liftime-limiting mechanism is beautifully illustrated in experiments of Imre and co-workers on ozone and methyl iodide.[8] Although these molecules dissociate in their excited electronic states, 1 out of $10^6$ molecules emits a photon and returns to the ground electronic state before dissociating. Imre and co-workers were able to disperse the tiny amount of emitted light and obtain its spectrum. The positions of the lines in the spectrum contain information about high-lying vibrational levels (near the dissociation threshold) on the ground-state potential energy surface. The intensities of the lines mark the pathway along which the wavepacket exits in the multidimensional vibrational space. Imre and co-workers observed a long progression of overtones in the emission from each of the molecules, corresponding to the large geometry changes a molecule experiences on its way to dissociation.

Imre and co-workers also illustrated the third lifetime-limiting mechanism.[8] By tuning their incident light frequency in the wings of the methyl iodide absorption spectrum, they observed the disappearance of all the higher overtones. This observation is entirely consistent with a shorter effective lifetime in the excited electronic state, and the molecule returning to the ground electronic state from a geometry nearly the same as that of the ground state, that is, from a geometry close to that corresponding to the one generated at the instant of excitation. The third lifetime mechanism has also been nicely illustrated in the experiments of Myers and co-workers, who used preresonance Raman intensities to infer the early time (femtosecond) wavepacket dynamics in the purple membrane protein, bacteriorhodopsin, and in the isomerization of cis- and trans-stilbene.[9] At these short times the wavepacket moves along the path of steepest descent on the excited-state potential energy surface. Hence, Myers and co-workers were able to recover information about the force constants of different coordinates on the excited-state potential energy surface from the preresonance Raman intensities. Numerical modeling of these intensities was carried out using semiclassical gaussian wavepacket propagation (a fully quantum-mechanical calculation including all the relevant degrees of freedom in a molecule as large as stilbene is at present intractable). Since the relevant dynamics last for less than a vibrational period, the gaussian wavepacket propagation, including spreading in a locally harmonic potential, is an excellent approximation, for any number of vibrational degrees of freedom.

If we reflect on the physical meaning of the variable $t = t_2 - t_1$, the integration variable in the expression for the Raman amplitude [Eq. (2.9)], we realize that it is precisely the time delay between the incident and scattered photons. The integral ranges continuously over all values of time delay, from 0, where the second photon is emitted instantaneously after the first photon is incident (scattering), to $\infty$, where the second photon is emitted long after the first photon is incident (separable absorption and emission process). If one thinks of the absorption of the photon as a half-collision process, and the emission as a second half-collision, the molecule and photons synchronize their clocks at each of these two instants. The time delay between incident and emitted photons is therefore commensurate with the effective lifetime of the molecule on the excited electronic state. Within the framework of the time-dependent wavepacket picture, this is the amount of time that the initial wavepacket propagates under the influence of the excited-electronic-state potential energy surface. Therefore, the shortest time emission corresponds to small excursions of the wavepacket from the initially populated region; the wavepacket returns to the ground electronic state near to where it originated and, consequently, primarily the initial vibrational state and low-lying states are populated (Rayleigh scattering and Raman scattering to fundamentals). Longer-time emission corresponds to larger excursions from the initially populated region; the wavepacket returns to the ground electronic state far from where it originated and higher vibrational levels of the ground electronic state are populated (Raman scattering to overtones and combinations). By the same reasoning, the three different situations previously discussed in which short-time wavepacket propagation is sufficient to calculate scattering intensities correspond to short time delays between the incident and scattered photon.

## III. TIME-DEPENDENT CARS

### A. Motivation

We now examine, as a first case of exploitation of the coherence properties of the coupled molecule–electromagnetic field system, a representation of CARS suitable to the description of time-dependent measurements. Our goal is to understand the conditions on controlling stimulated emission with pulses separated in time. It will be seen that the representation of time-dependent CARS includes the standard energy frame formulation as a special case when only CW fields are involved. The traditional CARS frequency matching condition, $\omega_0 = \omega_1 - \omega_2 + \omega_3$, generalizes for non-CW fields to the Fourier components at $\omega_0$ of the recurrences of a time evolved wavepacket.[10]

The CARS process involves four photons. Typically, two laser beams are employed, with frequencies $\omega_1$ and $\omega_2$, whose frequency difference is resonant with a vibrational transition of the ground electronic state. Photon three is generally of the same frequency as photon one; photon four is emitted spontaneously with a frequency equal to $\omega_1 - \omega_2 + \omega_3 = 2\omega_1 - \omega_2$, that is, such that the molecule returns to its ground vibrational state. The key to the usefulness of CARS is that, if certain momentum and energy phase-matching conditions are satisfied, the fourth photon in the process is emitted coherently. Because the signal beam is generated by a coherent mixing of three photons, it is extremely intense, often several orders of magnitude more intense than spontaneous Raman signals. Furthermore, it is emitted in the direction of the incident beams, whereas spontaneous Raman signals are distributed over $4\pi$ solid angle.

We are interested in the case when the frequency difference $\omega_1 - \omega_2$ is resonant with a "zeroth-order" state, that is, a nonstationary state of the molecule. Examples of such states are the C–H stretch overtones of benzene[11,12] and the fundamental of the ring breathing mode (vibration 12) of the alkylbenzenes.[13] Under low-resolution conditions, transitions to these vibrational states look like single lines. At higher resolution these "lines" are seen to consist of a tightly clustered set of transitions to molecular eigenstates. If this cluster of molecular eigenstates is excited coherently, it will evolve via intramolecular vibrational redistribution processes (IVR), which transfer amplitude of excitation from one part of the molecule to another. The time scale for this evolution, and the possible existence of quantum beats, or recurrences in the vibrational energy distribution, depend on both the particular molecule and the energy range studied.

### B. Theoretical Considerations

In traditional treatments, the amplitude of the CARS signal is proportional to the third-order molecular hyperpolarizability,

$$P_\rho^{(3)}(\omega_0) \propto \gamma_{\rho\sigma\tau\nu}(-\omega_0, \omega_1, \omega_2, \omega_3) E_\sigma(\omega_1) E_\tau(\omega_2) E_\nu(\omega_3), \tag{3.1}$$

where $\gamma_{\rho\sigma\tau\nu}$ is a fourth-rank tensor, and hence is comprised of 81 ($3^4$) Cartesian components and $E_\sigma$ is the field amplitude corresponding to photon $\sigma$.[14] Each of the 81 components contains 24 terms, corresponding to the number of permutations of the four photons in the process. $\gamma$ is usually displayed in the energy representation. In this representation each of the 24 terms contains a fourfold product of Franck–Condon matrix elements, formally involving all the molecular eigenstates. Clearly, if many states contribute to the scattering process, evaluation of these terms becomes prohibitively difficult. Furthermore, and more important for the present discussion, the conventional expression for $\gamma$ is useful only for the case of four CW photons. If, as a consequence of using pulsed radiation, there is a significant spread in the frequencies of the three incident fields, $\gamma$ must be evaluated for all combinations of the frequencies and integrated over $\omega_1$, $\omega_2$, and $\omega_3$.

We now examine formulation of a time-frame expression for CARS amplitude, valid for pulsed or CW light sources, which avoids all the previously stated difficulties. We deal directly with $P^{(3)}(\omega_0)$ and thereby circumvent the calculation of $\gamma$ at all combinations of frequencies.

The fourth-order time-dependent perturbation theory expression for $P^{(3)}(\omega_0)$ is

$$P_\rho^{(3)}(\omega_0) = \frac{k}{\hbar^3} \int_{-\infty}^{\infty} \int_{-\infty}^{t_4} \int_{-\infty}^{t_3} \int_{-\infty}^{t_2} dt_4\, dt_3\, dt_2\, dt_1 \langle g | e^{i\omega_g t_4} (\mu_\rho e^{i\omega_0 t_4}) \times e^{-iH(t_4-t_3)/\hbar} [\mu_\nu E_3(t_3)] e^{-iH(t_3-t_2)/\hbar} [\mu_\tau E_2(t_2)] e^{-iH(t_2-t_1)/\hbar} \times [\mu_\sigma E_1(t_1)] e^{-i\omega_g t_1} | g \rangle + 23 \text{ permutation terms.} \tag{3.2}$$

In Eq. (3.2), $\mu_\rho$, $\mu_\nu$, $\mu_\tau$, $\mu_\sigma$ are dipole moment operators and $E_1$, $E_2$, $E_3$, $E_4$ are the temporal field amplitudes corresponding to each of the four photons. Formally, $H$ is the full electronic-nuclear Hamiltonian. However, in practice, $H$ is the Born–Oppenheimer Hamiltonian of either the ground or excited electronic state, depending on the resonance condition satisfied by the preceding photons.

There are several different situations for which the fourfold integral in Eq. (3.2) can be simplified. For the common experimental situation where photons 1 and 3 are off-resonance, to a very good approximation $e^{-iH(t_2-t_1)/\hbar}$ and $e^{-iH(t_4-t_3)/\hbar}$ can be replaced by $\delta(t_2 - t_1)$ and $\delta(t_4 - t_3)$, respectively. Substitution of these approximations yields

$$P_\rho^{(3)}(\omega_0) = \frac{k}{\hbar^3} \int_{-\infty}^{\infty} \int_{-\infty}^{t_4} dt_4\, dt_2 \langle g | e^{i\omega_g t_4} (\mu_\rho e^{i\omega_0 t_4}) [\mu_\nu E_3(t_4)] e^{-iH(t_4-t_2)/\hbar} \times [\mu_\tau E_2(t_2)] [\mu_\sigma E_1(t_2)] e^{-i\omega_g t_2} | g \rangle + PT. \tag{3.3}$$

Equation (3.3) has the following interpretation: $\mu_\tau \mu_\sigma | g \rangle$ is an initial wavepacket; $E_2(t_2) E_1(t_2)$ prepares a zeroth-order vibrational state (or wavepacket) on the ground-state Born–Oppenheimer surface; and $e^{-iH(t_4-t_2)/\hbar}$ propagates this wavepacket on the ground-state Born–Oppenheimer surface. The projection is then taken onto $\langle g|\mu_\rho\mu_\nu$. Equation (3.3), with the preceding interpretation, is strongly reminiscent of the representation of two-photon resonance Raman scattering, Eq. (2.7). In our context, it is the Fourier components at $\omega_0$ of the wavepacket recurrences that determine $P_\rho^{(3)}(\omega_0)$.

The expression for the explicit time dependence of the CARS amplitude can also be simplified. This time-dependent amplitude is given by

$$P_\rho^{(4)}(t) = \int_{-\infty}^{\infty} P_\rho^{(3)}(\omega_0)e^{i\omega_0 t_4}\,d\omega_0, \tag{3.4}$$

which is simply the inverse Fourier transform of Eq. (3.2) or Eq. (3.3). The fourfold integral over time is seen to reduce to a threefold integral. We obtain, then, using Eq. (3.3) for $P_\rho^{(3)}(\omega_0)$,

$$\begin{aligned} P_\rho^{(3)}(t_4) = \frac{2\pi k}{\hbar^3}\int_{-\infty}^{t_4} & dt_2\langle g|e^{-i\omega_g t_4}(\mu_\rho)[\mu_\nu E_3(t_4)]e^{-iH(t_4-t_2)/\hbar} \\ & \times [\mu_\tau E_2(t_2)][\mu_\sigma E_1(t_2)]e^{-i\omega_g t_2}|g\rangle. \end{aligned} \tag{3.5}$$

We shall see that (3.5) predicts dramatically different decay times for the resonant and nonresonant contributions to the CARS signal. This effect has been illustrated beautifully in the experiments reported by Zinth et al.[15] and Kamga and Sceats.[16]

A third set of simplifications arise when the envelope in time of the light pulses can be considered $\delta$ functions on the time scale of the IVR (or other dynamical process). In the case where we can approximate $E_1(t_2)$ and $E_2(t_2)$ as $\delta(t_2^*)$ and $E_3(t_4)$ as $\delta(t_3^*)$, Eq. (3.5) simplifies to

$$P_\rho^{(3)}(t_4) \propto \bar{E}_1\bar{E}_2\bar{E}_3\langle g|\mu_\rho\mu_\nu e^{-iH(t_4^*-t_2^*)/\hbar}\mu_\tau\mu_\sigma|g\rangle\delta(t_4 - t_4^*) + PT. \tag{3.6}$$

These simplifications are valid when the time scale separation $\tau_{IVR} \gg \tau_{\text{pulse}} \gg \tau_{\Delta\omega}$ holds, where $\Delta\omega$ is the mismatch of the incident frequencies from an electronic transition.

Of course, in the fully general case, photon one and photon three may be resonant with an excited electronic state and Eq. (3.2) should be used to calculate the intensity; however, under resonance conditions only a few of the 24 terms make significant contributions to the CARS amplitude.

In order to recover the formula for $\gamma_{\rho\sigma\tau\nu}$ in Ref. 15, we return to Eq. (3.2) and set

$$
\begin{aligned}
E_1(t_1) &= E_1 e^{-i\omega_1 t_1}, \\
E_2(t_2) &= E_2 e^{i\omega_2 t_2}, \\
E_3(t_3) &= E_3 e^{-i\omega_3 t_3}.
\end{aligned}
\tag{3.7}
$$

We now rewrite Eq. (3.2) in an expanded form,

$$
\begin{aligned}
P_\rho^{(3)}(\omega_0) = \frac{k}{\hbar^3} &\int_{-\infty}^{\infty} \int_{-\infty}^{t_4} \int_{-\infty}^{t_3} \int_{-\infty}^{t_2} dt_4\, dt_3\, dt_2\, dt_1\, E_1 E_2 E_3 \\
&\times \langle g | \mu_\rho e^{-iH(t_4 - t_3)/\hbar} e^{i(\omega_3 - \omega_2 + \omega_1 + \omega_g)(t_4 - t_3)} \mu_\nu \\
&\times e^{-iH(t_3 - t_2)/\hbar} e^{i(-\omega_2 + \omega_1 + \omega_g)(t_3 - t_2)} \mu_\tau e^{-iH(t_2 - t_1)/\hbar} \mu_\sigma \\
&\times e^{i(\omega_1 + \omega_g)(t_2 - t_1)} | g \rangle e^{-i(\omega_g + \omega_1 - \omega_2 + \omega_3 - \omega_0 - \omega_g)t_4}.
\end{aligned}
\tag{3.8}
$$

The reader may easily verify the equivalence of this equation to Eq. (3.2). The additional phase factors in this expression cancel out in a dominolike fashion. We then transform to the new variables

$$
\begin{aligned}
W &= t_2 - t_1, \\
X &= t_3 - t_2, \\
Y &= t_4 - t_3, \\
t_4 &= t_4.
\end{aligned}
$$

Making use of the integral representation for the delta function, we obtain

$$
\begin{aligned}
P_\rho^{(3)}(\omega_0) = \frac{k}{\hbar^3} &\int_0^{\infty} \int_0^{\infty} \int_0^{\infty} dW\, dX\, dY \langle g | \mu_\rho e^{-iHY} \\
&\times e^{i(\omega_3 - \omega_2 + \omega_1 + \omega_g)Y} \mu_\nu e^{-iHX} e^{i(-\omega_2 + \omega_1 + \omega_g)X} \mu_\tau e^{-iHW} \\
&\times e^{i(\omega_1 + \omega_g)W} \mu_\sigma | g \rangle E_1 E_2 E_3 \delta(\omega_g + \omega_1 - \omega_2 + \omega_3 \\
&- \omega_0 - \omega_g) + PT.
\end{aligned}
\tag{3.9}
$$

Inserting three complete sets of states ($|j\rangle, |k\rangle, |l\rangle$) and integrating over $W$, $X$, and $Y$ we obtain

$$
\begin{aligned}
P_\rho^{(3)}(\omega_0) &= \left( \frac{k}{\hbar^3} \sum_{j,k,l} \frac{\langle g | \mu_\rho | l \rangle \langle l | \mu_\nu | k \rangle \langle k | \mu_\tau | j \rangle \langle j | \mu_\sigma | g \rangle}{(\omega_{jg} - \omega_1)(\omega_{kg} - \omega_1 + \omega_2)(\omega_{lg} - \omega_0)} + PT \right) \\
&\quad \times E_1 E_2 E_3 \delta(\omega_1 - \omega_2 + \omega_3 - \omega_0) \qquad (3.10a) \\
&= k \gamma_{\rho\sigma\tau\nu} E_\sigma(\omega_1) E_\tau(\omega_2) E_\nu(\omega_3) \delta(\omega_1 - \omega_2 + \omega_3 - \omega_0). \qquad (3.10b)
\end{aligned}
$$

The first term in parentheses in Eq. (3.10a) is precisely the second term in Eq. (37) of Ref. 14. The other terms may be obtained by permutation of $\omega_1$, $-\omega_2$, $\omega_3$, and $-\omega_0$. Note the $\delta$ function condition on $\omega_0$, which represents the conventional CARS frequency-matching condition.

### C. Some Examples

We now demonstrate some of the great variety of behavior that arises from different choices for resonance conditions for each of the photons and different possible intramolecular energy transfer pathways.

It is instructive to examine the behavior of $P_\rho^{(3)}(t_4)$ without including $E_3(t_3)$ in order to obtain a global picture for a variety of different delay times of photon three relative to photon two. We therefore examine the quantity

$$P_\rho'(t_4) \equiv \int_{-\infty}^{t_4} dt_2 \langle g|\mu_\rho \mu_\nu e^{-iH(t_4-t_2)/\hbar} \mu_\tau \mu_\sigma |g\rangle E_2(t_2) E_1(t_2) e^{-i\omega_g t_2}. \quad (3.11)$$

$P_\rho^{(3)}(t_4)$ may be obtained from $P_\rho'$ (for photon three nonresonant with an excited electronic state) by simply multiplying $P'$ by a narrow envelope in time at the desired delay. For definiteness we suppose that $E_2(t_2)E_1(t_2)$ has the form

$$E_2(t_2)E_1(t_2) = a(t_2) e^{-i(\omega_1-\omega_2)t_2}, \quad (3.12)$$

where the envelope function $a(t_2)$ is given by

$$a(t_2) = \frac{1}{2\pi\sigma^{1/4}} \exp\left(\frac{-t_2^2}{2\sigma^2}\right). \quad (3.13)$$

We are interested in a product pulse that is narrow in time (of order 1 psec) while at the same time relatively narrow in frequency (of order 10 wavenumbers), with a central frequency $\omega_1$. Furthermore, we assume that

$$\begin{aligned} C(t) &\equiv \langle g|\mu_\rho \mu_\nu e^{-iHt/\hbar} \mu_\tau \mu_\sigma |g\rangle \\ &\cong e^{-\Gamma_{IVR} t} e^{i\omega_s t}. \end{aligned} \quad (3.14)$$

The exponential decay of IVR on the ground-state potential surface is a crude but useful description of that process in many molecules. It may be justified on theoretical grounds by considering the Bixon–Jortner level coupling scheme. In (3.11), $\omega_s$ is the central frequency for the vibrational feature $s$.

Equation (3.8) may now be rewritten in the form

$$P_\rho'(t_4) = \int_{-\infty}^{t_4} a(t_2) e^{-\Gamma_{IVR}(t_4-t_2)} e^{i\Delta\omega(t_4-t_2)}\, dt_2, \quad (3.15)$$

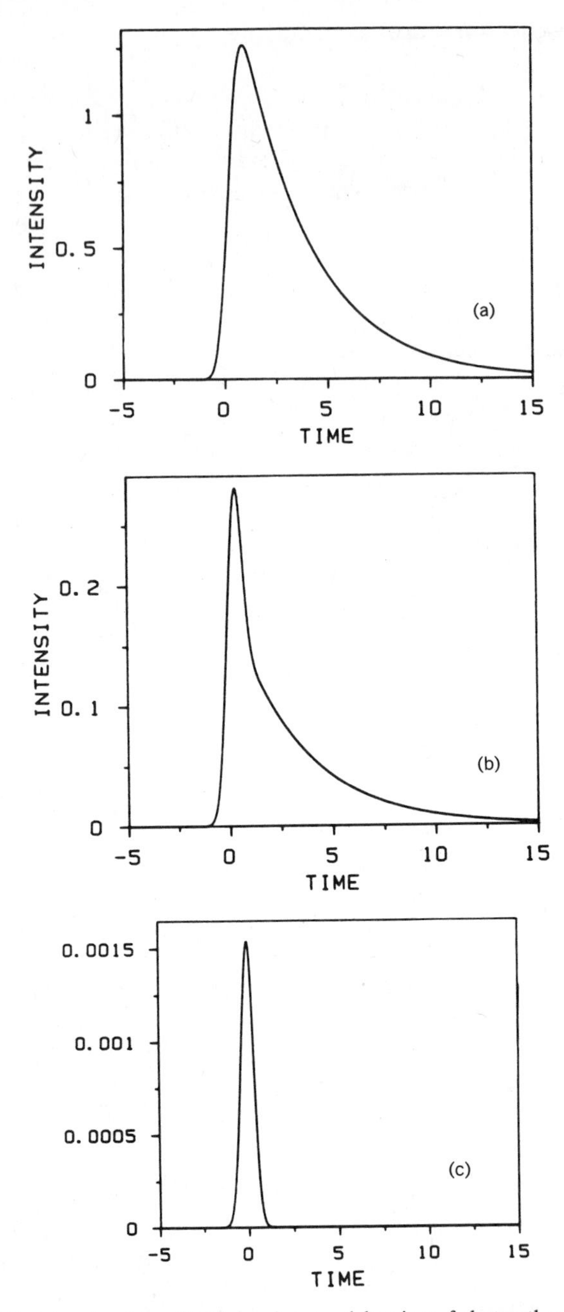


**Figure 6.** (*A*) The intensity of the CARS signal versus delay time of photon three (psec). The pulse shapes are gaussian with time constant $\sigma = 0.5$ psec, and $\Gamma_{IVR} = 0.15$ psec$^{-1}$. The frequency difference, $\Delta\omega \equiv \omega_g + \omega_1 - \omega_2 - \omega_s = 0$. The decay of the CARS signal is determined by the time constant for the IVR. (*B*) Same as *A* but $\Delta\omega = 3$ psec$^{-1}$. The overall intensity is greatly reduced, and the decay time is now much faster than the IVR time constant. (*C*) Same as *A* and *B* but $\Delta\omega = 30$ psec$^{-1}$. The overall intensity is two orders of magnitude smaller than in *B* and the CARS signal vanishes when the initial excitation is finished.

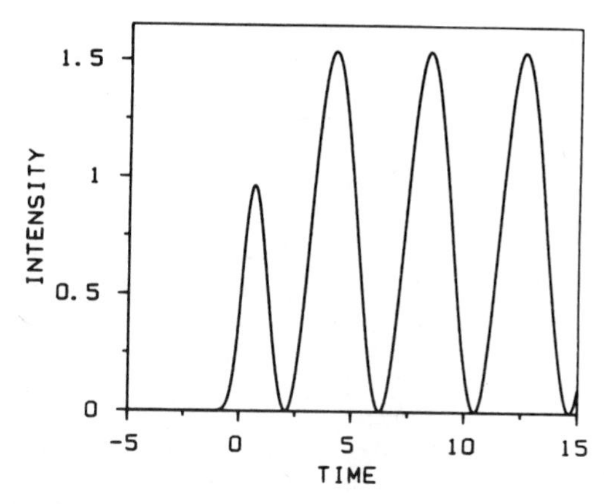


**Figure 7.** CARS intensity versus delay time. The pulse shapes are gaussian with time constant $\sigma = 0.5$ psec. IVR is that of a two-level system ($N = 2$) resulting in quantum beats with complete and regular recurrences. The coupling parameter, $v = 0.75\hbar$ psec$^{-1}$.

where $\Delta\omega \equiv \omega_g + \omega_1 - \omega_2 - \omega_s$ represents the detuning of the difference frequency from the vibrational transition.

Figure 6 displays a plot of $P'_\rho(t_4)$ versus $t_4$ when $\Gamma_{IVR} = 0.15$ psec$^{-1}$ and $\Delta\omega = 0$, 3, and 30 psec$^{-1}$. The CARS signal is observed to decay much more quickly as the detuning is increased, and the overall intensity of the nonresonant signal is much lower than the resonant signal. These observations are in agreement with the results of Zinth et al.[15] who demonstrate the effects mentioned both experimentally and numerically. Kamga and Sceats[16] used the effect gainfully to eliminate nonresonant background from the CARS signal from two component liquids.

In Figs. 7–9 we again consider $P'_\rho(t_4)$ as a function of $t_4$. We specialize to the case $\Delta\omega = 0$, and we now consider a different model for the IVR on the ground-state surface. The zeroth-order level, $s$, is coupled to its "nearest-neighbor" zeroth-order level only. This second zeroth-order level is in turn coupled to its nearest-neighbor level, and so on. The energy of each of the zeroth-order levels is taken to be the same, as is the coupling between levels. This is a crude model of a long-chain molecule or of a set of normal modes that are coupled in sequence. The number of coupled levels is varied from 2 to 5 to 10, simulating small, intermediate, and large molecules. Mathematically, the coupling matrix is tridiagonal, with all diagonal elements equal and all near-diagonal elements equal. The eigenvectors and eigenvalues for this matrix can be evaluated analytically; furthermore, the form of the decay for $N$ coupled levels is an $N$-point quadrature approximation to another analytic function, which represents the limiting decay for an infinite number

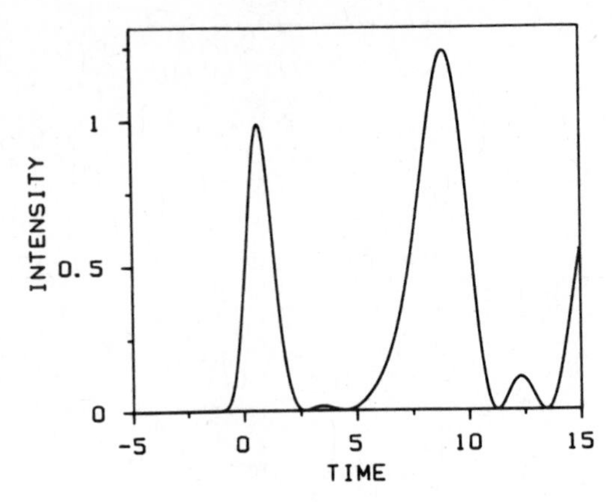


**Figure 8.** Same as Fig. 7 but for $N = 5$. The quantum beats are still strong but irregular.

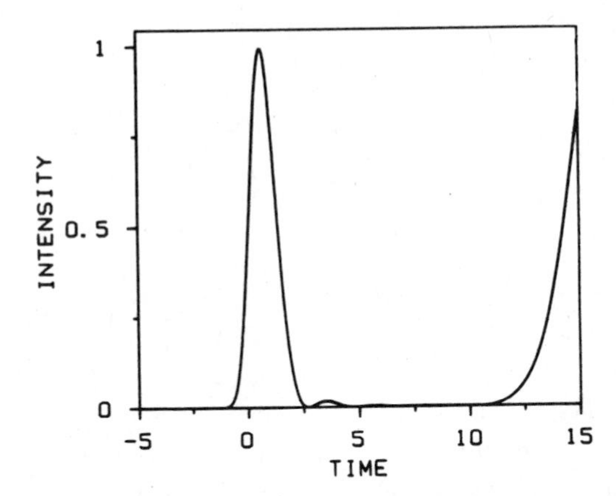


**Figure 9.** Same as Fig. 7 but for $N = 10$. Note the weak, early recurrences as well as the (truncated) strong, later recurrence.

of coupled levels. Interestingly, this limiting decay shows early time recurrences in the CARS signal despite irreversible long-time decay. Details of the calculation can be found in Ref. 10.

Figure 7 shows the time dependence of $P'_\rho(t_4)$ for the two-level case. Note the complete recurrences at times $n\tau = 2\pi n/\omega$. It is strange but true that the first feature ($t = 0$) has less amplitude than the others, because it represents the convolution of the waveform $a(t)$ with the one-sided function $C(t) \equiv \langle s|e^{-iH_g t}|s\rangle$ for $t > 0$. Figure 8 shows the five-level case. The initial decay is very similar to that in Fig. 7, but the recurrences are on a longer time scale and more irregular in the five-level case. Figure 9 shows the 10-level case. The

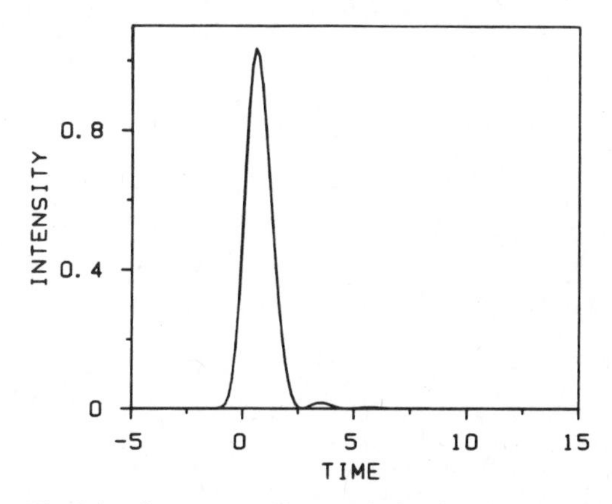


**Figure 10.** Same as Fig. 7 but for $N = \infty$. The correlation function $C(t)$ was determined analytically as the limit of an $N$-point quadrature approximation as $N \to \infty$, and is given by $(\hbar/vt)J_1(2vt/\hbar)$ (see text). Note that the weak, early recurrences persist while the strong, late recurrences are absent.

strong recurrences are, relative to those in Figs. 7 and 8, pushed farther out in time. There is, however, a weak sequence of recurrences at short times ($t = 3.5$, 7 psec), which persist in position and intensity as $N$ increases further.

As one might expect, $C(t)$ approaches a limiting form as $N$ goes to infinity. This form is

$$C(t) = \frac{\hbar}{vt} J_1\left(\frac{2vt}{\hbar}\right), \tag{3.16}$$

where $v$ is the matrix element coupling $s$ to its neighbors. Figure 10 shows the function $|P'_\rho(t_4)|^2$ versus $t_4$ corresponding to (3.16). Note that short-time recurrences persist while long-time strong recurrences do not.

Figures 11 and 12 show

$$\left| \int_{-\infty}^{t_4} \int_{-\infty}^{t_2} dt_1\, dt_2 \langle g | e^{-iH_g(t_4-t_2)} b(t_2) e^{-iH_{ex}(t_2-t_1)} a(t_1) e^{-i\omega_g t_1} | g \rangle \right|^2 \tag{3.17}$$

as a function of $t_4$. In both figures the overlap function is taken to have the functional form $C(t) = J_1(2\Gamma t)/\Gamma t$ on the excited- *and* the ground-state potential surfaces, as previously motivated. $\Gamma$ is taken to be 0.75 psec$^{-1}$ on the ground state surface in both plots, while it is 0.7 psec$^{-1}$ on the excited surface for Fig. 11 and 0.07 psec$^{-1}$ on the excited state surface for Fig. 12. Comparison of Figs. 11 and 12 yields the interesting result that the decay of the CARS

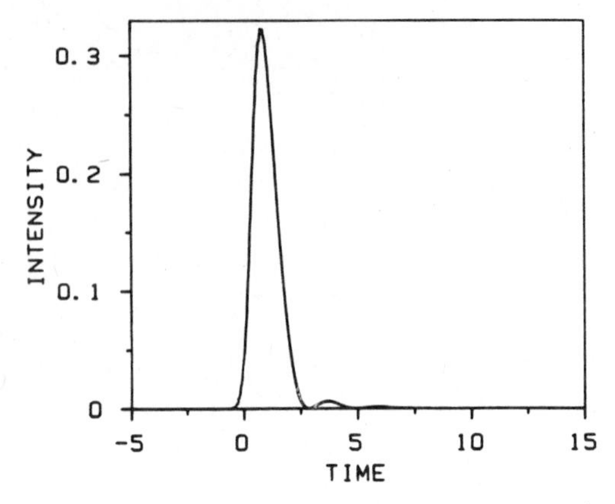


**Figure 11.** Resonant CARS intensity versus delay time: $\Gamma_{ex} = 0.7$ psec$^{-1}$, $\Gamma_g = 0.75$ psec$^{-1}$.

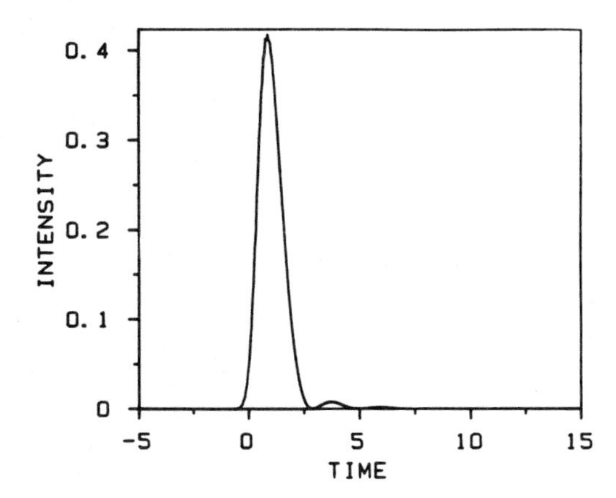


**Figure 12.** Resonant CARS intensity versus delay time: $\Gamma_{ex} = 0.07$ psec$^{-1}$, $\Gamma_g = 0.75$ psec$^{-1}$. Comparison with Fig. 11 shows that the resonant CARS signal is insensitive to the rate of IVR on the excited-state potential surface.

signal is insensitive to the rate of decay of IVR in the excited electronic state. This conclusion holds whether the IVR is much faster, comparable to, or much slower than the time envelope of the waveforms $a$ and $b$. Indeed, the only assumption required for this conclusion to be valid is that the centers of the waveforms $a$ and $b$ be coincident in time, for the following reasons. The convolution

$$\tilde{\gamma} \equiv \int_{-\infty}^{t_2} \langle g| e^{-iH_{ex}(t_2-t_1)/\hbar} a(t_1) |g\rangle \, dt_1 \tag{3.18}$$

is very sensitive to the IVR rate on the excited-state potential surface: If $\Gamma$ is very big, $\tilde{\gamma}$ will be very sharply peaked in time; while if $\Gamma$ is very small, $\tilde{\gamma}$ will be very broad in time. However, the convolution will have a minimum width equal to the width of $a(t_1)$. Since $b$ has the same width as $a$, and its center is coincident with the center of $a$, the product of $b$ with the convolution is just the width of $b$. It is interesting to note that one may insert an adjustable delay between photons one and two, now taking care that photon two is coincident with photon three. Then the IVR of a vibrational level in the excited electronic surface is probed with no complications arising from ground-electronic-state IVR. This small change in experimental setup allows, in principle, a ready comparison of IVR rates of the corresponding vibrational levels in the excited and the ground electronic states.

It is important to note that the preceding arguments do not apply to photons three and four. When photon three is resonant the emission of photon four may be significantly delayed, because photon four is emitted spontaneously. Hence, the quasicoincidence of the waveforms of photons three and four, which was assumed, is no longer guaranteed. Excited-state IVR may now contribute to the decay of the CARS signal, and no simple interpretation in terms of ground-state IVR will be possible.

## IV. PHOTON ECHOES IN MULTILEVEL SYSTEMS

### A. Motivation

We now examine, as the next step in understanding how to control the time evolution of an excitation in an isolated molecule, the photon echo phenomenon in multilevel systems. The key question to be addressed is how the complications introduced by the multilevel character of a system are manifest in the coherent dynamics that determine the photon echo.

The existence of a photon echo in a two-level optical system was predicted by analogy with the spin echo in a two-level magnetic system; it was first observed in 1966 by Abella et al.[17] The key ideas in generating this analogy are as follows. Consider an ensemble of two-level systems. In general, the members of the ensemble are in somewhat different environmental states, for example, at different sites of a lattice or moving with different velocities in a gas. Consequently, there is a distribution of splittings of the two levels. If the ensemble of systems just described is subjected to a strong electromagnetic field pulse, a macroscopic polarization is generated. Ordinarily, this macroscopic polarization decays with a time constant in the nanosecond range; the photon echo refers to the recurrence of the macroscopic polarization many decay lifetimes later. The decay of the macroscopic polarization arises from the dephasing of the initially prepared coherent superposition of the two levels in the ensemble, because the inhomogeneous distribution of level splittings leads to a distribution in the phases of the systems that broadens as time increases. Nevertheless, each system, and the ensemble, evolve in a deterministic fashion, so application of a series of field pulses at later times can remove the phase distribution in the ensemble of systems, recreating momentarily the initial macroscopic polarization, which in turn generates a burst of light.[18,19]

Warren and Zewail[20] have discussed the generalization of coherent optical effects, including photon echos, to multilevel systems. Their work is derived by analogy with the nuclear magnetic resonance (NMR) literature, and is formulated in terms of a multidimensional density matrix. The system studied consists of many coupled two-level systems. The coupled two-level systems can be described as a single multilevel system, capable of one- and multi-quantum transitions. Relative to the systems we will be dealing with, which have the irregular spacings associated with the vibrational or rotational manifold of an isolated molecule, the level structures of their systems display considerable symmetry.

In the weak-field regime the Born–Oppenheimer representation is valid (neglecting nonadiabatic effects) both while the light is on and when it is off. That being so, the complete spectrum of possibilities from continuous wave to pump-and-probe experiments can be described in a consistent dynamical wavepacket representation [Eq. (2.7)]. The formation of the amplitude on the excited-electronic-state surface is visualized as the result of interference between wavepackets continuously sent up from the ground-state surface. Each packet, after formation on the excited-state surface, propagates, so the interference is between all of the wavepackets displaced to various points on the surface, including the one at the point of introduction defined by the Franck–Condon principle. Clearly, as the duration of the light pulse increases and becomes infinite, the range from rapidly evolving nonstationary to stationary amplitudes can be produced on the excited-state surface, but the amplitudes so produced are always significantly smaller than that of the initial vibrational state on the ground-state surface.

In contrast, in the strong-field regime the Born–Oppenheimer approximation breaks down while the light is on, that is, the field is sufficiently strong that the isolated molecule Hamiltonian, for which the Born–Oppenheimer surfaces are approximate solutions, has no meaning. We note that if such a strong field remains on for a time on the order of a vibrational period, the dynamical processes that occur, including multiphoton absorption, are very complicated. We shall assume (although the situation is not completely realizable) that the strong-field pulse is "instantaneous" with respect to the time scale defined by the spread in Franck–Condon factors of the optical transition. This assumption implies that the Born–Oppenheimer surfaces are destroyed for the momentary duration of the light pulse; when the pulse is off, the Born–Oppenheimer surfaces are again good representations of the solutions of the isolated molecule Hamiltonian. In this limit a significant fraction of the nuclear vibrational amplitude is exchanged between the Born–Oppenheimer surfaces, on which surfaces said amplitudes evolve in time. There is, then, a clear analogy between this behavior in the strong-field regime and that of the two isolated levels in the traditional photon echo problem: the resonant photon mixes and thereby momentarily destroys the integrity of the two original levels of the isolated system.

### B. Wavefunction Representation of the Photon Echo in a Two-Level System

We consider first, for the purpose of establishing the language we use, the case of a photon echo in a two-level system.[18,19,21] When the field is on, the Hamiltonian takes the form

$$H_1 = \begin{pmatrix} E_a & 0 \\ 0 & E_b \end{pmatrix} - \mu E \begin{pmatrix} 0 & \cos(\omega t) \\ \cos(\omega t) & 0 \end{pmatrix}. \tag{4.1}$$

We seek a solution of the time-dependent Schrödinger equation for $H = H_1$,

$$i\hbar \frac{\partial \Psi}{\partial t} = H_1(t)\Psi. \tag{4.2}$$

It is convenient to transform into a rotating frame via the time-dependent unitary matrix

$$U = \begin{pmatrix} e^{i\omega t/2} & 0 \\ 0 & e^{-i\omega t/2} \end{pmatrix}, \tag{4.3}$$

$$\frac{\partial U}{\partial t} = \begin{pmatrix} i\frac{\omega}{2} e^{i\omega t/2} & 0 \\ 0 & -\frac{i\omega}{2} e^{-i\omega t/2} \end{pmatrix}, \tag{4.4}$$

$$\Psi' = U^{-1}\Psi. \tag{4.5}$$

Given $U$, the Schrödinger equation can be transformed by the operations

$$i\hbar U^{-1} \frac{\partial (UU^{-1})\Psi}{\partial t} = U^{-1}H(UU^{-1})\Psi, \tag{4.6}$$

which can be written

$$i\hbar U^{-1}\frac{\partial U\Psi^{-1}}{\partial t} = U^{-1}HU\Psi'. \tag{4.7}$$

After a little regrouping one obtains

$$i\hbar\frac{\partial\Psi'}{\partial t} = \begin{pmatrix} E_a & 0 \\ 0 & E_b \end{pmatrix} - \mu E\begin{pmatrix} 0 & \cos\omega t e^{i\omega t} \\ \cos\omega t e^{i\omega t} & 0 \end{pmatrix} + \begin{pmatrix} \frac{\hbar\omega}{2} & 0 \\ 0 & -\frac{\hbar\omega}{2} \end{pmatrix}\Psi'. \tag{4.8}$$

In the rotating wave approximation terms of the form $e^{2i\omega t}$ and $e^{-2i\omega t}$ are neglected with respect to unity, whereupon

$$i\hbar\frac{\partial\Psi'}{\partial t} = \begin{pmatrix} E_a & 0 \\ 0 & E_b \end{pmatrix} - (\mu E/2)\begin{pmatrix} 0 & 1 \\ 1 & 0 \end{pmatrix} + \begin{pmatrix} \frac{\hbar\omega}{2} & 0 \\ 0 & -\frac{\hbar\omega}{2} \end{pmatrix}\Psi' \tag{4.9}$$

$$= \begin{pmatrix} \bar{E} & 0 \\ 0 & \bar{E} \end{pmatrix} + \begin{pmatrix} \frac{\hbar\Delta\omega}{2} & 0 \\ 0 & -\frac{\hbar\Delta\omega}{2} \end{pmatrix} - (\mu E/2)\begin{pmatrix} 0 & 1 \\ 1 & 0 \end{pmatrix}\Psi' = H'\Psi',$$

with $\bar{E} = \frac{1}{2}(E_a + E_b)$ and $\Delta\omega = [(E_b - E_a)/\hbar] - \omega$. The field mixed states are simple when $\Delta\omega = 0$:

$$\Psi_1 = 2^{-1/2}[\psi_a + \psi_b], \tag{4.10}$$

$$\Psi_2 = 2^{-1/2}[\psi_a - \psi_b], \tag{4.11}$$

with eigenvalues

$$E_1 = \bar{E} + \frac{\mu E}{2}, \tag{4.12}$$

$$E_2 = \bar{E} - \frac{\mu E}{2}. \tag{4.13}$$

We are interested in the case where $\Delta\omega$ is small with respect to $\mu E$. In that case it is sufficiently accurate to adopt these simple linear combinations for all $\Delta\omega$.

Immediately before the exciting pulse we can represent the state of the system in the form

$$\Psi'(0) = \psi_a = 2^{-1/2}[\psi_1 + \psi_2]. \tag{4.14}$$

The system now evolves under the influence of the pulse until $\mu E t_1 = \pi/2$. For obvious reasons, this pulse is called a $\pi/2$ pulse. Immediately after the $\pi/2$ pulse the system wavefunction is given by

$$\begin{aligned}\Psi'(t_1) &= 2^{-1/2}[\psi_1 e^{i\pi/4} + \psi_2 e^{-i\pi/4}]e^{-i\bar{E}t_1/\hbar} \\ &= 2^{-1/2}[\psi_a + i\psi_b]e^{-i\bar{E}t_1/\hbar}.\end{aligned} \tag{4.15}$$

With the field off the system evolves freely until time $t_2$, at which time the wavefunction has the form

$$\Psi'(t_2) = 2^{-1/2}[\psi_a e^{i\Delta\omega(t_2-t_1)/2} + i\psi_b e^{-i\Delta\omega(t_2-t_1)/2}]e^{-i\bar{E}t_2/\hbar}. \tag{4.16}$$

An intense pulse is now applied for a period of time twice as long as the first pulse, $\mu E t = \pi$ ($\pi$-pulse). To see its effect on the wavefunction we rewrite Eq. (4.14) as

$$\begin{aligned}\Psi'(t_2) = 2^{-1/2}\bigg(&\frac{e^{i\Delta\omega(t_2-t_1)/2} + ie^{-i\Delta\omega(t_2-t_1)/2}}{2^{1/2}}\psi_1 \\ &+ \frac{e^{i\Delta\omega(t_2-t_1)/2} - ie^{-i\Delta\omega(t_2-t_1)/2}}{2^{1/2}}\psi_2\bigg)e^{-i\bar{E}t_2/\hbar}.\end{aligned} \tag{4.17}$$

After the pulse

$$\psi_1 \to \psi_1 e^{i\pi/2} = i\psi_1, \tag{4.18}$$

$$\psi_2 \to \psi_2 e^{-i\pi/2} = -i\psi_2, \tag{4.19}$$

which, after substitution into (4.17) and regrouping, leads to

$$\Psi'(t_3) = 2^{-1/2}(i\psi_b e^{i\Delta\omega(t_2-t_1)/2} - \psi_a e^{-i\Delta\omega(t_2-t_1)/2})e^{-i\bar{E}t_3/\hbar}. \tag{4.20}$$

We note that after the $\pi$ pulse the phase that was previously associated with $\psi_a$ is now associated with $\psi_b$, and vice versa with the field off. The system again evolves freely until time $t$, whereupon

$$\Psi'(t) = 2^{-1/2}(\psi_b i e^{i\Delta\omega(t_2-t_1)/2} e^{-i\Delta\omega(t-t_3)/2} - \psi_a e^{-i\Delta\omega(t_2-t_1)/2} e^{i\Delta\omega(t-t_3)/2}) e^{-i\bar{E}t/\hbar}. \tag{4.21}$$

Transforming the representation back to the nonrotating frame leads to

$$\begin{aligned} \Psi &= U\Psi' \\ &= 2(\psi_b i e^{-i\Delta\omega/2-[(t-t_3)-(t_2-t_1)]} e^{-i\omega t/2} \\ &\quad - \psi_a e^{(i\Delta\omega/2)[(t-t_3)-(t_2-t_1)]} e^{i\omega t/2}) e^{-i\bar{E}t/\hbar}. \end{aligned} \tag{4.22}$$

We now define the transition moment, as usual, by

$$\mu_{ab} \equiv \int d^3 r \psi_a^*(er)\psi_b, \tag{4.23}$$

and assume that $\mu_{aa} = \mu_{bb} = 0$. Then the polarization is given by

$$\begin{aligned} P' &\equiv \iint d^3 r d^3 R \Psi^*(er)\Psi \\ &= \mu_{ab} \sin(\Delta\omega[(t-t_3)-(t_2-t_1)] - \omega t) \\ &= \mu_{ab}\{\sin(\Delta\omega[(t-t_3)-(t_2-t_1)])\cos(\omega t) \\ &\quad - \cos(\Delta\omega[(t-t_3)-(t_2-t_1)])\sin(\omega t)\}. \end{aligned} \tag{4.24}$$

The echo amplitude is proportional to the coefficient of $\sin(\omega t)$ in (24):[2]

$$P = \mu_{ab} \cos(\Delta\omega[t_2 - t_1 + t_3 - t]). \tag{4.25}$$

The cosine dependence of the echo is a familiar result; for $t - t_3 = t_2 - t_1$ the echo has a maximum. In most discussions of photon echoes it is the inhomogeneous distribution of systems that gives rise to the echo. For a gaussian distribution of energy spacings $\Delta\omega$,

$$W(\Delta\omega) = \frac{1}{2^{1/2}\pi\sigma^2} e^{-\Delta\omega^2/2\sigma^2}, \tag{4.26}$$

we have

$$\int d(\Delta\omega) W(\Delta\omega) P(\Delta\omega) = \mu_{ab} e^{-[(t-t_3)-(t_2-t_1)]^2\sigma^2/2}. \tag{4.27}$$

### C. The Photon Echo in a Multilevel System

A convenient analogy between the multilevel and two-level cases is generated by separating the electronic and nuclear wavefunctions of the multilevel system.[21] We define $\psi_a(r, R)$ and $\psi_b(r, R)$ as the electronic wavefunctions associated with the ground and excited Born–Oppenheimer electronic states of the system. The Hamiltonian with the light on is then given by

$$H = \begin{pmatrix} H_a(R) & 0 \\ 0 & H_b(R) \end{pmatrix} + E \begin{pmatrix} 0 & \mu_{ab}(R) \cos \omega t \\ \mu_{ba}(R) \cos \omega t & 0 \end{pmatrix}, \tag{4.28}$$

where

$$H_a(R) \equiv \int d^3r \psi_a^*(r) H(r, R) \psi_a(r), \tag{4.29}$$

$$H_b(R) \equiv \int d^3r \psi_b^*(r) H(r, R) \psi_b(r), \tag{4.30}$$

and

$$\mu_{ab}(R) = \mu_{ba}(R) = \int d^3r \psi_a^*(r, R) \mu \psi_b(r, R). \tag{4.31}$$

We assume that the molecule has no permanent dipole moment, that is, $\mu_{aa}(R) = \mu_{bb}(R) = 0$. Moreover, we assume the Condon approximation: $\mu_{ab}(R) = \mu_{ba}(R) =$ constant. Non-Condon terms would introduce interesting new effects, particularly if $\mu(R)$ has a node. However, that is beyond the scope of the current treatment. We transform the representation to a rotating frame and adopt the rotating-wave approximation. Then

$$H' = \begin{pmatrix} H_a(R) + \dfrac{\hbar\omega}{2} & 0 \\ 0 & H_b(R) - \dfrac{\hbar\omega}{2} \end{pmatrix} + \mu_{ab}(R) E \begin{pmatrix} 0 & 1 \\ 1 & 0 \end{pmatrix}. \tag{4.32}$$

When the field is off

$$H' = \begin{pmatrix} H_a(R) + \dfrac{\hbar\omega}{2} & 0 \\ 0 & H_b(R) - \dfrac{\hbar\omega}{2} \end{pmatrix}. \tag{4.33}$$

Let

$$E_a \equiv \int d^3R\chi(0)^* H_a(R)\chi(0), \tag{4.34}$$

$$E_b \equiv \int d^3R\chi(0)^* H_b(R)\chi(0), \tag{4.35}$$

where $\chi(0)$ is the initial (ground) vibrational state of the ground Born–Oppenheimer surface. Then we rewrite Eq. (4.33) as

$$H' = \begin{pmatrix} \bar{E} & 0 \\ 0 & \bar{E} \end{pmatrix} + \begin{pmatrix} \dfrac{\hbar\Delta\omega}{2} & 0 \\ 0 & -\dfrac{\hbar\Delta\omega}{2} \end{pmatrix} + \begin{pmatrix} \bar{H}_a(R) & 0 \\ 0 & \bar{H}_b(R) \end{pmatrix}, \tag{4.36}$$

where $\bar{E} \equiv \frac{1}{2}(E_a + E_b)$ and $\Delta\omega = [(E_b - E_a)/\hbar] - \omega$. $\bar{H}_a$ and $\bar{H}_b$ are defined by Eq. (4.36). $\bar{E}_a$ is the zero-point energy on surface $a$, while $\bar{E}_b$ is approximately the vertical displacement of surface $b$ relative to surface $a$ in the Franck–Condon region. $\bar{H}_a$ and $\bar{H}_b$ are the ground- and excited-state Born–Oppenheimer Hamiltonians with the average energies subtracted. These Hamiltonians play a crucial role in defining the free induction decay and photon echo in a multilevel system. Indeed, they play a role analogous to $\Delta\omega$ for the ensemble of two-level systems. Their "size" is related to the spread in energy of the sizable Franck–Condon factors for the optical transitions we will be dealing with. The time conjugate to this energy spread is assumed to be short with respect to the time between pulses. The latter assumption allows us to neglect this part of the Hamiltonian while the pulses are on. This argument is developed more fully in the Appendix to this section. Although $\Delta\omega$ formally enters into the multilevel case, henceforth we specialize to $\Delta\omega = 0$ for ease of presentation.

We express the initial wavefunction as

$$\Psi'(0) = \Psi(0) = \psi_a\chi(0), \tag{4.37}$$

where $\chi$ is the (in general nonstationary) nuclear wavefunction. After the $\pi/2$ pulse we have

$$\begin{aligned} \Psi'(t_1) &= \frac{1}{2^{-1/2}}[\psi_a\chi_a(t_1) + i\psi_b\chi_b(t_1)]e^{-i\bar{E}t_1/\hbar} \\ &= 2^{-1/2}[\psi_a\chi_a(0) + i\psi_b\chi_b(0)]e^{-i\bar{E}t_1/\hbar} \end{aligned} \tag{4.38}$$

under the approximation that the time scale for wavepacket propagation is much longer than $t_1$, the duration of the pulse. Note that $\chi(0) = \chi_a(0) = \chi_b(0)$. With the light off, the system evolves freely and after time $t_2$ the wavefunction assumes the form

$$\Psi'(t_2) = 2^{-1/2}[\psi_a e^{-i\bar{H}t_a(t_2-t_1)/\hbar}\chi(0) + i\psi_b e^{-i\bar{H}t_b(t_2-t_1)/\hbar}\chi(0)]e^{-i\bar{E}t_2/\hbar}. \quad (4.39)$$

Since $\chi_a$ propagates on the ground-state surface, it is given in the rotating frame by

$$\chi_a(t_2) = e^{-i\bar{H}_a(t_2-t_1)/\hbar}\chi_a(0) = \chi_a(0). \quad (4.40)$$

However, $\chi_b$ propagates on the excited-state surface, for which it is not an eigenstate, but rather a wavepacket. The time-dependent form of $\chi_b$ is given semiclassically by[6]

$$\chi_b(t_2) = e^{-i\bar{H}t_b(t_2-t_1)/\hbar}\chi(0) = e^{-\alpha(x-x_{t_2})^2+(i/\hbar)p_t(x-x_{t_2})+(i/\hbar)\gamma_{t_2}}$$
$$\equiv g_b(t_2 - t_1)e^{(i/\hbar)\gamma_{t_2}}. \quad (4.41)$$

Here $x_t, p_t$ are the classical position and momentum, respectively, which satisfy Hamilton's equations of motion

$$\dot{p}_t = -\frac{\partial H_b}{\partial x_t}, \quad (4.42)$$

$$\dot{x}_t = \frac{\partial H_b}{\partial p_t}, \quad (4.43)$$

subject to the initial conditions $p_{t_1} = 0, x_{t_1} = 0$. $x = 0$ is defined as the position of the ground-state Born–Oppenheimer surface minimum. Moreover, $\gamma_t$ is the coordinate independent phase

$$\gamma_t = \int_0^t p_t \cdot \dot{q}_t \, dt - (E_{b,cl} + E_{0b} - E_b)t \cong S_b(t), \quad (4.44)$$

where $E_{b,cl}$ is the classical energy of the wavepacket on surface $b$ and $E_{0b}$ is the zero-point energy on surface $b$. In Eq. (4.44) we have used the definition of the classical action integral, $S_b$, and we have made the reasonable assertion that

$$E_{b,cl} + E_{0b} - E_b = 0. \quad (4.45)$$

The $\pi$ pulse begins at time $t_2$; its effect, analogous to that in the two-level system, is to exchange the amplitude between surfaces $a$ and $b$:

$$\Psi'(t_3) = 2^{-1/2}[-\psi_a \chi_a(t_3) + i\psi_b \chi_b(t_3)]e^{-i\bar{E}t_3/\hbar} \quad (4.46)$$
$$= 2^{-1/2}[-\psi_a e^{-i\bar{H}_b(t_2-t_1)/\hbar}\chi(0) + i\psi_b e^{-i\bar{H}_a(t_2-t_1)/\hbar}\chi(0)]e^{-i\bar{E}t_3/\hbar}.$$

At this stage of the evolution $\chi_a(t_3) = \chi_b(t_2)$ becomes a wavepacket on the *ground*-state surface and is given semiclassically, at later times, by

$$\chi_a(t) = e^{-\alpha(x-x_t)^2+(i/\hbar)p_t+(i/\hbar)\gamma_t} g_a(t - t_3)e^{(i/\hbar)\gamma_t}, \quad (4.47)$$

with initial conditions on the classical equations of motion

$$x_{t_3} = x_{t_2},$$
$$p_{t_3} = p_{t_2}, \quad (4.48)$$
$$\gamma_{t_3} = \gamma_{t_2}.$$

Similarly,

$$\chi_b(t_3) = \chi_a(t_2) = \chi(0) \quad (4.49)$$

becomes a wavepacket on the excited-state surface and is given semiclassically as before. The system now propagates freely; after a total time $t$ the wavefunction is given by

$$\Psi'(t) = 2^{-1/2}[-\psi_a \chi_a(t) + i\psi_b \chi_b(t)]e^{-i\bar{E}t/\hbar}, \quad (4.50)$$

where

$$\chi_a(t) = e^{-i\bar{H}_a(t-t_3)/\hbar}e^{-i\bar{H}_b(t_2-t_1)/\hbar}\chi(0), \quad (4.51)$$
$$\chi_b(t) = e^{-i\bar{H}_b(t-t_3)/\hbar}e^{-i\bar{H}_a(t_2-t_1)/\hbar}\chi(0). \quad (4.52)$$

Note the resemblance of Eqs. (4.51) and (4.52) to Eqs. (4.21) for the two-level system. Using the notation $g_a$, $g_b$ we may write the wavefunction in the form

$$\Psi'(t) = 2^{-1/2}[-\psi_a g_a(t - t_3; t_2 - t_1)e^{(i/\hbar)S_b(t_2-t_1)}e^{(i/\hbar)[S_a(t-t_3)-E_{a,cl}(t-t_3)]}$$
$$+ i\psi_b g_b(t - t_3)e^{(i/\hbar)S_b(t-t_3)}]e^{-i\bar{E}t/\hbar}, \quad (4.53)$$

where $E_{a,cl}$ is the classical energy of the wavepacket when it makes the vertical transition down from surface $b$ to surface $a$. In the nonrotating frame

$$\Psi(t) = 2^{-1/2}[-\psi_a g_b(t - t_3; t_2 - t_1)e^{(i/\hbar)S_b(t_2-t_1)} \times e^{(i/\hbar)[S_a(t-t_3)-E_{a,cl}(t-t_3)]}e^{i\omega t/2} + i\psi_b g_a(t - t_3)e^{(i/\hbar)S_b(t-t_3)}e^{(i/\hbar)\omega t/2}]e^{-i\bar{E}t}. \quad (4.54)$$

The inclusion of only two Born–Oppenheimer states in our derivation is warranted if the one-photon absorption process is resonant with the electronic transition frequency. (Two-photon and multiphoton absorption processes are assumed to be nonresonant.) However, at the same time the pulse must be broad enough in frequency to include the spread of Franck–Condon-allowed vibrational levels.

We display in Fig. 13 a schematic representation of the time evolution of the coupled pulsed field multilevel system. The $\pi/2$ pulse prepares half the

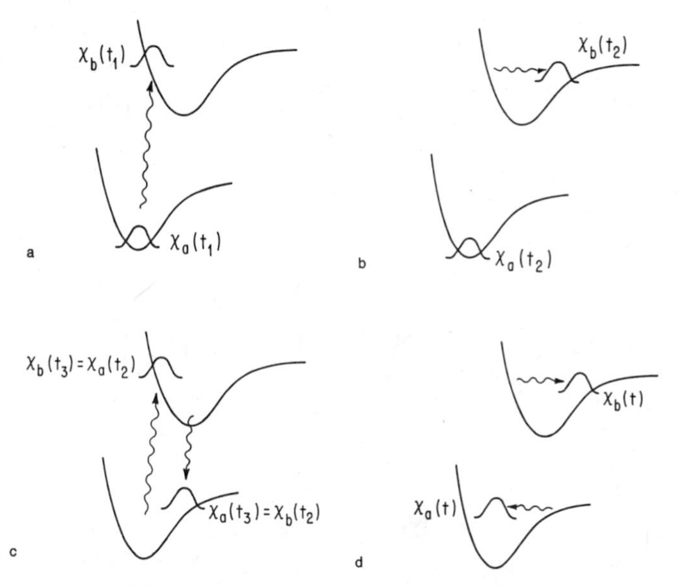


**Figure 13.** The (*a*) ground and (*b*) excited-electronic-state Born–Oppenheimer potential surfaces. The $\pi/2$ pulse moves half of the initial amplitude, $\chi(0)$, from a surface $a$ to surface $b$. After the pulse the nuclear wavefunctions of the ground-state and excited-state surfaces are denoted $\chi_a(t_1)$ and $\chi_b(t_1)$. (*b*) Wavepacket evolution of $\chi_b$. Motion of the wavepacket causes the overlap of the ground-state wavefunction and the excited-state wavefunction to decay, resulting in "free induction decay." $\chi_a$ remains in place in coordinate space. (*c*) The $\pi$ pulse exchanges the amplitude of surface $a$ with surface $b$. (*d*) Wavepacket evolution proceeds on both surface $a$ and surface $b$. When the two wavepackets overlap at some later time, a photon echo results.

excitation amplitude on one surface and half on the other, whereupon the nuclear wavefunction evolves on both surfaces. Starting at time $t_1$, and until time $t_2$, the amplitude on surface $b$, $\chi_b(t_2)$, moves away from the Franck–Condon region, as shown in Fig. 13$b$. In contrast, the amplitude on surface $a$, $\chi_a$, remains in place in coordinate space, while developing a simple phase factor. The $\pi$ pulse exchanges the amplitude of surface $a$ with that of surface $b$, which is the multilevel system analog of exchanging phases in the two-level system. Note that this exchange of amplitude is consistent with the properties of Franck–Condon and strong field transitions. Clearly, for $t_3 > t_2$ the new ground-state wavepacket, $\chi_a(t_3)$, evolves on the ground-state surface, moving away from where it came down.

Taking the final formula for $\Psi$ and substituting into the (generalized) expression for the polarization we obtain

$$P' = \int d^3r d^3R \Psi^* \hat{\mu} \Psi \tag{4.55}$$

$$= \int [e^{-iH_bt''/\hbar} e^{-iH_at'/\hbar} \psi_b \chi(0)]^* \hat{\mu} [e^{-iH_at''/\hbar} e^{-iH_bt'/\hbar} \psi_a \chi(0)] e^{i\omega t} d^3r d^3R + \text{C.C.}$$

$$= \Sigma(t''; t') e^{(i/\hbar)[S_b(t'') - S_b(t')]} e^{(i/\hbar)[S_a(t'') - E_{a,cl}t'']} e^{i\omega t} + \text{C.C.},$$

where $\Sigma(t''; t')$ is defined as

$$\Sigma(t''; t') = \int d^3R g_a^*(R, t''; t') \mu(R) g_b(R, t''; 0). \tag{4.56}$$

$\Sigma$ is the dynamical overlap of the wavepacket on the ground state with that on the excited state; it is a function of $t' \equiv t_2 - t_1$ and $t'' \equiv t - t_3$. It is not surprising that "two-surface" wavepacket overlap at time $t$ is a requirement for an echo at this time. In general, the wavepacket propagation on the ground-state surface reduces or destroys the echo. However, there are two circumstances for which the echo is strong. First, the transition to the ground-state surface may take place with zero momentum from a point directly above a ground-state minimum. The mostly likely example is the case for which the $\pi$ pulse generates the vertical transition down when the wavepacket returns to the Franck–Condon geometry. Second, the period for a wavepacket recurrence on the ground-state surface ($\tau = t''$) may be commensurate with the propagation time on the excited-state surface ($t'$). There may be one or several choices for $t'$ that satisfy this relationship. However, even if the wavepacket overlap is large, the echo amplitude can be severly diminished because of the additional phase factor in Eq. (4.54). If we take $t'' = t'$, this additional phase factor simplifies to

$$S_a(t'') - E_{a,cl}t'', \tag{4.57}$$

which vanishes (1) when $S$ and $E$ are independently zero (the case where the wavepacket makes the vertical transition down from the Franck–Condon geometry) or (2) when $t'' = \tau$ (the ground-state surface multidimensional period for a recurrence) if the ground-state surface is harmonic.

One of the new features of the photon echo phenomenon in the multilevel case is that not all initial pulse delays, $t' \equiv t_2 - t_1$, are equivalent. If $t'$ is chosen such that $g(t')$ has returned to the Franck–Condon region, for instance, the echo will be more intense than otherwise. Furthermore, for other values of $t'$ the most complete echo may not be at $t'' = t'$, but will depend on choosing $t'' \equiv t_4 - t_3$ such that $\Sigma(t''; t')$ and the extra phase factor are maximized.

## D. The Role of Rotations

So far we have concentrated attention on the vibrational manifold, without considering the rotational structure within this manifold. It should be noted at the outset that the rotational selection rules, $\Delta J = 0, \pm 1$, valid for one-photon processes, completely break down for multiphoton processes, although the selection rules $\Delta M = 0$, $\Delta K = 0$ for a parallel transition remain valid. Thus, a wide range of $J$ states associated with a particular vibrational state may become populated. Felker et al.[22] have recently reported the observation of rotational coherence in large molecules. They observe a coherent superposition of precisely three $J$ states, arising from $\Delta J = 0, \pm 1$ in a one-photon process. The multiphoton process prepares a similar coherent population of $J$ states, capable of exhibiting quantum interference phenomena, but many more $J$ levels may be involved.

During the $\pi/2$ pulse a coherent population in $J$ is prepared, associated with both the ground- and the excited-electronic-state vibrational wavepacket. Hence, a coherent ro-vibrational wavepacket is formed. In analogy with the case where non-Condon effects play a significant role in the vibrational portion of the wavefunction, the rotational wavepackets in general are not identical on the two surfaces. Nevertheless, the two rotational wavepackets should be sufficiently similar to be strongly coupled.

The rotational distribution does not, however, significantly qualify our conclusions as to the rephasing of vibrational states. The free induction decay and rephasing of the wavepacket due to rotations is on a much longer time scale than for the vibrations. Nevertheless, the rotational wavepacket has interesting properties of its own and it is instructive to consider a two-level vibrational problem, each vibration having many associated rotational states. Denoting the rotational superposition state on the ground electronic state

$P_1(\theta, t)$ and on the excited electronic state $P_2(\theta, t)$, the polarization $P$ will be proportional to

$$\langle P_1(\theta, t)|\cos\theta|P_2(\theta, t)\rangle. \tag{4.58}$$

It would be very interesting to explore the accessible forms of $P_1(\theta, t)$ and $P_2(\theta, t)$. For instance, how many $J$ states may be populated? How much anisotropy is there in the $\theta$ distribution? In spite of limited knowledge about $P_1$ and $P_2$, we may still proceed with our analysis:

$$\langle P_1(\theta, t)|\cos\theta|P_2(\theta, t)\rangle \tag{4.59}$$
$$= \sum_{J,J'} [\langle P_1(\theta, 0)|J', K, M\rangle e^{i2\pi(J'+1)J'Bt} \langle J'KM|\cos\theta|JKM\rangle$$
$$\times \langle JKM|P_2(\theta, 0)\rangle e^{i2\pi(J+1)JB} e^{-i2\Pi\nu_{ev}t}$$

$$= \sum_J \Bigg( \langle P_1(\theta, 0)|J+1, KM\rangle e^{i4\pi(J+1)Bt} \frac{(J-M+1)}{(2J+1)} \frac{(J+M+1)}{(2J+3)} \tag{4.60}$$
$$+ \langle P_1(\theta, 0)|J-1, KM\rangle e^{-i4\pi JBt} \frac{(J-M)}{(2J-1)} \frac{(J+M)}{(2J+3)} \Bigg) e^{i\phi} \langle JKM|P_2(\theta, 0)\rangle,$$

where

$$\phi = -2\pi\nu_{ev}t/\hbar \tag{4.61}$$

and $\nu_{ev}$ is the vibronic energy spacing. In writing Eq. (4.59) we have used the fact that the symmetric top parameters $A$ and $B$ are identical in the ground and the excited electronic state, and that $\Delta K = 0$ in the transition. It is apparent from Eq. (4.60) that regardless of the coefficients in the sum there will be complete recurrences of the sum at times $J = n/2B$, since $J$ is an integer. These recurrences give rise to coherent emission at the recurrence times.

In real systems the symmetric top parameters $A$ and $B$ may differ in the ground and excited states. Furthermore, when Coriolis coupling is taken into account, $K$ is no longer a good quantum number and the $J$ level spacing is not completely regular. These effects will tend to reduce, but not totally eliminate, recurrences of coherent transients and echoes.

### E. Further Comments

Many of the previously described conclusions were anticipated by Jortner and Kommandeur,[23] using a different formalism. Jortner and Kommandeur identify the initial decay of the wavepacket autocorrelation function with free induction decay and note that later recurrences of the autocorrelation function are a prerequisite for photon echoes. In contrast, we have emphasized the coordinate space wavepacket motion as a semiclassical guide to determining the recurrence times. Moreover, Jortner and Kommandeur's treatment is specialized to a single vibrational level on the ground electronic state. This eliminates are possibility of intramolecular dynamics, on the ground-electronic-state potential energy surface. The requirement for overlap between the dynamical wavepackets on the two potential energy surfaces for an echo to occur is absent from their work.

The photon echo is clearly one of a wide variety of optical coherence phenomena. All coherence phenomena are characterized by a directional signal and an intensity that is proportional to $N^2$, where $N$ is the number of emitters. In particular, we note the strong resemblance between the present formulation of the photon echo and the time-dependent formulation of CARS[16] described in the preceding section. We recall that a CARS signal is observed at a time when the convolution of the wavepacket autocorrelation function with the three incident photon fields is peaked. The same dynamical quantity—the wavepacket correlation function—enters into the description of both CARS and photon echoes. This may be understood as follows. The emission of coherent radiation arises from a macroscopic dipole moment, which requires both a microscopic transition dipole moment in a large number of emitters and phase matching of many such microscopie emitters. In two-level systems the emphasis is usually placed on the phase matching. When there are many inequivalent emitters, as in a solid, the phase matching, and hence the coherent radiation, will decay. In multilevel systems the individual microscopic transition dipole moments depend on the overlap integral between a time-dependent excited-state amplitude and a ground-state amplitude. In the case of CARS this overlap involves a third-order perturbation amplitude $\psi^{(3)}(t)$ with the ground vibrational level of the ground electronic state. For photon echoes it is the overlap of $\chi_a(t)$ with $\chi_b(t)$. The overlap may decay and show later recurrences induced solely by the intramolecular dynamics.

An additional result that emerges from our study concerns the extent to which wavepacket control is possible using coherent pulse sequences. In a two-level system one can exchange the phases of the two levels with a $\pi$ pulse and, in effect, achieve time reversal of the state of the system. In a multilevel system the extent of control is much more restricted. The center of the wavepacket evolves according to the Franck–Condon principle and Hamilton's equations of motion, which in turn are dictated by nature's potential energy surfaces. What can be controlled by the experimenter is the instant at which the wavepacket changes surfaces. This concept forms the basis for a scheme for controlling the selectivity of a reaction,[24,25] which we discuss in the next section.

For many molecules of interest there exist radiationless transitions that couple the levels of an electronically excited surface to a dense manifold of quasidegenerate levels on one or more other electronic surfaces, and these latter levels have vanishingly small transition dipole matrix elements with the initial level on the ground state surface. As shown in the preceding section, exponential decay of the amplitude of a wavepacket on an excited state surface via, say, a radiationless process, reduces the amplitude of a coherent emission signal but does not destroy the coherence.

### F. Appendix

The Hamiltonian we adopt is a $2 \times 2$ matrix of operators. It represents the ground and the excited electronic states within the Born–Oppenheimer approximation, coupled by the radiation field interacting with the dipole operators:

$$H = \begin{pmatrix} H_a & \mu E(t) \\ \mu E(t) & H_b \end{pmatrix}. \tag{A1}$$

The time-dependent Schrödinger equation reads

$$i\hbar \frac{\partial}{\partial t} \begin{pmatrix} \psi_a \\ \psi_b \end{pmatrix} = \begin{pmatrix} H_a & \mu E(t) \\ \mu E(t) & H_b \end{pmatrix} \begin{pmatrix} \psi_a \\ \psi_b \end{pmatrix}. \tag{A2}$$

At $t = 0$, $\psi_0 = \psi_a$, the ground vibrational state of $H_a$:

$$H_a \psi_0 = E_a \psi_0. \tag{A3}$$

Also, $\psi_b = 0$.

The two coupled differential equations in Eq. (A2) can be transformed to two coupled integral equations:

$$\psi_a(t) = e^{-iH_a t/\hbar} \psi_a(0) - \frac{i}{\hbar} \int_{-\infty}^{t} e^{-iH_a(t-t')/\hbar} \mu E(t') \psi_b(t')\, dt', \tag{A4a}$$

$$\psi_b(t) = -\frac{i}{\hbar} \int_{-\infty}^{t} e^{-iH_b(t-t')/\hbar} \mu E(t') \psi_a(t')\, dt'. \tag{A4b}$$

The reader may easily verify that Eqs. (A4) are formal solutions of Eq. (A2) by differentiating. We consider the strong-field limit, where we further assume that the duration of the pulses is short compared with wavepacket motion on either the ground or excited Born–Oppenheimer potential surface. This situation may be described in terms of $\pi$ pulses, familiar from the NMR and two-level optical literature. Implicitly, the strong-field case involves a multiphoton process: multiple absorption/emission processes lead to a cycling of amplitude between the two Born–Oppenheimer surfaces. It seems clear that if the pulse duration, $\tau_p$, is much shorter that the time scale for any molecular dynamics on the excited-state surface to take place, we should be able to treat the system as a two-level system. We will now prove this. We identify the time scale for wavepacket motion on the excited state surface with the decay time of the autocorrelation function of the wavepacket, $\tau_c$. This quantity, in turn, is inversely proportional to the width of the electronic absorption spectrum, $\Delta E$,

$$\tau_c = \frac{\hbar}{\Delta E} = \frac{\hbar}{[\langle\psi_a| H_b^2 |\psi_a\rangle - \langle\psi_a| H_b |\psi_a\rangle^2]^{1/2}}. \tag{A5}$$

We now proceed to derive the equivalence to a two-level system when the pulse satisfies the condition

$$\tau_p \ll \tau_c. \tag{A6}$$

Consider the probability for being in electronic state $b$ at time $t$:

$$\langle\psi_b(t)|\psi_b(t)\rangle = \frac{1}{\hbar^2}\int_{-\infty}^{t} dt' \int_{-\infty}^{t} dt'' \times \langle\psi_a(t'')|\,\mu E(t'')e^{iH_b(t-t'')/\hbar}e^{-iH_b(t-t')/\hbar}\mu E(t')|\psi_a(t')\rangle. \tag{A7}$$

For the sake of definiteness we choose a square pulse for $E(t)$

$$E(t) = A(t)\cos(\omega t), \tag{A8}$$

where

$$A(t) = \begin{cases} 1, & 0 < t < \tau_p, \\ 0, & t < 0, t > \tau_p. \end{cases} \tag{A9}$$

Substituting into Eq. (A7) we find

$$\langle\psi_b(t)|\psi_b(t)\rangle = \frac{1}{\hbar^2}\int_{0}^{\tau_p} dt' \int_{0}^{\tau_p} dt'' \times \langle\psi_a(t'')|\,\mu_{ab}e^{-iH_b(t'-t'')/\hbar}\mu_{ba}|\psi_a(t')\rangle\cos(\omega t')\cos(\omega t''). \tag{A10a}$$

In the rotating-wave approximation this becomes

$$\langle\psi_b(t)|\psi_b(t)\rangle = \frac{1}{\hbar^2}\int_0^{\tau_p} dt' \int_0^{\tau_p} dt'' \langle\psi_a(t'')\mu_{ab}e^{-iH_b(t'-t'')/\hbar}\mu_{ab}|\psi_a(t')\rangle e^{i\omega(t'-t'')}. \tag{A10b}$$

We next expand the matrix element in Eq. (A10b) in a cumulant expansion:

$$\langle\phi_a(t'')|e^{-iH_b(t'-t'')/\hbar}|\phi_a(t')\rangle = e^{\sum_{n=1}^{\infty} K_n(t'-t'')^n}\langle\phi_a(t'')|\phi_a t')\rangle, \tag{A11}$$

where

$$K_1(t'',t') = -\frac{i}{\hbar}\frac{\langle\phi_a(t'')|H_b|\phi_a(t')\rangle}{\langle\phi_a(t'')|\phi_a(t')\rangle}, \tag{A12}$$

$$K_2(t'',t') = -\frac{1}{2\hbar^2}\left[\frac{\langle\phi_a(t'')|H_b^2|\phi_a(t')\rangle}{\langle\phi_a(t'')|\phi_a(t')\rangle} - \frac{\langle\phi_a(t'')|H_b|\phi_a(t')\rangle^2}{\langle\phi_a(t'')|\phi_a(t')\rangle^2}\right],$$

and so forth, and $|\phi\rangle \equiv \mu_{ba}|\psi\rangle$. The $K_n$ are expected to be only weakly time dependent on time scales much shorter than $\tau_c$. For a gaussian absorption profile, $K_3(0,0)$ and higher $K_n(0,0)$ vanish. We will assume that these cumulants are significantly smaller than $K_1$, $K_2$. We next note that from Eq. (A5)

$$K_2(t'',t') \approx K_2(0,0) = -\frac{1}{2\tau_c^2}$$

provided $\mu_{ab}(R) = \mu_{ba}(R) = \mu$ (Condon approximation). Moreover, the largest value that $t' - t''$ can assume is $2\tau_p$. Hence, from Eq. (A6) the product

$$|K_2|(t' - t'')^2 \ll 1 \tag{A13}$$

for all $t'$, $t''$. Equation (A11) then becomes

$$\langle\phi_a(t'')|e^{-iH_b(t'-t'')/\hbar}|\phi_a(t')\rangle \approx \langle\phi_a(t'')e^{K_1(t'-t'')}|\phi_a(t')\rangle. \tag{A14}$$

Again assuming the Condon approximation, and the slow time dependence of $K_1(t'',t')$, Eq. (A12) reduces to

$$K_1(t'',t') \approx -\frac{i}{\hbar}\langle\psi_a(t'')|H_b|\psi_a(t')\rangle \approx -\frac{i}{\hbar}\langle\psi_a(0)|H_b|\psi_a(0)\rangle = E_b. \tag{A15}$$

Substituting Eqs. (A14) and (A15) into (A10b) we obtain

$$\langle\psi_b(t)|\psi_b(t)\rangle = \frac{1}{\hbar^2}\int_0^{\tau_p} dt' \int_0^{\tau_p} dt'' \langle\psi_a(t'')\mu_{ab}e^{-iE_b(t'-t'')/\hbar}\mu_{ba}|\psi_a(t')\rangle e^{-i\omega(t'-t'')}. \tag{A16}$$

Essentially the same procedure can be used to show that

$$\begin{aligned}
\langle\psi_a(t)|\psi_a(\mathrm{t})\rangle &\\
= 1 - \frac{i}{\hbar}\int_0^{\tau_p} & dt'\langle\psi_a(0)|e^{+iH_at'/\hbar}\mu_{ab}|\psi_b(t')\rangle\\
-\frac{1}{\hbar^2}\int_0^{\tau_p} & dt'\int_0^{\tau_p} dt''\langle\psi_b(t'')|\mu_{ba}e^{-iH_a(t'-t'')/\hbar}\mu_{ab}|\psi_b(t')\rangle e^{i\omega(t'-t'')}\\
= 1 - \frac{i}{\hbar}\int_0^{\tau_p} & dt'\langle\psi_a(0)|e^{-iE_at'/\hbar}\mu_{ab}|\psi_b(t')\rangle\\
-\frac{1}{\hbar^2}\int_0^{\tau_p} & dt'\int_0^{\tau_p} dt''\langle\psi_b(t'')|\mu_{ba}e^{-iE_a(t'-t'')/\hbar}\mu_{ab}|\psi_b(t')\rangle e^{i\omega(t'-t'')}.
\end{aligned} \tag{A17}$$

Hence, we have been able to show that for sufficiently short pulses the operators $H_b$, $H_a$ may be replaced by $E_b$, $E_a$, as argued in the text.

## V. COHERENT PULSE SEQUENCE CONTROL OF PRODUCT FORMATION IN CHEMICAL REACTIONS

### A. Motivation

We now examine whether or not it is possible to control the shape, duration, and temporal spacing of a coherent pulse sequence so as to control product formation in a chemical reaction. The ideas used are related to those described in the preceding sections; the method for achieving control of product formation that we propose can be thought of as a generalization of flash photolysis. As originally conceived, flash photolysis is a pump-and-probe technique for studying chemical reactions. Although the experimenter exerts some control over the system by virtue of choice of the pump, for example, selection of wavelength and pulse duration, the probe part of the experiment is only used to follow the natural evolution of the initially prepared system. Our conceptual extension of flash photolysis exploits the properties of the coupling of a molecule to the electromagnetic field so that, following preparation in some initial state, and evolution for a defined period of time, a second pulse directs the evolution of the system to some desired final state. The methodology we describe defines a theoretical paradigm, that is, shows that, for realistic model systems, control of selectivity of chemical reaction can be achieved.[24,25]

### B. General Remarks

We consider a system for which there can be a reaction on the ground-electronic-state potential energy surface, and ask how that reaction can be mediated by excitation to, evolution on, and stimulated deexcitation from, an excited electronic state. The excitation and stimulation pulse shapes, durations, and separations required to achieve selectivity of product formation depend on the properties of the excited-state potential energy surface. In the relevant time domain, which is defined by the shape of the excited-state potential energy surface, we shall show that it is possible to take advantage of the localization in phase space of the time-dependent quantum mechanical amplitude and thereby carry our selective chemistry.

Briefly put, it is assumed that the ground-electronic-state Born–Oppenheimer potential energy surface has two or more exit channels corresponding to the formation of two or more distinct chemical species. It is also assumed that there exists an excited-state potential energy surface whose minimum is displaced from that of the ground-state surface and whose normal coordinates are rotated from those of the ground-state surface. It is instructive to begin with a classical-mechanical description of the dynamics. Consider the hypothetical potential energy surface shown in Fig. 14; it has a central minimum and two nonequivalent exit channels separated from the minimum by saddle points. The trajectory that begins at rest at the minimum of the ground-state surface is projected vertically up to the excited-state surface (Fig. 15). It now evolves for some time on the excited-state surface, after which it is projected vertically back down to the ground-state surface. The time spent on the excited-state surface is one of the controllable variables in our scheme. The trajectory is now propagated on the ground-state surface long enough to determine its ultimate fate, that is, whether it leads to $A + BC$, $AB + C$, or $ABC$. Figure 16*a* shows a trajectory that exits from channel 1 ($A + BC$; excited state propagation time is 600 a.u.). Figure 16*b* shows a trajectory that exits from channel 2 ($AB + C$; excited state propagation time is 2100 a.u.). As may be seen in Figs. 17 and 18, in the classical-mechanical description there are windows of 50–100 a.u. width for exit out of a desired channel.

The quantum-mechanical description of the dynamics follows a very similar pattern. At the instant that the first photon is incident, the ground-state wavefunction makes a vertical (Franck–Condon) transition to the excited-state surface. The ground-state wavefunction is not a stationary state on the excited-state potential energy surface, so it must evolve as $t$ increases. There are some interesting analytical properties of this time evolution if the excited-state surface is harmonic. In that case a gaussian wavepacket remains

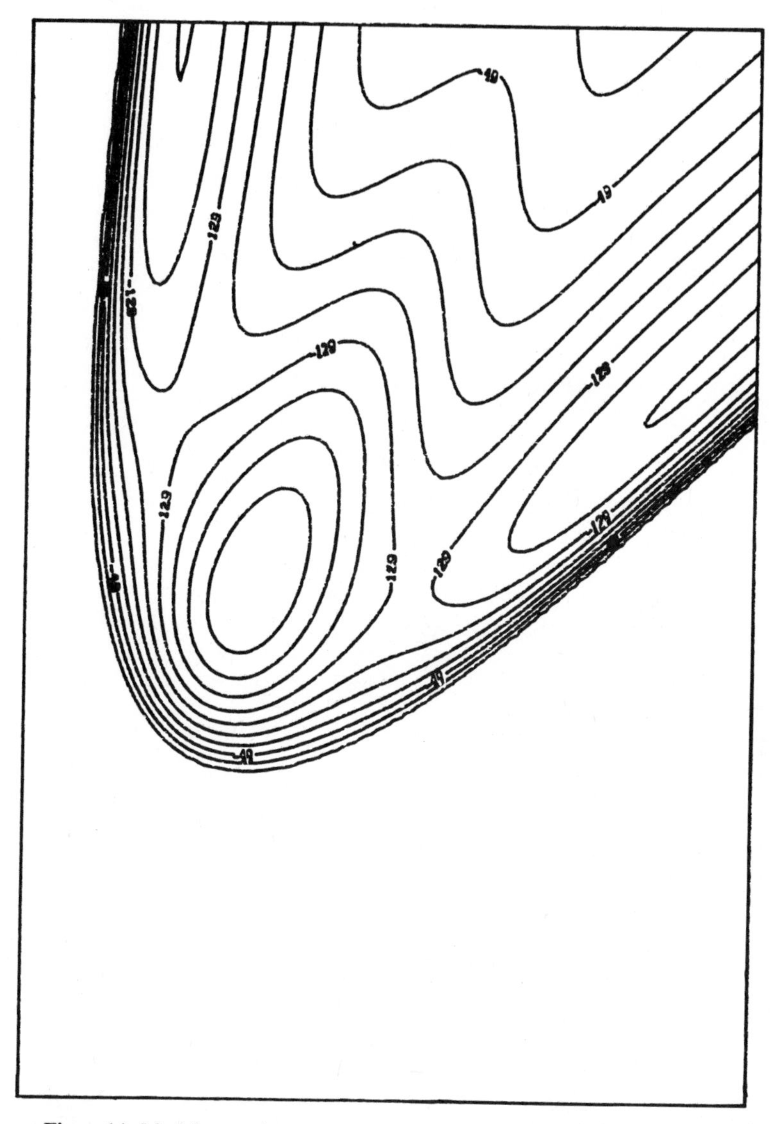

**Figure 14.** Model ground-state Born–Oppenheimer potential energy surface.

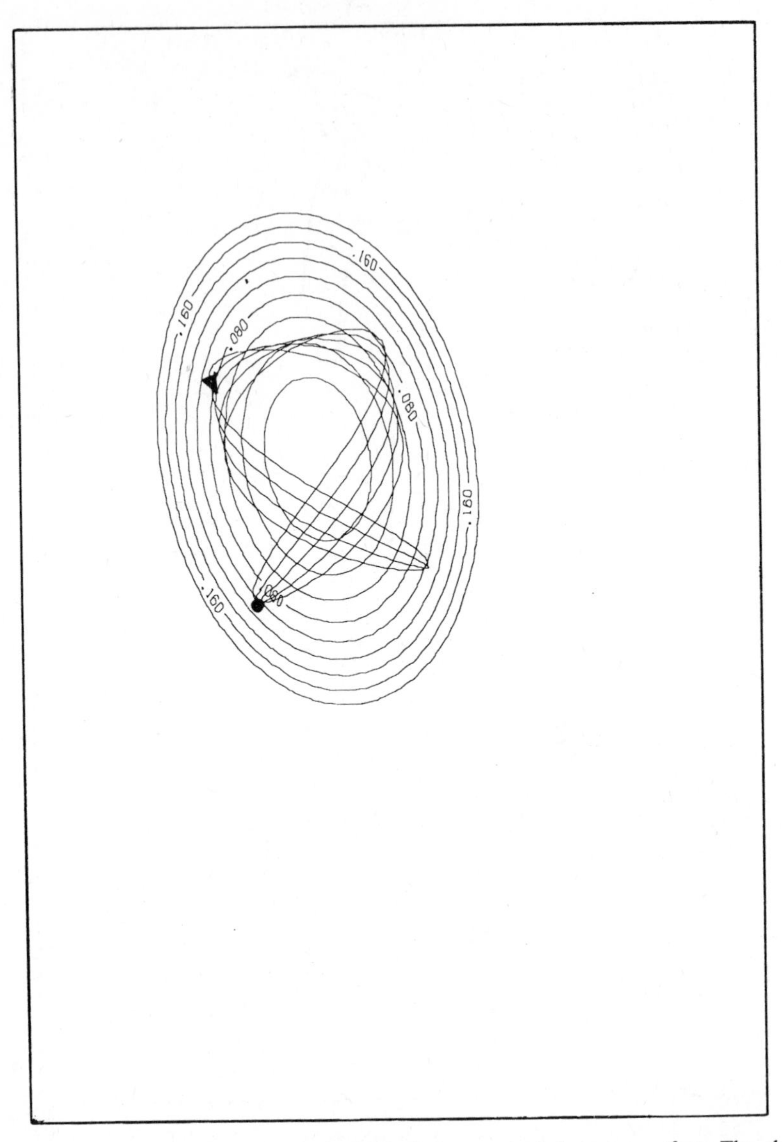


**Figure 15.** Harmonic-excited-state Born–Oppenheimer potential energy surface. The classical trajectory that originates at rest from the ground-state equilibrium geometry is shown superimposed.

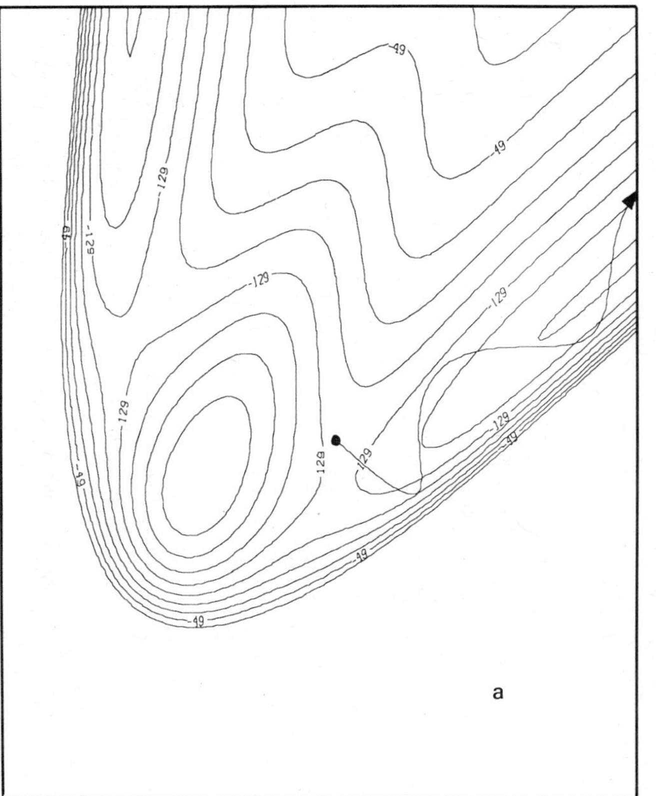


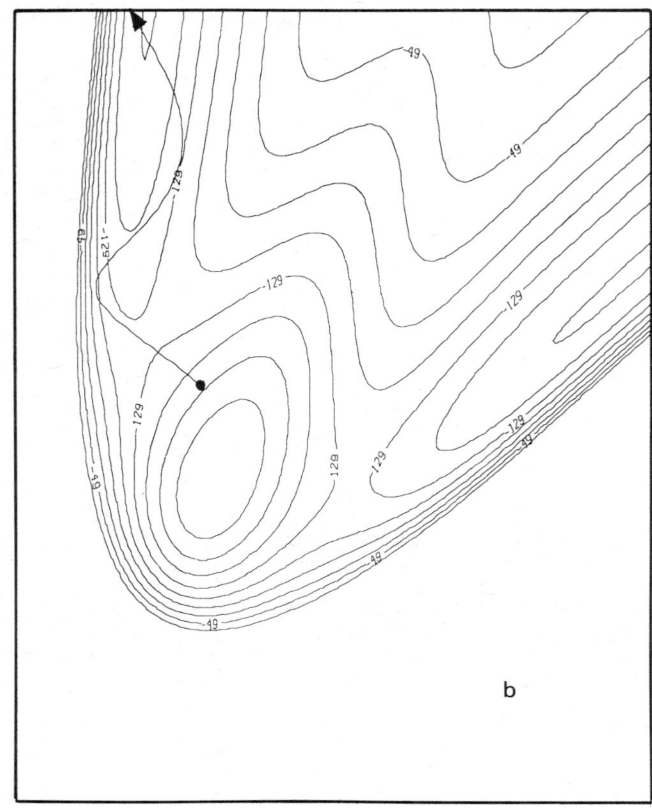


**Figure 16.** Classical trajectories on the ground-state surface that arise from a vertical transition down (coordinates and momentum unchanged) after propagation time $t_2 - t_1$ on the excited-state potential energy surface: (*a*) $t_2 - t_1 = 600$ a.u., (*b*) $t_2 - t_1 = 2100$ a.u.

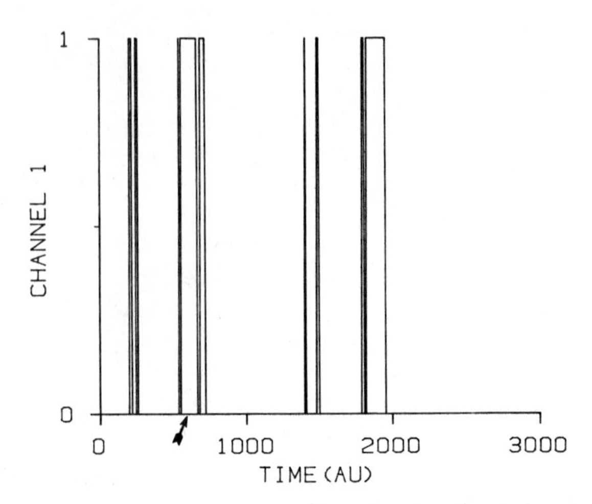


**Figure 17.** Probability (0 or 1) of exit from channel 1 as a function of excited-state potential energy surface propagation time.

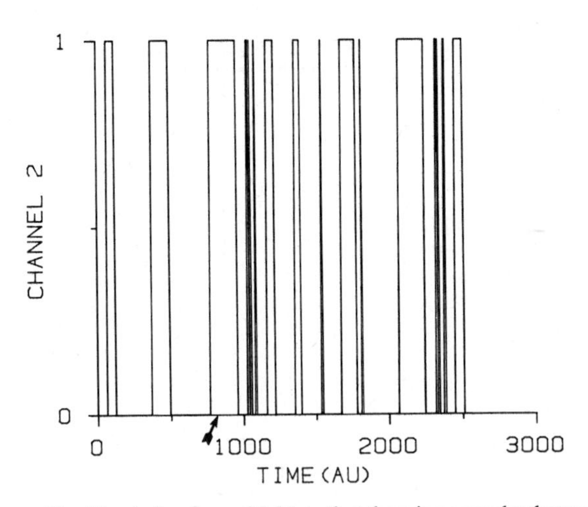


**Figure 18.** Same as Fig. 17 only for channel 2. Note that there is no overlap between the windows in Fig. 17 and Fig. 18, but there are time intervals that correspond to bound trajectories on the ground-state potential energy surface that appear as zero amplitude in both plots.

gaussian for all time, the center of the gaussian wavepacket follows the classical trajectory for harmonic oscillation, in both coordinate and momentum space, and the gaussian wavepacket develops a phase equal to the classical action integral for the same motion, namely, $\phi = \int_0^t (p\dot{q} - E)\,dt$. These properties are retained to a good approximation for smooth anharmonic potential energy surfaces. Moreover, Ehrenfest's theorem[26] ensures that the center of the wavepacket will obey the classical equations of motion for any potential surface, provided the wavepacket remains sufficiently localized. The duration of the propagation on the excited-state surface can be regulated by the delay of a second pulse relative to the initial pulse of light. The second pulse leads to a vertical (Franck–Condon) transition down to the ground-state surface. Note that the wavefunction amplitude is unchanged in the Franck–Condon transition. If the delay and width of the second pulse are chosen on the basis of the position and width of the windows in Figs. 17 and 18 it is plausible to expect the wavepacket amplitude on the ground surface to select one channel over the other. The results of quantum-mechanical calculations of wavepacket propagation on the excited-state and ground-state potential energy surfaces, for a variety of different excited-state potential energy surfaces and a range of pulse delays, show that this expectation is fulfilled.

## C. General Theory

### 1. *Second-Order Time-Dependent Perturbation Theory*

The Hamiltonian we adopt is a $2 \times 2$ matrix of operators. It represents the ground and the excited electronic states within the Born–Oppenheimer approximation, coupled by the radiation field interacting with the dipole operators ($\mu = \mu_{ab} = \mu_{ba}$):

$$H = \begin{pmatrix} H_a & \mu E(t) \\ \mu E(t) & H_b \end{pmatrix}. \tag{5.1}$$

The time-dependent Schrödinger equation reads

$$i\hbar \frac{\partial}{\partial t}\begin{pmatrix} \psi_a \\ \psi_b \end{pmatrix} = \begin{pmatrix} H_a \mu E(t) \\ \mu E(t) H_b \end{pmatrix}\begin{pmatrix} \psi_a \\ \psi_b \end{pmatrix}. \tag{5.2}$$

At $t = 0$, $\psi_0 = \psi_a$, the ground vibrational state of $H_a$, so that

$$H_a \psi_0 = E_a \psi_0. \tag{5.3}$$

Also, $\psi_b = 0$ (we are assuming the system is at 0 K).

The two coupled differential equations in Eq. (5.2) can be transformed to two coupled integral equations, namely,

$$\psi_a(t) = e^{-iHt/\hbar}\psi_a(0) - \frac{i}{\hbar}\int_{-\infty}^{t} e^{-iH(t-t')/\hbar}\mu E(t')\psi_b(t')\,dt', \tag{5.4a}$$

$$\psi_b(t) = -\frac{i}{\hbar}\int_{-\infty}^{t} e^{-iH_b(t-t')/\hbar}\mu E(t')\psi_a(t')\,dt'. \tag{5.4b}$$

The reader may easily verify that Eqs. (5.4) are formal solutions of Eq. (5.2) by differentiation. We consider first the weak-field regime.

The time-dependent perturbation theory expression for the first-order amplitude on the excited-state surface, $\psi_b^{(1)}(t)$, is given by

$$\psi_b^{(1)}(t_2) = -\frac{i}{\hbar}\int_{-\infty}^{t_2} e^{-iH_b(t_2-t_1)/\hbar}\mu E(t_1)\psi_a(0)e^{-i\omega_a t_1}\,dt_1. \tag{5.5a}$$

This equation is obtained from Eq. (4b) by substituting

$$\psi_a(0)e^{-i\omega_a t_1} \cong \psi_a(t_1) \tag{5.5b}$$

and identifying $t_1 \Leftrightarrow t'$ and $t_2 \Leftrightarrow t$.

The time-dependent perturbation theory expression for the second-order amplitude on the ground-state surface, $\psi_a^{(2)}(t)$, is

$$\psi_a^{(2)}(t) = -\frac{i}{\hbar}\int_{-\infty}^{t} e^{-iH_a(t-t_2)/\hbar}\mu E(t_2)\psi_b^{(1)}(t_2)\,dt_2 \tag{5.6a}$$

$$= \frac{-1}{\hbar^2}\int_{-\infty}^{t}\int_{-\infty}^{t_2} e^{-iH_a(t-t_2)/\hbar}\mu E(t_2)e^{-iH_b(t_2-t_1)/\hbar} \times \mu E(t_1)\psi_a(0)e^{-i\omega_a t_1}\,dt_1\,dt_2. \tag{5.6b}$$

Note that Eq. (5.6b) contains the field strength to second order only. Equation (5.6b) has the following simple interpretation: $\psi_a(0)$ evolves on the ground-state surface from $t = 0$ until $t = t_1$. At time $t_1$ it makes a vertical transition to the excited-state surface. The wavefunction propagates on the excited-state surface from time $t_1$ until time $t_2$. At time $t_2$ the wavefunction makes a vertical transition back to the ground-state surface. The wavefunction then evolves on the ground-state surface from time $t_2$ until time $t$. In general, the waveforms $E(t_1)$, $E(t_2)$ are extended; therefore, we must integrate over $t_1$ and $t_2$, all the instants at which the transition up and the transition down may take place. For some purposes it will be more useful to work with Eq. (5.6a), which also

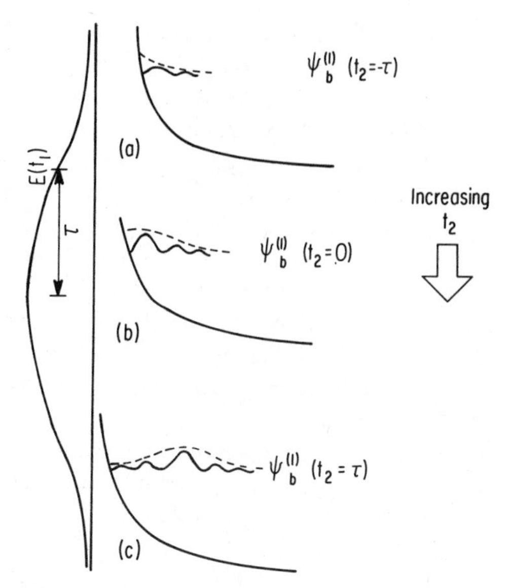


**Figure 19.** A series of pictures of the first-order amplitude on the excited state surface, $\psi_b^{(1)}(t_2)$. The amplitude is the convolution of the pulse $E(t_1)$, of half-width $\tau$ (shown on the left), with the freely evolving excited-state wavepacket $\phi(t)$.

has a simple interpretation. The first-order amplitude on the excited state surface, $\psi_b^{(1)}$, is a function of $t_2$. A series of schematic pictures of $\psi_b^{(1)}(t_2)$ is shown in Fig. 19, for a pulse that is long relative to the time it takes to begin to propagate on surface $b$: $\psi_b^{(1)}$ is simply the convolution of the waveform, $E(t_1)$, with the freely existing excited-state wavepacket, $\phi(t) = e^{-iHt/\hbar}\phi(0)$, where $\phi(0) = \mu\psi_a(0)$. If $E(t_1) = e^{i\omega_I t_1}$, then the reader may show from Eq. (5.5a),

$$\lim_{t_2\to\infty} \psi_b^{(1)}(t_2) \propto \psi_n \delta(E_n - \hbar(\omega_i + \omega_I)), \tag{5.7}$$

where $\psi_n$ is an eigenstate of surface $b$ with energy $E_n = \hbar(\omega_i + \omega_I)$. This wavefunction $\psi_b^{(1)}$ is projected down to the ground-state surface, where it continues to propagate until time $t$. Again, since $E(t_2)$ is extended, we must integrate over all $t_2$ at which the transition down may take place.

### 2. *Variational Formulation of Control of Chemical Products*

We now pose the question: Given the most general shape of an exciting and stimulating (deexciting) pulse, which shape is optimal in producing a desired chemical product? A little reflection shows that this is a problem in the calculus of variations, where the probability of obtaining product $\alpha$, $I_\alpha$, is a functional of the functions $a(s)$ and $b(t)$ (the exciting and deexciting pulse shapes). There are three reasons we believe the variational formulation is fruitful. (1) In principle, it will give optimal pulse shapes that can be constructed experimentally to obtain maximum yield for a given number of photons. (2) Different classes of solutions may be expected for different formulations of the variational problem, for example, for photodissociation versus photoisomerization, reaction in the presence of a curve crossing or multiphoton absorption. The solution to the variational problem could in principle be surprising, and an unexpected pulse shape may allow control of chemical products. (3) The solution should reflect the properties of the ground- and the excited-state potential energy surfaces, and thereby elucidate the partnership of surfaces and pulses in control of products.

To formulate the problem more precisely, we are interested in maximizing (or minimizing) flux out of a particular chemical arrangement channel, $\alpha$, relative to all others. This leads us to consider the quantity $I_\alpha$,

$$I_\alpha = \lim_{t\to\infty} \int_0^\infty dk \sum_{n_\alpha=0}^{\infty} |\langle n_\alpha, k_{n_\alpha} | \psi_a^{(2)}(t)\rangle|^2, \tag{5.8}$$

where $n_\alpha$ labels fragment internal states associated with arrangement channel $\alpha$ and $k_{n_\alpha}$ labels the associated translational states. We sum (integrate) over internal and translational states belonging to $\alpha$ because we are interested in the arrangement channel only.

Defining the projection operator for the $\alpha$th channel, $P_\alpha$, as

$$P_\alpha \equiv \int_0^\infty dk_{n_\alpha} \sum_{n_\alpha=0}^{\infty} |n_\alpha, k_{n_\alpha}\rangle \langle n_\alpha, k_{n_\alpha}|, \tag{5.9}$$

Eq. (5.8) can be rewritten as

$$I_\alpha = \lim_{t\to\infty} \langle \psi_a^{(2)}(t) | P_\alpha | \psi_a^{(2)}(t)\rangle \tag{5.10}$$

or, expressing $I_\alpha$ in terms of $\psi_b^{(1)}$,

$$I_\alpha = \lim_{t\to\infty} \int_{-\infty}^{\infty} dt_2' \int_{-\infty}^{\infty} dt_2 \langle \psi_b^{(1)}(t_2') | \mu b^*(t_2') e^{iH_a(t-t_2')/\hbar}$$
$$\times P_\alpha e^{-iH_a(t-t_2)/\hbar} \mu b(t_2) | \psi_b^{(1)}(t_2)\rangle \tag{5.11}$$

where we have substituted Eq. (5.6a) into Eq. (5.10).

As with any problem in the calculus of variations, the choice of constraints strongly affects the solution obtained. We choose the constraints

$$\int_{-\infty}^{\infty} |a(s)|^2\, ds = \int_{-\infty}^{\infty} |b(t)|^2\, dt = C, \tag{5.12}$$

where $C$ is a constant (we set $C = 1$ for convenience). The physical meaning of the constraint is that the total number of photons available to achieve product is fixed.

We will solve separately for the excitation pulse, $a(t_1)$, and the deexcitation pulse, $b(t_2)$, beginning with the latter. Beginning with Eq. (5.11) we have

$$I_\alpha[b] = \lim_{t\to\infty} \int_{-\infty}^{\infty} dt_2' \int_{-\infty}^{\infty} dt_2 \langle \psi_b^{(1)}(t_2')|\, \mu b^*(t_2') e^{iH_a(t-t_2')/\hbar}$$
$$\times P_\alpha e^{-iH_a(t-t_2)/\hbar} \mu b(t_2) | \psi_b^{(1)}(t_2)\rangle \tag{5.13}$$
$$= \int_{-\infty}^{\infty} dt_2' \int_{-\infty}^{\infty} dt_2\, b^*(t_2') b(t_2) A_\alpha(t_2, t_2'), \tag{5.14}$$

where

$$A_\alpha(t_2, t_2') \equiv \lim_{t\to\infty} \langle \psi_b^{(1)}(t_2')|\, \mu e^{iH_a(t-t_2')/\hbar} P_\alpha e^{-iH_a(t-t_2)/\hbar} \mu \psi_b^{(1)}(t_2)\rangle. \tag{5.15}$$

The form $I_\alpha[b]$ indicates that $I_\alpha$ is a functional of $b$.

Using the method of Lagrange multipliers we find

$$I_\alpha[b] = \int_{-\infty}^{\infty} dt_2' \int_{-\infty}^{\infty} dt_2 b^*(t_2') b(t_2) A_\alpha(t_2, t_2') + \lambda \int_{-\infty}^{\infty} |b(t)|^2\, dt. \tag{5.16}$$

We now write

$$I_\alpha = \bar{I}_\alpha + \delta I_\alpha,$$
$$b = \bar{b} + \delta b, \tag{5.17}$$

where $\bar{I}_\alpha$ is the optimal value of $I_\alpha$ and $\bar{b}(t)$ is the form of $b$ which optimizes $I$ while satisfying the constraints. $\delta I_\alpha$ and $\delta b$ are variations in $I_\alpha$ and $b$, respectively, away from the optimal values. Substituting Eq. (5.17) into Eq. (5.16), subtracting the optimal values from both sides, and neglecting second-order variations, yields

$$\delta I_\alpha = 2\,\mathrm{Re}\left(\int_{-\infty}^{\infty} dt_2' \int_{-\infty}^{\infty} dt_2\, b^*(t_2')\delta b(t_2) A_\alpha(t_2, t_2') + \delta \int_{-\infty}^{\infty} b^*(t)\delta b(t)\, dt\right)$$
$$= 0. \tag{5.18}$$

Since (5.18) holds for all variations $\delta b$, real or imaginary, we find that

$$\int_{-\infty}^{\infty} dt_2' \bar{b}^*(t_2') A_\alpha(t_2, t_2') + \lambda \bar{b}^*(t_2) = 0. \tag{5.19}$$

Eq. (5.19) is a Fredholm integral equation of the second kind. The Fredholm integral equation is easily solved if the kernel, $A_\alpha$, is separable, that is,

$$A_\alpha(t_2, t_2') = \sum_{i=1}^{N} \tilde{A}_i(t_2)\tilde{A}_i'(t_2'), \tag{5.20}$$

where $N$ is some finite number. In our case we want to allow formally for an infinite number of states, since the wavepacket has a dispersion in energy with, potentially, a long high-energy tail. More important, we want to integrate over the continuum of translational states, $k_{n_\alpha}$; hence, our kernel most definitely is not separable. We note, however, that the kernel $A_\alpha(t_2, t_2')$ is real and symmetric, since it is real and symmetric for all possible final states. According to the Hilbert–Schmit theory of integral equations there exists at least one solution of the integral equation (5.19), and possibly an infinite number. It is not clear under what conditions there will be more than one solution to Eq. (5.19).

Assuming we have solved for $\bar{b}(t)$, we will have $\bar{b}([a], t)$. That is, the optimal $\bar{b}(t_2)$ is a functional of $a(t_1)$. We now direct our attention to solving for the optimal $a(t_1)$.

Substituting Eq. (5.6b) into Eq. (5.10) we obtain

$$I_\alpha[a, \bar{b}[a]] = \lim_{t\to\infty} \int_{-\infty}^{\infty} dt_1' \int_{-\infty}^{\infty} dt_1\, e^{i\omega_i t_1'} \langle \psi_a(0) |\, \mu a^*(t_1') e^{iH_b(t-t_1')/\hbar}$$
$$\times\, g_\alpha([a], t) e^{-iH_b(t-t_1)/\hbar} \mu a(t_1) | \psi_a(0) \rangle e^{-i\omega_i t_1}$$
$$= \int_{-\infty}^{\infty} dt_1' \int_{-\infty}^{\infty} dt_1\, a^*(t_1') a(t_1) G_\alpha([a], t_1, t_1'), \tag{5.21}$$

where

$$g_\alpha([a], t) = \int_{-\infty}^{\infty} dt_2' \int_{-\infty}^{\infty} dt_2\, e^{-iH_b(t-t_2')/\hbar} \mu \bar{b}^*([a], t_2') e^{iH_a(t-t_2')/\hbar}$$
$$\times\, P_\alpha e^{-iH_a(t-t_2)/\hbar} \mu \bar{b}([a], t_2) e^{iH_b(t-t_2)/\hbar} \tag{5.22}$$

and

$$G_\alpha([a], t_1, t_1') = \lim_{t\to\infty} e^{i\omega_i t_1'} \langle \psi_a(0)| \mu e^{iH_b(t-t_1')/\hbar} g_\alpha([a], t) e^{-iH_b(t-t_1)/\hbar} \mu |\psi_a(0)\rangle e^{-i\omega_i t_1}. \tag{5.23}$$

Note that Eq. (5.21) has the same simple structural form as Eq. (5.14). The analysis of $I_\alpha[a]$ then proceeds much as for $I_\alpha[b]$, except that $G_\alpha(t_1, t_1')$ is a functional of $a$, whereas $A_\alpha(t_2, t_2')$ was not a functional of $b$ [cf. Eqs. (5.23) and (5.15)].

Using the method of Lagrange multipliers we write

$$I_\alpha[a] = \int_{-\infty}^{\infty} dt_1' \int_{-\infty}^{\infty} dt_1\, a^*(t_1') a(t_1) G_\alpha([a], t_1, t_1') + \lambda \int_{-\infty}^{\infty} |a(t)|^2\, dt. \tag{5.24}$$

Proceeding as before we set

$$I_\alpha = \bar{I}_\alpha + \delta I_\alpha,$$
$$a = \bar{a} + \delta a, \tag{5.25}$$

where $\bar{I}_\alpha$ is the optimal value of $I_\alpha$ with respect to variations in $a$, and $\bar{a}$ is the form of $a$ that optimizes $I$ while satisfying the constraints. $\delta I_\alpha$ and $\delta a$ are variations in $I_\alpha$ and $a$, respectively, away from $\bar{I}_\alpha$ and $\bar{a}$.

Substituting Eqs. (5.25) into Eq. (5.24), subtracting the optimal values from both sides, and neglecting second-order variations yields

$$\delta I_\alpha = 2\,\mathrm{Re}\left(\int_{-\infty}^{\infty} dt_1' \int_{-\infty}^{\infty} dt_1\, a^*(t_1') a(t_1) G_\alpha([a], t_1, t_1') a^*(t_1') a(t_1) \times \frac{\delta G_\alpha}{\delta a} \delta a(t_1) + \lambda \int_{-\infty}^{\infty} a^*(t) \delta a(t)\, dt\right) = 0, \tag{5.26}$$

where

$$\frac{\delta G_\alpha}{\delta a} = \lim_{t\to\infty} e^{i\omega_i t_1'} \langle \psi_a(0)| \mu e^{iH_b(t-t_1')/\hbar} \frac{\delta g_\alpha(t)}{\delta a} e^{-iH_b(t-t_1)/\hbar} \mu |\psi_a(0)\rangle e^{-i\omega_i t_1} \tag{5.27}$$

and

$$\frac{\delta g_\alpha}{\delta a} = \int_{-\infty}^{\infty} dt_2' \int_{-\infty}^{\infty} dt_2\, e^{-iH_b(t-t_2')/\hbar} \mu \bar{b}^*([a], t_2') e^{-iH_a(t-t_2')/\hbar} \times P_\alpha e^{-iH_a(t-t_2)/\hbar} \mu \frac{\delta \bar{b}(t_2)}{\delta a} e^{-iH_b(t-t_2)/\hbar}. \tag{5.28}$$

The quantity $\delta\bar{b}/\delta a$ can be found, in principle, if $\bar{b}[a]$ is known. Since Eq. (5.26) holds for all variations $\delta a$, we find the following equation for $a(t_1)$:

$$\bar{a}^*(t_1) = \frac{1}{\lambda}\int_{-\infty}^{\infty} dt_1' \left( \bar{a}^*(t_1')G_\alpha([\bar{a}], t_1, t_1') + \bar{a}^*(t_1')\bar{a}(t_1)\frac{\delta G_\alpha}{\delta a}([\bar{a}], t_1, t_1') \right). \tag{5.29}$$

Equation (5.29) provides a route to calculating $a(t_1)$. It is a nonlinear integral equation and, therefore, formidable to solve analytically or numerically.

### D. Numerical Calculation of Wavepacket Propagation

The wavepacket propagation procedure used is an adaptation of several existing grid methods.[5b] The wavefunctions on the ground- and excited-state potential energy surfaces are discretized; they are represented by their values at a set of $64 \times 64$ or $128 \times 128$ grid points. The grid we use is a rectangular lattice in bond length coordinates, $R_1$ and $R_2$, chosen to encompass the coordinate space region of interest, namely, the bound-state region and the adjacent portion of the two exit channels on the ground-state potential energy surface. The region on the excited-state potential energy surface involves the identical portion of coordinate space. This propagation method is in principle exact if the grid is allowed to extend to infinity and the grid point spacing and time step are reduced to zero.

The equations integrated numerically were the full nonperturbative expressions [Eq. (5.2)], since this more accurate calculation required no additional effort. The price paid for following the full amplitude $\psi_a(t)$ rather than $\psi_a^{(2)}(t)$ is that much stronger fields are necessary in order for the amplitude that exits to be visible, relative to the unperturbed amplitude. For sequences consisting of short pulses, the excitation and evolution processes are separable; strong fields give the same pattern of amplitude in the exit channels as weak fields that have the same time dependence, only more flux. We used strong fields in the calculations below. The reader is referred to the original articles for details of the theoretical calculations.[25]

### E. Quantum-Mechanical Results

We show below results for four different model systems. All the systems use the same ground-state potential energy surface (Fig. 14), but different excited-state potential energy surfaces. All the surfaces are models for the coupled symmetric and antisymmetric stretch vibrations in a collinear molecule. The masses for the system are 1823, 1823, and 2646 a.u., corresponding to HHD. Although the ground-state potential energy surface of true HHD is not bound, there are excited states that are bound. Our primary reason for choosing hydrogenic masses was to facilitate the quantum-mechanical calculations, which are substantially more difficult for larger masses. In general, one expects the classical–quantum correspondence, upon which our selectivity scheme is based, to improve for larger masses.

### 1. *Harmonic Excited-State Surface*

Equipotential contours for the excited-state potential energy surface are shown in Fig. 15. The surface is harmonic and rotated relative to the ground-state potential energy surface. The classical-mechanical motion of a single trajectory originating at the Franck–Condon geometry with zero momentum on this excited-state potential energy surface was already presented earlier in this section (Fig. 15). The winding motion in that figure is a consequence of the rotation of the excited-state normal coordinates relative to those of the ground-state surface (Duschinsky effect). Figures 17 and 18 depict the windows for dissociation out channel 1 and channel 2, respectively, and show consistent alternation that is a consequence of the magnitude of the rotation of one set of normal coordinates relative to the others. Rotation of normal modes may not, strictly speaking, be necessary for alternation of exit channels; however, anharmonicities on the ground-state surface (and on the excited-state surface, as in the following examples) must then contribute to strong mode mixing.

We now exploit the dissociation windows to choose a suitable pulse sequence for quantum-mechanical calculations. We choose

$$\mu E(t) = A(t)\cos(\omega_e t) + B(t)\cos(\omega_d t),$$

where $A(t)$ and $B(t)$ are gaussian pulses, with a delay between them chosen on the basis of the classical windows. The parameters of the pulse sequence are given elsewhere.[25] The pulse delay $t_d - t_e = 600$ a.u. was chosen to correspond to the broad window in Fig. 17 of $t_2 - t_1 = 600$ a.u. The first pulse was deliberately chosen to be very narrow in time, so the excitation and evolution processes are almost separable. The values for $A$ and $B$ were chosen so that

$$\int_{-\infty}^{\infty} A(t)\,dt = \pi$$

and

$$\int_{-\infty}^{\infty} B(t)\,dt = \pi.$$

Figures 20*a*–20*c* represent the excited-state wavefunction at $t = 200, 400$, and 600 a.u., before the second pulse. (In this figure, as in all subsequent figures,

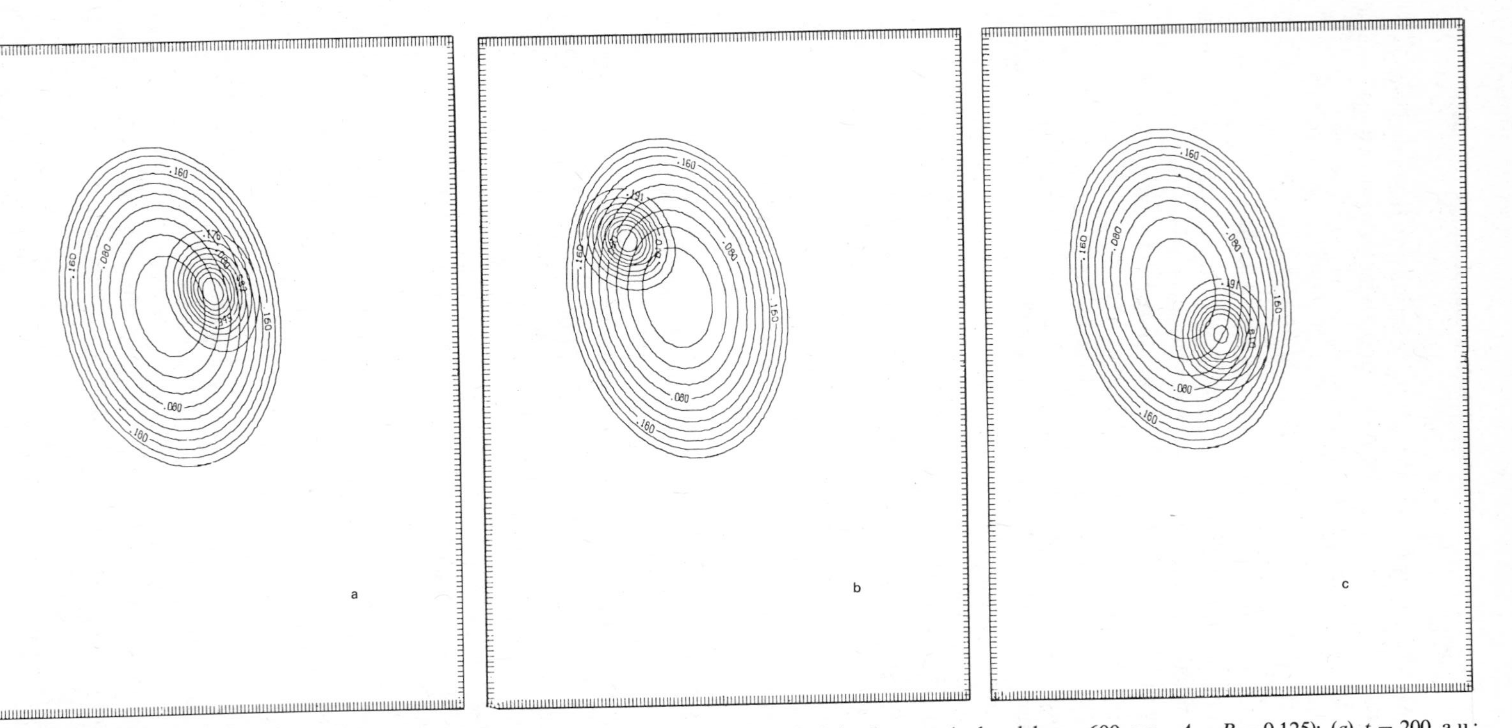


**Figure 20.** Magnitude of the excited state wavefunction for the pulse sequence described in the text (pulse delay = 600 a.u., $A = B = 0.125$): (*a*) $t = 200$ a.u.; (*b*) $t = 400$ a.u.; (*c*) $t = 600$ a.u. Note the agreement with the results obtained for the classical trajectory (Fig. 15).

it is the magnitude of the wavefunction that is plotted.) Note how the wavepacket tracks the classical trajectory (Fig. 16*a*). Figures 21*a*–21*c* show the ground-state wavefunction at $t = 0$, 800, and 1000 a.u. Although not all the amplitude exits to form product, that amplitude that does exit leaves virtually completely from channel 1. Figure 20 demonstrates that in favorable cases, as long as the vertical transition back to the ground-state surface is timed to be close enough to a classical exit window, much of the wavepacket amplitude on the ground state will exit out the corresponding channel.

We now want to establish that use of a different delay time from that cited in the previous paragraph will lead to amplitude exiting exclusively from channel 2. The new delay and width were suggested by the exit window in Fig. 18 centered at $t = 825$ a.u. Parameters for the second pulse sequence are given elsewhere.[25] An examination of the excited-state wavefunction at $t =$ 800 a.u. (Fig. 22) shows, as expected, that it tracks the classical trajectory that has propagated for the same length of time. Figures 23*a*–23*c* show the ground-state wavefunction at $t = 0$, 1000, and 1200 a.u. following stimulation from the excited state. As before, although most of the amplitude remains in the bound region, the amplitude that does exit does so exclusively from channel 2. Thus, we observe excellent control of products in this case: for one choice of delay time between pulses we generate an exit from channel 1, while with a second delay time we generate an exit from channel 2.

This example is particularly favorable for our approach. The harmonic excited-state potential energy surface ensures that the wavepacket retains its near-gaussian shape, at least up to the deexcitation time. Of course, after that time the anharmonic ground-state potential energy surface leads to a highly non-gaussian distribution. We provide in the following a more severe test of our approach, using anharmonic excited-state surfaces, where wavepacket delocalization and bifurcation may begin immediately after the initial excitation pulse.

### 2. *Anharmonic Excited-State Surfaces*

*a. Broad and Shallow Surface.* Model II has a very broad and shallow excited-state potential energy surface. It was initially believed that use of this surface as an intermediary would favor slow alternation (as a function of pulse delay) in the flux out of channel 1 or channel 2 because of the low antisymmetric stretch frequency. We find, instead, that wavepacket spreading is particularly dramatic in this example because of the flatness of the surface.

Equipotential contours for the excited-state surface are shown in Fig. 24. As before, we show the initial vibrational wavefunction as well as the classical trajectory on the excited-state surface that originates from rest on the ground-state surface. The parameters for the excited-state surface are given elsewhere.[25] Figures 25*a*–25*c* show the excited-state wavefunction at $t = 200$

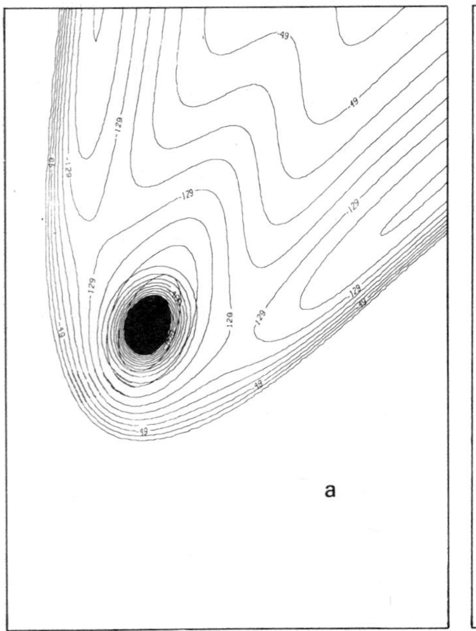


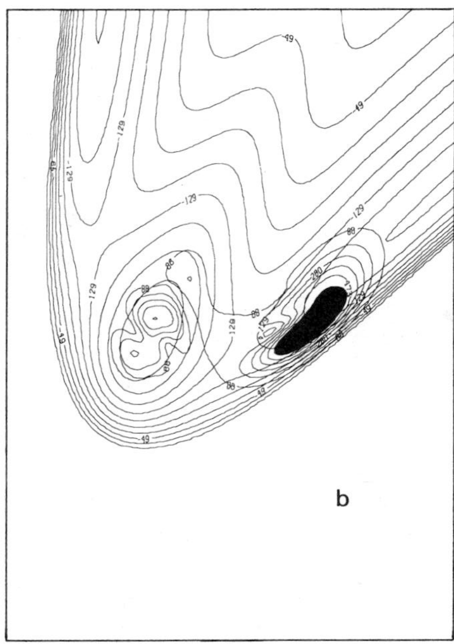


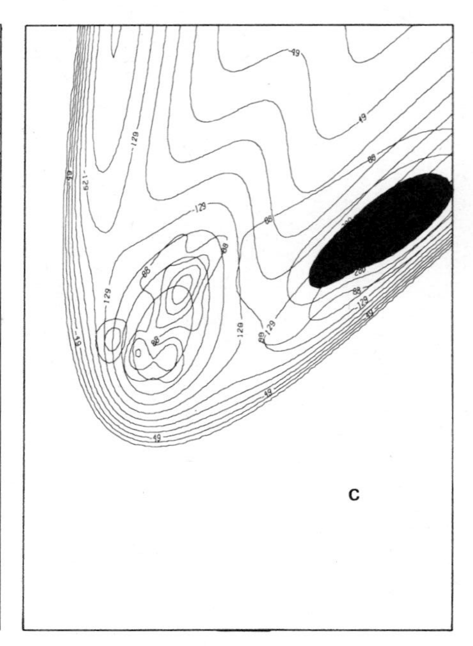


**Figure 21.** Magnitude of the ground-state wavefunction for the pulse sequence in Fig. 20 (*a*) $t = 0$; (*b*) $t = 800$ a.u.; (*c*) $t = 1000$ a.u. Note the agreement with the results obtained for the classical trajectory (Fig. 16*a*). Although some of the amplitude remains in the bound region, that which does exit does so exclusively from channel 1.

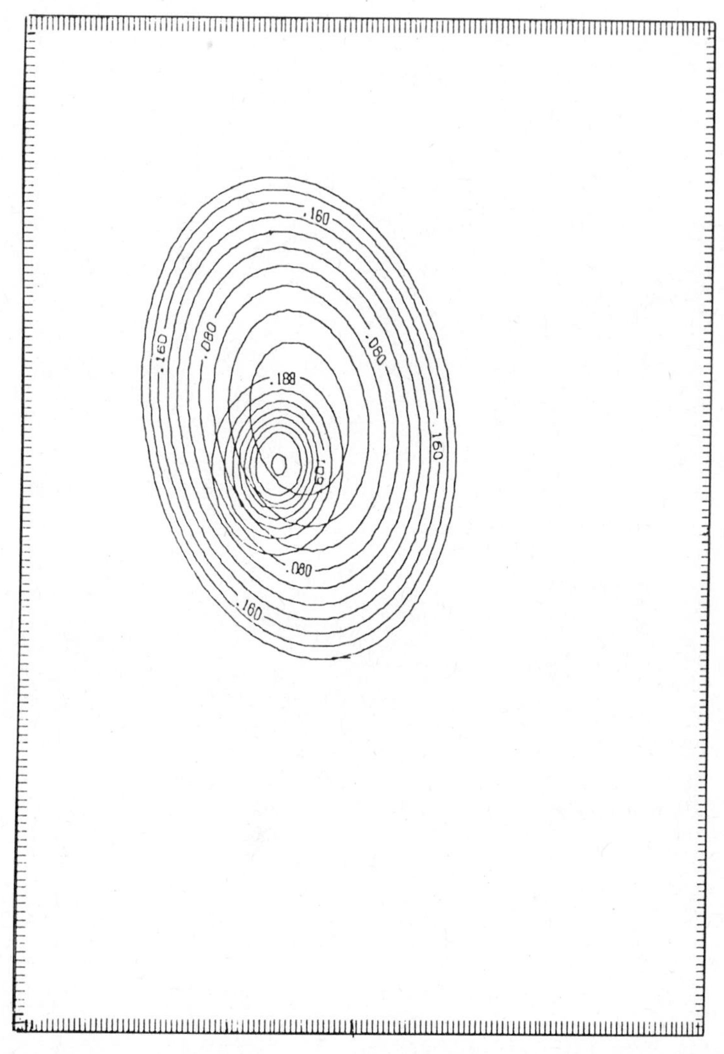


**Figure 22.** Magnitude of the excited-state wavefunction at $t = 800$ a.u., for the second pulse sequence described in the text (pulse delay $= 825$ a.u., $A = B = 0.125$). The excited state wavefunction at $t = 200$, 400, and 600 a.u. is virtually identical to that in Figs. 20*a*–20*c*.

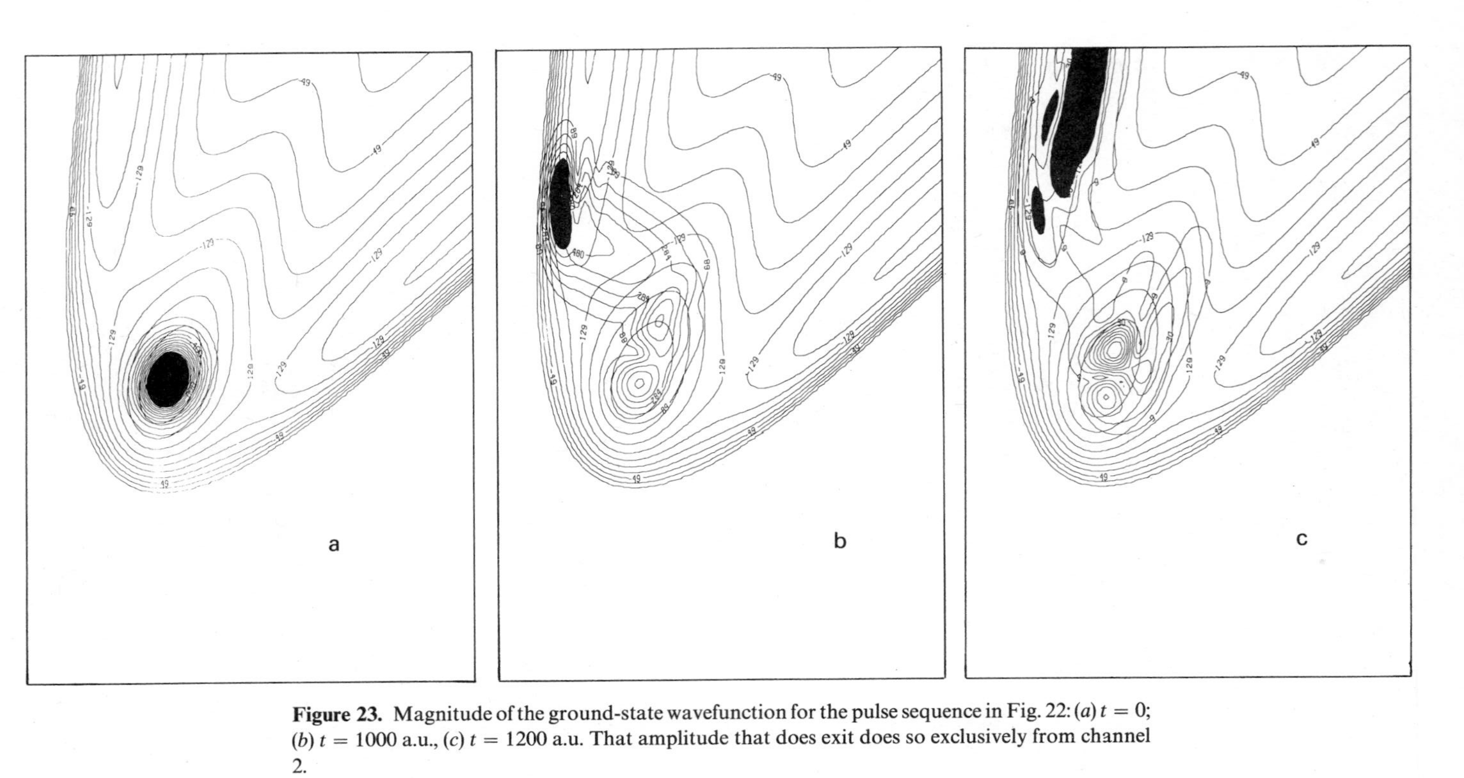


**Figure 23.** Magnitude of the ground-state wavefunction for the pulse sequence in Fig. 22: (*a*) $t = 0$; (*b*) $t = 1000$ a.u., (*c*) $t = 1200$ a.u. That amplitude that does exit does so exclusively from channel 2.

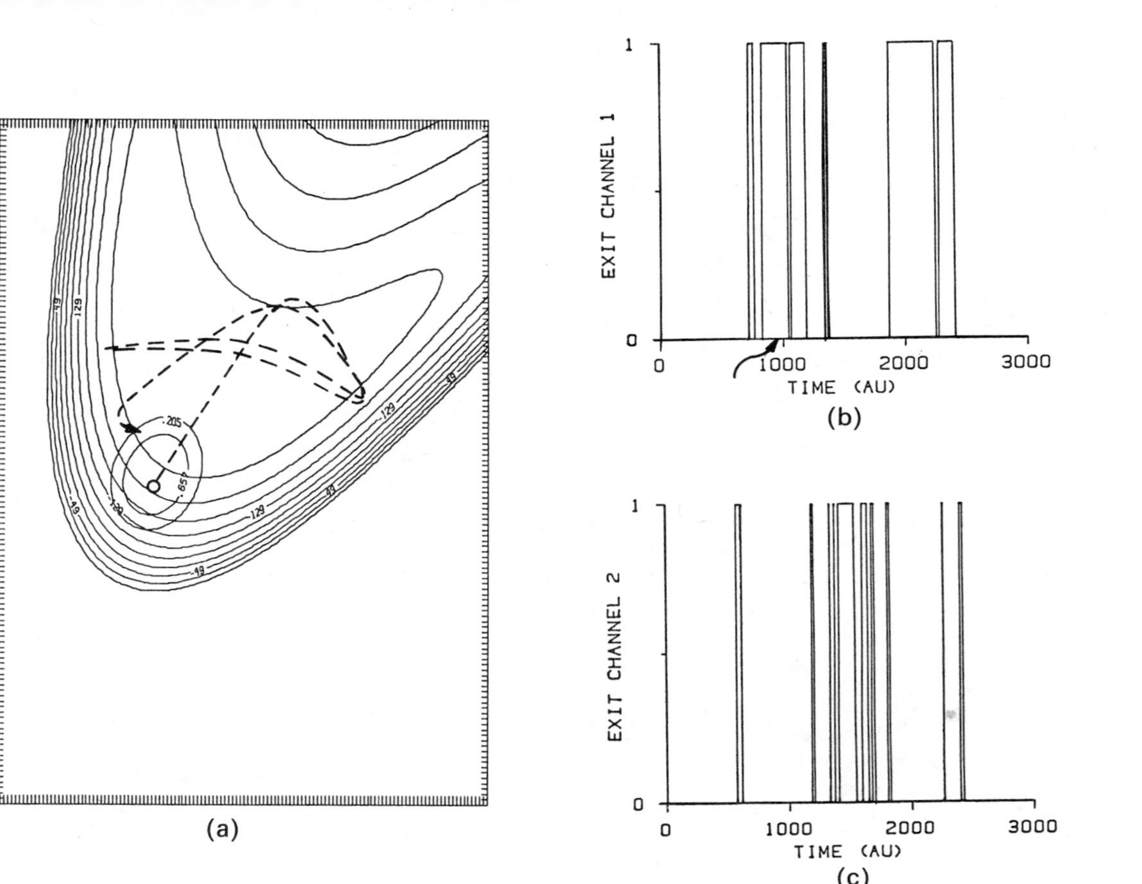


**Figure 24.** (*a*) Anharmonic excited-state potential energy surface. The classical trajectory that originates from rest from the ground-state equilibrium geometry is shown superposed. (*b*) Probability (0 or 1) of exit from channel 1 as a function of excited-state propagation time. (*c*) Same as (*b*), only for exit channel 2.

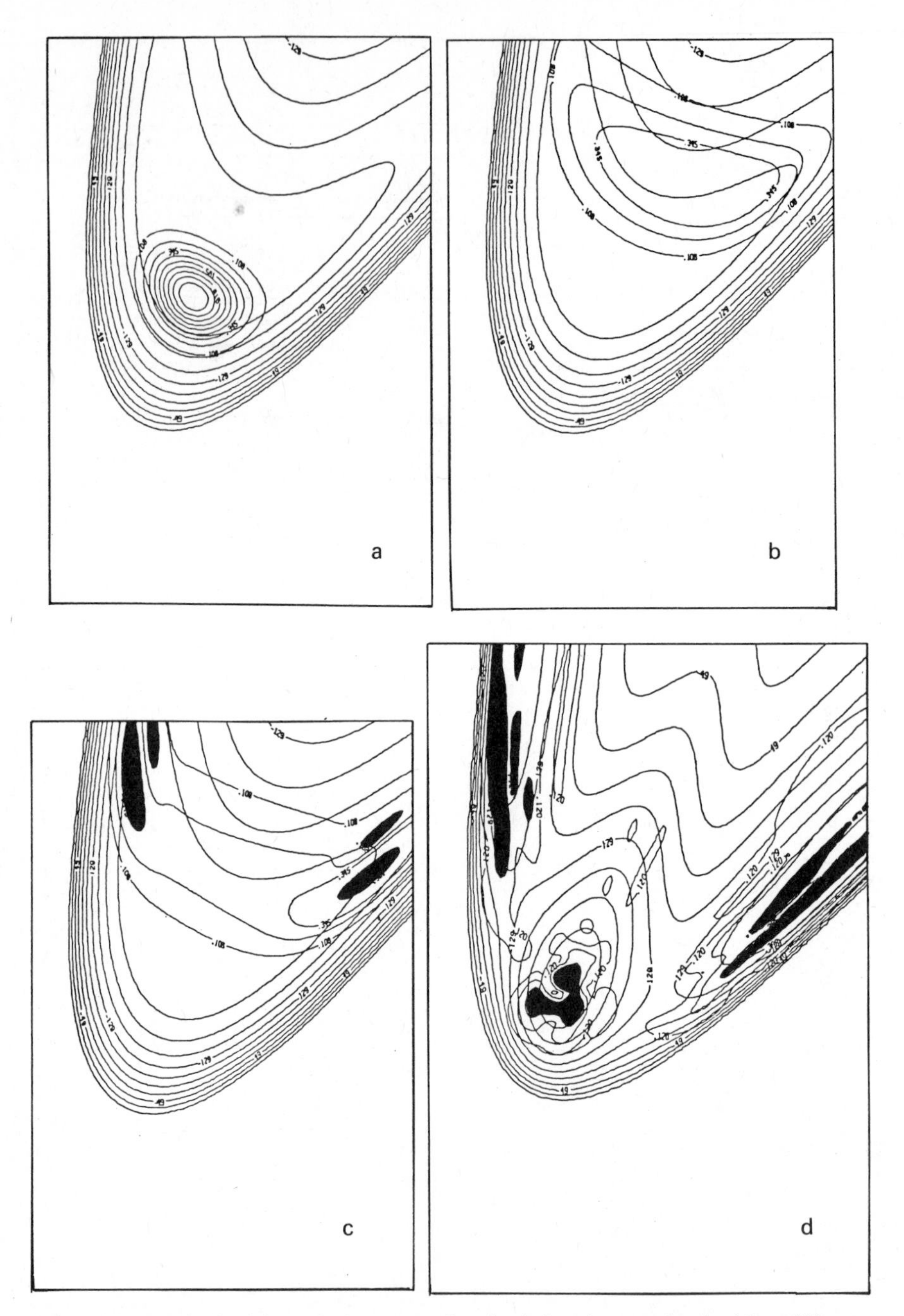


**Figure 25.** Magnitude of the excited-state wavefunction before the second pulse: (*a*) $t = 200$ a.u.; (*b*) $t = 600$ a.u. (*c*) $t = 800$ a.u. Note the extensive wavepacket spreading because the surface is so flat. This spreading will undermine the selectivity of products. (*d*) Ground-state wavefunction at $t = 1200$ a.u., after the second pulse. The poor selectivity of products is apparent.

a.u., 600 a.u., and 800 a.u. respectively, before the second pulse. Clearly, the quantum-mechanical amplitude is spreading severely. Figure 25 *d* shows the amplitude on the ground-state surface at $t = 1200$ a.u., after the second pulse. The poor selectivity is apparent from the figure.

The failure to achieve selectivity in this model system can be traced to the dynamics on the anharmonic excited-state surface, and, in particular, the wavepacket bifurcation. This observation motivated us to explore more systematically the features of the excited-state potential energy surface and excited-state wavepacket dynamics that are compatible with the proposed selectivity scheme.

*b. Typical Excited-State Surface.* Model III has a "typical" excited-state potential energy surface. Specifically, the excited-state minimum is displaced to larger distance than the ground-state minimum, the frequencies in the symmetric and asymmetric stretch coordinates are roughly equal, and the force constants are in the same range as their ground-state values. This model was intended to represent the case in which the wavepacket stays close to the harmonic region of the excited-state potential energy surface, and thus be reasonably well localized. In fact, as will be seen, the wavepacket spreading is pronounced. The parameters defining this surface are given elsewhere.[25]

Equipotential contours for the excited state surface are shown in Fig. 26*a*. Figures 27*a*–27*c* show the excited-state wavefunction at $t = 200$ a.u., 400 a.u., and 600 a.u., respectively, before the second pulse. Again, the quantum-mechanical amplitude is spreading severely, as the wavepacket migrates toward the soft part of the Morse potential. Figure 27*d* shows the amplitude on the ground-state surface at $t = 1000$ a.u., after the second pulse. The selectivity out channel 2 is virtually complete (no amplitude exits from channel 1). This result was unexpected: The classical window predicts an exit from channel 1.

*c. Excited-State Minimum at Smaller Distance than the Ground-State Minimum.* The special feature of this excited-state surface, model IV, is that the excited-state minimum is displaced to smaller distance than the ground-state minimum. The difference in bond lengths leads to a variety of effects not seen with the other potential energy surfaces, since the initial transition is to the soft part of the Morse potential, while the second transition (depending on the instant) is to the strongly repulsive part of the Morse potential. This model was designed to explore the possibility of the wavepacket returning to the original potential energy surface with considerable potential energy, which could be converted to the kinetic energy required for dissociation. A related feature is that there is dramatic wavepacket contraction, or focusing, as the wavepacket evolves on the excited-state surface from the soft to the hard part of the potential. The parameters for the surface are given elsewhere.[25]

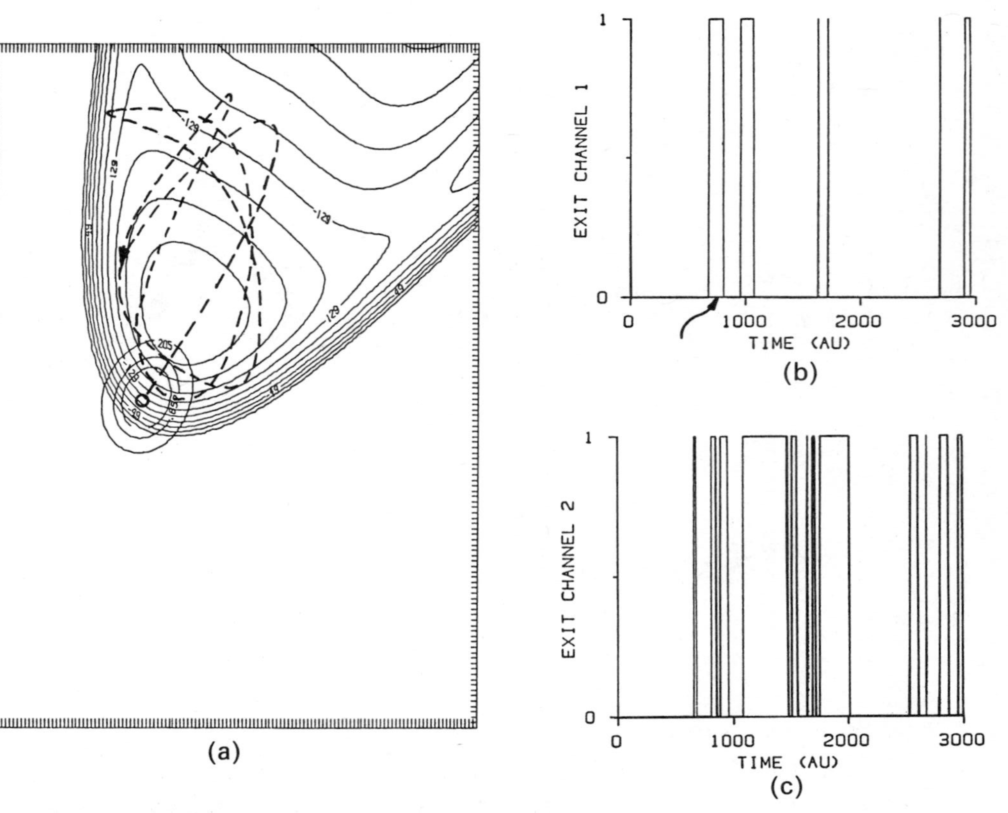


**Figure 26.** (*a*) Anharmonic excited-state potential energy surface. The classical trajectory that originates from rest from the ground-state equilibrium geometry is shown superposed. (*b*) Probability (0 or 1) of exit from channel 1 as a function of excited state propagation time. (*c*) Same as (*b*) only for exit channel 2.

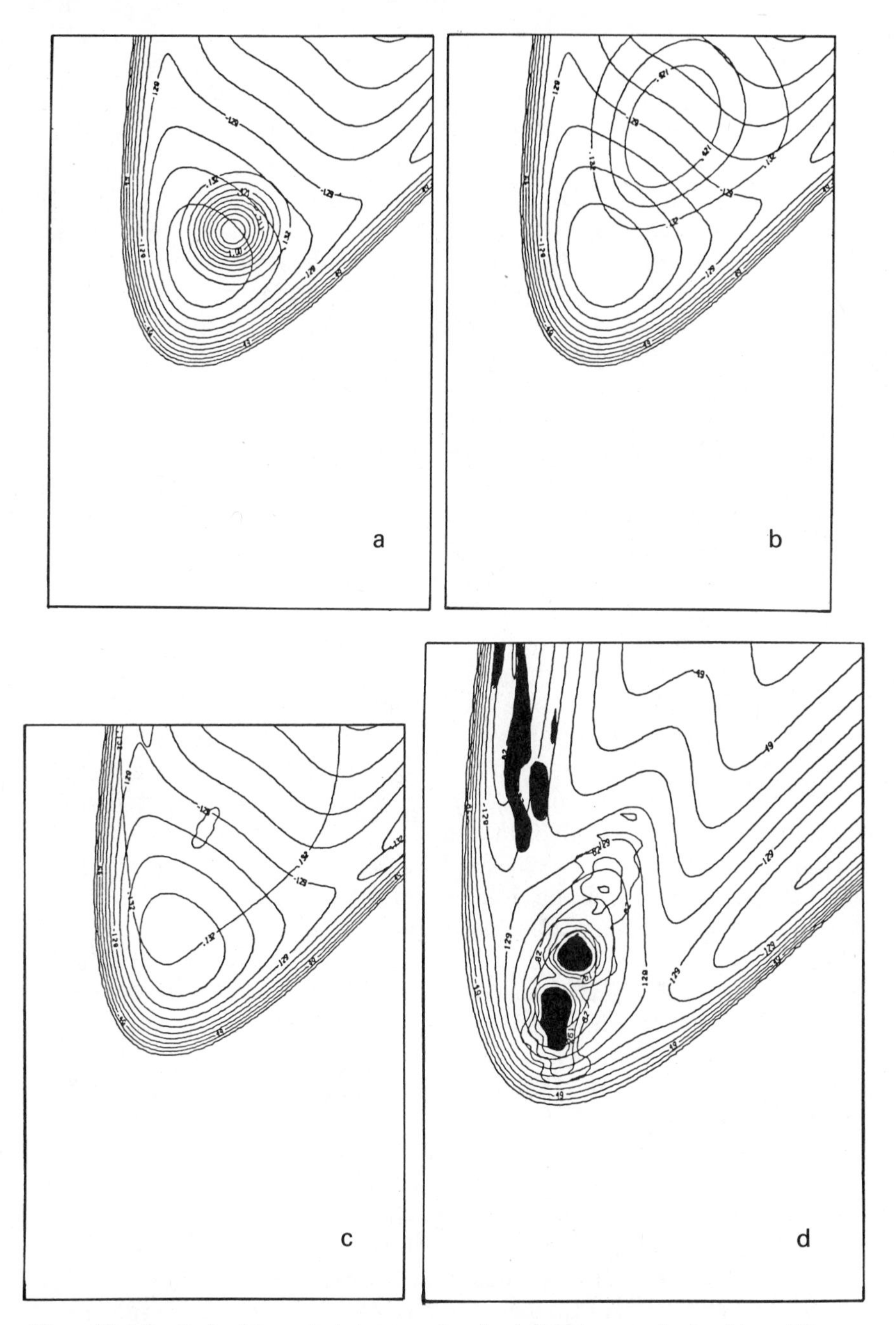

**Figure 27.** Magnitude of the excited-state wavefunction before the second pulse: (*a*) $t = 200$ a.u.; (*b*) $t = 400$ a.u.; (*c*) $t = 600$ a.u. Note the wavepacket spreading is still significant, as the wavepacket approaches the soft part of the Morse potential. (*d*) Ground-state wavefunction at $t = 1000$ a.u., after the second pulse. There is complete selectivity out of channel 2, while the classical mechanics predicts selectivity for exit out of channel 1.

Equipotential contours for the excited-state surface are shown in Fig. 28. As before, we show the initial vibrational wavefunction as well as the unique classical trajectory on the excited-state surface that originates from rest on the ground-state surface. The first pulse sequence examined was at $t = 30, 610$ a.u. Figures 29*a*–29*c* show the excited-state wavefunction at $t = 200$ a.u., 400 a.u., and 600 a.u., respectively, before the second pulse. The wavepacket begins on the soft part of the Morse potential, and initially begins to spread. However, as the wavepacket migrates to the hard part of the Morse potential, it contracts very dramatically. Figures 29*d*–29*f* show the amplitude on the ground-state surface at $t = 800$, 1000, and 1200 a.u., respectively, after the second pulse. It is apparent that a substantial fraction of wavepacket amplitude exits from channel 2, while virtually no amplitude exits from channel 1.

The next pulse sequence examined was at $t = 30, 1010$ a.u. Figures 30*a* and 30*b* show the excited-state wavefunction at $t = 800$ and 1000 a.u., respectively, before the second pulse. Figures 30*c*–30*e* show the amplitudes on the ground-state surface at $t = 1200$, 1400, and 1600 a.u., respectively, after the second pulse. The wavepacket breaks up on the ground-state surface with roughly equal amplitudes escaping from channel 1 and channel 2.

Figure 31 shows the branching ratio as a function of stimulation time, for stimulation pulses centered 200 a.u. apart, from 210 up to 1010 a.u. Note the dramatic differences between the branching ratios at different times relative to each other as well as relative to the amplitude that remains bound.

Thus, an excited-state potential energy surface with shorter equilibrium bond lengths than the ground-state surface proves to be the most useful for our selectivity of reactivity scheme, because it focuses the motion of the wavepacket. Deeper wells and/or higher barriers on the excited-state surface than on the ground-state surface are probably helpful as well. In general, the vibrational energy acquired on the steep excited-state surface is used to break a bond on the shallower ground-state surface. Although these changes in molecular parameters on excitation are not common, there are cases for which they occur. Alternatively, one may use our scheme with the roles of excited state and ground state reversed. Consider starting out in the ground vibrational state of the excited electronic state. This is the initial condition for ordinary emission spectroscopy. Then one may use a two-pulse sequence to stimulate amplitude down to the ground electronic state and back up to the original electronic state. Now the steepness of the ground-state surface barriers accelerate the nuclear motion so that enough kinetic energy is acquired for dissociation on the excited-state surface; also, the tighter bonds on the ground-state surface serve to focus the wavepacket. With this interpretation, the parameters used in the last example are reasonable for many molecules.

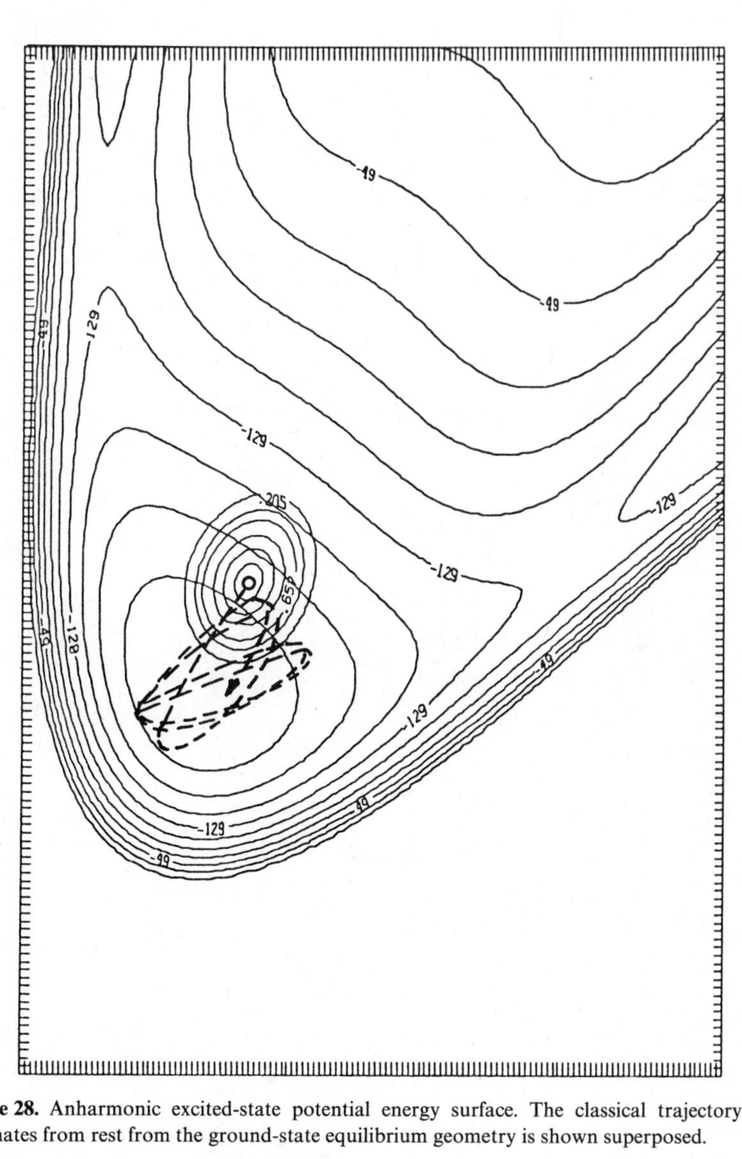


**Figure 28.** Anharmonic excited-state potential energy surface. The classical trajectory that originates from rest from the ground-state equilibrium geometry is shown superposed.

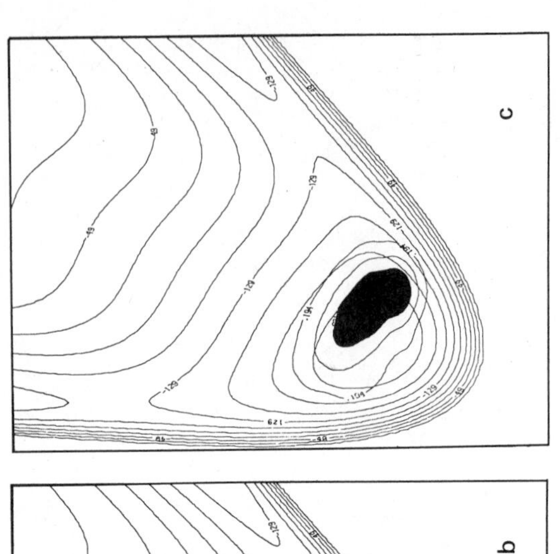
c

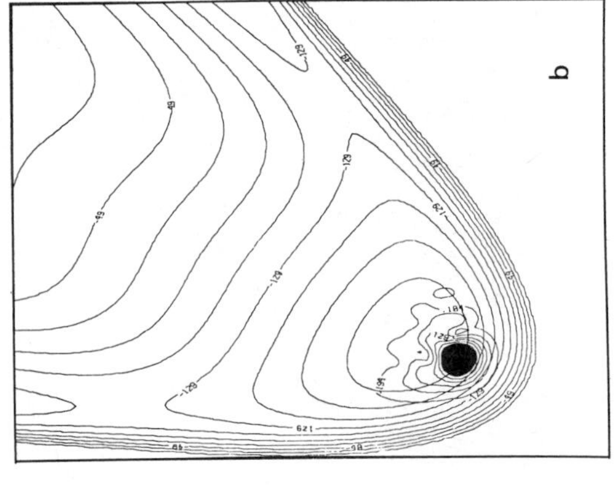
b

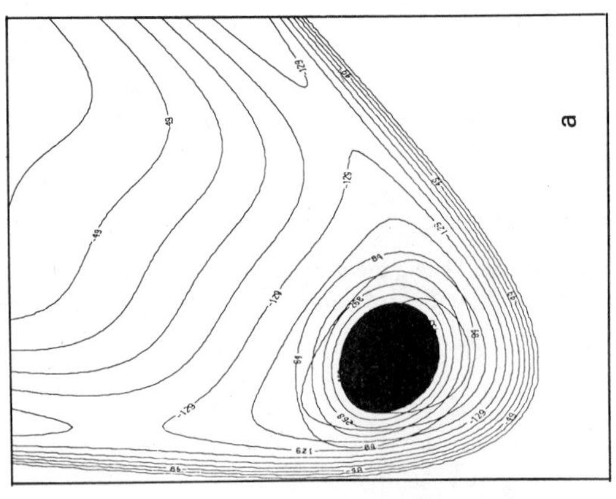
a

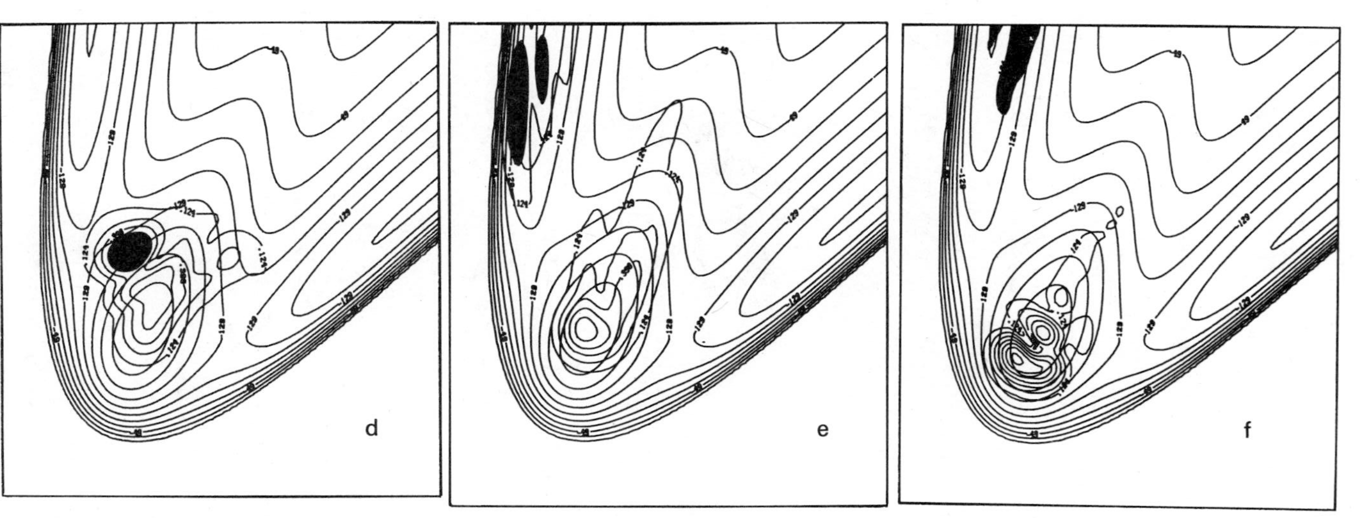


**Figure 29.** Magnitude of the excited-state wavefunction before the second pulse at $t = 610$ a.u.: (*a*) $t = 200$ a.u.; (*b*) $t = 400$ a.u., (*c*) $t = 600$ a.u. Note the dramatic wavepacket contraction as the wavepacket approaches the hard part of the Morse potential. (*d*) Ground-state wavefunction at $t = 800$ a.u., after the second pulse. (*e*) Ground-state wavefunction at $t = 1000$ a.u. (*f*) Ground-state wavefunction at $t = 1200$ a.u. A significant fraction of the wavepacket amplitude is exiting from channel 2, while virtually no amplitude exits from channel 1.

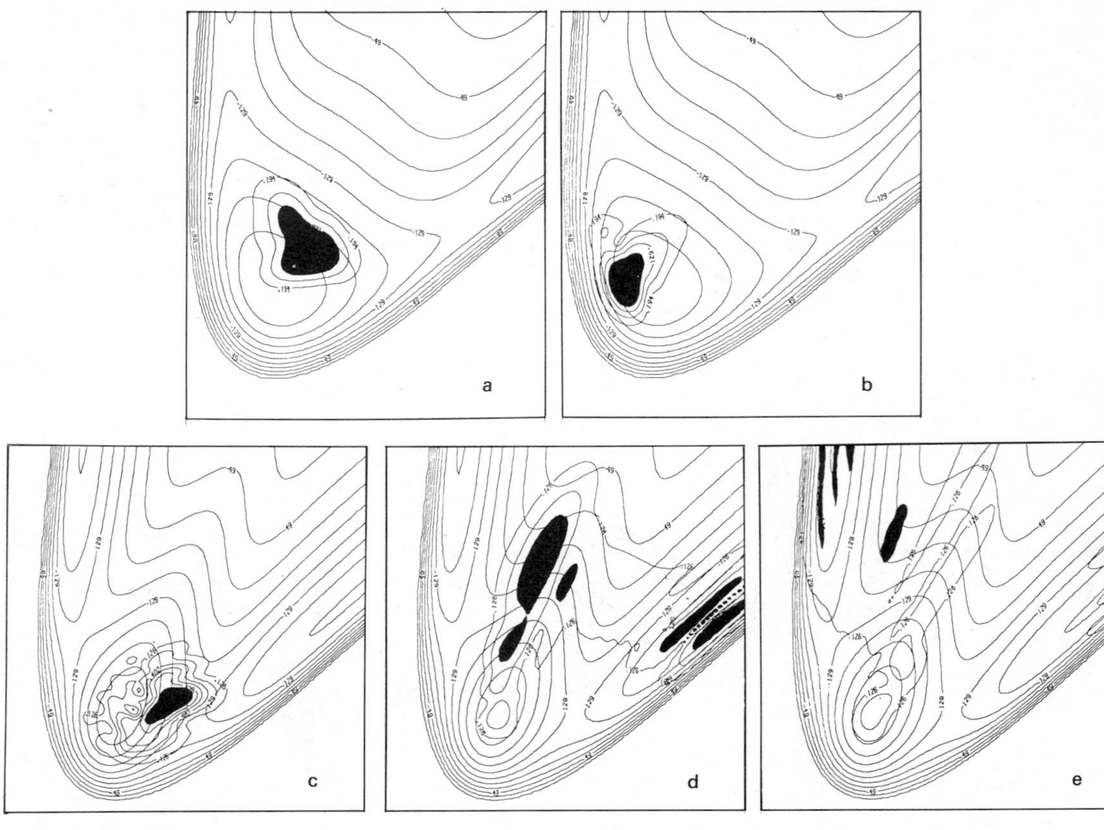


**Figure 30.** Magnitude of the excited-state wavefunction before the second pulse at $t = 1010$ a.u.: (*a*) $t = 800$ a.u., (*b*) $t = 1000$ a.u. (*c*) Ground-state wavefunction at $t = 1200$ a.u. (*d*) Ground-state wavefunction at $t = 1400$ a.u. (*e*) Ground-state wavefunction at $t = 1600$ a.u. Note the wavepacket breakup on the ground-state surface, with roughly equal amplitudes exiting from channel 1 and channel 2.

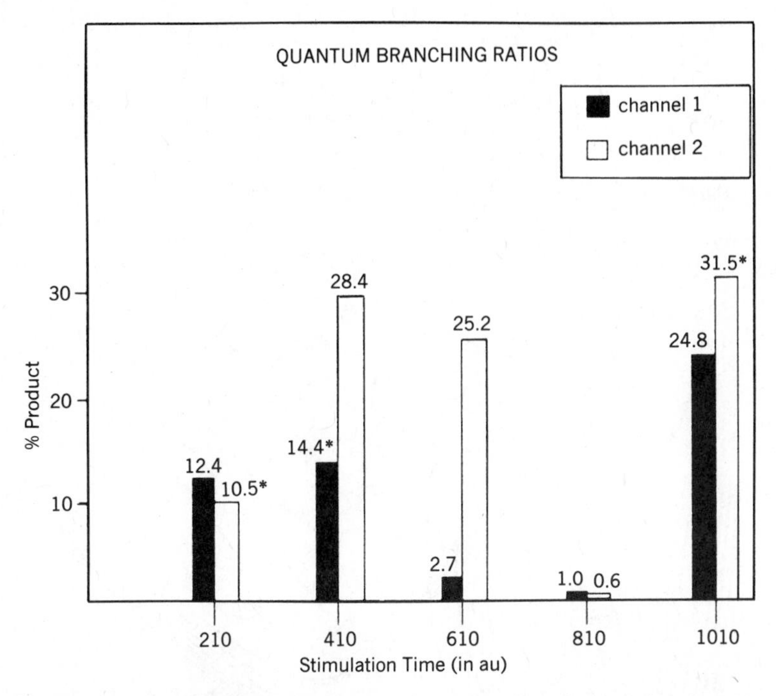


**Figure 31.** Quantum-mechanical branching ratio as a function of stimulation time. Note the dramatic differences between the branching ratios at different times, relative to each other as well as relative to the amplitude that remains bound.

### 3. *Classical Propagation of the Wigner Distribution*

A study of the last anharmonic system previously, discussed was conducted usng a single classical trajectory originating at the Franck–Condon geometry with zero momentum. The trajectory failed to reproduce many of qualitative features of Fig. 31. A single classical trajectory cannot exhibit the tendency of the wavepacket to bifurcate. Moreover, the single trajectory we used had no zero-point energy. We, therefore, also examined a swarm of trajectories with an initial gaussian distribution in **p** and **x**, which corresponds to the Wigner transform of the ground vibrational state of the ground-state potential energy surface,[27] where the ground-state surface is expanded up to quadratic terms about the equilibrium geometry. This distribution was discretized with 81 sets of $p$, $x$ initial conditions. The initial distribution in coordinate space is shown in Fig. 32. There are nine initial values for $p$, $x$ corresponding to each initial value for **p**, **x**. We point out that since the initial conditions for $p$ and $x$ are independent, there is a range of zero point energies in the swarm.

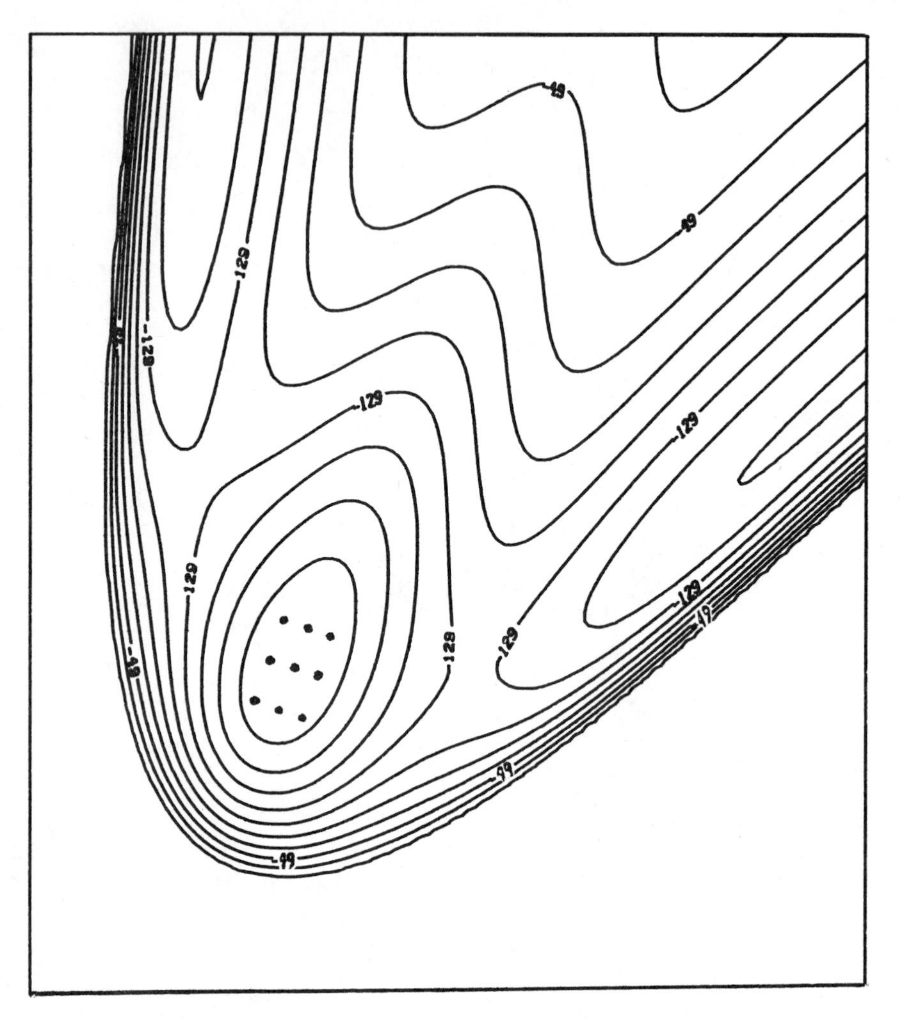


**Figure 32.** Initial swarm of classical trajectories on the ground-state potential energy surface. The swarm consists of 81 trajectories (there are nine different momentum combinations for each of the coordinate combinations).

Figures 33*a*–33*c* show the swarm on the excited-state potential energy surface, for the same pulse sequence as Fig. 29 (second pulse at 610 a.u.). The swarm mimics closely the quantum wavepacket, including the sequence of contraction and spreading. Figures 33*d*–33*f* shows the swarm on the ground-state potential energy surface, after the second pulse. Those trajectories that do exit do so from channel 2.

Figures 34*a* and 34*b* show the swarm on the excited-state potential energy

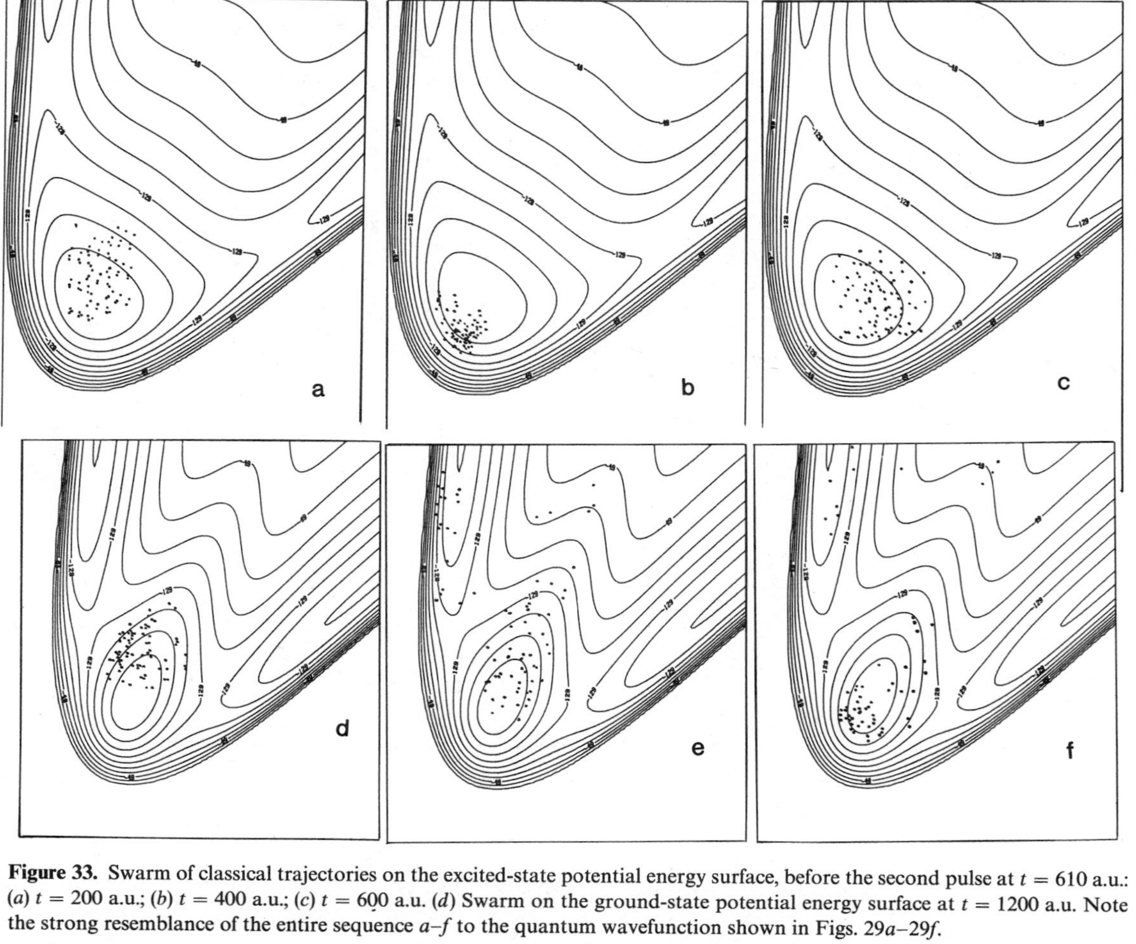


**Figure 33.** Swarm of classical trajectories on the excited-state potential energy surface, before the second pulse at $t = 610$ a.u.: (*a*) $t = 200$ a.u.; (*b*) $t = 400$ a.u.; (*c*) $t = 600$ a.u. (*d*) Swarm on the ground-state potential energy surface at $t = 1200$ a.u. Note the strong resemblance of the entire sequence *a*–*f* to the quantum wavefunction shown in Figs. 29*a*–29*f*.

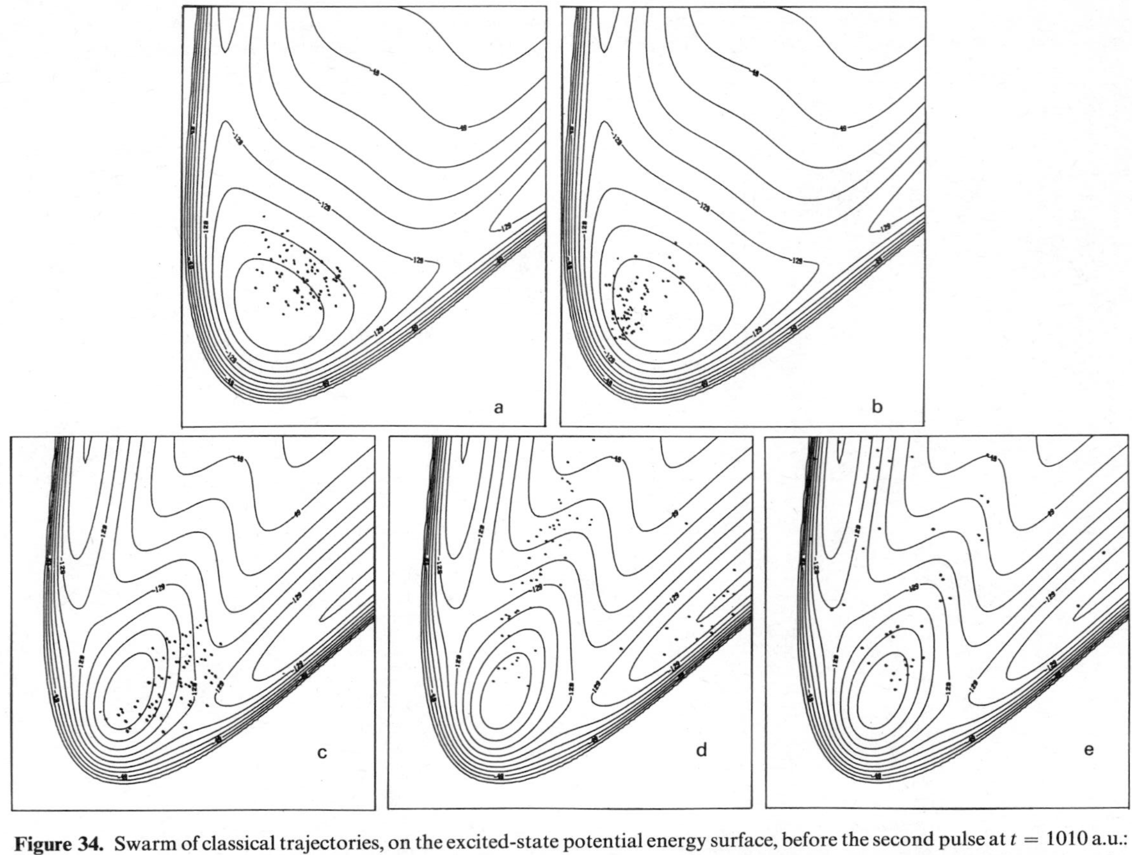


**Figure 34.** Swarm of classical trajectories, on the excited-state potential energy surface, before the second pulse at $t = 1010$ a.u.: (*a*) $t = 800$ a.u.; (*b*) $t = 1000$ a.u. (*c*) Swarm on the ground-state potential energy surface at $t = 1200$ a.u. (*d*) Swarm on the ground state at $t = 1400$ a.u. (*e*) Swarm on the ground state at $t = 1600$ a.u. Note the breakup of the swarm on the ground-state surface. The entire sequence *a*–*e* is in close agreement with the quantum wavefunction, Figs. 30*a*–30*e*.

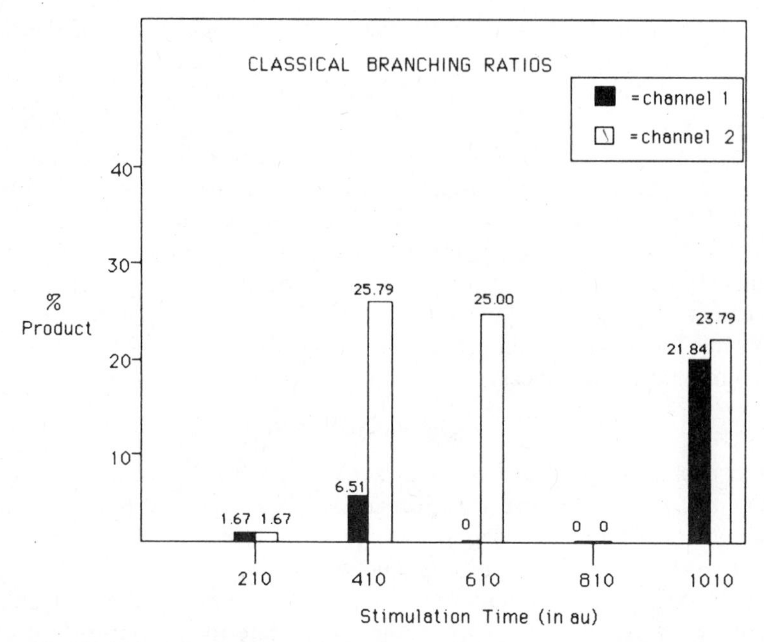


**Figure 35.** Classical mechanical branching ratio as a function of stimulation time. Note the qualitative agreement with the quantum branching ratio, Fig. 31.

surface for the same pulse sequence as Fig. 30 (second pulse at $t = 1010$ a.u.). Figures 34*c*–34*e* show the swarm on the ground-state potential energy surface after the second pulse. The swarm breaks up on the ground-state surface with a substantial number of trajectories exiting from channel 1, followed by an approximately equal number exiting from channel 2. The entire sequence is in close agreement with the quantum-mechanical results, Figs. 20*a*–20*e*.

Figure 35 shows the classical branching ratio as a function of stimulation time, for the same stimulation pulses as in Fig. 31. Note the qualitative agreement with the quantum-mechanical results in both a relative and absolute sense.

Clearly, a Wigner swarm of classical trajectories is a valuable exploratory tool for estimating quantum-mechanical branching ratios. The computer times involved are more than 100 times shorter than for the quantum-mechanical calculations. For larger masses and more degrees of freedom the disparity between classical and quantum calculation times will become even more pronounced. At the same time, the classical–quantum correspondence should be even better than illustrated here, for larger mass systems.

The classical–quantum correspondence is particularly good if the time-evolving state is localized relative to the scale on which anharmonicities in the potential become pronounced. If the state is delocalized, the classical propagation neglects interference terms, which are likely to be significant.[27] Hence, one must be somewhat cautious about calculating product branching ratios that originate from excited vibrational states and, more important for our purposes, when the pulse duration is long compared with respect to the time it takes for the wavefunction to begin propagating on the excited-state surface. In the latter case the convolution of the pulse with the propagating wavefunction gives rise to an extended wavefunction. During the free evolution stage following the pulse, the classical swarm may not be faithful to the quantum interference effects.

### F. Alternative Schemes

The two choices of initial and final surfaces previously mentioned can be thought of as examples from a spectrum of possibilities inherent in a more general scheme for achieving selectivity of reactivity. That more general scheme involves use of some electronic state to assist selectivity of product formation but allows the initial and final states to be different. Imagine a Franck–Condon transition from some initial state to an intermediate electronic state followed, after a controlled delay, by a transition to a third electronic state (which could be the initial state). If the final-state potential energy surface and the intermediate-state potential energy surface have the right properties, use of shaped pulses and control of pulse separation will permit selectivity of reactivity on the final-state potential energy surface. It is also possible to imagine the use of detuning from resonance with the intermediate electronic state as a tool to augment control of the time scale for evolution in that state.

Clearly, there are many ways in which the ideas we have proposed must be extended. Among the more important extensions we cite variational optimization of the shape, duration, and separation of the pulses used to generate the selectivity of reactivity, and analysis of the changes induced by the inclusion of all degrees of freedom of the molecule (say in the sense of a reaction path Hamiltonian, or a dynamical path Hamiltonian). For studies involving more degrees of freedom, a swarm of classical trajectories should be a very useful tool.

Recently, several independent theoretical studies, notably by Holme and Hutchinson (HH) and by Brumer and Shapiro (BS), have concluded that two-photon (or more precisely, two-frequency) processes can afford some control over the preparation of an initial state or the resulting product distributions.[28,29] Both of these sets of workers emphasize the use of monochromatic lasers and all also emphasize the importance of creating a superposition of vibrational states in controlling final arrangement channels. This is in accord with the Tannor–Rice (TR) scheme, the wavepacket described in the previous sections being a superposition of a multitude of vibrational levels. The classical–quantum correspondence principle, so central to the TR scheme, is not invoked in the work of HH and BS.

There is another interesting contrast between the TR scheme and these other two-photon/two-frequency schemes. In order to direct amplitude out of competing chemical channels, one requires significant control over the momentum of the prepared state, which is what one is controlling via the pulse delay in the TR scheme. In contrast, the initially prepared states of HH and BS have little or no momentum. This may account for the emphasis in the TR scheme on chemically distinct arrangement channels, while the emphasis in the HH and BS schemes is on chemically identical but electronically distinct arrangement channels. With that introduction we proceed to review the recent work of Holme and Hutchinson, and Brumer and Shapiro.

Holme and Hutchinson suggest tuning two, or several, lasers to nearly degenerate molecular eigenstates. By adjusting the strength of the field at each frequency, they are able to combine these two molecular eigenstates with arbitrary coefficients, and thereby prepare a desired superposition state. These workers have focused on applications to local mode overtones, where field-free evolution would lead to relaxation into an intramolecular bath.

A second aspect of the HH scheme is the use of strong fields. Mukamel and Shan have shown that strong fields may be used to suppress intramolecular vibrational relaxation (IVR) if the frequency of Rabi cycling ($\Omega$) is much larger than the rate of IVR ($K_{IVR}$).[30] The energy remains localized so long as the laser field interacts with the molecule. Holme et al. are presumably in the regime where $K_{IVR} > \Omega$, so that multiple laser frequencies are still required to suppress IVR. HH envision using two lasers to selectively excite a local mode for times long enough to be chemically significant. Recently, these workers have applied their approach to the control of singlet–triplet and triplet–triplet branching ratios in a realistic model for the diatomic CS.

Brumer and Shapiro suggest the use of three lasers to gain control over product arrangement. The first laser is tuned to a transition to a low-lying vibrational level of the ground electronic state, creating a coherent superposition of this excited vibrational state and the ground vibrational state. The second and third lasers are tuned to excite the two initial states in the coherent superposition to the same total energy. These workers are concerned with the case where there is a degeneracy at the final energy (as there always is in dissociation with more than one open channel). The relative amplitude and phase of lasers two and three are adjustable parameters in the BS scheme;

they are used to control the composition of the state that is populated at the final energy $E_f$. Brumer and Shapiro have shown that there is no way, using a single laser, to gain control over this composition, hence the need for a two-step process.

As an explicit example, BS consider the dissociation of methyl iodide, in which two product channels are accessible at energy $E$:

$$CH_3I \begin{cases} \nearrow CH_3 + I \\ \searrow CH_3 + I^* \end{cases} \tag{5.30}$$

Studies of this system show a broad range of control over the I to I* product ratio. For example, a superposition of $|E_1\rangle$ and $|E_3\rangle$ (the first and third vibrational states of the ground-electronic-state potential energy surface) allows an increase of the yield of I from 30%, the value attained by excitation with one frequency, to more than 70%. Furthermore, using a diatomic model for $CH_3I$, BS were able to define conditions which reduce the I yield to zero or increase it fully to one.

Figure 36 summarizes the relationship between the Tannor–Rice, Holme–Hutchinson, and Brumer–Shapiro schemes for control of photochemical products.

Sleva and Zewail (private communication) have recently reported "photon locking" in a system that ordinarily undergoes pure dephasing ($T_2$ type processes). In a single $\pi$-pulse experiment their observed $T_2$ is $\sim 1$ nsec. If a second pulse, phase shifted by $\pi/2$ from the original pulse, is applied continuously after the initial $\pi$ pulse, the phases of the ensemble of systems are "locked". The dephasing time is now $\approx 1\ \mu$sec. Mukamel and Shan recently reported a calculation with essentially the same conclusion, that is, that intramolecular vibrational redistribution can be suppressed by application of suitable fields to a system.[30] In particular, Mukamel and Shan show that if the Rabi frequency is large compared with the energy spread of states that have a component of the optically allowed state,

$$\Omega \gg V_{sl}, \tag{5.31}$$

and if a strong field is allowed to operate continuously, IVR will be suppressed on a time scale much longer than that characteristic of IVR in the absence of the strong field.

We have applied these ideas to the case described in Figs. 24 and 25, where wavepacket spreading on the broad, anharmonic, excited-state potential energy surface destroys the selectivity. A square pulse was used for

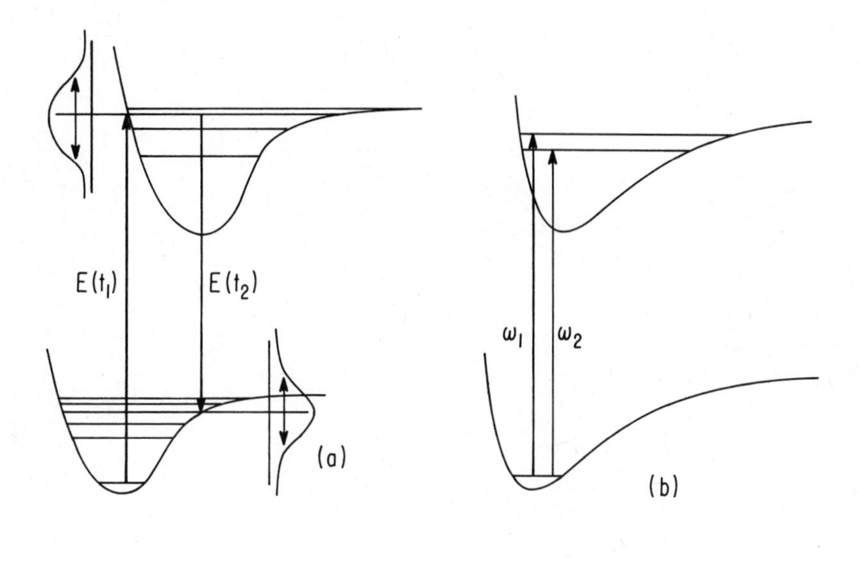


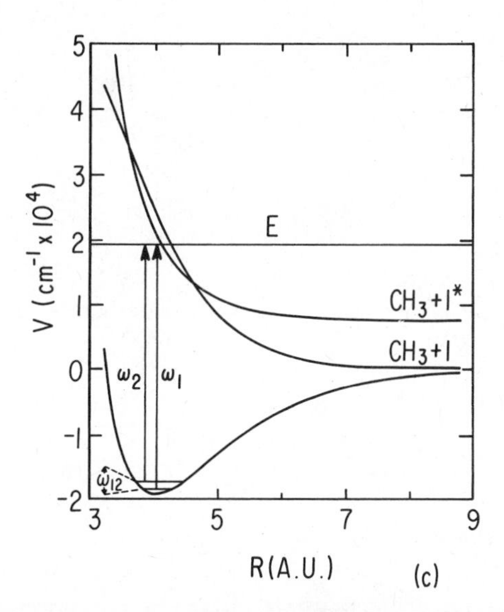


**Figure 36.** Comparison of Tannor–Rice, Holme–Hutchinson, and Brumer–Shapiro selectivity schemes. (*a*) Tannor–Rice scheme uses two *pulses*, where each pulse is wide enough in frequency to excite a superposition of many vibrational levels. (*b*) Holme–Hutchinson scheme uses two monochromatic photons to prepare a superposition state on the excited-state surface (*c*) Brumer–Shapiro scheme uses one photon to prepare a superposition state on the ground-state surface; then two additional photons to excite the superposition state to the excited state surface.

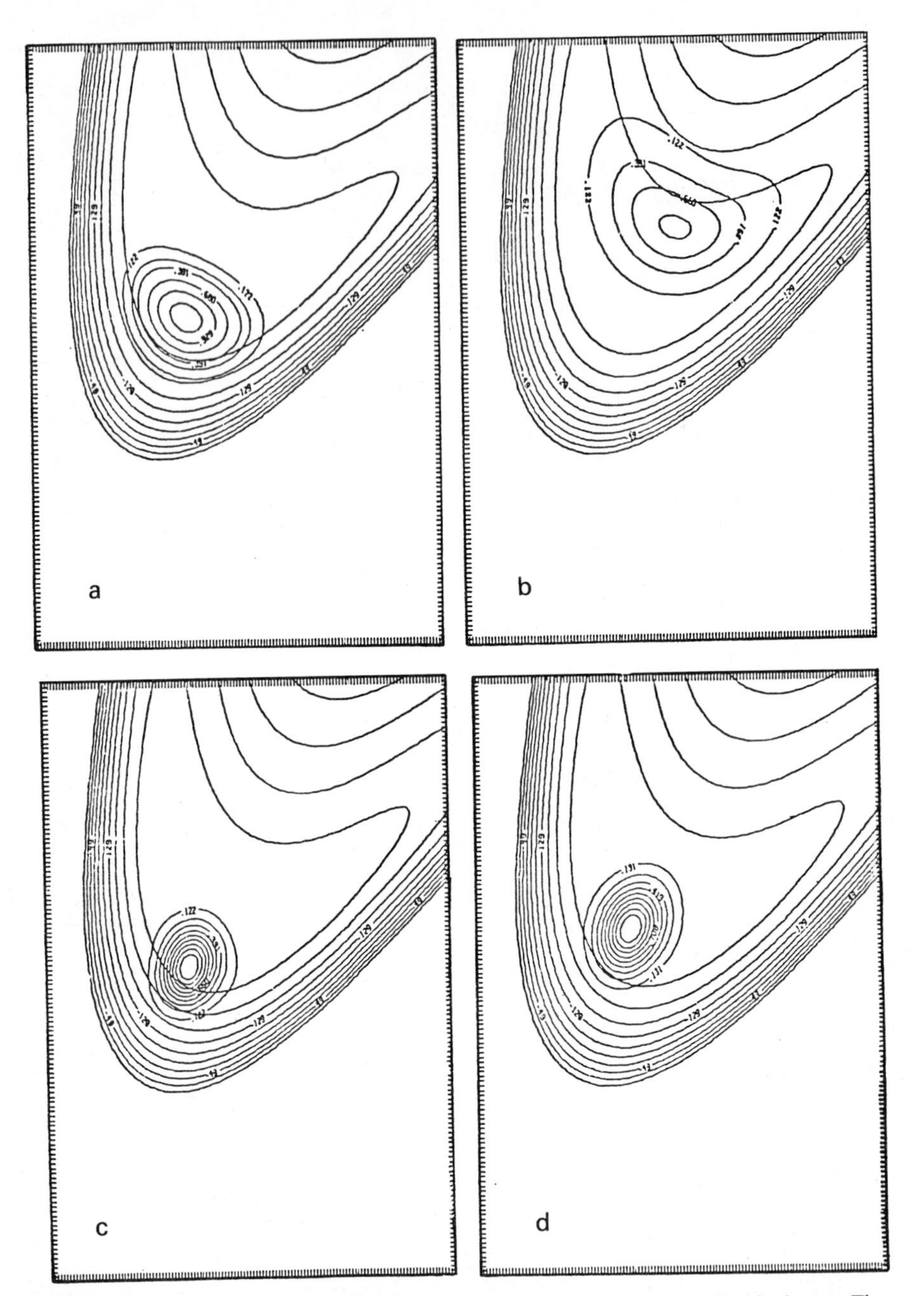


**Figure 37.** Excited-state wavefunction for the photon locking sequence described in the text. The wavefunction for the pulse sequence without locking is presented for comparison. The potential energy surface is the same as in Figs. 25 and 26. (*a*) $t = 200$ a.u., without locking; (*b*) $t = 400$ a.u., without locking; (*c*) $t = 200$ a.u. with locking, (*d*) $t = 400$ a.u. with locking. Note that the locking pulse curbs the wavepacket spreading.

excitation:

$$\begin{gathered}\mu E(t) = A\cos(\omega_e t),\\ A = 0.125, \quad 20 \lesssim t \lesssim 40,\\ A = 0, \quad t \lesssim 20, \quad t \gtrsim 40,\\ \omega_e = 0.10.\end{gathered} \tag{5.32}$$

An additional continuous pulse was applied, namely,

$$\begin{gathered}C(t) = C\cos(\omega_c t + \pi/2),\\ C = 0.125, \quad 40 \lesssim t \lesssim 1000,\\ C = 0, \quad t \lesssim 40, \quad t \gtrsim 1000,\\ \omega_c = 0.10.\end{gathered} \tag{5.33}$$

Figures 37*a* and 37*b* show the wavefunction on the excited state potential surface at $t = 200$ a.u. and 400 a.u., without the locking pulse. Figures 37*c* and 37*d* show the wavefunction at the corresponding times with the locking pulse. The motion of the center of the wavepacket is greatly reduced. More important, with respect to selectivity, there is almost no wavepacket spreading. This example suggests that strong fields may be used in conjunction with the carefully tailored waveforms we have described above to achieve selectivity of reaction.

## VI. CONCLUDING REMARKS

We have been concerned with describing a scheme for controlling the selectivity of a chemical reaction that avoids the use of assumptions concerning the localization of energy in particular chemical bonds or the directed transfer of energy to particular bonds. The existence of such a scheme establishes a theoretical paradigm, but does not guarantee the translation of the paradigm into a practical methodology. The key to that translation, for the scheme we propose, is the generation of very short (femtosecond) tunable and shapable pulses of light. Pulse shaping and phase shifting of laser pulses as short as 100 psec has been demonstrated by Haner, Spano, and Warren (private communication), while the generation of tunable femtosecond unshaped pulses has been demonstrated by Shank[31] and others.[33] The marriage of these two technologies will permit exploitation of coherent pulse sequence control of product formation in chemical reactions.

The ideas discussed in this article are not restricted to control of the reactivity of isolated molecules by the use of pulsed lasers. Gadzuk (private communication) has proposed a realization of the Tannor–Rice scheme in the case of surface-induced dissociative scattering of a polyatomic molecule. In this case it is imagined that on both incoming and outgoing legs of the scattering trajectory there is a crossing of potential energy surfaces that permits charge transfer between the molecule and the material surface, generally in the form of creating a short-lived negative ion state of the molecule. The time of propagation on the negative ion potential energy surface is controlled by varying the work function of the material surface, which in turn controls the crossing point of the potential energy surfaces, hence the timing of the charge transfer. Gadzuk shows, for a simple case, that this scheme does lead to preferential formation of one or the other products of a reaction.

We expect many more inventive suggestions for controlling the selectivity of product formation in a chemical reaction to be proposed.

**Note added in proof:** It has come to our attention after this review was completed that the work by Holme and Hutchinson, (Ref. 28), accidentally neglected the time-dependent phase of the initial state prepared by the laser. This has been confirmed through private communication with the authors. As far as we know there is no physical justification for this neglect, which played a crucial role in the selectivity which they observed.

## Acknowledgment

The research described in this paper has been supported by the NSF and the AFOSR.

## References

1. E. J. Heller, *Acc. Chem. Res.* **14**, 368 (1981).
2. E. J. Heller, *J. Chem. Phys.* **68**, 2066 (1978); **68**, 3891 (1978); K. C. Kulander and E. J. Heller, *J. Chem. Phys.* **69**, 2439 (1978); R. C. Brown and E. J. Heller, *J. Chem. Phys.* **75**, 186 (1981).
3. E. J. Heller and W. Gelbart, *J. Chem. Phys.* **73**, 626 (1980), E. J. Heller, E. B. Stechel, and M. J. Davis, *J. Chem. Phys.* **73**, 4720 (1980); L. Tutt, J. Schindler, D. J. Tannor, E. J. Heller, and J. I. Zink, *J. Phys. Chem.* **87**, 3017 (1983).
4. S.-Y. Lee and E. J. Heller, *J. Chem. Phys.* **71**, 4777 (1979); D. J. Tannor and E. J. Heller, *J. Chem. Phys.* **77**, 202 (1982); A. B. Myers, R. A. Mathies, D. J. Tannor, and E. J. Heller, *J. Chem. Phys.* **77**, 3857 (1982); E. J. Heller, R. L. Sundberg, and D. J. Tannor, *J. Phys. Chem.* **86**, 1822 (1982); R. L. Sundberg and E. J. Heller, *Chem. Phys. Lett.* **93**, 586 (1982).
5. (*a*) E. A. McCullough and R. E. Wyatt, *J. Chem. Phys.* **51**, 1253 (1969); **54**, 3578 (1971); A. Askar and S. Cakmak, *J. Chem. Phys.* **68**, 2794 (1978); M. D. Feit and J. A. Fleck, Jr., *J. Chem. Phys.* **78**, 301 (1983); (*b*) D. Kosloff and R. Kosloff, *J. Comput. Phys.* **52**, 35 (1983); R. Kosloff and D. Kosloff, *J. Chem. Phys.* **79**, 1823 (1983).
6. E. J. Heller, *J. Chem. Phys.* **62**, 1544 (1975); **65**, 4979 (1976); R. D. Coalson and M. Karplus, *Chem. Phys. Lett.* **90**, 301 (1982).
7. R. A. Covaleskie, D. A. Dolson, and C. S. Parmenter, *J. Chem. Phys.* **72**, 5774 (1980).
8. D. Imre, J. Kinsey, R. Field, and D. Katayama, *J. Phys. Chem.* **86**, 2564 (1982); D. Imre, J. L. Kinsey, A. Sinha, and J. Krenos, *J. Phys. Chem.* **88**, 3956 (1984).

9. A. B. Myers, R. A. Harris, and R. A. Mathies, *J. Chem. Phys.* **79**, 603 (1983); A. B. Myers and R. A. Mathies, *J. Chem. Phys.* **81**, 1552 (1984).
10. D. J. Tannor, S. A. Rice, and P. M. Weber, *J. Chem. Phys.* **83**, 6158 (1985).
11. K. V. Reddy, D. F. Heller, and M. J. Berry, *J. Chem. Phys.* **76**, 2814 (1982).
12. E. L. Sibert, III, W. P. Reinhardt, and J. T. Hynes, *Chem. Phys. Lett.* **92**, 455 (1982).
13. J. B. Hopkins, P. R. R. Langridge-Smith, and R. E. Smalley, *J. Chem. Phys.* **78**, 3410 (1983).
14. J. J. Valentini, in *Spectrometric Techniques*. Academic, New York, 1985, Vol. 4.
15. W. Zinth, A. Laubereau, and W. Kaiser, *Opt. Commun.* **26**, 457 (1978).
16. F. M. Kamga and M. G. Sceats, *Opt. Lett.* **5**, 126 (1980).
17. I. D. Abella, N. A. Kurnit, and S. R. Hartmann, *Phys. Rev.* **141**, 391 (1966).
18. M. J. Burns, W. K. Lin, and A. H. Zewail, in *Spectroscopy and Excitation Dynamics of Condensed Molecular Systems* (V. M. Agranovich and R. M. Hochstrasser, eds.). North-Holland, Amsterdam, 1983, Chap. 7.
19. M. Sargent, III, M. O. Scully, and W. E. Lamb, Jr., *Laser Physics*. Addison-Wesley, Reading, MA, 1974.
20. W. S. Warren and A. H. Zewail, *J. Chem. Phys.* **78**, 2298 (1983).
21. D. J. Tannor and S. A. Rice, in *Understanding Molecular Properties*. D. Reidel, Dortrecht, 1986.
22. P. M. Felker, J. S. Baskin, and A. H. Zewail, *J. Chem. Phys.* **90**, 724 (1986).
23. J. Jortner and J. Kommandeur, *Chem. Phys.* **28**, 273 (1978).
24. D. J. Tannor and S. A. Rice, *J. Chem. Phys.* **83**, 5013 (1985).
25. D. J. Tannor, R. Kosloff, and S. A. Rice, *J. Chem. Phys.* **85**, 5805 (1986).
26. A. Messiah, *Quantum Mechanics*. Wiley, New York, 1958, Vol I.
27. E. J. Heller, *J. Chem. Phys.* **65**, 1289 (1977); E. J. Heller and R. C. Brown, *J. Chem. Phys.* **75**, 1048 (1981).
28. T. A. Holme and J. S. Hutchinson, *Chem. Phys. Lett.* **124**, 181 (1986), *J. Chem. Phys.* **86**, 42 (1987).
29. M. Shapiro and P. Brumer, *J. Chem. Phys.* **84**, 4103 (1986); P. Brumer and M. Shapiro, *Chem. Phys. Lett.* **126**, 54 (1986).
30. S. Mukamel and K. Shan, *Chem. Phys. Lett.* **117**, 489 (1985).
31. C. V. Shank, R. L. Fork, R. Yen, R. H. Stolen, and W. J. Tomlinson, *Appl. Phys. Lett.* **40**, 761 (1982).
32. R. A. Engh, J. W. Petrich, and G. R. Fleming, *J. Phys. Chem.* **89**, 618 (1985).

# AUTHOR INDEX

Numbers in parentheses are reference numbers and indicate that the author's work is referred to although his name is not mentioned in the text. Numbers in *italic* show the pages on which the complete references are listed.

# SUBJECT INDEX